MARINE BIOLOGY

MICROALGAE: BIOTECHNOLOGY, MICROBIOLOGY AND ENERGY

MARINE BIOLOGY

Additional books in this series can be found on Nova's website under the Series tab.

Additional E-books in this series can be found on Nova's website under the E-books tab.

MARINE BIOLOGY

MICROALGAE: BIOTECHNOLOGY, MICROBIOLOGY AND ENERGY

MELANIE N. JOHANSEN
EDITOR

Nova Science Publishers, Inc.
New York

Copyright © 2012 by Nova Science Publishers, Inc.

All rights reserved. No part of this book may be reproduced, stored in a retrieval system or transmitted in any form or by any means: electronic, electrostatic, magnetic, tape, mechanical photocopying, recording or otherwise without the written permission of the Publisher.

For permission to use material from this book please contact us:
Telephone 631-231-7269; Fax 631-231-8175
Web Site: http://www.novapublishers.com

NOTICE TO THE READER

The Publisher has taken reasonable care in the preparation of this book, but makes no expressed or implied warranty of any kind and assumes no responsibility for any errors or omissions. No liability is assumed for incidental or consequential damages in connection with or arising out of information contained in this book. The Publisher shall not be liable for any special, consequential, or exemplary damages resulting, in whole or in part, from the readers' use of, or reliance upon, this material. Any parts of this book based on government reports are so indicated and copyright is claimed for those parts to the extent applicable to compilations of such works.

Independent verification should be sought for any data, advice or recommendations contained in this book. In addition, no responsibility is assumed by the publisher for any injury and/or damage to persons or property arising from any methods, products, instructions, ideas or otherwise contained in this publication.

This publication is designed to provide accurate and authoritative information with regard to the subject matter covered herein. It is sold with the clear understanding that the Publisher is not engaged in rendering legal or any other professional services. If legal or any other expert assistance is required, the services of a competent person should be sought. FROM A DECLARATION OF PARTICIPANTS JOINTLY ADOPTED BY A COMMITTEE OF THE AMERICAN BAR ASSOCIATION AND A COMMITTEE OF PUBLISHERS.

Additional color graphics may be available in the e-book version of this book.

Library of Congress Cataloging-in-Publication Data

Microalgae : biotechnology, microbiology, and energy / editor, Melanie N. Johansen.
  p. cm.
 Includes bibliographical references and index.
 ISBN 978-1-61324-625-2 (hardcover : alk. paper) 1. Microalgae. 2. Microalgae--Biotechnology. 3. Microalgae--Microbiology. 4. Biomass energy. I. Johansen, Melanie N.
 QK568.M52M53 2011
 579.8--dc22

2011014563

Published by Nova Science Publishers, Inc. †New York

CONTENTS

Preface		**vii**
Chapter 1	Microalgae Biotechnological Applications: Nutrition, Health and Environment *A. E. Marques, J. R. Miranda, A. P. Batista, and L. Gouveia*	**1**
Chapter 2	Assessing the Renewability of Biodiesel from Microalgae via Different Transesterification Processes *Ehiaze Ehimen, Zhifa Sun, and Gerry Carrington*	**61**
Chapter 3	Toxicity and Removal of Organic Pollutants by Microalgae: A Review *Lin Ke, Yuk Shan Wong, and Nora F. Y. Tam*	**101**
Chapter 4	Microalgal Engineering: The Metabolic Products and the Bioprocess *Jorge Alberto Vieira Costa, Michele Greque de Morais and, Michele da Rosa Andrade*	**141**
Chapter 5	Hydrothermal Carbonization of Microalgae and Other Low Cellulosic Biomass Materials *Steven M. Heilmann, Marc G. von Keitz, and Kenneth J. Valentas*	**171**
Chapter 6	Investigations on the Use of Microalgae for Aquaculture *José Antonio López Elías, Luis Rafael Martínez Córdova, and Marcel Martínez Porchas*	**201**
Chapter 7	Microalgae: The Future of Green Energy *K. K. I. U. Arunakumara*	**227**
Chapter 8	Real-Time Spectral Techniques for the Detection of Buildup of Valuable Compounds and Stress in Microalgal Cultures: Implications for Biotechnology *Alexei Solovchenko, Inna Khozin-Goldberg, and Olga Chivkunova*	**251**
Chapter 9	Microalgae as an Alternative Feed Stock for Green Biofuel Technology *G. S. Anisha and Rojan P. John*	**277**

Chapter 10	A Critical Review: Microalgal CO_2 Sequestration, Which Strain Is the Best? *Yanna Liang*	295
Chapter 11	Use of Microalgae as Biological Indicators of Pollution: Looking for New Relevant Cytotoxicity Endpoints *Ángeles Cid, Raquel Prado, Carmen Rioboo, Paula Suárez-Bregua, and Concepción Herrero*	311
Chapter 12	Application of Green Technology on Production of Eyes-Protecting Algal Carotenoids from Microalgae *Chao-Rui Chen, Chieh-Ming J. Chang, Chun-Ting Shen, Shih-Lan Hsu, Bing-Chung Liau, Po-Yen Chen, and Jia-Jiuan Wu*	325
Chapter 13	Microalgae as Biodeteriogens of Stone Cultural Heritage: Qualitative and Quantitative Research by Non-Contact Techniques *Ana Zélia Miller, Miguel Ángel Rogerio-Candelera, Amélia Dionísio, Maria Filomena Macedo, and Cesareo Saiz-Jimenez*	345
Chapter 14	Astaxanthin Production in Cysts and Vegetative Cells of the Microalga *Haematococcus Pluvialis* Flotow *C. Herrero, M. Orosa, J. Abalde, C. Rioboo, and A. Cid*	359
Chapter 15	Nitrogen Solubility, Antigenicity, and Safety Evaluation of an Enzymatic Protein Hydrolysate from Green Microalga *Chlorella Vulgaris* *Humberto J. Morris, Olimpia Carrillo, María E. Alonso, Rosa C. Bermúdez, Alfredo Alfonso, Onel Fong, Juan E. Betancourt, Gabriel Llauradó, and Ángel Almarales*	373
Chapter 16	Heterotrophic Microalgae in Biotechnology *Niels Thomas Eriksen*	387
Chapter 17	Microalgae Growth and Fatty Acid Composition Depending on Carbon Dioxide Concentration *C. Griehl, H. Polhardt, D. Müller, and S. Bieler*	413
Index		455

PREFACE

Microalgae are microscopic algae, typically found in freshwater and marine systems. Microalgae, capable of performing photosynthesis, are important for life on earth; they produce approximately half of the atmospheric oxygen and use simultaneously the greenhouse gas carbon dioxide to grow photoautotrophically. The biodiversity of microalgae is enormous and they represent an almost untapped resource. In this book, the authors present current research in the study of microalgae, including microalgal biotechnological applications in nutrition, health and the environment; using microalgae biomass for biodiesel and biofuel production and microalgae for aquaculture.

Chapter 1 - Microalgae (prokaryotic and eukaryotic) consist of a wide range of autotrophic organisms which grow through photosynthesis just like land based plants. Their unicellular structure allows them to easily convert solar energy into chemical energy through CO_2 fixation and O_2 evolution, being well adapted to capture CO_2 and store it as biomass. Microalgae and cyanobacteria have an interesting and not yet fully exploited potential in biotechnology. They can be used to enhance the nutritional value of food and animal feed due to their chemical composition, playing a crucial role in aquaculture. Highly valuable molecules like natural dyes (e.g. carotenoids), polyunsaturated fatty acids, polysaccharides and vitamins from algal origin are being exploited and can be applied in the nutritional supplements; cosmetics (e.g. phycocyanin) and pharmaceuticals. In fact, microalgae and cyanobacteria are able to produce several biologically active compounds with reported antifungal, antibacterial, anticancer, antiviral (e.g. anti-HIV), immunosuppressive, anti-inflamatory and antioxidant activity. Nowadays, there is a focus on using microalgae in renewable energy sources and environmental applications. Microalgae are a potential source for biofuels production such as biodiesel, bioethanol, biohydrogen and biogas. These can be produced through a biorefinery concept, in which every component of the biomass is used to produce usable products. This strategy can integrate several different conversion technologies (chemical, biochemical, termochemical and direct combustion) providing a higher cost effective and environmental sustainability for the biofuels production. Environmental applications can include CO_2 sequestration and wastewater treatment. This can be achieved by coupling microalgae production systems with industrial polluting facilities.

Chapter 2 - Using process modelling tools, the conventional and in-situ transesterification processes for biodiesel production from microalgae biomass was modelled.The raw material and process energy requirements of the up-scaled process were obtained for the different transesterification processes, and a renewability assessment of the various schemes was carried out. The biomass cultivation and biodiesel production process renewability was assessed by comparing the minimum work required to restore the non-renewable resources degraded in the biomass and biodiesel production process with the useful work available from the main process products. If the maximum work obtained from the process products is larger than the restoration work, the process is considered as renewable. In a present day scenario (with the use of fossil fuel sources for the production of the process raw materials, such as for methanol and sulphuric acid production, and electricity), all the transesterification processes were shown to be non-renewable. The influence of the choice of the electricity generation scheme, raw material source and the type of heating fuels (including heating and drying technology) on the process renewability was also examined. The process renewability of the in-situ transesterification of microalgae lipids to biodiesel was found to significantly improve with the use of renewable electricity, reacting alcohols from biomass fermentation and heat pump technology to facilitate the biomass drying and process heating.

Chapter 3 - The ubiquity and persistence of organic pollutants in the aquatic environment are of potential risk to aquatic habitats and human health due to their highly toxic, mutagenic and carcinogenic properties. Bioremediation, a cost-effective technology to remove organic pollutants from contaminated water bodies, involves a number of biological processes, including accumulation, transformation and degradation, mediated mainly by microorganisms such as bacteria, fungi and microalgae. Eukaryotic microalgae are dominant primary producers and play a central role in the fixation and turnover of carbon and other nutrient elements. However, their role in the remediation of aquatic contaminants and the relative importance of the processes involved are much less understood, as compared to bacteria and fungi. Further, most of the studies on the novel remediation technology using microalgae have concentrated on metals and nutrients and much less on organic pollutants. Screening of tolerant species is a crucial step in bioremediation, and the understanding of the response and adaptive changes in microalgal cells to toxicity induced by organic pollutants is equally important as it serves as a scientific basis of remediation practices. However, there has been very little published information on the toxicity, resistance and adaptations of microalgae to persistent organic pollutants (POPs). This paper reviews the recent research on the sensitivity, tolerance and adaptations of microalgae to the toxicity of various POPs, including organochlorine pesticides (OCPs), polychlorinated biphenyls (PCBs), polycyclic aromatic hydrocarbons (PAHs), polybrominated diphenyl ethers (PBDEs) and some other emerging environmental endocrine disrupting compounds. The review focuses on the physiological and biochemical changes in microalgae and their relationships with tolerance. The mechanisms and factors affecting the capacity of microalgal cells to remove POPs, as well as the feasibility, limitations and future research directions of employing microalgae in POPs remediation, are also addressed.

Chapter 4 - Many analyses have been carried out about the future possibility of exhausting the planet's resources and its ability to sustain its inhabitants. The use of microorganisms and their metabolic products by humans is one of the most significant fields of biotechnology activities. Microalgae are descendants of the first photosynthetic life forms. More than 3,500 million years ago the oxygen atmosphere was made up by cyanobacteria and other forms of life could evolve. Since then, microalgae have contributed to regulating the planet's biosphere. The use of solar energy through photosynthesis in microalgae cultivation is a clean, efficient and low cost process, because the sun's energy is virtually free and unlimited. The biomass of microalgae and its products are employed in many fields: feed and food additives in agriculture, fertilizers, in the food industry, pharmaceutics, perfume making, medicine, in biosurfactants, biofuels and in science. This wide use is due to its fast growth, non-toxicity, assimilability (85–95%), high protein content (60–70%), well-balanced amino acid composition, richness in vitamins, minerals, fatty acids, biopigments, biopolymers, and the fact that is has a great variety of biologically active agents in appreciable amounts. Microalgae are considered to be efficient immunopotentiators and have anticarcinogenic and antiviral effects. Producing biocompounds from microalgae has the additional advantage of simultaneously fixing large amounts of carbon dioxide. The use of these alternative sources reduces costs and can generate carbon credits. Microalgae can be grown on land that is unsuitable for agriculture and farming, or on inhospitable land such as deserts, using brackish water and/or wastes from the desalination process. The composition and rates of photosynthesis and growth of these organisms are strongly dependent on growth conditions. Manipulating these conditions can result in higher yield metabolites that are of interest. At the end of the process, according to the characteristics of the microalgal biomass obtained, it can be used to produce different compounds. The objective of this study was to present traditional and advanced bioproducts obtained from microalgal biomass and to describe the characteristics of the cultivation process.

Chapter 5 - Hydrothermal carbonization is a process in which biomass is heated in water under pressure at temperatures below 250 °C to create a char product. A significant advantage of the process is that water is not removed from the char by evaporation but by filtration, providing a favorable energy input to output ratio.

With higher plants, the chemistry derives primarily from lignin, hemicellulose and cellulose components. Cellulose is the most recalcitrant of these and requires relatively high temperatures and long reaction periods for conversion into a highly carbonized char product.

The author's approach has been to examine biomass materials having relatively low cellulose contents, applying reaction temperatures generally below 225 °C, and for reaction periods of less than 2 hours. These conditions are believed to be conducive to continuous processing and to increase the carbon content in the char primarily via a dehydration mechanism, rather than by loss of carbon dioxide. Substrates that have been examined in this manner included microalgae, cyanobacteria, and fermentation residues such as distiller's grains, brewer's grains and others.

It was determined during the course of these investigations that fatty acids created by hydrolysis of lipids during the process do not chemically contribute to char formation but are adsorbed onto the char and can be recovered by solvent extraction in high yield. Therefore, the process provides fatty acid, char, and aqueous filtrate products, all of which have utility.

Chapter 6 - Despite of its high cost, the use of live feed is essential and frequently irreplaceable in the aquaculture of mollusks, fishes, crustaceans, and some other aquatic organisms, especially in the larviculture and nursery phases (Lin et al. 2009). The use of microalgae during these phases seems to be a universal practice, because some microalgae species have adequate physical and nutritional characteristics for the early development of aquatic organisms, and the operative costs for their production is commonly lower compared to the production of other organisms or formulated feeds (Martínez Córdova et al. 1999; Lovatelli et al. 2004).

Chapter 7 - Carbon neutral renewable source of energy is needed to displace petroleum-derived fuels, which contribute to global warming and are of limited availability. Biodiesel and bioethanol, in this context, are the two potential renewable fuels that have gained substantial attraction. However, sustainability of biodiesel and bioethanol production from conventional agricultural crops is still questionable. Microalgae, a source of biodiesel are at the center of new research conducted with the aim of completely displacing fossil-based diesel. With special reference to biodiesel, the present article reviews prospects and constraints of microalgae as a source of biofuel.

There are at least 30,000 known species of microalgae, of which only a handful are currently of commercial significance due to their non-energy products such as nutraceuticals, pigments, proteins and functional foods. Though may vary with the species, microalgal biomass can be rich in proteins or rich in lipids or have a balanced composition of lipids, sugars and proteins. Under laboratory conditions, some microalgae strains were reported to generate 70 % lipid in their biomass. The fundamental chemical reaction required to produce biodiesel is the esterification of lipids, either triglycerides or oil, with alcohol, which results in a fatty acid alkylester called biodiesel (Fatty acid methyl-ester). As the fastest growing photosynthesizing organisms, biomass harvest of microalgae (158 tons/ha) is significantly higher than that of crop species such as sugarcane (75 tons/ha) used for bioethanol production. Under optimum growing conditions, a hectare of microalgae may potentially yield about 8,000 liters of biodiesel, which is 10 to 1000 times as much liquid fuel per year per hectare as conventional crops.

However, achieving the capacity to inexpensively produce biodiesel from microalgae is still challengeable. It could therefore be concluded that though microalgae are considered to be a potential source of green energy, the sustainability will largely depend on development of cost effective culture and processing techniques. Screening and collecting strains of algal species to access their potential for high oil production with high biomass productivity, investigating the physiology and biochemistry of the algae, use of molecular-biology and genetic engineering techniques to enhance the oil yield and development of advance processing techniques of cost competitive are considered to be the priority areas of research concern.

Chapter 8 - Single-cell algae (microalgae) are among the most promising resources for the production of biofuels and bioactive compounds, as well as for CO_2 biomitigation and bioremediation. Improvement of microalgal photobiotechnologies for the production of value-added products such as long-chain polyunsaturated fatty acids, storage triacylglycerols and carotenoids, requires fast and reliable, and preferably non-destructive techniques for on-line monitoring of the target product's content and the physiological condition of the algal culture. These techniques can provide essential information for timely and informed decisions on adjusting illumination conditions and medium composition, and on the optimal time for biomass harvesting. Often, such decisions must be taken within hours, and mistakes can lead to a significant reduction in productivity or a total loss of the culture. A promising approach for real-time non-destructive monitoring of laboratory and upscaled microalgal cultures is based on measuring the optical properties of algal suspensions, such as absorption, scattering and reflection of light by microalgal cells in certain spectral regions. To this aim, the following criteria should be met: i) reliable spectral measurements, ii) efficient algorithms for the processing of spectral data, and iii) a thorough understanding of the relationships between changes in physiological condition and/or biochemical composition of the algal culture and accompanying changes in its optical properties. This chapter presents a review of recent experimental work in this area, with an emphasis on investigations conducted by the authors and their colleagues in the fields of physiology, biochemistry and spectroscopy which have implications for the cultivation of biotechnologically important microalgal species.

Chapter 9 – The worldwide fossil fuel reserves are on the decline but the fuel demand is increasing remarkably. The combustion of fossil fuels needs to be reduced due to several environmental concerns. Biofuels are receiving attention as alternative renewable and sustainable fuels to ease our reliance on fossil fuels. Biodiesel and bioethanol, the two most successful biofuels in the transport sector, are currently produced in increasing amounts from oil or food crops, but their production on a large scale competes with world food supply and security. Microalgae offer a favourable alternative source of biomass for biofuel production without compromising land and water resources since they can be easily cultured on waste land which cannot support agriculture. This chapter focuses on the potential of microalgal biomass for production of the transport fuels, biodiesel and bioethanol and the bottlenecks and prospects in algal fuel technology.

Chapter 10 - While human beings are combating against global warming, fuel shortage, resource depletion, and economic downturn, microalgae, the oldest plants on earth, are gaining intensive and unprecedented attentions. The broad variety, wide distribution, and versatile growth conditions allow microalgae to be used in various fields. To be more specific, microalgae can assist humans in solving many of the challenges we are facing. But taking advantages of their unique capabilities requires better knowledge of them. Different microalgae thrive in different environment. This review focuses on identifying the best species/strains for sequestering CO_2 from flue gas released from stationary sources. Though no complete studies have been conducted for selected strains, this review helps to narrow the range and pave the way for future in-depth investigations of well-suited microalgal species in terms of capturing CO_2 and developing value-added products.

Chapter 11 - An important amount of the applied load of pesticides enter into aquatic ecosystems from agricultural runoff or leaching and, as a consequence, have become some of the organic pollutants that appear most frequently on aquatic ecosystems. The assessment of toxic potential in surface water is one of the main tasks of environmental monitoring for the control of pollution. Animal organisms such as fishes or mussels have been examined intensively whereas little information is available on the susceptibility of water plants and plankton organisms.

As primary producers, microalgae constitute the first level of aquatic trophic chains. Due to its microscopic size, it is possible to get sample at population and community levels. Some species can be cultivated in photobioreactors under controlled conditions. Because of their short generation times, microalgae respond rapidly to environmental changes, and any effect on them will affect to higher trophic levels. In addition, microalgae offer the possibility to study the trans-generational effects of pollutant exposure, being a model of choice for the study of the long term effects of pollutant exposure at population level. Furthermore, microalgal tests are generally sensitive, rapid and low-cost effective. For all these reasons, the use of microalgal toxicity tests is increasing, and today these tests are frequently required by authorities for notifications of chemicals and are also increasingly being used to manage chemical discharges. For example, algal toxicity tests of chemicals are mandatory tests for notification of chemicals in the European Union countries. Others fields of use for algae in toxicity assessment are industrial wastewaters and leachates from waste deposits.

Cytotoxic effects of aquatic pollutants on microalgae are very heterogeneous, and they are influenced by the environmental conditions and the test species. Growth, photosynthesis, chlorophyll fluorescence and others parameters reflect the toxic effects of pollutants on microalgae; however, other relevant endpoints are less known because experimental difficulties, especially under in vivo conditions.

During the last two decades, our research group has a high priority scientific objective: study the effect of different aquatic pollutants on freshwater microalgae, with the aim to develop new methods for the detection of contaminants based on the physiological response of microalgae, with the purpose of providing an early warning signal of sublethal levels of pollution.

Chapter 12 - This study investigated co-solvent modified supercritical carbon dioxide ($SC-CO_2$) extraction of lipids and carotenoids from the microalgal species of *Nannochloropsis oculata*. The changes in content of zeaxanthin in submicronized precipitates generated from the supercritical anti-solvent (SAS) process were also examined. The effect of operational conditions on amount, recovery of the zeaxanthin and mean size, morphology of the precipitates was obtained from experimentally designed SAS process. The mean size of particles falls within several hundreds of nano meters and is highly dependent on the injection time, the content of zeaxanthin in the particulates ranged from 65 to 71%. Finally, the biological assays including antioxidant and anti-tyrosinase abilities were tested to evidence the bioactivity of zeaxanthin. This study demonstrates that elution chromatography coupled with a SAS process is an environmentally benign method to recover zeaxanthin from *N. oculata* as well as to produce nanosize particles containing zeaxanthin from algal solutions.

Chapter 13 - Biological colonisation of stone is one of the main problems related to monuments and buildings conservation. It is amply recognised that microalgae have the greatest ecological importance as pioneer colonisers of stone materials, conducting to aesthetic, physical and chemical damages. Their deterioration potential is related with their photoautotrophic nature, using the mineral components of stone substrates and sunlight as energy source without any presence of organic matter.

Stone biodeterioration by microalgae has been assessed by several authors. Most of the employed methodologies for microbial identification and monitoring are time-consuming and require extensive sampling. In addition, the scaffolding and sampling procedures required may also transform the researcher in a biodeteriorating agent itself. In this chapter, non-contact techniques for colonisation detection and monitoring are proposed in order to fulfil the mission of heritage preservation. *In vivo* chlorophyll *a* fluorescence and digital image analysis were applied to estimate microalgal biomass and to quantify coverage of limestone samples artificially colonised by algal communities. The results showed that *Ançã* and especially *Lecce* limestones were extensively colonised on their surfaces revealing significant epilithic growth, whereas *Escúzar* and *San Cristobal* limestones were endolithically colonised by photoautotrophic microorganisms.

The easily handled, portable and non-destructive techniques proposed allow the understanding of stone biodeterioration processes avoiding contact and damaging of the objects, which ensures a wide field of application on cultural heritage studies and the design of appropriate conservation and maintenance strategies.

Chapter 14 - Carotenoids are isoprenoid polyene pigments widely distributed in nature. They are the main source of the red, orange or yellow colour of many edible fruits (lemons, peaches, apricots, oranges, strawberries, cherries, and others), vegetables (carrots and tomatoes), mushrooms (milk-caps), and flowers. They are also found in animal products:

eggs, crustaceans (lobsters, crabs and shrimps) and fish (salmonids) (De Saint Blanquat, 1988).

Chapter 15 - Green microalgae biomass would represent in tropical countries an innovative proteinaceous bioresource for developing protein hydrolysates. The proteolytic modification could have special importance for the improvement of solubility of algal protein and for decreasing its residual antigenicity. This chapter examined the nitrogen solubility, residual antigenicity and safety of *Chlorella vulgaris* protein hydrolysate (Cv-PH). A high increase of nitrogen solubility in Cv-PH, with respect to *Chlorella* aqueous extract (Cv-EA) was observed over a wide pH range (2-8). Residual antigenicity of Cv-PH was measured using male guinea pigs sensitized with Cv-EA. Neither mortality nor positive anaphylaxis symptoms were observed in Cv-PH challenged animals. The safety of Cv-PH was evaluated in an oral acute toxicity study (OECD Guideline 423) and in a 28-day repeated dose oral toxicity study (OECD Guideline 407) using mice as an experimental animal model. In the acute toxicity study (at a dose of 2 000 mg/kg) neither mortality nor changes in general condition were observed over a 14-days observation period. In the repeated dose oral toxicity study (a limit test at a dose of 2 000 mg/kg) no clinical changes were found in the experimental animals. The increased hemoglobin levels and leukocyte counts, particularly neutrophils, observed in Cv-PH groups may be related to the hemopoiesis stimulatory effect reported previously in *Chlorella* protein hydrolysates. Organ weights at the end of experimentation and histopathological tests revealed no significant influence of Cv-PH. These findings indicate the safety of Cv-PH in preclinical studies. Since there were no observed adverse effects of Cv-PH in these studies, the NOAEL (no observed adverse effect level) for *Chlorella* protein hydrolysate is 2 000 mg/kg/day administered orally for 28 days. An extended knowledge of the functional properties and safety of microalgae hydrolysates can be useful in understanding their potential use in the food and pharmaceutical industries.

Chapter 16 - Heterotrophic microalgal species can be grown in processes and in bioreactors resembling what is used to grow the more common types of industrial microorganisms, bacteria, yeast, and fungi. This opportunity gives heterotrophic microalgal cultures some advantages over phototrophic microalgal cultures in terms of productivity and hygienic standard. Phototrophic microalgal and cyanobacterial cultures are typically 1-2 orders of magnitude less productive than what is often obtained in heterotrophic cultures, partly because only limited amounts of light can be supplied to these cultures, and partly because inhomogeneous light intensities inside the cultures result in low photosynthetic yields near culture surfaces and no photosynthetic activity in central zones too deep to be reached by light. It is also less problematic to maintain cultures axenic in ordinary bioreactors with more compact designs than large-scale photobioreactors, where large surface areas are needed to maximise the collection of light. Heterotrophic cultures are not influenced by climate and weather, in contrast to sunlight dependent, large-scale phototrophic cultures located outdoors.

Cultivation of heterotrophic microalgae is, however, also not unproblematic. Heterotrophic microalgae grow more slowly than many bacteria and yeasts, and heterotrophic microalgae are therefore mainly of interest if they produce something that is not made by other types of microorganisms. Only a limited number of microalgal species will grow heterotrophically, and the number of heterotrophic microalgae synthesising valuable products that cannot be obtained also from other sources is low. Still, heterotrophic species from a phylogenetically highly diverse selection chlorophytes, rhodophytes, cyanidiophytes, diatoms, heterokontophytes, euglenoids, and dinoflagellates have been or are being developed for productions of food, feed, lipids, pigments, and more. A few heterotrophic microalgal processes have also matured to commercialisation. Green algae of the genus *Chlorella* are produced heterotrophically and used as health food, and docosahexaenoic acid, an essential ω-3 poly-unsaturated fatty acid is produced in the dinoflagellate *Crypthecodinium cohnii* and the thraustochytrids *Schizochytrium* sp. and *Ulkenia* sp. and added to infant formula and foods.

Chapter 17 - The increase of atmospheric carbon dioxide is considered to be one of the main causes of global warming. Between 1990 and 2008, atmospheric CO_2 rose from 280 ppm (20.541 mill. t) to 400 ppm (29.381 mill. t). At the same time, fossil oil resources are said to be depleted within a few decades if fuel consumption remains at current levels. It is, therefore, crucial to explore alternatives to oil producing sources and also to reduce atmospheric CO_2 concentration.

Microalgae have been suggested as excellent candidates meeting the requirements: they are able to fix large amounts of carbon dioxide and transform it into biomass with high content of lipids.

The lipid content of microalgae is characteristic of their genus and species and also depends on the different growth phases and culture conditions like nutrient supply (especially nitrogen and phosphate amounts in culture medium), light intensity, temperature, pH and carbon dioxide concentration.

Different microalgae species of the division of Chlorophyta (*Scenedesmus* sp. and *Chlorella* sp.) were investigated according to their biomass productivity, lipid content, fatty acid profile and tolerance of high levels of carbon dioxide. The study showed that a higher CO_2 level leads to a decrease in biomass concentration and an increase in lipid content of the analysed species. For the most part, lipids contain saturated and unsaturated fatty acids with a chain length between C14 and C22. With increasing CO_2 concentration, the content of unsaturated fatty acids with 1 and 2 double bonds increases, whereas the content of linolenic acid, an acid with 3 double bonds, decreases.

In: Microalgae: Biotechnology, Microbiology and Energy
Editor: Melanie N. Johnsen

ISBN 978-1-61324-625-2
© 2012 Nova Science Publishers, Inc.

Chapter 1

MICROALGAE BIOTECHNOLOGICAL APPLICATIONS: NUTRITION, HEALTH AND ENVIRONMENT

A. E. Marques, J. R. Miranda, A. P. Batista, and L. Gouveia

Unidade de Bioenergia - Laboratório Nacional de Energia e Geologia (LNEG).
Estrada do Paço do Lumiar, 1649-038 Lisboa, Portugal

ABSTRACT

Microalgae (prokaryotic and eukaryotic) consist of a wide range of autotrophic organisms which grow through photosynthesis just like land based plants. Their unicellular structure allows them to easily convert solar energy into chemical energy through CO_2 fixation and O_2 evolution, being well adapted to capture CO_2 and store it as biomass. Microalgae and cyanobacteria have an interesting and not yet fully exploited potential in biotechnology. They can be used to enhance the nutritional value of food and animal feed due to their chemical composition, playing a crucial role in aquaculture. Highly valuable molecules like natural dyes (e.g. carotenoids), polyunsaturated fatty acids, polysaccharides and vitamins from algal origin are being exploited and can be applied in the nutritional supplements; cosmetics (e.g. phycocyanin) and pharmaceuticals. In fact, microalgae and cyanobacteria are able to produce several biologically active compounds with reported antifungal, antibacterial, anticancer, antiviral (e.g. anti-HIV), immunosuppressive, anti-inflamatory and antioxidant activity. Nowadays, there is a focus on using microalgae in renewable energy sources and environmental applications. Microalgae are a potential source for biofuels production such as biodiesel, bioetanol, biohydrogen and biogas. These can be produced through a biorefinery concept, in which every component of the biomass is used to produce usable products. This strategy can integrate several different conversion technologies (chemical, biochemical, termochemical and direct combustion) providing a higher cost effective and environmental sustainability for the biofuels production. Environmental applications can include CO_2 sequestration and wastewater treatment. This can be achieved by coupling microalgae production systems with industrial polluting facilities.

Keywords: microalgae; nutrition; biomolecules; health; bioenergy; biofuels; CO_2 mitigation; wastewater treatment; genetic engineering.

1. INTRODUCTION

Microalgae are an extremely heterogeneous group of organisms, described as a life-form, not a systematic unit. They are regarded as unicellular photoautotrophic (contain chlorophyll *a*) microorganisms, than can be eukaryotic or prokaryotic. The diversity of the microalgae is very broad and is reflected in an equally wide range of metabolisms and biochemical properties as a diversity of pigments, photosynthetic storage products, cell walls and mucilages, fatty acids and lipids, oils, sterols and hydrocarbons, and bioactive compounds, including secondary metabolites (Gouveia et al., 2010). This biodiversity implies that groups of organisms are differentiated by some measure of the extent to which their gene pools are separated and how this is expressed phenotypically.

The phylogeny of algae and related organisms has evolved dramatically in recent years. Molecular and ultrastructural evidence of evolutionarily conserved features (e.g. ribosomal RNA gene sequencing, flagellar hairs and roots, plastid and mitochondrial structure, the mitotic apparatus) has combined to create an exciting, dynamic field of inquiry.

For purposes of classification and for understanding biological and evolutionary relatedness (phylogeny), the specie is the fundamental unit for classifying groups of organisms. With theincreasing availability of molecular information (e.g. protein and nucleic acid sequence data), there has been a movement toward reconciliation of taxonomic and phylogenetic approaches. This is leading to classification systems that reflect some biological reality, such as the degree to which groups of populations are genetically similar, with implications for evolutionary history and speciation (Metting Jr, 1996).

Historically, species of microalgae were recognized on the basis of phenotypic properties, such as whole organism morphology, cellular anatomy and ultrastructure, metabolism and physiology and were described and categorized according to the International Code of Botanical Nomenclature. More recently, it has been recommended by some bacteriologists that the taxonomy of prokaryotic blue-green algae (cyanobacteria or cyanophytes) and prochlorophytes be treated under the International Code of Nomenclature of Bacteria, with the botanical system serving as the baseline (Castenholz and Waterbury, 1989). However, very few species have been described under the bacteriological code and most researchers agree to the pressing needs for resolution of this issue. Figure 1.1 illustrates the diversity among major lineages, where it can be seen that algae are phylogenetically more diverse than either plants or animals. Among the algae, fossil records show the blue-green lineage arising early in the Precambrian. The red and green lineages date from the mid to late-Precambrian. Brown algae are first seen in the Paleozoic while most other lineages date from the early to late Mesozoic (Anderson, 1996). Figure 1.2 shows different species of algae and the diversity of phenotypes.

These microorganisms occur in terrestrial environments and, although most species require at least a film of liquid water to be metabolically active, they can be found in a remarkable range of habitats, from snowfields to the edges of hot springs, and from damp earth and the leaves of plants to sun-baked desert soils. Eventhe insides of rocks are open to colonization, with endolithic algae occupying tiny cracks in rocks in alpine and arctic environments that show no obvious signs of vegetation.

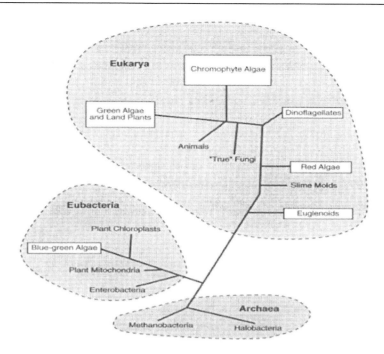

Figure 1.1. Depiction of the phylogenetic relatedness of some groups of algae. Lengths of line segments are proportional to evolutionary distance based on analysis of ribosomal RNA gene sequences (modified from Radmer and Parker, 1994).

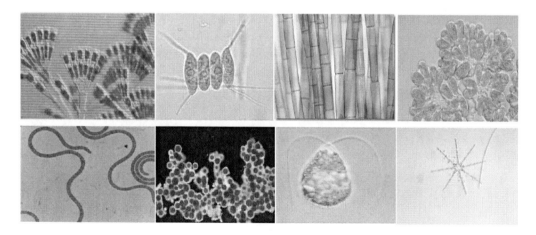

Figure 1.2. Phenotypical diversity of microalgae. Different forms and shapes according to their genetic and environmental characteristics.

Because they are key primary producers, algae play a vital role in the Earth's carbon cycle, and the Earth's atmosphere would not contain free oxygen at all if it were not for the activities of algae and cyanobacteria (Graham and Wilcox, 2000; van den Hoek et al., 1995). In fact, as the predominant component of the marine and freshwater plankton, microalgae are primarily responsible for the 40-50% of total global photosynthetic primary production contributed by all algae (Harlin and Darley, 1988).

Microalgae have been the subject of applied research for their commercial and industrial potential since the early 1950's when productivity and yield were first studied in mass culture (Burlew, 1953). More recently, microalgae have been targeted as a source of bioactive compounds and pharmaceuticals, specialty chemicals, health foods, aquaculture feeds, and for waste treatment, agriculture and biofuels (Akatsuka,1990; Borowitzka and Borowitzka,1988; Lembi and Waaland, 1989; Radmer, 1996; Spolaore et al., 2006).

Microalgal biotechnology is a form of biomass production similar to conventional agriculture, presenting some advantages namely because algae are more photosynthetically efficient than terrestrial plants. In addition, microalgae can reach higher biomass productivities, faster growth rate, higher CO_2 fixation and O_2 production rates when compared with higher plants. They are feasibly grown in liquid medium which can be handled easily, and can be cultivated in variable climates and non-arable land, including marginal areas unsuitable for agricultural purposes (e.g. desert and seashore lands). They can use far less water than traditional crops, with the advantage of using non-potable water, including waste waters that can be treated by them. Microalgae cultivation can avoid environmental impacts, such as soil desertification and deforestationand do not displace food crop cultures. Moreover, their production is not seasonal; there is no need for pesticides or herbicides and does not produce contaminants (Chisty, 2007; Rodolfi et al., 2009).

Commercial production of phototrophic microbial biomass is limited to a few microalgal species such as *Arthrospira* and *Dunaliella*,that are cultivated in open ponds (Figure 1.3) by means of a selective environment (e.g. high pH or salinity) or a high growth rate (Tredici, 2004).

Closed photoautotrophic culture systems (e.g. tubular, flat, vertical cylinders and sleeves photobioreactors), with transparent walls (glass or plastic), have been developed in the last years, to overcome the limitations of open systems, mainly the low productivity and risk of contamination (e.g. microorganisms, heavy metals), and to enable the culture of specific microalgae that do not grow in highly selective environments (e.g. high salinity and alcalinity) (Borowitzka, 1999).

Figure 1.3. Microalgae cultivated in open ponds systems. *Dunaliella* by Cognis at Hutt Lagoon (Australia) (a); *Arthrospira* in raceway ponds by Cyanotech (Hawaii, USA) (b) and Earthrise Farms (California, USA) (c).

Some attempts have been made to develop commercial-scale photobioreactors, but most were closed after a few months of operation, including Photo Bioreactors Ltd plant in Santa Ana (Murcia, Spain). The first truly successful large-scale industrial production of microalgae in a closed photobioreactor has been accomplished by the system developed by Prof. Otto Pulz (2001) in a plant built in Klötze (Germany) by Ökologische Produkte Altmark GmbH

(ÖPA) and run by IGV Ltd. The plant consists of a 700 000 L glass tubular reactor (500 km total length), divided in 20 subunits, installed in a 12 000 m² greenhouse (Figure 1.4).

In this chapter a brief overview on various biotechnological applications of microalgae studied so far will be presented.

Figure 1.4. *Chlorella* sp. growing in tubular photobioreactors (Klötze, Germany).

2. MICROALGAE BIOTECHNOLOGY AND NUTRITION

Microalgae use as natural food by indigenous populations has occurred for centuries, however, the cultivation of microalgae is only a few decades old (Borowitzka, 1999). Edible blue-green microalgae, including *Nostoc*, *Spirulina*, and *Aphanizomenon* species, have been used as a nutrient-dense food for many centuries in Asia, Africa and Mexico (Figure 2.1) (Hallman, 2007; Abdulqader et al., 2000).

Figure 2.1. Aztecs harvesting algae from lakes in the Valley of Mexico (a); Kanembu women gathering *Spirulina* from area around Lake Chad (b).

Among the thousands of species that are believed to exist (Chaumont, 1993; Radmer and Parker, 1994), only a few thousands strains are kept in collections, a few hundred are investigated for chemical content and just a handful are cultivated in industrial quantities

(Olaizola, 2003). Some of the most biotechnologically relevant microalgae are the Cyanobacteria *Arthrospira* (*Spirulina*) and the green algae (Chlorophycea) *Chlorella vulgaris*, *Haematococcus pluvialis* and *Dunaliella salina* which are already widely commercialized and used, mainly as nutritional supplements for humans and as animal feed additives.

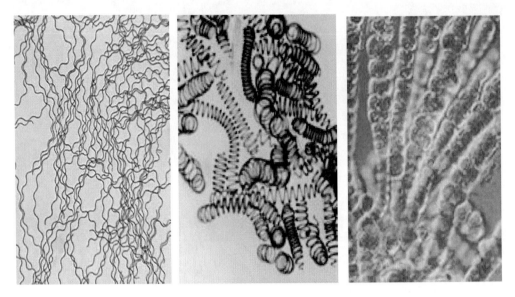

Figure 2.2. *Arthrospira* (*Spirulina*) sp.

Arthrospira (*Spirulina*) is an ancient microscopic filamentous cyanobacteria (prokaryotic) that belongs to the Class Oscilatoriacea (Figure 2.2). It is classified as a microalga (blue-green alga) due to its chlorophyll *a* content and ability to do photosynthesis (photoautotrophic). *Spirulina* grows profusely in certain alkaline lakes in Mexico and Africa, forming massive blooms, and has been used as food by local populations since ancient times (Yamaguchi, 1997). Since the late 1970s, when the first large-scale *Spirulina* production plant was established in Mexico, it has been extensively produced around the world (Hawaii, California, China, Taiwan, Japan) using open raceway ponds (Borowitzka, 1999). It is estimated a total production of 3000 tons/year, being broadly used in food and feed supplements, due of its high protein content and its excellent nutritive value, such as high γ-linolenic acid (GLA; 18:3ω6) and vitamin B_{12} level (Ötles and Pire, 2001; Shimamatsu, 2004). *Spirulina* is also the main source of natural phycocyanin, a valuable blue pigment used as a natural food and cosmetic colouring (Ötles and Pire, 2001; Kato, 1994; Shimamatsu, 2004).

Chlorella was the first microalga to be isolated and cultivated in laboratory by Beijerinck in 1890. Belongs to the Chlorophyta (green algae) family, and presents Chlorophyll *a* and *b* and several carotenoids, that may be synthesized and accumulated outside the chloroplast under conditions of nitrogen deficiency and/or other stress, colouring the alga orange (Figure 2.3).

Chlorella has been used as an alternative medicine in the Far East since ancient times and it is known as a traditional food in the Orient. The commercial production of *Chlorella* as a novel health food commodity started in Japan in the 1960s, under the scientific supervision of the Microalgae Research Institute of Japan (*Chlorella* Institute), and by 1980 there were 46 large-scale factories in Asia (Borowitzka, 1999). Nowadays, *Chlorella* is widely produced and marketed as a health food supplement in many countries, including China, Japan, Europe and the US, being estimated a total production around 2000 ton/year in the 1990s (Lee, 1997).

Chlorella is considered as a potential source of a wide spectrum of nutrients (*e.g.* carotenoids, vitamins, minerals) being widely used in the healthy food market as well as for animal feed and aquaculture.

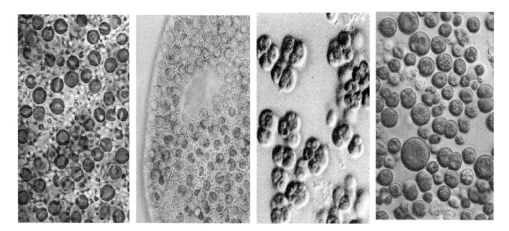

Figure 2.3. *Chlorella vulgaris* green and orange (carotenogenic) microalga.

Haematococcus is a freshwater, unicellular, green alga (Chlorophyceae, order Volvocales) that is extensively used for the production of the orange-red pigment astaxanthin (Figure 2.4). When green vegetative cells come across stress conditions (e.g. nitrogen deficiency, high light intensity, salt stress) the alga rapidly differentiates into encysted cells that accumulate the ketocarotenoid astaxanthin (3,3'-dihydroxy-β,β-carotene-4,4'-dione) in globules outside the chloroplast. It has been suggested that the accumulated astaxanthin might function as a protective agent against oxidative stress damage (Kobayashi et al., 1997).This carotenoid pigment is a potent radical scavenger and singlet oxygen quencher, with increasing amount of evidence suggesting that surpasses the antioxidant benefits of β-carotene, vitamin C and vitamin E (Todd-Lorenz and Cysewski, 2000).

In the 1990s in the USA and India, several plants started with large-scale production of *Haematococcus pluvialis*, which is currently the prime natural source of astaxanthin for commercial exploitation, particularly as pigmentation source in farmed salmon, trout and poultry industries.

Dunaliella salina is an halotolerant microalga, naturally occurring in salted lakes, that is able to accumulate very large amounts of β-carotene, a valuable chemical mainly used as natural food colouring and provitamin A (retinol). The *D. salina* community in Pink Lake, Victoria (Australia) was estimated to contain up to 14% of this carotenoid in their dry weight (Aasen et al., 1969), and in culture some *Dunaliella* strains may also contain up to 10% and more β-carotene, under nutrient-stressed, high salt and high light conditions (Ben-Amotz and Avron, 1980; Oren, 2005). Apart from β-carotene *Dunaliella* produces another valuable chemical, glycerol.

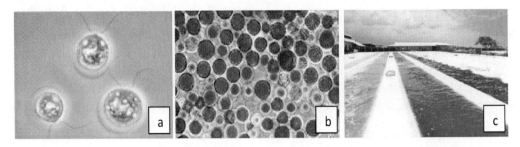

Figure 2.4. *Haematococcus pluvialis* a) green vegetative cells; b) encysted carotenogenic cells; c) cultivated in open raceway ponds.

2.1. Microalgae in Human Nutrition

In early 1950's microalgae were considered to be a good supplement and/or fortification in diets for malnourished children and adults, as a single cell protein (SCP) source, in response to concerns of an increasing world population and a predicted insufficient protein supply.

Although microalgae are consumed since ancient times, they are considered as unconventional food items and have to undergo a series of toxicological tests to prove their harmlessness. In fact, some algae have been tested under all possible aspects much more carefully than most of any conventional food commodities (Becker, 2004).

Some of the prerequisites for the utilization of algal biomass for humans and animals include the determination of proximate chemical composition; biogenic toxic substances; non-biogenic toxic compounds; protein quality studies; biochemical nutritional studies; supplementary value of algae to conventional food sources; sanitary analysis; safety evaluations (feeding trials with animals); clinical studies (test for safety and suitability of the product for human consumption) and acceptability studies (Becker, 2004).

Some human nutritional studies were done with humans and the authors suggest that the algae daily consumption should be restricted to about 20 g, with no harmful side effects occurrence, even after a prolonged period of intake (Becker, 1988). While some studies report that people have lived solely on algae for prolonged periods of time without developing any negative symptoms, in other studies, discomfort, vomiting, nausea, and poor digestibility of even small amounts of algae were reported.

Powell et al. (1961) performed one of the first studies, in which a meal containing up to 500 g of a mixture of *Chlorella* and *Scenedesmus* was given to young healthy men. Subjects tolerated well 100 g incorporation levels, but above this some gastrointestinal disorders were observed.

Gross et al. (1982) performed a study feeding algae (*Scenedesmus obliquus*) to children (5 g/daily) and adults (10 g/daily), incorporated into their normal diet, during four-week test period. Haematological data, urine, serum protein, uric acid concentration and weight changes were measured, and no changes in the analyzed parameters were found, except a slight increase in weight, especially important for children. The same authors also carried out a study (Gross et al., 1978) with slightly (group I) and seriously (group II) malnourished infants during three weeks. The four-years-old children of group I (10 g algae/daily) showed a significant increase in weight (27 g/day) compared with the other children of the same group who received a normal diet, and no adverse symptoms were recorded. The second group was nourished with a diet enriched with 0.87 g algae/kg body weight, substituting only 8% of the total protein and the daily increase in weight was about sevenfold (in spite of a low protein contribution) and all anthropogenic parameters were normal. The authors concluded that the significant improvement in the state of the health was attributed not only to the algal protein but also to therapeutic factors.

Almost no adverse symptoms have been revealed so far in connection with the consumption of microalgae and unwanted side effects appear to be extremely rare (Becker, 2004). However, there are still some health concerns remaining regarding the ingestion of microalgae. Several strains of cyanobacteria have been identified with the production of biogenic toxins (Cox et al., 2005). However, these cases are associated to wild algal blooms and no such cases have ever been reported in connection to mass cultured algae (Becker, 2004). Non-biogenic toxins, such as heavy metals and other contaminants, can also be avoided by proper cultivation techniques and non-polluted cultivation areas. The content in nucleic acids (RNA and DNA) is another concern, since these are sources of purines which are uric acid precursors that when accumulated in the serum may increase the risk of gout and kidney stones. In fact, this has been the major limitation for SCP use as food or food ingredient. Although, microalgae have relatively low nucleic acid contents (4-6%) as compared to yeasts (8-12%) and bacteria (20%), so an intake below 20 g of algae per day or 0.3 g of algae per kg of body weight should present no harm (Becker, 2004).

2.1.1. Novel Foods Regulation

Authorization of novel foods and novel food ingredients is harmonised in the European Union (EU) by the Regulation EC 258/97. Foods commercialised in at least one Member State before the entry into force of the Regulation on Novel Foods on 15 May 1997, are on the EU market under the *"principle of mutual recognition"*. This is the case of the microalgae *Arthrospira* (*Spirulina*) *platensis*, *Chlorella pyrenoidosa*, and *Aphanizomenon flosaquae* (filamentous blue-green algae from Klamath Lake, Oregon USA), according to the DG Health and Consumer Protection, Novel Foods Catalogue (http://ec.europa.eu/food/food/biotechnology/novelfood/).

In order to ensure the highest level of protection of human health, novel foods must undergo a safety assessment before being placed on the EU market.

The application of DHA-rich algal oil from *Schizochytrium* sp. for additional food uses by Martek Biosciences Corporation (USA) is currently under evaluation. The microalga *Odontella aurita* from Innovalg (France) was considered substantially equivalent (simplified procedure in article 5th) to other authorized algae in December 2002, as well as DHA (docosahexahenoic acid)-rich microalgal oil (DHActive™) from Nutrinova (Germany) in November 2003. Novel foods notifications of Astaxanthin-rich extracts derived from *Haematococcus pluvialis* have been approved for several companies such as US Nutra (USA), AstaReal AB (Sweden), Alga Technologies Ltd (Israel) and Cyanotech (USA). The successful authorization of these microalgal based foods and food ingredients broaden perspectives for a wider inclusion of these valuable microrganisms in the diet.

2.1.2. Examples of Microalgae Food Applications

Commercial large-scale production of microalgae started in the early 1960s in Japan with the culture of *Chlorella* as a food additive, which was followed in the 1970s and 1980s by expanded world production in countries such as USA, India, Israel, and Australia (Spolaore et al., 2006; Borowitzka, 1999). In 2004, the microalgae industry had grown to produce 7000 tonnes of dry matter per year (Table 2.1) with *Chlorella, Spirulina* and *Dunaliella* dominating the market (Pulz and Gross, 2004).

Nowadays, microalgae are mainly marketed as health food or food supplement and commonly sold in the form of tablets, capsules, and liquids. Much attention is being diverted to algae as ingredient factories, particularly of nutritional ingredients such as omega-3 fatty acid DHA and astaxanthin carotenoid. In fact, while *Chlorella*, *Spirulina* and *Haematococcus* biomass are sold at 36-50€/kg for human and animal nutrition (aquaculture), fine chemical compounds reach far most expensive prices: β-carotene 215-2150 €/kg, DHA oil 43€/g, astaxanthin 7 €/mg and phycocyanin 11€/mg (Brennan and Owende, 2010).

Table 2.1. Major microalgae commercialized for human nutrition (adapted from Pulz and Gross, 2004; Spolaore et al., 2006 and Hallmann, 2007)

Microalga	Major Producers	Products	World Production (ton/year)
Spirulina (*Arthrosphira*)	Hainan Simai Pharmacy Co. (China) Earthrise Nutritionals (California, USA) Cyanotech Corp. (Hawaii, USA) Myanmar Spirulina factory (Myanmar)	powders, extracts tablets, powders, extracts tablets, powders, beverages, extracts tablets, chips, pasta and liquid extract	3000
Chlorella	Taiwan Chlorella Manufacturing Co. (Taiwan) Klötze (Germany)	tablets, powders, nectar, noodles powders	2000
Dunaliella salina	Cognis Nutrition and Health (Australia)	powders β-carotene	1200
Aphanizomenon flos-aquae	Blue Green Foods (USA) Vision (USA)	capsules, crystals powder, capsules, crystals	500
Haematococcus pluvialis	Cyanotech (USA), Mera Pharmaceuticals (USA), Parry's Pharmaceuticals (India), Algatech (Israel)	astaxanthin	300
Chrypthecodinium cohnii	Martek (USA)	DHA oil	240
Schizochytrium	Martek (USA)	DHA oil	10

Additionally, there is an increasingly growing market for food products with microalgae addition (Figure 2.5) such as pastas, biscuits, bread, snack foods, candy bars or gums, yoghurts, drink mixes, soft drinks, etc., either as nutritious supplement, or as source of natural food colorant (Becker, 2004).

In some countries (Germany, France, Japan, USA, China, Thailand), food production and distribution companies have already started serious activities to market functional foods with microalgae and cyanobacteria (Pulzand Gross, 2004). Although, the biotechnological exploitation of microalgae resources for human nutrition purposes is restricted to very few species due to the strict food safety regulations, commercial factors, market demand and specific preparation (Pulz and Gross, 2004). Potential consumers may have some reluctance to use algal or algal products related to conservative ethnic factors including religious and socio-economic aspects (Becker, 2004).

But foods supplemented with microalgae biomass might be sensorially more convenient and variable, thus combining health benefits with attractiveness to consumers, namely in terms of colour.

Also, several microalgae when correctly processed have an attractive or piquant taste and could be thus well incorporated into many types of foods, adding not only nutritional value, but also new, unique and attractive tastes (Richmond, 2004).

Figure 2.5. Examples of commercial microalgae based food products.

In the last years, some research has been carried out by our team regarding the development a range of novel attractive healthy foods, prepared from microalgae biomass, rich in carotenoids and polyunsaturated fatty acids with antioxidant effect. At the same time toxicological studies involving all the microalgae to be incorporated are being conducted (Bandarra et al., 2010). Traditional foods, such as mayonnaises/salad dressings, puddings/gelled desserts, biscuits/cookies and pasta, largely consumed on daily basis on different European diets, were studied. The effect of microalgal concentration on the products colour parameters was investigated, as well as its stability throughout the processing conditions and during storage time.

Pea protein stabilized oil-in-water emulsions (mayonnaise type) were coloured with *Chlorella vulgaris* (green/carotenogenic) and *Haematococcus pluvialis* biomass (carotenogenic) (Raymundo et al., 2005; Gouveia et al., 2006). Pea protein isolate was also used, in combination with kappa-carrageenan and starch polysaccharides, to develop "animal-free" gelled desserts, as an alternative to dairy-desserts (Nunes et al., 2006). The gels were coloured with different microalgae such as *Chlorella vulgaris* (green/carotenogenic), *Haematococcus pluvialis* (carotenogenic), *Spirulina maxima* and *Diacronema vlkianum* (Batista et al., 2008; 2011; Gouveia et al., 2008a).

Chlorella vulgaris (green) (Gouveia et al., 2007) and *Isochrysis galbana* (Gouveia et al., 2008b) have been studied as colouring and PUFA-ω3 sources in short dough butter cookies, previously optimized (Piteira et al., 2004).

More recently, the addition of microalgal biomass *Chlorella vulgaris* (green/carotenogenic), *Spirulina maxima*, *Isochrysis galbana* and *Diacronema vlkianum* on durum wheat semolina pasta products was studied (Fradique et al., 2010).

In general, the developed products presented appealing and stable colours (Figure 2.6) with added value in terms of health benefits, considering the antioxidant properties and PUFA-ω3 content of the microalgae. The results obtained are promising since it was possible to obtain common food products enriched with microalgae, resulting stable, attractive and healthier foods with enormous potential in the functional food market.

Another example of emerging applications of microalgae in novel food products is the study of Valencia and co-workers (2007) that developed dry fermented sausages rich in docosahexaenoic acid (DHA) with oil from the microalgae *Schizochytrium* sp. The influence on nutritional properties, sensorial quality and oxidation stability was evaluated, with promising results.

However, in this study only the oil fraction was added while in the above cited studies by our team full microalgal biomass was added. This approach allows saving costs related to microalgae cell rupture and extraction, as at the same time provides natural encapsulation of the bioactive compounds (e.g. pigments, antioxidants, PUFAs) that could otherwise be degraded throughout food processing operations.

Figure 2.6. Oil-in-water emulsions (a); pastas (b); desserts (c); and short-dough biscuits with microalgal biomass addition.

2.2. Microalgae in Animal Nutrition

Several microalgae (e.g. *Chlorella, Tetraselmis, Spirulina, Nannochloropsis, Nitzchia, Navicula, Chaetoceros, Scenedesmus, Haematococcus, Crypthecodinium*), macroalgae (e.g. *Laminaria, Gracilaria, Ulva, Padina, Pavonica*) and fungi (*Mortierella, Saccharomyces, Phaffia, Vibrio marinus*) can be used in both terrestrial (e.g. poultry, ruminants and pigs) and aquatic animal feed (Harel and Clayton, 2004).

Using even very small amounts of microalgal biomass can positively affect the physiology of animals by improved immune response, resulting in growth promotion, disease resistance, antiviral and antibacterial action, improved gut function, probiotic colonization stimulation, as well as by improved feed conversion, reproductive performance and weight control (Harel and Clayton, 2004). The external appearance of the animals may also be improved, resulting in healthy skin and a lustrous coat, for both farming animals (poultry, cows, breeding bulls) and pets (cats, dogs, rabbits, ornamental fishes and birds) (Certik and Shimizu, 1999).

The large number of nutritional and toxicological evaluations already conducted has demonstrated the suitability of algae biomass as a valuable feed supplement (Becker, 1994a). In fact, 30% of the current world algal production is sold for animal feed applications (Becker, 2004).

Microalgae play a key role in aquaculture, since they constitute the basis of the natural food chain, being the food source for larvae of many species of mollusks, crustaceans and fish. In addition, microalgae serve as a food source for zooplankton production (rotifers, copepods), which in turn are used as feed for rearing fish larvae (Lavens and Sorgeloos 1996; Pulz and Gross, 2004).

The marine microalgae *Isochrysis galbana* and *Diacronema vlkianum* (Haptophyceae) are recognized by their ability to produce long chain polyunsaturated fatty acids (LC-PUFA), mainly eicosapentaenoic acid (EPA, 20:5ω3) and also docosahexaenoic acid (DHA, 22:6ω3), that are accumulated as oil droplets in prominent lipid bodies in the cell (Liu and Lin, 2001; Bandarra et al., 2003; Donato et al., 2003). These microalgae are widespread in marine environment, forming a major part of marine phytoplankton, and have been used as feed species for commercial rearing of many aquatic animals, particularly larval and juvenile molluscs, crustacean and fish species (Fidalgo et al., 1998).

Haematococcus pluvialis, identified as the organism which can accumulate the highest level of astaxanthin in nature (1.5-3.0% dry weight) (Todd-Lorenz and Cysewski, 2000), has been widely used in the rearing of some marine animals, such as salmonids, trouts, shrimps, lobsters and crayfish, providing a pinkish-red hue similar to the wild species.

3. MICROALGAE BIOTECHNOLOGY AND PHARMACEUTICALS

Pharmaceutical industry is growing at a CAGR (Compound Annual Growth Rate) of around 8% while the global pharmaceutical market is forecasted to reach 764,144€ billion in 2012 (www.oilgae.com).

Macroalgae, microalgae and cyanobacteria represent a very large, relatively unexploited reservoir of novel compounds, many of which are likely to show biological activity, presenting unique and interesting structures and functions (Febles et al., 1995; Yamaguchi, 1997; Siddhanta et al., 1997). In the past four decades, algae have been shown as a potential source of these compounds such as organic acids, carbohydrates, amino acids, peptides, vitamins, antibiotics, enzymes, pigments, toxic compounds. They are part of the normal cell growth, related to the interactions with the environment, and their production is usually favoured by sub-optimal growth conditions (Ghasemi et al., 2004; Ghasemi et al., 2007; Sukumaran and Thevanathan, 2011).

Crude extracts of marine macroalgae have been extensively studied (more than 20 research papers) for their antimicrobial properties (e.g. Allen and Dawson, 1960; Burkholder et al.,1960; Duff et al.,1966; Hornsey and Hide, 1974; Glombitza, 1979; Henriquez et al., 1979; Pesando et al., 1979; Pesando and Caram, 1984; Caccamese and Azzolina, 1979; Calvo et al., 1986; Alam et al., 1994; Horikawa et al., 1966; Robles Centeno et al., 1996; Rovirosa and San Martin, 1997; Robles Centeno and Ballantine, 1999; Walter and Mahesh, 2000;

Dhamotharan, 2002; Salvadoret al., 2007) however, reports on fresh water algae are only a few (e.g.Prattet al., 1944; Harder and Opperman, 1954; Davidson, 1961; Gupta and Shrivastava, 1965; Stangenberg, 1968; Pandey and Gupta, 1977; Debro and Ward, 1979; Mason et al., 1982; Prashantkumar et al., 2006).

Cyanobacteria have also been identified as one of the most promising group of organisms from which novel and biochemically active natural products are isolated. Cyanobacteria such as *Microcystis, Anabaena, Nostoc* and *Oscillatoria* produce a great variety of secondary metabolites. The only comparable group is *actinomycetes*, which has yielded a tremendous number of metabolites. The rate of discovery from traditional microbial drug producers like *actinomycetes* and *hyphomycetes*, which were in the focus of pharmaceutical research for decades, is decreasing and it is the time to turn to cyanobacteria and exploit their potential. This is of paramount importance to fight increasingly resistant pathogens and newly emergent diseases (Hayashi et al., 1994).

Unfortunately, most of the studies have only used *in vitro* assays and, it is likely that most of these compounds have little or no application in medicine as they are either too toxic or inactive *in vivo* (Borowitzka, 1995).

Although there has been some interest in the exploitation of cultured algae, microalgae and cyanobacteria to develop pharmaceutical compounds, the study of these secondary metabolites and their controlled long-term production is still in its infancy. One of the major reasons why drug candidates do not make it to the world pharmaceutical market is the relatively small yield of the compounds available from natural stock, usually between 1 and 50 g on a yearly basis. There is a need to develop detailed procedures for the production of biochemically active secondary metabolites from these organisms.

Bioprocess intensification strives to overcome this shortfall by developing detailed mechanistic growth kinetics for particular organisms and thereby design bioreactors based on an ecological approach. Bioprocess intensification involves optimizing fermentation yield via media composition and field strategies, dynamic control of physical conditions, induction genetics, immobilization, and bioreactor engineering. Since, most of these organisms grow under photoautotrophic condition with carbon dioxide and light as a carbon and energy source respectively, it leaves relatively fewer, perhaps more critical, culture parameters to be manipulated for control of secondary metabolite production (Burja et al., 2001; 2002). In this section, some microalgae and cyanobacteria bioactive compounds will be addressed.

3.1. Cyanobacteria as a Producer of Bioactive Compounds

Cyanobacteria have been screened for new pharmaceuticals such as antibiotics, antiviral, anticancer, enzyme inhibitory agents and therapeutic applications in the treatment of cancer. Published data until 1996 revealed 208 cyanobacterial compounds with biological activity while in 2001 the number of compounds screened was raised to 424 (Figure 3.1) (Burja et al., 2001). The reported biological activities comprise cytotoxic, antitumor, antibiotic, antimicrobial (antibacterial, antifungal, antiprotozoa), antiviral (*e.g.* anti-HIV) activities as well as biomodulatory effects like immunosuppressive and anti-inflammatory (Burja et al., 2001; Singh et al., 2005).

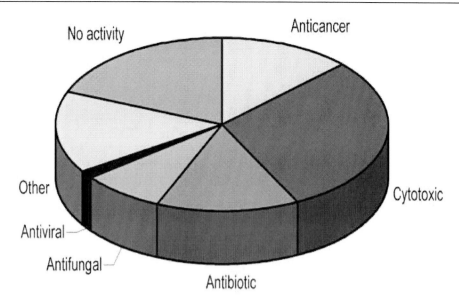

Figure 3.1. Reported biological activity of cyanobacterial compounds – 2001 (424 compounds) (Nagle et al., 1996; Burja et al., 2001).

Some examples of cyanobacterial derived compounds with antiviral, anticancer, anti-inflammatory, antimicrobial among other activities are presented in the following subsections.

3.1.1. Antiviral Activity

Cyanovirin-N (CV-N) is a unique, 101 amino acid long, 11 kDa protein (Figure 3.2). CV-N is directly virucidal. It inactivates, potently and irreversibly, diverse primary strains of HIV-1, including M-tropic forms involved in sexual transmission of HIV, and also blocks cell-to-cell transmission of HIV infection.

This protein was discovered as a constituent of a cultured cyanobacterium, *Nostoc ellipsosporum*, and both the sequence and the 3-D structure of CV-N are unprecedented. (Burja et al., 2001).

CV-N is extremely resistant to physicochemical degradation and can withstand treatment with denaturants, detergents and organic solvents, multiple freeze-thaw cycles, and heat with no apparent loss of antiviral activity.

CV-N interacts in an unusual manner with the viral envelope and inhibits fusion of virus with CD4 cell membrane (Yang et al., 1997). It has a potent activity against all immunodeficiency viruses (HIV-1, M- and T-tropic strains of HIV-1, HIV-2, SIV (simian) and FIV (feline) (Burja et al., 2001). Several patents have been filed to protect this method of HIV prevention (Boyd, 2001; 2002; 2004).

Gustafson et al. (1989) reported the protection of human lymphoblastoid T cells from the cytopathic effect of HIV infection with the extract of the blue-green algae *Lyngbya lagerheimeii* and *Phormidium tenue*. A new class of HIV inhibitors called sulfonic acid, containing glycolipid, was isolated from the extract of blue-green algae and the compounds were found to be active against the HIV virus. Calcium spirulan (Ca-SP), a novel sulphated polysaccharide, is an antiviral agent. This compound selectively inhibits the entry of enveloped virus (Herpes simplex, human cytomegalovirus, measles virus) into the cell (Hayashi et al., 1996; Hayashi and Hayashi, 1996; Ayehunie et al., 1998).

```
NH2-L G K F S Q T C Y N
    S A I Q G S V L T S
    T C E R T N G G Y N
    T S S I D L N S V I
    E N V D G S L K W Q
    P S N F I E T C R N
    T Q L A G S S E L A
    A E C K T R A Q Q F
    V S T K I N L D D H I
    A N I D G T L K Y E —COOH
```

Figure 3.2. Cyanovirin N aminoacid sequence (Burja et al., 2001).

3.1.2. Anticancer Effect

Borophycin (Figure 3.3) is a boron containing metabolite isolated from marine strains of cyanobacteria *Nostoc linckia* and *Nostoc spongiaeforme var. tenue*. It exhibits potent cytotoxicity against human epidermoid carcinoma and human colorectal adenocarcinoma cell lines (Burja et al., 2001).

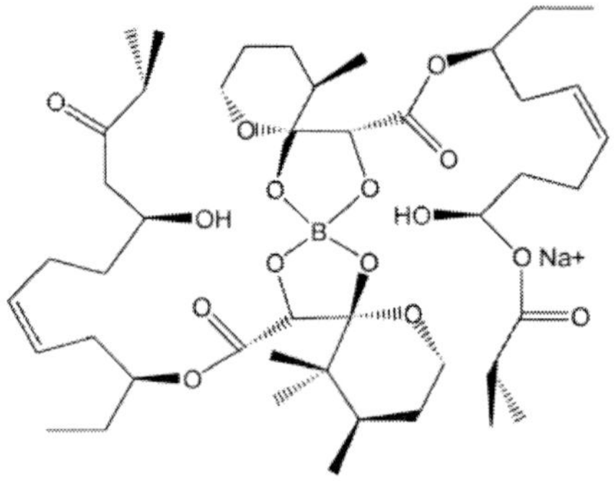

Figure 3.3. Chemical structure of Borophycin (Singh et al., 2005).

Cryptophycin (Figure 3.4) first isolated from *Nostoc* sp. ATCC 53789. It has also been isolated from *Nostoc* sp. GSV 224 and has exhibited potent cytotoxicity against human tumor cell lines. It shows good activity against a broad spectrum of drug-sensitive and drug-resistant murine and human solid tumors (Burja et al., 2001). The mechanism of cytotoxicity of the cryptophycins is tubulin-interaction, with a disruption of tubulin-dynamics, resulting in apoptosis of tumor cells (Panda et al., 1998).

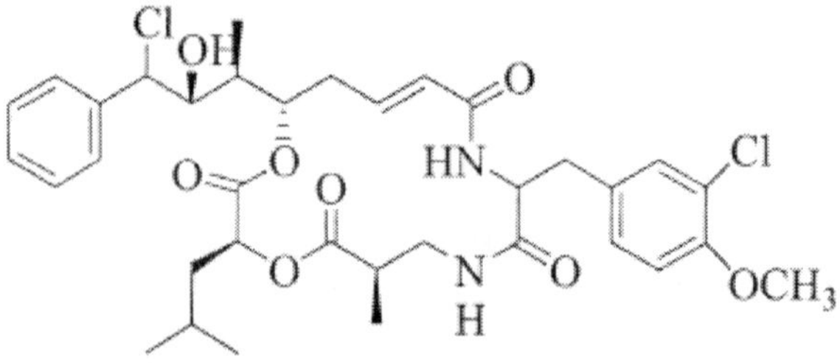

Figure 3.4. Chemical structure of Cryptophycin (Singh et al., 2005).

Mathew et al. (1995) reported that the oral supplementation of *Spirulina fusiformis* resulted in regression of subjects with homogenous leukolakia. In addition, extracts of *Spirulina* and *Dunaliella* inhibited the chemically induced carcinogenesis in model hamster buccal pouches (Schwartz et al., 1987; Schwartz et al., 1988).

Studies have also showed that sulphated polysaccharide, calcium spirulans appears to inhibit tumor invasion and metastasis of melanoma cells and inhibit the tumor invasion of basement membrane (Mishima et al., 1998). *Aphanizomenon flos-aquae* extract containing a high concentration of phycocyanin inhibited the *in vitro* growth of one of four tumor cell lines tested, indicating the sensitivity of cell lines to the phycocyanin. *Phormidium tenue* contain several diacylglycerols that inhibit chemically induced tumors on mice (Tokuda et al., 1996).

Curacin was isolated from the Curaso strain marine Cyanobacterium *Lyngbya majuscula* by Gerwick's group in 1994. It is an important lead compound for a new type of anticancer drugs. It is an antimitotic agent that inhibits microtubule assembly and binding of cholchicine to tubulin (Gerwick et al., 1994).

3.1.3. Anti-Inflammatory Activity

Cyanobacteria contain significant amounts of carotenoids (β-carotene, lycopene, lutein) having antioxidant properties. By the quenching action on the reactive oxygen species, these carotenoids also have anti-inflammatory activities. The anti-inflammatory activity of blue-green algae is also due to phycocyanin. C-phycocyanin is a free radical scavenger (Bhat and Madyastha, 2000) and has a significant hepatoprotective effect (Vadiraja et al., 1998). The anti-inflammatory effect seemed to be a result of phycocyanin inhibiting the formation of leucotriene, an inflammatory metabolite derived from arachidonic acid (Romay et al., 1999). *Aphanizomenon flos-aquae* decrease the level of arachidonic acid (Kushak et al., 2000) and also contain significant amounts of ω-3-α linolenic acid which inhibit the formation of inflammatory postaglandins and arachidonate metabolite.

3.1.4. Antimicrobials

Some species of Cyanobacteria produce antibiotic compounds (Kulik, 1995; Noaman et al., 2004). Thillairajasekar et al. (2009) tested extracts of *Trichodesmium erythraeum* against bacteria and fungi. Hexane extracts showed activity against *B. subtilis, S. aureus, E. faecalis, S. typhi, P. aeruginosa, P. vulgaris, Erwinia*sp. and inhibited the growth of *Tricophyton simii, T.mentagrophytes, T.rubrum, Aspergillus flavus, Aspergillus niger, Scopulariopsis*sp *and Botrytis cinera*.

Abedin and Taha (2008) also tested cyanobacteria *Anabaena oryzae, Tolypothrix ceytonica* and *Spirulina platensis* against various organisms that incite diseases of humans and plants such as *E. coli, B. subtilis, S. aureus, P. aeruginosa, A. niger, A. flavus, Penicillium herquei, Fusarium moniliforme, Helminthosporium sp., Alternaria brassicae, S. cerevisae* and *Candida albicans*.

Martins et al. (2008) tested nine cyanobacteria strains (*Synechocystis* and *Synechococcus*) that have showed antibiotic activity against two Gram-positive bacteria, *Clavibacter michiganensis* subs. *insidiosum* and *Cellulomonasuda*. These cyanobacteria have not revealed effects against fungus *Candida albicans* and Gram-negative bacteria.

3.1.5. Toxins

The cytotoxic activity, important for anticancer drugs development, is likely related to defence strategies in the highly competitive marine environment, since usually only those organisms lacking an immune system are prolific producers of secondary metabolites such as toxins (Burja et al., 2001).

Hepatotoxins are the most commonly encountered toxins involving cyanobacteria and include the cyclic peptides microcystin and nodularin. *Microcystis aeruginosa* and *Nodularia spumigena* synthesize toxins destructive to liver cells (Burja et al., 2001). To date over 50 different variants of microcystins have been isolated from the species of *Anabaena, Hapalasiphon, Microcystis, Nostoc and Oscillatoria*. One of the most toxic genera of cyanobacteria belonging to the order *Oscillatoroales* is *Lyngbya*. They are responsible for the synthesis of cytotoxic compounds such as antillatoxin, aplysiatoxin, debromoaplysiatoxin and lyngbyatoxin A, B and C (Shimizu, 2003).

Kalkitoxin, discovered and purified from organic extracts of *Lyngbya majuscula*, is a neurotoxin that blocks sodium channels preventing the nerves from firing off their electrical signals. Topiramate helps to suppress epileptic attacks largely by blocking sodium channels. Painkillers like lidocaine are sodium channel blockers. Kalkitoxin could treat these disorders including neurodegenerative diseases by selectively activating and blocking sodium channels. This toxin is a useful pharmaceutical compound and a valuable tool to understand the working of sodium channels and the effect of disease on them (Wu et al., 2000).

L. majuscula also contains Antillatoxin. This lipodesipeptide toxin is an extremely potent ichthyotoxin. Its activity is comparable to that of brevetoxin and involves the activation of voltage-gated sodium channels. It is interesting that out of two potent ichthyotoxins from *L. majuscula* one is a sodium channel blocker and other an activator (Burja et al., 2002; Li et al., 2001).

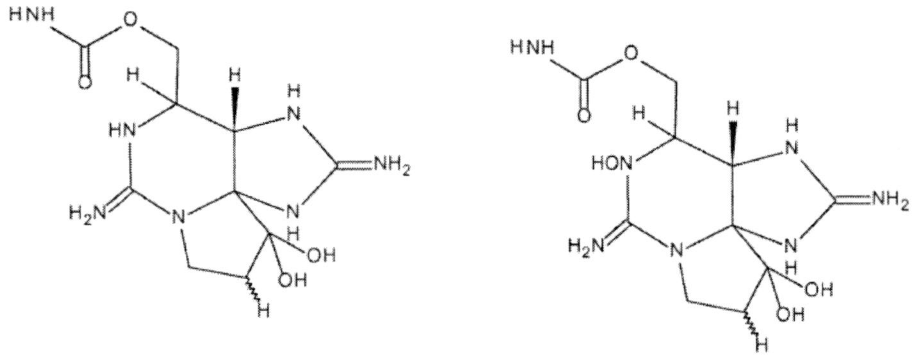

Figure 3.6. Chemical structures of Saxitoxin and Neosaxitoxin.

Saxitoxins (Figure 3.6) are neurotoxic alkaloids, which are known as paralytic shellfish poisons. The name saxitoxin was derived from the mollusk in which it was first identified. *Alexandrium catenella, A. minutum, A. ostenfeldi, A. tamarense, Gymnodinium catenatum* and *Pyrodinium bahamense* secrete saxitoxins. This toxin blocks neuronal transmission by binding to the voltage gated Na^+ channels in nerve cells, thus causing a neurotoxic effect. Saxitoxin is highly toxic and kills a guinea pig at only 5µg/kg when injected intramuscularly.

The oral LD50 for human is 5.7µg/kg. This neurotoxin specifically and selectively binds to the sodium channel in neural cells. Thus, it physically occludes the opening of Na$^+$ channels and prevents any sodium cation from going in or out of the cell. Since, neuronal transmittance of impulse and messages depends on depolarization of the cell the action potential is stopped, impairing a variety of body functions, including breathing. The diaphragm may stop working and death may occur after cardiorespiratory failure.

Anatoxin-*a* and homoanatoxin-*a* (Figure 3.7) are secondary amines and are postsynaptic depolarizing neuromuscular blocking agents (Carmichael et al., 1977) that bind strongly to the nicotinic acetylcholine receptor (Spivak et al., 1980). These compounds are potent neurotoxins, which cause rapid death due to respiratory arrest (Devlin et al., 1977). Anatoxin-*a* produced by *Anabaena flos-aquae* (Gorham et al., 1964) enters the body by inhalation, injection and exposure to high concentration through the skin. Homoanatoxin-*a* is structurally similar to anatoxin-*a* found in *Oscillatoria formosa* (Lilleheil et al., 1997). Anatoxin-*a* is a potent neurotoxin, a cholinesterase inhibitor (Mahmood and Carmichael, 1986; 1987) and induces hypersalivation in mammals.

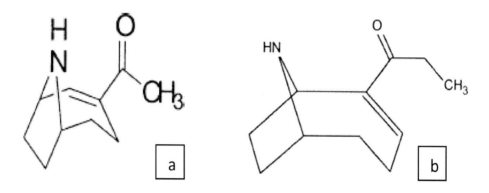

Figure 3.7. Chemical structures of Anatoxin-*a* (a) and Homoanatoxin-*a* (b).

3.1.6. Protease Inhibitors

Five classes of protease inhibitors have been reported from the toxic genera of cyanobacteria: they are micropeptins, aerugenosins, microginins, anabaenopeptins and microverdins. Serine protease inhibitors of micropeptin type are the most common inhibitors from cyanobacteria with more than fifty compounds. Some cyanopeptolins are specific inhibitors of serine proteases, including elastase, which is of critical importance in a number of diseases like lung emphysema, which is mediated by excessive action of elastase. Furthermore, it has been proposed that unphysiologically high levels of elastase activity are involved in myocardial damage and may cause a particular form of psoriasis. Cyanopeptolins are subjected to inhibition assays with commercial proteases, which are of medicinal relevance, like trypsin, thrombin, plasmin, papain and elastase (Grach-Pogrebinsky et al., 2003; Matern et al., 2001).

Scyptolin are cyclic desipeptides with elastase inhibiting activity, isolated from terrestrial cyanobacterium *Scytonema hofmanni* PCC 7110. These metabolites significantly inhibited porcine pancreatic elastase in invitro assays (Figure 3.8).

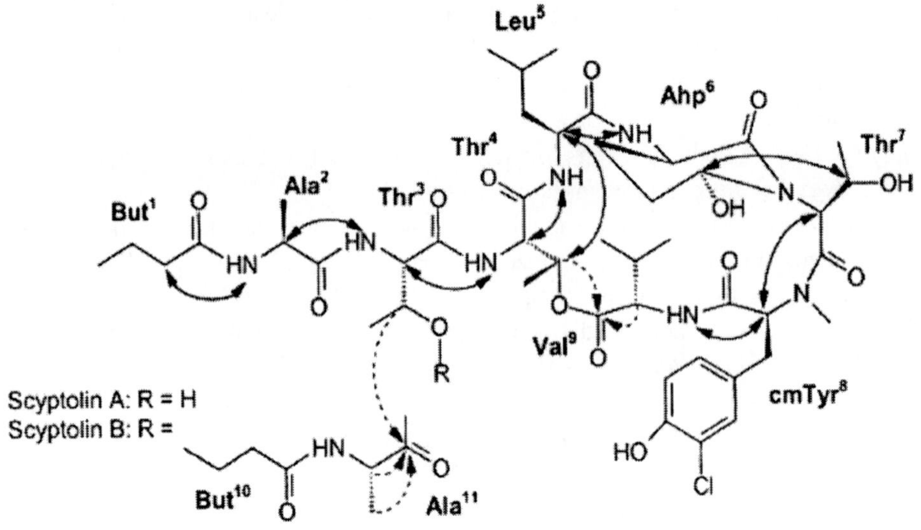

Figure 3.8. Chemical structure of Scyptolin.

Natural elastase inhibitors might serve as valuable lead structures in pharmaceutical research dedicated to the development of more effective drugs (Matern et al., 2001). Three new protease inhibitors were isolated from *Planktothrix rubescens*planktopeptin (BL1125, planktopeptin BL843 and planktopeptin BL1061). They are micropeptin type serine protease inhibitors and they were also found to be elastase and chymotrypsin inhibitors (Grach-Pogrebinsky et al., 2003).

3.2. Microalgae as a Producer of Bioactive Compounds

As mentioned above, microalgae are able to produce a wide range of active substances with antibacterial, antiviral, antifungal, enzyme inhibiting, immunostimulant, cytotoxic and antiplasmodial activities. Most of the isolated substances belong to groups of polyketides, amides, alkaloids and peptides (Ghasemi et al., 2004).

Pratt et al. (1944) were the first to isolate an antibacterial substance from *Chlorella*. A mixture of fatty acids, named chlorellin, exhibited inhibitory activity against both Gram-positive and Gram-negative bacteria.The effect of antimicrobial activity of *Chlorella* species has been reported in other studies such as Kellam and Walker (1989), Ördög et al. (2004), Debro and Ward (1979).

The microalgae such as *Chlorella* spp., *Scenedesmus* spp. (Ördöget al., 2004), *Chlamydomonas* spp. (Kellam and Walker, 1989), *Euglenaviridis* (Das et al., 2005), *F. ambigua* (Ghasemi et al., 2004), have been reported as the main groups of microalgae to produce antimicrobial substances. Also *Ochromonas* sp. and *Prymnesium parvum* produce toxins that may have potential pharmaceutical applications (Katircioglu et al., 2006; Borowitzka and Borowitzka, 1988). The ability to produce antimicrobial agents may be significant not only as a defensive instrument for the algal strains (Mundt et al., 2001) but also as a good source of the new bioactive compounds from a pharmaceutical point-of-view. Temperature of incubation, pH of the culture medium, incubation period, medium constituents and light intensity are the important factors influencing antimicrobial agent production (Noaman et al., 2004).

Pithophora oedogonia was tested for antibacterial activity against clinical isolates of common human pathogenic bacteria namely, *Escherichia coli, Klebsiella pneumoniae, Proteus mirabilis, Salmonella typhi, Pseudomonas aeruginosa, Vibrio cholerae, Shigella flexnerii, Staphylococcus aureus, Streptococcus pyogenes* and *Streptococcus faecalis*. Methanolic extract residue dissolved in diethylether exhibited good activity against *Streptococcus pyogenes, Streptococcus faecalis* and *E. coli*. Activity of silica gel column fractions is significant and comparable to that of standard antibiotics. Chromatatron fractions recorded very low MIC values for *Streptococcus pyogenes, Streptococcus faecalis* and *Escherichia coli* as compared to that of standard antibiotics. The findings reported by Sukumaran and Thevanathan (2011) suggested that the 'nuisance alga' *Pithophora oedogonia*, could serve as a potential source of biologically active natural products for pharmaceutical application.

Katircioglu (2006) have studied ten microalgae strains isolated from different freshwater reservoirs situated in various topographies in Turkey. Their antimicrobial agent production was carried out on various organisms such as *Bacillus subtilis, Bacillus thuringiensis, Bacillus cereus, Bacillus megaterium, Yersinia enterocolitica, E. coli, Staphylococcus aureus, Micrococcus luteus, Micrococcus flavus, Pseudomonas aeruginosa, Saccharomyces cerevisiae, Candida albicans*. The findings in this study reveal that activity is maintained against antimicrobial activity of acetone and ether extracts on Gram-negative bacteria; methanol extracts on Gram-positive bacteria; ethanol extracts on both Gram-positive and Gram-negative organisms. Chloroform extracts, on the other hand, were not found to reveal any antimicrobial activity. Schwartz et al. (1990) reported that an algal extract from the green alga *Caulerpa vanbosseae* was active in inhibiting prenyl transferase, 5-lipoxygenase and contained an antibacterial agent with good activity against a strain of methicillin-resistant *Staphylococcus aureus*.

Ghasemi et al. (2007) have screened 60 strains of microalgae from paddy fields of Iran in order to study the antimicrobial effect against six strains of bacteria and four strains of fungi. The culture supernatants of 21 strains of microalgae and methanolic extracts of 8 strains exhibited significant antibacterial effect and 17 strains showed antifungal effect. *Chroococcus dispersus, Chlamydomonas reinhardtii* and *Chlorella vulgaris* appeared to be the most promising strains and it was shown that they excreted a broad spectrum of antimicrobial substances in the culture medium. Although among all of the species studied in this investigation (*S. aureus, S. epidermis, B. subtilis, E. coli, S. tiphy, P. aeruginosa, C. albicans, C. kefyr, A. Fumigatus, A. niger*) for antibacterial and antifungal activity, *Chroococcus dispersus* indicated widespread spectrum of antimicrobial activities.

Soltani et al. (2005) also describes the biological effects of *Chroococcus* spp. In this study the methanolic extract of *Chroococcus* sp. was found to be biologically active against *Staphylococcus epidermidis*. In another study (Mian et al., 2003), it was shown that *Chroococcus* sp. produced bioactive substances with antibiotic activity against *Bacillus cereus, S. aureus* and *S. epidermidis*.

3.3. Algae in Cosmeceuticals

Cosmeceuticals, a term derived from the words 'cosmetic and pharmaceutical', have drug-like benefits and contain active ingredients such as vitamins, phytochemicals, enzymes, polysaccharides, antioxidants, and essential oils. Cosmeceuticals have attracted increased attention because of their beneficial effects on human health (Kim et al., 2008).

In general, algae have some economical potential because they contain high amounts of minerals and bioactive compounds. Among them, brown and red algae are the common types used in cosmeceutic products. The largest market for algae cosmetics is France, with an estimated five thousand tons of wet algae used to cater the demand. Some components of algae extracts react with diverse skin proteins and form a protective gel on the surface and thereby reduce moisture loss (Malakoff, 1997). Kelp (brown alga) found in offshore contains many essential vitamins, minerals, and essential fatty acid like ω-3 and -6, which are known to facilitate cell regeneration and skin health. Also, extracts from Kelp act as a suntan stimulator when the skin is exposed to UV radiation and stimulate the activity of tyrosinase which could be used as thickeners in many cosmetics and hair products, especially conditioners (Mungo, 2005).

Chondrus cripsus (red alga) is rich in polysaccharides and minerals, including manganese, zinc, calcium, and magnesium which are related to hydrating, soothing, healing, moisturizing, and conditioning effects. The extract from *Codium tomentosum* (green alga) is a good source of glucuronic acid, which regulates water distribution within the skin, and protects the skin from the damagingeffects of a dry environment. The extract from *Laminaria saccharina* contains proteins, vitamins, minerals, and carbohydrates which regulates sebaceous gland activity, and has anti-inflammatory and healing properties (Fitton et al., 2007). The extract from *Crithmum maritimum* contains minerals, essential oils, polyphenols, flavonoids, and vitamin C and activates the protein synthesis of the connective tissues such as collagen and elastin resulting in the improvementof tone and elasticity in the skin.

The extract from *C. maritimum* contains bioactive components for skin protection (Majmudar, 2007). Extracts from *Ascophyllum nodosum* (brown algae) and *Asparagopsis armata* (red algae) contain anti-irritant components and reduce the level of vascular endothelial growth factor (VEGF) which stimulates the growth and dilation of tiny blood vessels. If VEGF is too high, it may cause problems with dilated capillaries, resulting in a very sensitive red skin. The extract from *Entermorpha compressa* (green alga) contains glycosamine, hydroxyproline, and polysaccharides and increases theblood circulation.

Some microalgal and cyanobacterial species are established in the skin caremarket, being the main ones *Arthrospira* and *Chlorella* (Stolz and Obermayer, 2005). Microalgae extracts can be mainly found in face and skin care products (e.g., anti-agingcream, refreshing or regenerant care products, emollient andas an anti-irritant in peelers). Microalgae are also represented in sun protection and hair care products. Two examples of commercially available products and their properties claimed by their companies are: a protein-rich extract from *Arthrospira* which repairs the signs of early skin aging, exerts a tightening effect and prevents stria formation (Protulines, Exsymol S.A.M., Monaco); and an extract from *Chlorella vulgaris* stimulates collagen synthesis in skin, thereby supporting tissue regeneration and wrinkle reduction (Dermochlorella, Codif, St. Malo, France). Recently, two new products have been launched by Pentapharm (Basel, Switzerland) (Stolz and Obermayer, 2005): an ingredient from *Nannochloropsis oculata* with excellent skin-tightening properties (short and long-term effects) (Pepha-Tight) and an ingredient from *D. salina*, which shows the ability to marked lystimulate cell proliferation and turnover and to positively influence the energy metabolism of skin (Pepha-Ctive).

In table 3.1 are presented some companies which commercialize pharmaceutical products from algae.

Table 3.1. Examples of companies deriving pharmaceutical products from algae

Companies deriving Pharmaceuticals from algae	
Rincon Pharmaceuticals	rinconpharma.com
Rallis	rallis.co.in
Monsanto	monsanto.com/default.asp
Jubilant Organosys	jubl.com
Piramal Healthcare	piramalhealthcare.com
AstraZeneca	astrazenecaindia.com
Idec Pharmaceuticals	idecpharm.com
National Facility for Marine Cyanobacteria	nfmc.res.in
Novo Nordisk India Private Ltd	novonordisk.co.in
Agri Life SOM Phytopharma	somphyto.com

4. Microalgae Biotechnology in Environmental Applications

Climate change is one of the defining issues of the 21st century. Humans are challenged to find a set of policies, practices and standards of behaviour that will provide long-term economic opportunities and improved quality of life around the world, while maintaining a sustainable climate and viable ecosystems (Pinto et al., 2002; Betenbough and Bentley, 2008; The National Academies, 2008; Atsumi et al., 2009; Energy Information Administration, 2009). Nowadays there is a great dependence on fossil fuels, causing the emission of pollutants, e.g. CO_x, NO_x, SO_x, C_xH_x, ash, among others (Debabrata and Veziroglu, 2001), which are responsible for most of the global warming observed over the last 50 years, according to the *Intergovernmental Panel on Climate Change* (IPCC) (Houghton et al., 2001; Brennan and Owende, 2010; Ghirardi and Mohanty, 2010). Algae (macroalgae, microalgae and cyanobacteria) show to be promising in some industrial applications with environmental benefits, such as biofuels production, CO_2 sequestration and wastewater treatment.

4.1. Biofuels Production Using Microalgae and Cyanobacteria

The majority of the current commercially available biofuels are bioethanol, derived from sugar cane or corn starch, or biodiesel derived from oil crops, including soybean and oilseed rape. Although biofuels have the potential to be environmentally beneficial compared to fossil fuels, there is some dispute as to whether these crop-based biofuels are economically competitive compared to fossil fuels. Furthermore, there is even more concern over the impact that the use of these crops for biofuels might have on food availability (Demirbas, 2009; Hill et al., 2006). Biofuels derived from the cultivation of algae have therefore been proposed as an alternative approach that does not have a negative impact on agriculture.

Algae, particularly green unicellular microalgae have been proposed for a long time as a potential renewable fuel source (Benemann et al., 1977; Oswald and Golueke, 1960). Microalgae have the potential to generate significant quantities of biomass and oil suitable for conversion to biodiesel. These organisms have been estimated to have higher biomass productivity than plant crops in terms of land area required for cultivation, can use non-arable land and non-potable water, are predicted to have lower cost per yield, and have the potential to reduce GHG emissions through the replacement of fossil fuels (Benemann and Oswald, 1996; Brennan and Owende, 2010; Brune et al., 2009; Chisti, 2008; Dismukes et al., 2008; Huntley and Redalje, 2007; Rittmann, 2008; Schenk et al., 2008; Sheehan et al., 1998; Stephens et al., 2010). As with plant-derived feedstocks, algae can be used directly or processed into liquid fuels and gas by a variety of biochemical conversion or thermochemical conversion processes (Amin, 2009; Brennan and Owende, 2010; Demirbas, 2009; Rittmann, 2008). The conversion of light to chemical energy can be responsible for a wide range of fuel synthesis: protons and electrons (for biohydrogen), sugars and starch (for bioethanol), oils (for biodiesel) and biomass (for BTL and biomethane) (Figure 4.1).

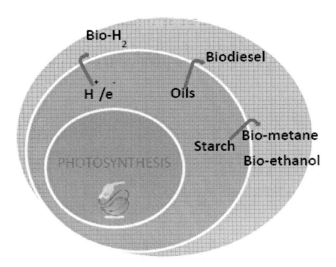

Figure 4.1. The role of algae photosynthesis in biofuels production (Gouveia, 2011).

Dried algae biomass may be used to generate energy by direct combustion (Kadam, 2002) but this is probably the least attractive use for algal biomass. Thermochemical conversion methods include gasification, pyrolysis, hydrogenation and liquefaction of the algal biomass to yield gas- or oil-based biofuels (McKendry, 2002a; 2002b; Miao and Wu, 2004). Biochemical conversion processes include fermentation and anaerobic digestion of the biomass to yield bioethanol or methane (McKendry, 2002a). In addition, hydrogen can be produced from algae by bio-photolysis (Melis, 2002). Finally, lipids, principally triacylglycerol lipids can be separated and isolated from harvested microalgae and then converted to biodiesel by transesterification (Figure 4.2) (Chisti, 2007; Hu et al., 2008; Miao and Wu, 2006).

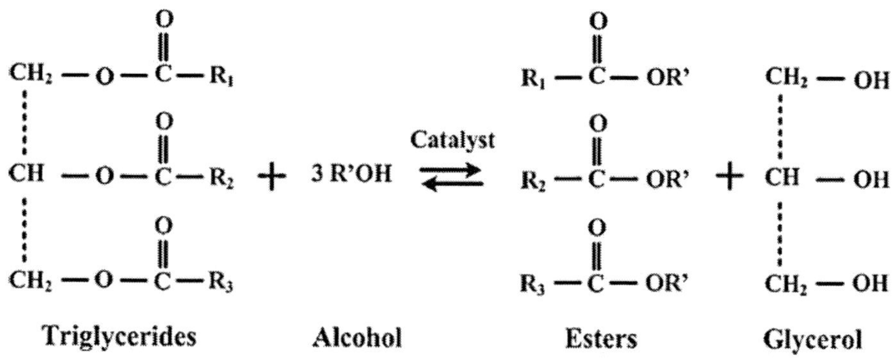

$$\underset{\text{Triglycerides}}{\begin{array}{c}CH_2-O-\overset{O}{\underset{\|}{C}}-R_1\\ CH-O-\overset{O}{\underset{\|}{C}}-R_2\\ CH_2-O-\overset{O}{\underset{\|}{C}}-R_3\end{array}} + \underset{\text{Alcohol}}{3\,R'OH} \underset{}{\overset{\text{Catalyst}}{\rightleftarrows}} \underset{\text{Esters}}{\begin{array}{c}R_1-\overset{O}{\underset{\|}{C}}-OR'\\ R_2-\overset{O}{\underset{\|}{C}}-OR'\\ R_3-\overset{O}{\underset{\|}{C}}-OR'\end{array}} + \underset{\text{Glycerol}}{\begin{array}{c}CH_2-OH\\ CH-OH\\ CH_2-OH\end{array}}$$

Figure 4.2. Transesterification of triglycerides (overall reaction) (adapted from Mata et al., 2010).

One of the attractions of microalgae as a biofuel feedstock is that they can be effectively grown in conditions which require minimal freshwater input unlike many plant-based biofuel crops, and utilize land which is otherwise non-productive to plant crops, thus making the process potentially sustainable with regard to preserving freshwater resources. For example, microalgae could be cultivated near the sea to use saline or brackish water. There has therefore been significant interest in the growth of microalgae for biofuels under saline conditions (e.g. Rodolfi et al., 2009; Takagi et al., 2006).

A number of studies have argued that biofuel production from algae, particularly biodiesel production is both economically and environmentally sustainable (Brune et al., 2009; Chisti, 2008; Huntley and Redalje, 2007; Stephens et al., 2010), although there have been some skeptical views of the long term viability and economics of biofuels from algae (Reijnders, 2008; van Beilen, 2010; Walker, 2009). One frequent criticism is that the requirement of fossil fuels in the biofuel production process, in the construction of algal growth facilities, supply of nutrients for algal growth, harvesting of algae and biomass processing, is not often considered in the evaluation of algal biofuel viability and would in fact give rise to a net negative energy output.

In this chapter the authors will emphasize biodiesel, bioethanol, biohydrogen and biomethane production by microalgae and cyanobacteria.

4.1.1. Biodiesel

The oil production capabilities of microalgae and the suggested potential productivity of oil from microalgae may be significantly greater than oilseed crops such as soybean (Sheehan et al., 1998) (Table 4.1). Nowadays, the focus in research is to identify microalgae strains and culture conditions, as well as genetic engineering modifications, to provide the greatest lipid productivities (Griffiths and Harrison, 2009; Hu et al., 2008). The synthesis and accumulation of a large amount of triacylglycerols (TAG) accompanied by considerable alterations in lipid and fatty acid composition, occurs under stress imposed by chemical (pH, salinity, nutrient starvation) and physical environmental stimuli (light and temperature), either acting individually or in combination) (Hu et al., 2008; Converti et al., 2009; Dean et al., 2010; Gouveia et al., 2009; Li et al., 2008; Rodolfi et al., 2009).

Microalgae strains, such as *Chlamydomonas reinhartii, Nannochloropsis* sp., *Neochloris oleoabundans, Scenedesmus obliquus, Nitschia* sp., *Schzochytrium* sp., *Chlorella protothecoides, Dunaliella tertiolecta* among others, were screened by many authors in order to choose the best lipid producers in terms of quantity (combination of biomass productivity and lipid content) and quality (fatty acid composition) as an oil source for biodiesel production (Xu et al., 2006; Miao and Wu, 2006; Chisti, 2007; Rodolfi et al., 2009; Lopes da Silva et al., 2008; Gouveia and Oliveira, 2009; Gouveia et al., 2009; Morowvat et al., 2010).

Typical microalgal biomass fatty acid composition is mainly composed of a mixture of unsaturated fatty acids, such as palmitoleic (16:1), oleic (18:1), linoleic (18:2) and linolenic (18:3) acid. Saturated fatty acids, such as palmitic (16:0) and stearic (18:0) are also present to a small extent (Meng et al., 2009; Gouveia and Oliveira, 2009).

Fatty acid composition can also vary both quantitatively and qualitatively with their physiological status and culture conditions (Hu et al., 2008).

In addition to chemical and physical factors, growth phase and/or aging or senescence of the culture also affects the TAG content and fatty acid composition (Hu et al., 2008). Lipid content and fatty acid composition is also subject to variation during the growth cycle, usually with an increase in TAGs in the stationary phase. The aging culture also increases the lipid content of the cells, with a notable increase in the saturated and mono-unsaturated fatty acids, and a decrease in PUFAs (Liang et al., 2006).

Table 4.1. The oil content of some microalgae species (adapted from Chisti, 2007)

Microalgae	Oil content (% dry wt)
Botryococcus braunii	25-75
Chlorella sp.	28-32
Cryptothecodinium cohnii	20
Cylindrotheca sp.	16-37
Dunaliella primolecta	23
Isochrysis sp.	25-33
Monallanthus salina	20
Nannochloris sp.	20-35
Nannochloropsis sp.	31-68
Neochloris oleoabundans	35-54
Nitzschia sp.	45-47
Phaeodactylum tricornutum	20-30
Schizochytrium sp.	50-77

4.1.2. Bioethanol

Bioethanol is already well established as a fuel most notably in Brazil and US (Goldemberg, 2007). It is usually obtained by alcoholic fermentation from starch (cereal grains, such as corn, wheat and sweet sorghum), sugar (sugar cane and sugar beet) and lignocellulosic feedstocks (Antolin et al., 2002).

S. cerevisiae is the universal organism for fuel ethanol production from glucose. Nevertheless, *Z. mobilis* is considered the most effective organism for production of ethanol, although it is not currently used commercially (Drapcho et al., 2008). Starch processing is a mature industry, and commercial enzymes required for starch hydrolysis are currently an available low cost technology.

Microalgal bioethanol can be produced through via dark fermentation, which is a low energy intensive process, but with very low yields: 1% (w/w) for *Chlamydomonas reinhardii* (Hirano et al., 1997) and 2,1% (w/w) for *Chlorococum littorale* (Ueno et al., 1998), which does not make this process appealing to the industry. The microalgae *Chlamydomonas perigranulata* has also been reported to produce intracellular bioethanol (Hon-Nami and Kunito, 1998; Hon-Nami, 2006).

Some microalgae have a high carbohydrate content (Table 4.2), and therefore a high potential for bioethanol production through yeast fermentation (e.g. Schenk et al., 2008; Hankamer et al., 2007) however only some research (Huntley and Redalje, 2007; Rosenberg et al., 2008; Subhadra and Edwards, 2010) has been done on this subject (Douskova et al., 2008). Hirano et al. (1997) conducted an experiment with *Chlorella vulgaris* microalga (37% starch content) through fermentation and yielded a 65% ethanol conversion rate, when compared to the theoretical conversion rate from starch. It has been demonstrated that supplementing the medium with iron can increase threefold the carbohydrate content (He et al. 2010). Douskova et al. (2008) have shown that the microalgae *Chlorella vulgaris* in phosphorus, nitrogen or sulphur limiting conditions, the starch content of the cells increased 83, 50 and 33%, respectively.

Table 4.2. Amount of carbohydrates from various species of microalgae on a dry weight basis (%) (Adapted from Becker 1994a; Harun et al. 2010)

Algae strains	Carbohydrates (%)
Scenedesmus obliquus	10–17
Scenedesmus dimorphus	21–52
Chlamydomonas rheinhardii	17
Chlorella vulgaris	12–17
Chlorella pyrenoidosa	26
Spirogyra sp.	33–64
Dunaliella bioculata	4
Dunaliella salina	32
Euglena gracilis	14–18
Prymnesium parvum	25–33

4.1.3. Biohydrogen

Hydrogen is an energy carrier with great potential in the transport sector, for domestic and industrial applications, where it is being explored for liquefaction of coal and upgrading of heavy oils in order to use in combustion engines, and fuel-cell electric vehicles (Balat, 2005), generating no air pollutants. Hence, in both the near and long term, hydrogen demand is expected to increase significantly (Balat, 2009).

Hydrogen can be produced in a number of ways (Madamwar et al., 2000). However currently, the developing H_2 economy is almost entirely dependent upon the use of carbon-based non-renewable resources, such as steam reforming of natural gas (~48%), petroleum refining (~30%), coal gasification (~18%) and nuclear powered water electrolysis (~4%) (Gregoire-Padro, 2005); from water through thermal and thermochemical processes, such as electrolysis and photolysis; and through biological production, such as steam reforming of bio-oils (Wang et al., 2007), dark and photo fermentation of organic materials and photolysis of water catalyzed by special microalga and cyanobacteria species (Kapdan and Kargi, 2006).

Algal biomass (whole or after oil and/or starch removal), can be converted in bio-H_2 by dark-fermentation, that is one of the major bio-processes using anaerobic organisms for bio-

H_2 production. *Enterobacter* and *Clostridium* bacteria strains are well known as good producers of bio-H_2 that are capable of utilizing various types of carbon (Angenent et al., 2004; Das, 2009; Cantrell et al., 2008).

On the other way, cyanobacteria and green algae are the only organisms currently known to be capable of both oxygenic photosynthesis and bio-H_2 production.

In cyanobacteria, hydrogen is produced by a light-dependent reaction catalyzed by nitrogenase or in dark-anaerobic conditions by a hydrogenase (Rao and Hall, 1996; Hansel and Lindblad, 1998), while in green algae, hydrogen is produced photosynthetically by the ability to harness the solar energy resource, to drive H_2 production, from H_2O (Melis et al., 2000; Ghirardi et al., 2000; Melis and Happe, 2001).

To date, H_2 production has been observed in only 30 genera of green algae (Boichenko and Hoffmann, 1994), highlighting the potential to find new H_2 producing eukaryotic phototrophs with higher H_2 production capacities.

For photobiological H_2 production, cyanobacteria, formerly called "blue-green algae" and "nitrogen-fixing" bacteria, are among the ideal candidates, since they have the simplest nutritional requirements. They can grow using air, water and mineral salts, with light as their only source of energy (Tamagnini et al., 2007; Lindblad et al., 2002). Table 4.3 shows some examples of green-algae and cyanobacteria that have the ability to produce molecular hydrogen.

Table 4.3. Some microalgae and cyanobacteria that are able to produce molecular hydrogen

Hydrogen-evolving microrganisms	
Geen algae	*Scenedesmus obliquus, Chlamydomonas reinhardii* and *Chlamydomonas moewusii*
Heterocystous cyanobacteria	*Anabaena azollae, Anabaena* CA, *Anabaena variabilis, Anabaena cylindrical, Nostoc muscorum, Nostoc spongiaeforme* and *Westiellopsis prolific*
Cyanobacteria nonheterocystous	*Plectonema boryanum, Oscillotoria Miami* BG7, *Oscillotoria limnetica, Synechococcus* sp., *Aphanothece halophytico, Mastidocladus laminosus* and *Phormidium valderianum*

In what concern molecular and physiology H_2 production by cyanobacteria two enzymes are involved: the nitrogenase(s) and the bi-directional hydrogenase. In N_2-fixing strains the net H_2 production is the result of H_2 evolution by nitrogenase and H_2 consumption mainly catalysed by an uptake hydrogenase. Consequently, the production of mutants deficient in H_2 uptake activity is necessary.

Moreover, the nitrogenase has a high ATP requirement and this lowers considerably its potential solar energy conversion efficiency. On the other hand, the bi-directional hydrogenase requires much less metabolic energy, but is extremely sensitive to oxygen (Das and Veziroglu, 2001; Schütz et al., 2004).

Masukara et al. (2001) demonstrated that uptake deficient mutants of *Anabaena* strains produce considerably more H_2 compared to the wild types. The maximum light driven-H_2 production rates for cyanobacteria have been reported to be 2.6 mmol H_2/g dry weight.h cultured in an Allen and Arnon culture medium, with nitrate molybdenium replaced by vanadium at 30°C, at 6500 lux of irradiance and 73 vol. % Ar, 25 vol. % N_2 and 2 vol. % CO_2 (Stage I).

At a stage II, gas atmosphere was changed to 93 vol. % Ar, 5 vol. % N_2 and 2 vol. % CO_2 (Sveshnikov et al., 1997). These rates compare well with hydrogen production from the most active green algae at 0.7-1 mmol H_2/(g dry mass / h) sustainable for hours to days in wild type strains of *Chlamydomonas reinhardtii*, *Chlorella vulgaris* and *Scenedesmus obliquus* (Boichenko et al., 2004).

4.1.4. Biogas

Organic material like crop biomass or liquid manure can be used to produce biogas via anaerobic digestion and fermentation.

Mixtures of bacteria are used to hydrolyze and break down the organic biopolymers (i.e. carbohydrates, lipids and proteins) into monomers, which are then converted into a methane-rich gas via fermentation (typically 50-75% CH_4).

Carbon dioxide is the second main component found in biogas (approximately 25-50%) and, like other interfering impurities, has to be removed before the methane is used for electricity generation.

Microalgae biomass is a source of a vast array of components that can be anaerobically digested to produce biogas. The use of this conversion technology eliminates several of the key obstacles that are responsible for the current high costs associated with algal biofuels, including drying, extraction, and fuel conversion, and as such may be a cost-effective methodology.

Several studies have been carried out that demonstrate the potential of this approach. According to Sialve et al., (2009) the methane content of the biogas from microalgae is 7 to 13% higher compared to the biogas from maize.

A recent study indicated that biogas production levels of 180 mL/g.d of biogas can be realized using a two-stage anaerobic digestion process with different strains of algae, with a methane concentration of 65% (Vergara-Fernandez et al., 2008). Calorific value are directly correlated with the microalgae lipid content, and under nitrogen starvation, results in a significant increase in the caloric value of the biomass with a decrease of the protein content and a reduction in the growth rate (Illman et al., 2000).

Sialve et al. (2009) evaluated the energetic content (normal and N-starvation growth) of the microalgae *Chlorella vulgaris*, *C. emersonii* and *C. protothecoides*, in two scenarios, such as the anaerobic digestion of the whole biomass and of the algal residues after lipids extraction.

From the latter process, biodiesel and methane could be obtained with a higher energetic value. However, the energetic cost of biomass harvesting and lipid recovery is probably higher than the recovery energy, especially since most of the techniques involve biomass drying (Carlsson et al., 2007).

When the cell lipid content does not exceed 40%, anaerobic digestion of the whole biomass appears to be the optimal strategy on an energy balance basis, for the energetic recovery of cell biomass, as concluded by the authors (Sialve et al., 2009).

Another study performed by Mussgnug et al. (2010) screened some microalgae for biogas production, namely *Chlamydomonas reinhardtii*, *Chlorella kessleri*, *Euglena gracilis*, *Spirulina (Arthrospira) platensis*, *Senedesmus obliquus* and *Dunaliella salina*, and it was demonstrated that the quantity of biogas potential is strongly dependent on the species and on the pre-treatment. *C. Reinhardtii* revealed to be the more efficient with a production of 587mL biogas/g volatile solids.

Anaerobic digestion well explored in the past, will probably re-emerge in the coming years either as a mandatory step to support large scale microalgae cultures or as a standalone bioenergy producing process (Sialve et al., 2009). This technology could be very effective for situations like integrated wastewater treatment, where algae are grown under uncontrolled conditions using strains which are not optimized for lipid production.

In table 4.4 are presented some companies which are directing research, pilot-scale and larger-scale facilities through the production of biofuels from algae.

Table 4.4. Examples of companies deriving biofuels from algae

Companies deriving biofuels from algae			
GGASS	gold-green.com	A2BE Carbon Capture	algaeatwork.com
Green Plains Renewable Energy	gpreinc.com	Algae Link	algaelink.com
Greenbelt resources	greenbeltresources.com	Algawheel	algaewheel.com
GreenStar	greenstarusa.com	Algenol Biofuels	algenolbiofuels.com
International Energy	iea.org	Aurora Biofuels	aurorabiofuels.com
Neptune Industry	seafood-norway.com	Bioking	biodieseltec.de
Neste Oil	nesteoil.com	Scipio Biofuels, Inc	scipiobiofuels.com
PetroAlgae	petroalgae.com	SequesCO$_2$	sequesco.com
PetroSun	petrosuninc.com	Sustainable Power	bionomicfuel.com
Pure Powe	pure-power.de	UOP Honeywell	uop.com
Renewable Energy Group	regfuel.com	XL Renewables	xldairygroup.com
Sapphire Energy	sapphireenergy.com	Bionavitas	bionavitas.com
Virent Energy Systems	virent.com	Blue Marble Energy	bluemarbleenergy.net
Solazyme	solazyme.com	Bodega Algae	bodegaalgae.com
Green Star Products, Inc.	greenstarusa.com	Cellana	cellana.com
Icon Energy	iconenergy.com	Chevron	chevron.com
Inventure Chemicals	inventurechem.com	Columbia Energy	Columbia-energy.com
Aquaflow	aquaflowgroup.com	Community Fuels	communityfuels.com
Enhanced Biofuels and Technologies	ebtiplc.com	Diversified Energy	diversified-energy.com
Kwikpower International	kwikpower.com	ForeverGreen	Forevergreen.org
Biofuel Systems	biofuelsystems.com	Algoil	Algenergy.eu
Solix Biofuels	solixbiofuels.com	OriginOil	originoil.com
Seambiotic	seambiotic.com	Alga Technologies	algatech.com
LiveFuels Inc	livefuels.com	AXI	alliedminds.com
Infinifuel Biodiesel	infinifuel.com	GlobalGreen Solutions	globalgreensolutions.com

4.2. CO_2 Sequestration Using Microalgae

The chemical reaction-based approaches and the biological mitigation are the two main CO_2 mitigation strategies normally used (Wang et al., 2008). In one hand, the chemical reaction-based CO_2 mitigation approaches are energy-consuming, use costly processes, and have disposal problems because both the captured CO_2 and the wasted absorbents need to be disposed of. In other hand, the biological CO_2 mitigation has attracted much attention in the last years since it leads to the production of biomass energy in the process of CO_2 fixation through photosynthesis (Pulz and Gross, 2004). Flue gases from power plants are responsible for more than 7% of the total world CO_2 emissions from energy use (Kadam, 1997). Also, industrial exhaust gases contains up to 15% CO_2 (Kadam, 2001; Maeda et al., 1995), providing a CO_2-rich source for microalgae cultivation and a potentially more efficient route for CO_2 bio-fixation.

Direct utilization of power plant flue gas has been considered for CO_2 sequestration systems (Benemann, 1993). The advantage of utilizing flue gas directly is the reduction of the cost of separating CO_2 gas. Since power plant flue gas contains a higher concentration of CO_2, identifying high CO_2 tolerant species is important. Also, the screening of algae that are tolerant to the high levels of SO_x and NO_x that are present in flue gases is important.

Several species have been tested under CO_2 concentrations of over 15%. For example, *Chlorococcum littorale* could grow under 60% CO_2 using the stepwise adaptation technique (Kodama et al., 1993). Another high CO_2 tolerant species is *Euglena gracilis*, whose growth was enhanced under 5-45 % concentration of CO_2 (Nakano et al., 1996).

There are reports of *Chlorella* sp. successful growth under 10% CO_2 (Hirata et al., 1996a; 1996b), 40% CO_2 (Hanagata et al., 1992), and even 100% CO_2 (Maeda et al., 1995) although the maximum growth rate occurred under a 10% concentration. *Scenedesmus* sp. could grow under 80% CO_2 conditions but the maximum cell mass was observed in 10-20% CO_2 concentrations (Hanagata et al., 1992). *Cyanidium caldarium* (Seckbach et al., 1971) and some other species of *Cyanidium* can grow in pure CO_2 (Graham and Wilcox, 2000). Table 4.5 summarizes the CO_2 tolerance of various species.

Table 4.5. CO_2 tolerance of various species

Species	Maximum CO_2 concentration	References
Cyanidium caldarium	100%	Seckbach et al., (1971)
Scenedesmus sp.	80%	Hanagata et al., (1992)
Chlorococcum litorale	60%	Kodama et al., (1993)
Synechococcus elongatus	60%	Myari, (1995)
Euglena gracilis	45%	Nakano et al., (1996)
Chlorella sp.	40%	Hanagata et al., (1992)
Eudorina spp.	20%	Hanagata et al., (1992)
Dunaliella tertiolecta	15%	Nagase et al., (1998)
Nannochloris sp.	15%	Yoshihara et al., (1996)
Chlamydomonas sp.	15%	Miura et al., (1993)
Tetraselmis sp.	14%	Matsumoto et al., (1996)

C. vulgaris grown on wastewater discharge from a steel plant successfully sequestered 0.624g CO_2 L^{-1} per day (Yun et al., 1997). Doucha et al. (2005) recorded 10–50% reduction in CO_2 concentration in flue gases using with *Chlorella* sp., with the efficacy decreasing with increasing rate of flue gas injection into microalgae culture. De Morais and Costa (2007), using *Spirulina* sp. obtained a maximum daily CO_2 biofixation of 53% for 6% CO_2 and 46% for 12% CO_2 in the injected flue gas, with the highest mean fixation rate being 38% for 6% CO_2. With *S. obliquus*, de Morais and Costa (2007) achieved biofixation rates of 28% and 14% for 6% and 12% CO_2, respectively.

Since the temperature of waste gas from thermal power stations is around 120°C, the use of thermophilic (T_{op}=42-100°C), or high temperature tolerant species (e.g. *Cyanidium caldarium*) are also being considered (Seckbach et al., 1971; Bayless et al., 2001). Otherwise, gases also need to be cooled prior to injection into the growth medium, increasing processing costs. In addition, some thermophiles produce unique secondary metabolites (Edwards, 1990), which may reduce overall costs for CO_2 sequestration. Miyairi (1995) examined the growth characteristics of *Synechococcus elongatus* under high CO_2 concentrations, being the upper limit of CO_2 and temperature, 60% CO_2 and 60°C, respectively (Miyairi, 1995). Although less tolerant than thermophiles, some mesophiles can still be productive under relatively high temperature (Edwards, 1990), being good candidates for the direct use of flue injection.

4.3. Wastewater Treatment with Microalgae

Wastewater treatment with microalgae has been investigated since the 1960s by Oswald et al. (Golueke et al., 1965; Oswald et al., 1959; Oswald and Gotaas, 1957). It was performed in a system called Advanced Integrated Wastewater Ponding System (AIWPS) infull scale operation (Oswald, 1978, 1988, 1991) and in research scale (high-rate pond - HRP) in California USA (Green et al., 1995a, 1995b, 1996; Oswald, 1991; Oswald et al., 1994).

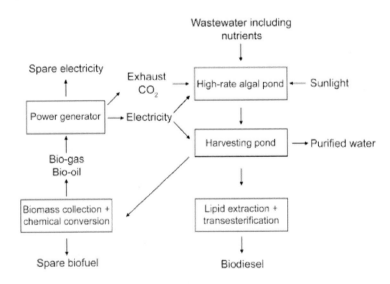

Figure 4.3. A flow-diagram showing how wastewater resources could be used for sustainable algal-based biofuel production (Pittman et al., 2011).

Algae can be applied to sequester, remove or transform pollutants such as excess nutrients, xenobiotics, and heavy metals from wastewater. These applications are known as phycoremediation. The treatment processes yield an output in the form of algal biomass that can be used to produce various compounds (Munoz and Guieysse, 2006).

For example, the production of microalgae biomass, biodiesel and other bio-products can be more environmentally sustainable, cost-effective and profitable, if combined with processes such as wastewater and flue gas treatments. In fact various studies demonstrated the use of microalgae for production of valuable products combined with environmental applications (Hodaifa et al., 2008; Jacob-Lopes et al., 2008, 2009; Murakami et al., 1998). Figure 4.3 demonstrate how wastewater can be used to produce biofuels and biomass.

4.3.1. Nutrients (N, P) Removal

Systems involving microalgae production and wastewater treatment, allow nutrition of microalgae by using organic compounds (nitrogen and phosphorous) available in some manufactures wastewater, not containing heavy metals and radioisotopes. Additionally, microalgae can mitigate the effects of sewage effluent and industrial sources of nitrogenous waste such as those originating from water treatment or fish aquaculture and at the same time contributing to biodiversity. Moreover, by removing nitrogen and carbon from water, microalgae can help the reduction of eutrophication in the aquatic environment (Mata et al., 2010).

Gonzales et al. (1997), Lee and Lee (2001) and Aslan and Kapdan (2006) studied *Chlorella vulgaris* as a nitrogen and phosphorus removal from wastewater. Aslan and Kapdan (2006) found an average removal efficiency of 72% for nitrogen and 28% for phosphorus. Other widely used microalgae for nutrient removal are *Scenedesmus* (Martínez et al., 2000) and *Spirulina* species (Olguín et al., 2003), *Nannochloris* (Jimenez-Perez et al., 2004), *Botryococcus brauinii* (An et al., 2003), and cyanobacterium *Phormidium bohneri* (Dumas et al., 1998; Laliberte et al., 1997).

Godos et al. (2010) studied the ability of two green microalgae (*Scenedesmus obliquus* and *Chlorella sorokiniana*), one cyanobacterium *(Spirulina platensis)*, one euglenophyt (*Euglena viridis*) and two wild microalgae consortia to photosynthetically support carbon, nitrogen and phosphorous removal from diluted piggery wastewaters. The results from the batch biodegradation tests, the batch oxygenation tests and the continuous piggery wastewater biodegradation operation confirmed that tolerance towards ammonia was the most important criterion for microalgae selection.

Kim et al. (2007) studied the newly-developed KEP Imedium,to which fermentation-treated swine urine has been added, and have determined to be an optimal growing medium for *Scenedesmus* cells in a batch mass culture, maintained for a minimum of 5 months, without any culture medium addition. The KEP I culture should improve the cost efficiency of industrial mass batch cultures and microalgal stock for species reservation, and may also prove to ameliorate certain environmental damages, via the recycling of animal wastewater. The KEP I medium containing fermented swine urine harbors organic materials (organic acids, enzymes, and hormones) generated by bacteria during the fermentation process. These materials serve to accelerate the physiological and biochemical activities of the growing cells, and also effect a delay in the onset of stationary phase of the cell divisions of *Scenedesmus*, despite the shortage of inorganic nutrients (N, P, and C) within the medium.

A study conducted by Chinnasamy et al. (2010) using a wastewater containing 85–90% carpet industry effluents with 10–15% municipal sewage, evaluated the feasibility of algal biomass and biodiesel production. Native algal strains were isolated from carpet wastewater. Preliminary growth studies indicated both fresh water and marine algae showed good growth in wastewaters. A consortium of 15 native algal isolates showed >96% nutrient removal in treated wastewater. Biomass production potential and lipid content of this consortium cultivated in treated wastewater were about 9.2–17.8 ton ha^{-1} year^{-1} and 6.82%, respectively. About 63.9% of algal oil obtained from the consortium could be converted into biodiesel.

4.3.2. Removal of Heavy Metals

Algae biomass can be used as an inexpensive biomaterial for the passive removal of toxic heavy metals. The relative affinity of raw *Sargassum* biomass for various divalent metal cations was determined at environmentally relevant concentrations to be Cu > Ca > Cd > Zn > Fe (Davis et al., 2003). Also microalgae have been used to remove heavy metals from wastewater (Wilde and Benemann, 1993; Perales-Vela et al., 2006).

Worms et al. (2010) studied the uptake of Cd(II) and Pb(II) by microalgae in the presence of colloidal organic matter from wastewater treatment plant effluents. The uptake of Cd by *Chlorella kessleri* was consistent with the speciation and measured free metal ion concentrations while Pb uptake was much greater than that expected from the speciation measurement.

Brown algae have proven to be the most effective and promising substrates due to the biochemical composition of the cell wall, which contains alginate and fucoidan. *Ascophyllum nodosum* (brown algae) has proven to be the most effective algal species to remove metals of cadmium, nickel, and zinc from monometallic solutions compared to green and red algae (Shi et al., 2007). A similar study proved that brown algae *Fucus vesiculosus* gave the highest removal efficiency of chromium (III) at high initial metal concentrations (Murphy et al., 2008). Another algal species, *S. obliquus* was examined for degrading cyanide from mining process wastewaters (Gurbuz et al., 2009). It was observed that cyanide was degraded up to 90% after introduction of algae into the system. *Spirogyra condensata* and *Rhizoclonium hieroglyphicum* also employed as biosorption substrates to remove chromium from tannery wastewater (Gurbuz et al., 2009). The pH and concentration of algae were concluded to have significant effect on removal of chromium thus indicating potential of algae for removal hazardous heavy metals in wastewater (Gurbuz et al., 2009).

5. GENETICS IN ALGAE

Genetic engineering is a method to produce valuable biomolecules to medicine or industry, that are difficult or even impossible to produce in another way, or which require prohibitively high production costs in other systems.

Genetic engineering in algae is a complex and fast-growing technology that has advanced significantly over the past two decades (Fuhrmann, 2002). Normally, more than just one species generates a desired product or shows another trait of interest. Therefore, careful selection of an appropriate target organism stands at the beginning of every algae transformation project. Aside from important issues like product quality and quantity, additional points have to be considered (Hallman, 2007). Selectable marker genes, promoters, reporter genes, transformation techniques, and other genetic tools and methods are already available for various species and are accessible to genetic transformation. Large-scale sequencing projects are also planned, in progress, or completed for several of these species; the most advanced genome projects are those for the red alga *Cyanidioschyzon merolae*, the diatom *Thalassiosira pseudonana*, and the three green algae *Chlamydomonas reinhardtii*, *Volvox carteri* and *Ostreococcus tauri* (Hallman, 2007).

A powerful driving force in algae transgenics is the prospect of using genetically modified algae as bioreactors. In general, nowadays non-transgenic, commercial algae biotechnology produces food additives, cosmetics, animal feed additives, pigments, polysaccharides, fatty acids, and biomass. But recent progress in algae transgenics promises a much broader field of application: molecular farming, the production of proteins or metabolites that are valuable to medicine or industry, seems to be feasible with transgenic algae systems (Figure 5.1). Indeed, the ability of transgenic algae to produce recombinant antibodies, vaccines, insecticidal proteins, or bio-hydrogen has already been demonstrated. Genetic modifications that enhance physiological properties of algae strains and optimization of algal production systems should further improve the potential of this promising technology in the future (Teng et al., 2002; Geng et al., 2003; Léon-Banares et al., 2004; Tan et al., 2005; Hallman, 2007; Rosenberg et al., 2008).

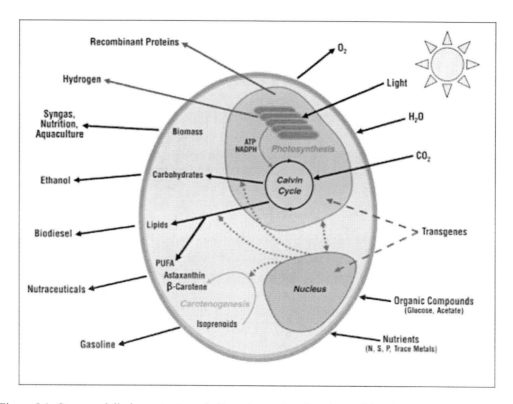

Figure 5.1. Commercially important metabolic pathways in microalgae. This schematic representation depicts simplified cellular pathways involved in the biosynthesis of various products derived from microalgae (adapted from Rosenberg et al., 2008).

5.1. Transformation Experiments

One of the most obvious arguments that stand for a certain species is the ease of cultivating the organism, especially under laboratory conditions. With respect to this point, small algae that grow with a short life cycle in liquid, axenic culture in a synthetic medium under defined environmental conditions are preferred for transformation experiments.

Ideally, the target species is also easily satisfied by inexpensive culture media and it grows to high densities, even under varying conditions. To allow for selection of transformed algae, a method must be established or be available that allows for regeneration of the target species from single cells. Most algae are photoautotrophs, so they require only light, water and basic nutrients for growth; other algae are heterotrophs and can also be grown in the dark if sugars are present in the culture medium (Borowitzka and Borowitzka, 1988; Hallman, 2007). For transformation experiments it might also be advantageous to use species with separate sexes and a well-known sexual life cycle that can be triggered under laboratory conditions. In this way, controlled genetic crosses can be made. It is also helpful to use a alga in which mutants can be easily generated or mutant collections exist (Hallman, 2007).

Previous molecular, biochemical, physiological or ecological knowledge of the target species and a somewhat developed molecular toolbox is desirable. Moreover, the existence of extensive sequence information is of extreme importance for algae transgenics.

So far, no algae species fulfills all of these requirements, but several microalgae come quite close to this ideal of a target organism (Hallman, 2007). Due to their small sized genomes, more than 30 Cyanobacteria complete genome sequences are available in different databases such as GenBank such as *Synechococcus elongatus*, *Anabaena* sp., *Nostoc punctiforme*, *Synechocystis* sp., *Thermosynechococcus elongatus*, *Gloeobacter violaceus*, and *Prochlorococcus marinus*.

The existence of extensive sequence information for a given target algae and the fulfillment of most of the other mentioned general requirements is an excellent basis for the realization of successful and diversified transformation experiments. Before these experiments can begin, some questions about the organism to be transformed, have to be raised. For example, if the organism have any close relatives that have been transformed before; which DNA constructs might be the most suitable for selection of the transgenic algae; which promoters should work and result in a high level of expression; how should the DNAs be transferred into the cells; how can expression of a given gene be monitored.

The use of selectable marker genes is normally required in all experiments that aim to generate stable transgenic algae, since only a very low percentage of treated organisms are successfully transformed. Selectable markers are often antibiotic resistance genes which are dominant markers as they confer a new trait to any transformed target strain of a certain species, no matter of the respective genotype (Ciferri et al., 1989; Léon-Banares et al., 2004).

The basis of almost all algal transformation methods is to cause, by various means, temporal permeabilization of the cell membrane, enabling DNA molecules to enter the cell. Entrance of the DNA into the nucleus and integration into the genome occurs without any external help. DNA integration mainly occurs by illegitimate recombination events, resulting in ectopic integration of the introduced DNA and, thus, culminates in stable genetic transformation. Nowadays, it is not difficult to permeabilize a cell membrane in order to introduce DNA. However, the affected reproductive cell must survive this life-threatening damage and DNA invasion and resume cell division.

There is a couple of working transformation methods for algae systems that enable the recovery of viable transformants (Ciferri et al., 1989). The most popular method is micro-particle bombardment, also referred to as micro-projectile bombardment, particle gun transformation, gene gun transformation, or simply biolistics (Schiedlmeier et al., 1994; Johnson et al., 2007; Sanford et al., 1993).

This course of action, which appears like an anti-algae military operation using carpet bombing, was successfully applied in *Chlamydomonas reinhardtii* (Kindle et al., 1989), *Volvox carteri* (Schiedlmeier et al., 1994), *Dunaliella salina* (Tan et al., 2005), *Gracilaria changii* (Gan et al., 2003), *Laminaria japonica* (Qin et al., 1999; Jiang et al., 2003), *Phaeodactylum tricornutum* (Apt et al., 1996), *Navicula saprophila* (Dunahay et al., 1995), *Cyclotella cryptica* (Dunahay et al., 1995), *Euglena gracilis* (Doetsch et al., 2001), *Porphyridium* sp. (Lapidot et al., 2002), *Cylindrotheca fusiformis* (Fischer et al., 1999), *Haematococcus pluvialis* (Steinbrenner and Sandmann, 2006), *Chlorella kessleri* (El-Sheekh, 1999), and *Chlorella sorokiniana* (Dawson et al., 1997).

Another less complex and less expensive transformation procedure involves preparation of a suspension of microalgae that is then agitated in the presence of micro- or macro-particles, polyethylene glycol and DNA. Several investigators have used silicon carbide (SiC) whiskers as micro-particles. These hard and rigid micro-particles allowed transformation of cells with intact cell walls including *Chlamydomonas reinhardtii* (Dunahay, 1993), *Symbiodinium microadriaticum* (ten Lohuis and Miller, 1998), and *Amphidinium* sp. (ten Lohuis and Miller, 1998).

Additionally, electroporation relies on pulses of electric current to permeabilize the cell membrane and has shown great success with cell wall deficient organisms (Shimogawara et al., 1998; Chen et al., 2001; Sun et al., 2005). Naked cells, protoplasts, cell wall reduced mutants and other cells with thin walls can be transformed through this technology, in which specially designed electrodes effect voltage across the plasma membrane that exceeds its dielectric strength. This large electronic pulse temporarily disturbs the phospholipid bilayer of the cell membrane, allowing molecules like DNA to pass. Cells of *Chlamydomonas reinhardtii* (Brown et al., 1991), *Cyanidioschyzon merolae* (Minoda et al., 2004), *Dunaliella salina* (Geng et al., 2003), and *Chlorella vulgaris* (Chow and Tung, 1999) have been transformed by this procedure.

In the following subsections there are some examples of genetic changes in cyanobacteria and microalgae with different applications as pharmaceutical and biofuels.

5.2. Genetic Improvement of Algae for Pharmaceuticals

Chlamydomonas reinhardtii is currently the most favored single cell algae for the expression of human genes, being already demonstrated successfully expression and assembly of a recombinant human monoclonal IgA antibody (Mayfield et al., 2003). Achieving high expression in transgenic algae and simplification of antibody purification required optimization of the codons of the corresponding gene and fusion of the IgA heavy chain to the variable region of the light chain using a flexible linker. In this way, antibody production can become not only much more convenient, but also much cheaper than expression in other systems. In addition, expression in an organism without an immune system allows expression of antibodies that would otherwise interfere with the immune system of the host animal used in conventional antibody production. Algae have also demonstrated suitability for synthesizing vaccines. In this regard, stable expression of the hepatitis B surface antigen gene has been shown in *Dunaliella salina* (Sayre et al., 2001; Geng et al., 2003; Sun et al., 2003). Since *Dunaliella* is otherwise used for nutrition, there is no need for purification of the antigen, so the intact algae could be used to deliver a vaccine.

A further project aims at the application of antigen producing algae in the fish industry. It is intended to use an alga-produced antigen to vaccinate fish against the hematopoietic necrosis virus (IHNV) which causes an infectious disease that kills 30% of the US trout population each year. Vaccination is realized simply by feeding the fish with the algae (Banicki, 2004).

Microalgae have also been shown to be useful for expressing insecticidal proteins. Because the green alga *Chlorella* is one possible food for mosquito larvae, the mosquito hormone trypsin-modulating oostatic factor (TMOF) was heterologously expressed in *Chlorella*. TMOF causes termination of trypsin biosynthesis in the mosquito gut. After feeding mosquito larvae with these recombinant *Chlorella* cells the larvae died within 72 h (Borovsky, 2003). Because diseases such as malaria, dengue and west Nile fever are transmitted via mosquitoes, mosquito abatement is an expensive requirement in tropical countries. Use of such transgenic algae might be a much cheaper alternative.

The use of algae as an expression system is not restricted to antibodies, antigens, or insecticidal proteins. Most notably, *Chlamydomonas reinhardtii* offers a general, attractive alternative to the traditional but costly mammalian-based expression systems (Franklin and Mayfield, 2004).

5.3. Genetic Improvement of Algae for Biofuels

Industrial methods for the production of biofuels using energy- rich carbon storage products, such as sugars and lipids are well established. These methods are currently being used on a large scale in the production of bioethanol from corn grain and biodiesel from oil seed crops. However, it might be possible to introduce biological pathways in microalgae cells that allow for the direct production of fuel products that require very little processing before distribution and use (Figure 5.2).

Several biological pathways have been described for the production of fatty acid esters, alkanes, and alcohols. However, the introduction of metabolic pathways for the direct production of fuels faces many challenges. The product yields for pathways that lead to the accumulation of compounds that are not necessarily useful for the cell are unlikely to be economically viable without comprehensive engineering of many aspects of microalgae metabolism.

In addition, many types of fuel products are potentially toxic, and tolerant species of microalgae may have to be generated.

Other factors required to lower the production costs of algae biofuel include maximizing the content of lipids and other biofuel precursors, increasing the rate of cell growth, identifying chemical inducers of these metabolic changes, and implementing multistage growth systems (Rosenberg et al., 2008; Radakovits et al., 2010).

Understanding microalgae lipid metabolism is of great interest for the ultimate production of diesel fuel surrogates. Both the quantity and the quality of diesel precursors from a specific strain are closely linked to how lipid metabolism is controlled.

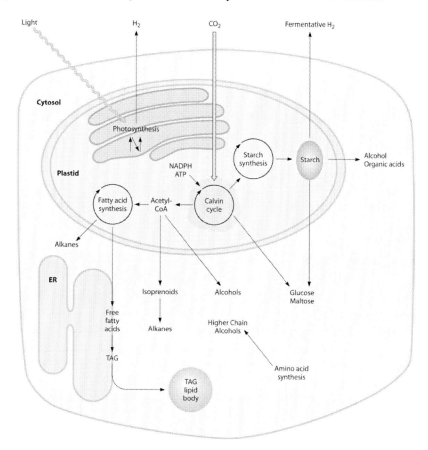

Figure 5.2. Microalgal metabolic pathways that can be leveraged for biofuel production. ER, endoplasmic reticulum (Radakovits et al., 2010).

Lipid biosynthesis and catabolism, as well as pathways that modify the length and saturation of fatty acids, have not been as thoroughly investigated for algae as they have been for terrestrial plants. However, many of the genes involved in lipid metabolism in terrestrial plants have homologs in the sequenced microalgal genomes. Therefore, it is probable that at least some of the transgenic strategies that have been used to modify the lipid content in higher plants will also be effective with microalgae.

Another possible approach to increasing the cellular lipid content is blocking metabolic pathways that lead to the accumulation of energy-rich storage compounds, such as starch.

When the goal is the production of carbohydrates from biomass to processing into bioethanol, one genetic strategy may be the decrease of the degradation of these compounds by microalgae (Radakovits et al., 2010).

In 1994, researchers were able to isolate the gene coding for acetyl-CoA carboxylase (ACCase), an enzyme that catalyzes an initial metabolic step in lipid biosynthesis in the diatom *Cyclotella cryptica* (Roessler et al., 1994). With this, they developed expression vectors and a transformation protocol that enabled the first attempt at metabolic engineering of a microalga (Dunahay et al., 1996). Even though the overexpression of ACCase did not have a significant effect on lipid synthesis, the tools and methodologies established by this work have set the stage for future genetic engineering efforts in microalgae (Sheehan et al., 1998).

Lindberg et al. (2010) applied the concept "photosynthetic biofuels". This consists of a system where the same organism serves both as photo-catalyst and producer of ready-made-fuel. This concept was applied upon genetic engineering of the cyanobacterium *Synechocystis*, conferring the ability to generate volatile isoprene hydrocarbons from CO_2 and H_2O. Heterologous expression of the *Pueraria montana* (kudzu) isoprene synthase (*IspS*) gene in *Synechocystis* enabled photosynthetic isoprene generation in these cyanobacteria.

Results of this work showed that oxygenic photosynthesis can be re-directed to generate useful small volatile hydrocarbons, while consuming CO_2, without a prior requirement for the harvesting, dewatering and processing of the respective biomass.

Concerning hydrogen metabolism, one strategy adopted to improve hydrogen yield is the inhibition or depletion of the gene that encodes for uptake hydrogenase, the enzyme that have the ability to consume hydrogen produced by some microalgae and cyanobacteria (Masukawa et al., 2002). These authors showed that by knocking out *hupL* gene of *Anabaena* sp. PCC 7120, H_2 production increased to 4–7 times that of wild-type.

Genetic techniques have also been applied with the aim of increasing H_2 photoproduction activity by decreasing light-harvesting antenna size, inhibiting state transitions, and other hydrogenase engineering (Beckmann et al., 2009; Ghirardi et al., 2009; Hankameret al., 2007; Melis, 2007).

In combination with physiological and biochemical approaches, these studies allowed important advancesin the understanding of H_2 metabolism and enzyme maturation (Bock et al., 2006; Posewitz et al., 2004).

In addition, numerous strategies are emerging to further advance our ability to optimize H_2 production in eukaryotic and prokaryotic phototrophs.

CONCLUSIONS

Microalgae are the tiniest plants on earth living mainly in aquatic environments and requiring only light, CO_2, N and P for their growth. Their enormous biochemical diversity enables their application in a vast range of biotechnological processes. Microalgae have been widely used in animal (food chain basis) and human nutrition, being expected to increase in the next years, considering their potential as a source of natural well-balanced nutrients. Major research is being conducted concerning microalgae health applications, namely by the isolation of pharmacologically active compounds.

An emerging area for microalgae biotechnology is environmental applications. This is mainly due to their ability of CO_2 mitigation, reducing green-house gas emissions that are related to global warming and climate changes; and their capability to growth in liquid effluent enabling wastewater treatment. Nowadays, there is a focus on using microalgae in renewable energy as a potential source for biofuels production such as biodiesel, bioetanol, biohydrogen and biogas.

Genetic and metabolic engineering are valuable tools which can improve even more the potential of microalgae as a natural source of nutraceutical, pharmaceutics and biofuels.

ACKNOWLEDGEMENTS

The authors would like to acknowledge FCT-Fundação para a Ciência e Tecnologia through the national projects "Microalgae as a sustainable raw material for biofuels production (biodiesel, bioethanol, bio-H2 and biogas)" (PTDC/AAC-AMB/100354/2008) and "Biohydrogen production from the cyanobacteria *Anabaena* sp. and its mutants" (PTDC/ENR/68457/2006).

REFERENCES

Aasen, A.J., Eimhjellen, K.E., and Liaaen-Jensen, S. (1969). An extreme source of □-carotene. *Acta Chemica Scandinavica*, 23, 2544-2545.

Abdulqader, G., Barsanti, L.,and Tredici, M.R. (2000).Harvest of *Arthrospira platensis* from Lake Kossorom (Chad) and its household usage among the Kanembu. *Journal of Applied Phycology*, 12, 493-498.

Abedin, R.M.A., and Taha, H.M. (2008). Antibacterial and antifungal activity of cyanobacteria and green microalgae.Evaluation of medium components by Placket-Burman design for antimicrobial activity of *Spirulina platensis*. *Global Journal of Biotechnology and Biochemistry*, 3, 22-31.

Akatsuka, I. (1990). *Introduction to Applied Phycology*. SPB Academic Publishing, The Hague.

Alam, K., Aguna, T., Maven, H., Taie, R., Rao, K.S., Burrows, I., Huber, M.E., and Rali,T.(1994). Preliminary screening of seaweeds, sea grass and Lemongrass oil from Papua New Guinea for antimicrobial and antifungal activity. *International Journal of Pharmacognosy*,32, 396-99.

Allen, M.B., and Dawson, E.Y. (1960).Production of antimicrobial substances by benthic tropical marine algae. *Journal of Bacteriology*,79, 459-460.

Amin, S. (2009). Review on biofuel oil and gas production processes from microalgae. *Energy Conversion and Management*, 50, 1834–1840.

An, J.Y., Sim, S.J., Lee, J.S., and Kim, B.K. (2003).Hydrocarbon production from secondarilytreated piggery wastewater by the green algae, *Botryococcus braunii*. *Journal of Applied Phycology*, 15, 185–91.

Andersen, R.A. (1996). Algae. In J.C. Hunter-Cevera and A. Belt (Eds.), *Maintaining Cultures for Biotechnology and Industry* (pp. 29-64).San Diego: Academic Press.

Angenent, L.T., Karim, K., Al-Dahhan, M.H., Wrenn, B.A., and Rosa, D.E. (2004) Production of bioenergy and biochemicals from industrial and agricultural wastewater.*Trends in Biotechnology*, 22, 477-485.

Antolin, G., Tinaut, F.V., Briceno, Y., Castano, V., Perez, C., and Ramirez, A.I. (2002). Optimization of biodiesel production by sunflower oil transesterification. *BioresourceTechnology*, 83, 111–114.

Apt, K.E., Kroth-Pancic, P.G., and Grossman, A.R.(1996) Stable nuclear transformation of the diatom *Phaeodactylum tricornutum. Molecular and General* Genetics, 252, 572-579.

Aslan, S., and Kapdan, I.K. (2006).Batch kinetics of nitrogen and phosphorus removalfrom synthetic wastewater by algae.*EcologicalEngineering*, 28, 64–70.

Atsumi, S., Higashide, W., and Liao, J.C. (2009).Direct photosynthetic recycling of carbon dioxide to isobutyraldehyde. *Nature Biotechnology*, 27, 1177-1182.

Ayehunie, S., Belay, A., Baba, T.W.,and Ruprecht, R.M. (1998). Inhibition of HIV-1 replication by an aqueous extract of *Spirulina platensis.Journal of Acquired Immune Deficiency Syndromes and Human Retrovirology*, 18, 7-12.

Balat, M.(2005). Current alternative engine fuels. *Energy Sources*, 27, 569–577.

Balat, M. (2009). Possible methods for hydrogen production. *Energy Sources*,31, 39-50.

Bandarra, N.M., Pereira, P.A, Batista, I.,and Vilela, M.H. (2003). Fatty acids, sterols and □-tocopherol in *Isochrysis galbana.Journal of Food Lipids*, 10, 25-34.

Bandarra, N.M., Duarte, D., Pinto, R., Sampayo, C., Ramos, M., Batista, I., Nunes, M.L., Batista, A.P., Raymundo, A., Gouveia, L., and Silva-Lima, B. (2010). Effect of dietary *n*-3 PUFA from microalgae on blood, liver, brain, kidney and heart lipids.*8th Euro Fed Lipid Congress – "Oils, Fats and Lipids: Health and Nutrition, Chemistry and Energy*. Munich 21-24th November.

Banicki, J.J.(2004) An alga a day keeps the doctor away. Engineered algae as a new means to vaccinate fish.*Twine Line*,26, 1-2.

Batista, A.P., Gouveia, L., Nunes, M.C., Franco, J.M., and Raymundo, A. (2008). Microalgae biomass as a novel functional ingredient in mixed gel systems. In P.A. Williams and G.O. Phillips (Eds),*Gums and Stabilisers for the Food Industry*(14th Edition, pp. 487-494). London: Royal Society of Chemistry.

Batista, A.P., Nunes, M.C., Gouveia, L., Sousa, I., Raymundo, A., Cordobés, F., Guerrero, A., and Franco, J.M. (2011). Microalgae biomass interaction in biopolymer gelled systems. *Food Hydrocolloids*, 25, 817-825.

Bayless, D.J., Kremer, G.G., Prudich, M.E., Stuart, B.J., Vis-Chiasson, M.L., Cooksey, K., and Muhs, J. (2001) Enhanced practical photosynthetic CO_2 mitigation. *Proceedings of the First National Conference on Carbon Sequestration*, 5A4, 1-14.

Becker, E.W. (1988). Micro-algae for human and animal consumption.In M.A. Borowitzka and L.J. Borowitzka(Eds.), *Micro-algal Biotechnology* (pp. 222-256).Cambridge University Press.

Becker, E.W. (1994a). *Microalgae: biotechnology and microbiology*. Cambridge University Press.

Becker, E.W. (1994b). Oil production. In J. Baddiley, N.H. Carey, I.J. Higgins and W.G. Potter (Eds),*Microalgae: Biotechnology and Microbiology*. Cambridge University Press.

Becker, E.W. (2004). Microalgae in human and animal nutrition. In A. Richmond (Ed.), *Handbook of microalgal culture* (pp. 312-351). Oxford: Blackwell Publishing.

Beckmann, J., Lehr, F., Finazzi, G., Hankamer, B., Posten, C., Wobbe, L., and Kruse, O. (2009). Improvement of light to biomass conversion by de-regulation of light-harvesting protein translation in *Chlamydomonas reinhardtii. Journal of Biotechnology*,142, 70–77.

Ben-Amotz, A., and Avron, M. (1980).Glicerol, □-carotene and dry algal meal production by commercial cultivation of *Dunaliella*. In G. Shelef, and C.J. Soeder (Eds.), *Algae Biomass* (pp. 603-610). Amsterdam: Elsevier/North Holland Biomedical Press.

Benemann, J.R. (1993) Utilization of carbon dioxide from fossil fuel - burning power plants with biological system.*Energy Conversion and Management*, 34, 999-1004.

Benemann, J.R., Weissman, J.C., Koopman, B.L. and Oswald, W.J. (1977).Energy production by microbial photosynthesis.*Nature*, 268, 19–23.

Benemann, J.R.,and Oswald, W.J. (1996).*Systems and economic analysis of microalgae ponds for conversion of CO_2 to biomass*.Department of Energy, Pittsburgh Energy Technology Center.Final Report, Grant No.DE-FG22-93PC93204.

Betenbough, M., and Bentley, W. (2008). Metabolic engineering in the 21st century: meeting global challenges of sustainability and health. *Current Opinion in Biotechnology*, 19, 411-413.

Bhat, V.B., and Madyastha, K.M. (2000).*C*-phycocyanin: a potent peroxyl radical scavenger *in vivo* and *in vitro*. *Biochemical and Biophysical Research Communications*, 275, 20-25.

Bock, A., King, P.W.; Blokesch, M., and Posewitz, M.C.(2006).Maturation of hydrogenases.*Advances in Microbial Physiology*,51,1–71.

Boichenko, V.A., and Hoffmann, P. (1994).Photosynthetic hydrogen-production in prokaryotes and eukaryotes - occurrence, mechanism, and functions.*Photosynthetica*, 30, 527-552.

Boichenko, V.A., Greenbaum, E., and Seibert, M. (2004).Hydrogen production by photosynthetic microorganisms. In M.D. Archer and J. Barber (Eds.), *Photoconversion of Solar Energy: Molecular to Global Photosynthesis* (Volume 2, pp. 397-452). London: Imperial College Press.

Borovsky, D.(2003) Trypsin-modulating oostatic factor: a potential new larvicide for mosquito control. *Journal of Experimental Biology*,206, 3869-3875.

Borowitzka, M.A. (1995). Microalgae as sources of pharmaceuticals and other biologically active compounds. *Journal of Applied Phycology*, 7, 3-15.

Borowitzka, M.A. (1999). Commercial production of microalgae: ponds, tanks, tubes and fermenters. *Journal of Biotechnology*, 70, 313-321.

Borowitzka, M.A., and Borowitzka, L.J.(1988).*Micro-algal Biotechnology*.Cambridge University Press.

Boyd, M.R. (2001). Anti-cyanovirin antibody with an internal image of gp 120, a method of use thereof, and a method of using a cyanovirin to induce an immune response to gp 120.United States Patent n° 6193982.

Boyd M.R. (2002). *Methods of using cyanovirins topically to inhibit viral infection*. United States Patent n° 6420336.

Boyd, M.R. (2004). Methods of using cyanovirins to inhibit viral infection. United States Patent n° 6743577.

Brennan, L., and Owende, P. (2010).Biofuels from microalgae – a review of technologies for production, processing, and extractions of biofuels and co-products.*Renewable and Sustainable Energy Reviews*, 14, 557-577.

Brown, L.E., Sprecher, S.L., and Keller, L.R.(1991) Introduction of exogenous DNA into *Chlamydomonas reinhardtii* by electroporation. *Molecular and Cellular Biology*,11, 2328-2332.

Brune, D.E., Lundquist, T.J.,and Benemann, J.R. (2009). Microalgal biomass for greenhouse gas reductions: potential for replacement of fossil fuels and animal feeds. *Journal of Environmental Engineering*, 135, 1136–1144.

Burja, A.M. Banaigs, B. Abou-Mansour, E. B., J.G. Wright, P.C. (2001).Marine cyanobacteria - a prolific source of natural products. *Tetrahedron.* 57, 9347-9377.

Burja A.M., Abou-Mansour, E., Banaigs, B., Payri C., Burgess J.G., and Wright P.C. (2002). Culture of marine cyanobacterium, *Lyngbya majuscule* (*Oscillatoriace*), for bioprocess intensified production of cyclic linear lipopeptides. *Journal of Microbiological Methods*, 48, 207-219.

Burkholder, P.R., Burkholder, L.M., and Almodover, L.R. (1960).Antibiotic activity of some marine algae of Puerto Rico.*Botanica Marina*,2, 149-154.

Burlew, J.S. (1953). *Algal Culture from Laboratory to Pilot Plant*. Carnegie Institute, Washington DC.

Caccamese, S., and Azzolina, R. (1979).Screening for antimicrobial activities in marine algae from Eastern Sicily. *Planta medica*,37, 333-339.

Calvo, M.A., Cabanes, F.J., and Abarca, L. (1986). Antifungal activity of some Mediteranean algae. *Mycopathologia,*93, 61-63.

Cantrell, K.B., Ducey, T., Ro, K.S., and Hunt, P.G. (2008) Livestock waste-to-bioenergy generation opportunities. *Bioresource Technology*, 99, 7941–7953.

Carlsson, A.S., van Beilen, J.B., Moller, R., and Clayton, D. (2007) Micro- and macro-algae utility for industrial applications.In D. Bowles (Ed.), *Outputs from the EPOBIO Project*. UK: CPL Press.

Carmichael, W.W., Biggs, D.F., and Gorham, P.R. (1977).Toxicology and pharmacological action of *Anabaena flos-aquae* toxin. *Science,* 187, 542-544.

Castenholz, R.W., and Waterbury, J.B. (1989). Oxygenic photosynthetic bacteria. Group I. Cyanobacteria. In J.T. Staley et al. (Eds.), *Bergey's Manual of Systematic Bacteriology*(Volume 3, pp. 1710-1798). Baltimore, Maryland:Williams and Wilkens.

Certik, M., and Shimizu, S. (1999).Biosynthesis and regulation of microbial polyunsaturated fatty acid production. *Journal of Biosciences and Bioengineering,* 87, 1-14.

Chaumont, D. (1993). Biotechnology of algal biomass production: a review of systems for outdoor mass culture. *Journal of Applied Phycology*, 5, 593-604.

Chen Y., Wang, Y., and Sun, Y. (2001).Highly efficient expression of rabbit neutrophil peptide-1 gene in *Chlorella ellipsoidea* cells. *Current Genetics*,39, 365–370.

Chinnasamy, S., Bhatnagar, A., Hunt, R.W., and Das, K.C. (2010). Microalgae cultivation in a wastewater dominated by carpet mill effluents for biofuel applications. *Bioresource Technology*, 101, 3097–3105.

Chisti, Y. (2007). Biodiesel from microalgae. *Biotechnology Advances*, 25, 294–306.

Chisti, Y. (2008). Biodiesel from microalgae beats bioethanol. *Trends in Biotechnology*, 26, 126–131.

Chow, K.C., and Tung, W.L.(1999) Electrotransformation of *Chlorella vulgaris*. *Plant Cell Reports*,18, 778-780.

Ciferri, O., Tiboni, O., and Sanangelatoni, A.M. (1989).The genetic manipulation of cyanobacteria and its potential uses.InR.C. Cresswell, T.A.V. Rees, N. Shah (Eds.),*Algal and Cyanobacterial Biotechnology*(1st Edition). Avon: Bath Press.

Converti, A., Casazza, A.A., Ortiz, E.Y., Perego, P.,and del Borghi, M. (2009). Effect of temperature and nitrogen concentration on the growth and lipid content of *Nannochloropsisoculata* and *Chlorellavulgaris* for biodiesel production. *Chemical Engineering and Processing*, 48, 1146–1151.

Cox, P.A., Banack, S.A., Murch, S.J., Rasmussen, U., Tien, G., Bidigare, R.R., Metcalf, J.S., Morrison, L.F., Codd, G.A.,and Bergman, B. (2005). Diverse taxa of cyanobacteria produce □-*N*-methylamino-*L*-alanine, a neurotoxic amino acid. *Proceedings of the National Academy of Sciences*, 102, 5074-5078.

Das, B.K., Pardhan, J., Pattnaik, P., Samantaray, B.R., and Samal, S.K. (2005).Production of antibacterials from the freshwater alga *Euglena viridis* (Ehren). *World Journal of Microbiology and Biotechnology*, 21, 45-50.

Das, D. (2009) Advances in biological hydrogen production processes: an approach towards commercialization. *International Journal of Hydrogen Energy*, 34, 7349-7357.

Das, D., and Veziroglu, T. (2001) Hydrogen production by biological processes: a survey of literature. *International Journal of Hydrogen Energy*, 26, 13–28.

Davidson, F.F. (1961). Antibacterial activity of *Oscillatoriaformosa* extract. *American Journalof Botany.*48, 542.

Davis, T.A., Volesky, B., and Mucci, A. (2003).A review of the biochemistry of heavy metalbiosorption by brown algae. *Water Research*, 37, 4311-4330.

Dawson, H.N., Burlingame, R., and Cannons, A.C.(1997) Stable transformation of *Chlorella*: rescue of nitrate reductase-deficient mutants with the nitrate reductase gene. *Current Microbiology*,35, 356-362.

de Morais, M.G., and Costa, J.A.V. (2007).Biofixation of carbon dioxide by *Spirulina* sp. and*Scenedesmus obliquus* cultivated in a three-stage serial tubular photobioreactor. *Journal of Biotechnology*, 129, 439–445.

Dean, A.P., Sigee, D.C., Estrada, B., and Pittman, J.K. (2010).Using FTIR spectroscopy for rapid determination of lipid accumulation in response to nitrogen limitation in freshwater microalgae. *Bioresource Technology*, 101, 4499–4507.

Debabrata, D., and Veziroglu, T.N. (2001).Hydrogen production by biological processes: a survey of literature. *International Journal of Hydrogen Energy*, 26, 13-28.

Debro, L.H., and Ward, H.B. (1979).Anibacterial activity of fresh water green algae.*Planta Medica*,36, 375 – 378.

Demirbas, A. (2009). Biofuels securing the planet's future energy needs. *Energy Conversion and Management*, 50, 2239–2249.

Devlin, J.P., Edwards, O.E., Gorham, P.R., Hunter, N.R., Pike, R.K., and Starvick, B. (1977).Anatoxin-*a*, a toxic alkaloid from *Anabaena flos-aquae* NCR-44h. *Canadian Journal of Chemistry*, 5, 1367-1371.

Dhamotharan, R. (2002). An investigation on the bioactive principles of *Padina tetrastromatica* Hauck and *Stoechospermum marginatum* (C.Ag.)Kuetz with respect to antimicrobial and biofertilizer properties.*PhD Thesis*.University of Madras.

Dismukes, G.C., Carrieri, D., Bennette, N., Ananyev, G.M., and Posewitz, M.C.(2008). Aquatic phototrophs: efficient alternatives to land-based crops for biofuels. *Current Opinion in Biotechnology*, 19, 235–240.

Doetsch, N.A., Favreau, M.R., Kuscuoglu, N., Thompson, M.D., and Hallick, R.B.(2001). Chloroplast transformation in *Euglena gracilis*: splicing of a group III twintron transcribed from a transgenic *psbK* operon. *Current Genetics*,39, 49-60.

Donato, M., Vilela, M.H.,and Bandarra, N.M. (2003).Fatty acids, sterols, □-tocopherol and total carotenoids composition of *Diacronema vlkianum*. *Journal of Food Lipids*, 10, 267-276.

Doucha, J., Straka, F., and Lı́vansky´, K. (2005).Utilization of flue gas for cultivation ofmicroalgae (*Chlorella* sp.) in an outdoor open thin-layer photobioreactor. *Journal of Applied Phycology*, 17, 403–12.

Douskova I, Doucha J, Machat J, Novak P, Umysova D, Vitova M, Zachleder V (2008) Microalgae as a means for converting flue gas CO_2 into biomass with a high content of starch. Bioenergy: Challenges and Opportunities International Conference and Exhibition on Bioenergy. Guimarães, Portugal, April 6th-9th.

Drapcho, C.M., Nhuan, N.P., and Walker, T.H. (2008).*Biofuels Engineering Process Technology*.Mc- Graw Hill, New York.

Duff, D.C.B., Bruce, D.L., and Anita, N.J. (1966).The antibacterial activity of marine plantonic algae. *Canadian Journal of Microbiology*,12, 877-884.

Dumas, A., Laliberte, G., Lessard, P., and Noue, J. (1998).Biotreatment of fish farm effluentsusing the cyanobacterium *Phormidium bohneri*. *Aquaculture Engineering*,17, 57–68.

Dunahay, T.G.(1993).Transformation of *Chlamydomonas reinhardtii* with silicon carbide whiskers. *BioTechniques*,15, 452- 460.

Dunahay, T.G., Jarvis, E.E., and Roessler, P.G.(1995) Genetic transformation of the diatoms *Cyclotella cryptica* and *Navicula saprophila*. *Journal of Phycology*,31, 1004-1012.

Dunahay, T.G., Jarvis, H.H., Dais, S.S., and Roessler, P.G. (1996).Manipulation of microalgal lipid production using genetic engineering, *Applied Biochemistry and Biotechnology*,57/58, 223–231.

Edwards, C. (1990). Microbiology of Extreme Environments. Milton Keynes, Great Britain: Open University Press.

El-Sheekh, M.M.(1999).Stable transformation of the intact cells of *Chlorella kessleri* with high velocity microprojectiles. *Biologia Plantarum*,42, 209-216.

Energy Information Administration(2009).International Energy Outlook 2009.US Department of Energy, Washington, DC. Available at: http://www.eia.doe.gov (last accessed on July 2010).

Febles, C. I., Arias, A., Gil-Rodriguez, M. C., Hardisson, A., and Sierra-Lopez, A. (1995).*In vitro* study of antimicrobial activity in algae (*Chlorophyta*, *Phaeophyta* and *Rhodophyta*) collected from the coast of Tenerife (in Spanish). *Anuario del Estudios Canarios*, 34, 181-192.

Fidalgo, J.P., Cid, A., Torres, E., Sukenik, A., and Herrero, C. (1998). Effects of nitrogen source and growth phase on proximate biochemical composition, lipid classes and fatty acid profile of the marine microalga *Isochrysis galbana*. *Aquaculture*, 166, 105-116.

Fischer, H., Robl, I., Sumper, M., and Kröger, N.(1999) Targeting and covalent modification of cell wall and membrane proteins heterologously expressed in the diatom *Cylindrotheca fusiformis*. *Journal of Phycology*,35, 113-120.

Fitton, J. H., Irhimeh, M., and Falk, N. (2007) Macroalgal fucoidan extracts: a new opportunity for marine cosmetics. *Cosmetics and Toiletries*, 122, 55-64.

Fradique, M.; Batista, A.P.; Nunes, M.C.; Gouveia, L.; Bandarra, N.M.; Raymundo, A. (2010). *Chlorellavulgaris* and *Spirulinamaxima* biomass incorporation in pasta products. *Journal of the Science of Food and Agriculture*, 90, 1656-1664.

Franklin, S.E., and Mayfield, S.P.(2004).Prospects for molecular farming in the green alga *Chlamydomonas*. *Current Opinion in Plant Biology*,7, 159-165.

Fuhrmann, M. (2002).Expanding the molecular toolkit for *Chlamydomonas reinhardtii* from history to new frontiers. *Protist*,153, 357–364.

Gan, S.Y, Qin, S., Othman, R.Y., Yu, D., and Phang, S.M.(2003). Transient expression of *lacZ* in particle bombarded *Gracilaria changii* (Gracilariales, Rhodophyta). *Journal of Applied Phycology* 15, 345-349.

Geng, D., Wang, Y., Wang, P., Li, W., and Sun, Y.(2003) Stable expression of hepatitis B surface antigen gene in *Dunaliella salina* (Chlorophyta). *Journal of Applied Phycology*,15, 451-456.

Gerwick, W.H., Proteau, P.J., Nagle, D.G., Hamel, E., Blokhin, A.,and Slate, D.L. (1994).Structure of curacin A, a novel antimitotic, antiproliferative, and brine shrimp toxic natural products from the marine Cyanobacterium *Lyngbya majuscula*. *Journal of Organic Chemistry*, 59, 1243–1245.

Ghasemi, Y., Tabatabaei-Yazdi, M., Shafiee, A., Amini, M., Shokravi, Sh., and Zarrini, G. (2004).Parsiguine, a novel antimicrobial substance from *Fischerella ambigua*. *Pharmaceutical Biology*, 2, 318-322.

Ghasemi, Y., Moradian, A., Mohagheghzadeh, A., Shokravi, S. and Morowvat, M.H. (2007). Antifungal and antibacterial activity of the microalgae collected from paddy fields of Iran: characterization of antimicrobial activity of *Chroococus dispersus*. *Journal of Biological Sciences*, 7, 904-910.

Ghirardi, M.L., Kosourov, S., Tsygankov, A., and Seibert, M. (2000).Two-phase photobiological algal H$_2$-production system. *Proceedings of the 2000 U.S. DOE Hydrogen Program Review*. National Renewable Energy Laboratory, Golden, Colorado (pp. 1-13).

Ghirardi, M.L., Dubini, A.,Yu, J. and Maness, P.C.(2009). Photobiological hydrogen-producing systems. *Chemical Society Reviews*,38, 52–61.

Ghirardi, M. L., and Mohanty, P. (2010).Oxygenic hydrogen photoproduction – current status of the technology. *Current Science*, 98, 499-507.

Glombitza, K.W. (1979). Antibiotics from algae. InH.AHoppe, T. Levring and Y. Tanaka (Eds.), *MarineAlgae in the Pharmaceutical Science* (pp. 303-343).Berlin: Walter de Gruyter.

Godos, I., Vargas, V.A., Blanco, S., González, M.C.G., Soto, R., García-Encina, P.A., Becares, E., and Muñoz, R. (2010). A comparative evaluation of microalgae for the degradation of piggery wastewater under photosynthetic oxygenation. *Bioresource Technology*, 101, 5150–5158.

Goldemberg, J. (2007). Ethanol for a sustainable energy future. *Science*,315, 808-810.

Golueke, C.G., Oswald, W.J., and Gee, H.K.(1965).Harvesting and processing sewage-grown planktonic algae. *Journal Water Pollution Control Federation*, 37, 471–498.

Gonzales, L.E., Canizares, R.O., and Baena, S. (1997). Efficiency of ammonia and phosphorusremoval from a Colombian agroindustrial wastewater by the microalgae *Chlorellavulgaris* and *Scenedesmusdimorphus*. *Bioresource Technology*, 60, 259–262.

Gorham, P.J., McLachlan, J., Hammer, U.T., and Kim, W.K. (1964).Isolation and toxic strains of *Anabaena flos-aquae*. *De Breb. Verh. Inetrnat. Verein. Limnol.*, 15, 796-804.

Gouveia, L., Marques, A.E., Sousa, J.M., Moura, P. and Bandarra, N.M. (2010). Microalgae – source of natural bioactive molecules as functional ingredients. *Food Science and Technology Bulletin: Functional Foods*, 7, 21-37.

Gouveia, L. (2011). *Microalgae as a feedstock for biofuels*.Springer Briefs in Microbiology.Springer. Heidelberg, Germany.

Gouveia, L., Batista, A.P., Raymundo, A., Sousa, I., and Empis, J. (2006). *Chlorella vulgaris* and *Haematococcus pluvialis* biomass as colouring and antioxidant in food emulsions. *European Food Research and Technology*, 222, 362-367.

Gouveia, L., Batista, A.P., Miranda, A., Empis, J., and Raymundo, A. (2007). *Chlorella vulgaris* biomass used as colouring source in traditional butter cookies. *Innovative Food Science and Emerging Technologies*, 8, 433-436.

Gouveia, L., Batista, A.P., Raymundo, A., and Bandarra, N.M. (2008a).*Spirulina maxima* and *Diacronema vlkianum* microalgae in vegetable gelled desserts. *Nutrition and Food Science*, 38, 492-501.

Gouveia, L., Coutinho, C., Mendonça, E., Batista, A.P., Sousa, I., Bandarra, N.M.,and Raymundo, A. (2008b). Sweet biscuits with *Isochrysis galbana* microalga biomass as a functional ingredient. *Journal of the Science of Food and Agriculture*, 88, 891-896.

Gouveia, L., and Oliveira, A.C. (2009). Microalgae as a raw material for biofuels production. *Journal of Industrial Microbiology and Biotechnology*,36, 269-274.

Gouveia, L., Marques, A.E., Lopes da Silva, T., and Reis, A. (2009) *Neochloris oleabundans* UTEX #1185: a suitable renewable lipid source for biofuel production. *Journal of Industrial Microbiology and Biotechnology*, 36, 821-826.

Grach-Progrebinsky, O., Sedmak, B., and Carmeli, S. (2003). Protease inhibitors from a Slovenian Lake Bled toxic water bloom of the cyanobacterium *Planktothrix rubescens*. *Tetrahedron*. 59, 8329-8336.

Graham, L.E., and Wilcox, L.W. (2000).*Algae*.Upper Saddle River, NJ: Prentice-Hall, Inc.

Green, F.B., Lundquist, T.J., and Oswald, W.J.(1995ab). Energetics of advanced integrated wastewater pond systems. *Water Science and Technoogy*, 31, 9–20.

Green, F.B., Bernstone, L., Lundquist, T.J., Muir, J., R.B. Tresan, and Oswald, W.J.(1995b). Methane fermentation, submerged gas collection, and the fate of carbon in advanced integrated wastewater pond systems. *Water Science and Technoogy*, 31, 55–65.

Green, F.B., Bernstone, L., Lundquist, T.J., and Oswald, W.J.(1996).Advanced integrated wastewater pond systems for nitrogenremoval. *Water Science and Technoogy*,33, 207–217.

Gregoire-Padro, C.E. (2005).Hydrogen basics. *First annual International Hydrogen Energy Implementation Conference*. The New Mexico Hydrogen Business Council, Santa Fe, NM.

Griffiths, M.J. and Harrison, S.T.L. (2009).Lipid productivity as a key characteristic for choosing algal species for biodiesel production. *Journal of Applied Phycology*, 21, 493–507.

Gross, R., Gross, U., Ramirez, A., Cuadra, K., Collazos, C., and Feldheim, W. (1978). Nutritional tests with green *Scenedesmus* with health and malnourished children. *Archiv fur Hydrobiologie, Beihefte Ergebnisse der Limnologie*, 11, 161-173.

Gross, R., Schneberger, H., Gross, U.,and Lorenzen, H. (1982). The nutritional quality of *Scenedesmus acutus* produced in a semi industrial plant in Peru. *Berichte der Deutschen Botanischen Gesellschaft*, 95, 323-327.

Gupta, A.B., and Shrivastava, G.C. (1965).On antibiotic properties of some fresh wateralgae. *Hydrobiologia*,25, 285-288.

Gurbuz, F., Ciftci, H., and Akcil, A. (2009).Biodegradation of cyanide containing effluents by*Scenedesmus obliquus*. *Journal of Hazardous Materials*, 162, 74–79.

Gustafson, K., Cardellina, J., Fullar, R.W., Weislow, O.S., Kiser, R.F., Snader, K., Patterson, G.M., and Boyd, M.R. (1989).AIDS-antiviral sulfolipids from cyanobacteria (blue-green algae). *Journal of the National. Cancer Institute*, 81, 1254-1258.

Hallmann, A. (2007). Algal transgenics and biotechnology. *Transgenic Plant Journal*, 1, 81-98.

Hanagata, N., Takeuchi, T., Fukuju, Y., Barnes, D. J., and Karube, I. (1992).Tolerance of microalgae to high CO_2 and high temperature. *Phytochemistry*, 31, 3345-3348.

Hankamer, B., Lehr, F., Rupprecht, J., Mussgnug, J.H., Posten, C., Kruse, O. (2007). Photosynthetic biomass and H_2 production by green algae: from bioengineering to bioreactor scale-up. *Physiologia Plantarum*, 131, 10-21.

Hansel, A., and Lindblad, P. (1998).Towards optimization of cyanobacteria as biotechnologically relevant producers of molecular hydrogen, a clean and renewable energy source. *Applied Microbiology Biotechnology*, 50, 153-160.

Harder, R., and Opperman, A. (1954).Antibiotics from green algae *Stichococcus bacillaris* and *Protosiphon botrydies*. *Arch. Mikrobiol.*,19, 398-401.

Harel, M., and Clayton, D. (2004). Feed formulation for terrestrial and aquatic animals. US Patent 20070082008 (WO/2004/080196).

Harlin, M.M., and Darley, W.M.(1988). The algae: an overview. InC.A. Lembi and RA Waaland (Eds.),*Algae and Human Affairs* (pp 3-27).Cambridge University Press.

Harun, R., Singh, M., Forde, G.M., Danquah, M.K. (2010) Bioprocess engineering of microalgae to produce a variety of consumer products.*Renewable and Sustainable Energy Reviews*, 14, 1037–1047.

Hayashi, O., Katoh, T.,and Okuwaki Y. (1994).Enhancement of antibody production in mice by dietary *Spirulina platensis*. *Journal of Nutritional Science and Vitaminology*, 40, 431-441.

Hayashi, T., and Hayashi, K. (1996).Calcium spirulan, an inhibitor of enveloped virus replication from a Blue-Green Alga *Spirulina platensis*. *Journal of Natural Products*, 59, 83-87.

Hayashi, K., Hayashi, T.,and Kojima, I. (1996). A natural sulfated polysaccharide, calcium spirulan, isolated from *Spirulina platensis*: *in vitro* and *ex vivo* evaluation of anti-herpes simplex virus and anti-human immunodeficiency virus activities. *AIDS Research and Human Retroviruses*, 12, 1463-1471.

He, H., Feng, C., Huashou, L., Wenzhou, X., Yongjun, L., and Yue, J. (2010).Effect of iron on growth, biochemical composition and paralytic shellfish poisoning toxins production of *Alexandriumtamarense*. *Harmful Algae*, 9, 98-104.

Henriquez, P., Candia, A., Norambuena, P., Silva, M., and Zemelman R. (1979). Antibiotic properties of Marine algae II.Screening of Chilean marine algae for antimicrobial activity. *Boanica Marina*,22, 451-453.

Hill, J., Nelson, E., Tilman, D., Polasky, S.,and Tiffany, D. (2006).Environmental, economic, and energetic costs and benefits of biodiesel and ethanol biofuels. *Proceedings from the National Academy of Sciences USA*, 103, 11206–11210.

Hirano, A., Ryohei, U., Shin, H., and Yasuyuki, O. (1997).CO_2 fixation and ethanol production with microalgal photosynthesis and intracellular anaerobic fermentation. *Energy*, 22, 137-142.

Hirata, S., Hayashitani, M., Taya, M., and Tone, S. (1996a).Carbon dioxide fixation in batch culture of *Chlorella* sp. using a photobioreactior with a sunlight-collection device. *Journal of Fermentation and Bioengineering*, 81, 470-472.

Hirata, S., Taya, M., and Tone, S. (1996b). Characterization of *Chlorella* cell cultures in batch and continuos operations under a photoautotrophic condition. *Journal of Chemical Engineering of Japan*, 29, 953-959.

Hodaifa, G., Martínez, M.E., Sánchez, S.(2008).Use of industrial wastewater from oliveoilextraction for biomass production of *Scenedesmusobliquus*. *Bioresource Technology*, 99, 1111–1117.

Hon-Nami, K. (2006).A unique feature of hydrogen recovery in endogenous starch-to-alcohol fermentation of the marine microalga. *Applied Biochemistry and Biotechnology*, 129-132, 808-828.

Hon-Nami, K., and Kunito, S. (1998) Microalgae cultivation in a tubular bioreactor and utilization of their cells. *Chinese Journal of Oceanology and Limnology*, 16, 75-83.

Horikawa, M., Noro, T., and Kamei, Y. (1966).Characterization and purification of antibacterial substances from rd alga *Laurencia okamurae*. *Marine and Highland Bioscience Center Report*.3, 45-52.

Hornsey, I.S., and Hide, D. (1974). The production of antimicrobial compounds by british marine algae. I. Antibiotic producing marine algae. *British Phycological Journal*,9, 353-361.

Houghton, J., Ding, Y., Griggs, D., Noguer, M., van der Linden, P., Dai, X., Maskell, K.,and Johnson, C. (2001). *Climate Change 2001. The Scientific Basis*.Cambridge University Press.

Hu, Q., Sommerfeld, M., Jarvis, E., Ghirardi, M., Posewitz, M., Seibert, M., and Darzins, A. (2008). Microalgal triacylglycerols as feedstocks for biofuel production: perspectives and advances. *The Plant Journal*, 54, 621–639.

Huntley, M.E., and Redalje, D.G. (2007). CO_2 mitigation and renewable oil from photosynthetic microbes: A new appraisal. *Mitigation and Adaptation Strategies for Global Change*, 12, 573–608.

Illman, A.M., Scragg, A.H., and Shales, S.W. (2000). Increase in *Chlorella* strains calorific values when grown in low nitrogen medium. *Enzyme and Microbial Technology*, 27, 631–635.

International Code of Botanical Nomenclature (2006) (Vienna Code).Regnum Vegetabile 146. A.R.G. Gantner Verlag KG(http://ibot.sav.sk/icbn/main.htm).

International Code of Nomenclature of Bacteria (1992).International Union of Microbiological Societies.Washington (DC): ASM Press.

Jacob-Lopes, E., Lacerda, L.M.C.F., and Franco, T.T. (2008).Biomass production and carbondioxide fixation by *Aphanothece microscopica* Nageli in a bubble columnphotobioreactor. *Biochemical Engineering Journal*, 2008, 40, 27–34.

Jacob-Lopes, E., Scoparo, C.H.G., Lacerda, L.M.C.F., and Franco, T.T. (2009).Effect of lightcycles (night/day) on CO_2 fixation and biomass production by microalgaein photobioreactors. *Chemical Engineering and Processing*, 48,306–10.

Jiang, P., Qin, S., and Tseng, C.K.(2003).Expression of the *lacZ* reporter gene in sporophytes of the seaweed *Laminaria japonica* (Phaeophyceae) by gametophyte-targeted transformation. *Plant Cell Reports*,21, 1211-1216.

Jimenez-Perez, M.V., Sanches-Castillo, P., Romera, O., Fernandez-Moreno, D., and Perez-Martinez, C. (2004).Growth and nutrient removal in free and immobilized planktonic green algae isolated from pig manure. *Enzyme and Microbial Technology*, 34, 392–398.

Johnson, E.A., Rosenberg, J., and McCarty, R.E. (2007). Expression by *Chlamydomonas reinhardtii* of a chloroplast ATP synthase with polyhistidine-tagged beta subunits, *Biochimica etBiophysica Acta*,1767, 374–380.

Kapdan, I.K, and Kargi, F. (2006).Bio-hydrogen production from waste materials. *Enzyme and Microbial Technology*, 38, 569–582.

Kadam, K.L. (1997). Power plant flue gas as a source of CO_2 for microalgae cultivation:economic impact of different process options. *Energy Conversion and Management*, 38(Suppl.), 505–510.

Kadam, K.L. (2001). Microalgae production from power plant flue gas: environmentalimplications on a life cycle basis. NREL/TP-510-29417. Colorado, USA: National Renewable Energy Laboratory.

Kadam, K.L. (2002). Environmental implications of power generation via coal–microalgae cofiring. *Energy*, 27, 905–922.

Katircioglu, H., Beyatli,Y., Aslim, B., Yüksekdag, Z. and Atici, T. (2006).Screening for antimicrobial agent production of some microalgae in freshwater. *The Internet Journal of Microbiology.*2.

Kato, T. (1994).Blue pigment from *Spirulina*. *New Food Industry*, 29, 17-21.

Kellam, S.J., and Walker, J.M. (1989).Antibacterial activity from marine microalgae in laboratory culture. *British Phycological Journal*, 24, 191-194.

Kim, M.K., Park, J.W., Park, C.S., Kim, S.J, Jeune, K.H, Chang, M.U, and Acreman, J. (2007).Enhanced production of *Scenedesmus* sp. (green microalgae) using a new medium containing fermented swine wastewater. *Bioresource Technology*, 98, 2220–2228.

Kim, S., Ravichandran, Y.D., Khan, S.B.,and Kim, Y.T. (2008). Prospective of the cosmaceuticals derived from marine organisms. *Biotechnology and Bioprocess Engineering*, 13, 511-523.

Kindle, K.L., Schnell, R.A., Fernandez, E., and Lefebvre, P.A.(1989).Stable nuclear transformation of *Chlamydomonas* using the *Chlamydomonas* gene for nitrate reductase. *Journal of Cell Biology*,109, 2589-2601.

Kobayashi, M., Kakizono, T., Nishio, N., Nagai, S., Kurimura, Y., and Tsuji, Y. (1997).Antioxidant role of astaxanthin in the green alga *Haematococcus pluvialis*. *Applied Microbiology and Biotechnology*, 48, 351-356.

Kodama, M., Ikemoto, H., and Miyachi, S. (1993).A new species of highly CO_2-tolreant fast-growing marine microalga suitable for high-density culture. *Journal of Marine Biotechnology*,1, 21-25.

Kulik, M.M. (1995). The potential for using cyanobacteria (blue green algae) and algae in the biological control of plant pathogenic bacteria and fungi. *European Journal of Plant Pathology*, 101,585-599.

Kushak, R. I., Drapeau, C.,van Cott, E.M., and Winter, H.H. (2000). Favorable effects of blue green algae *Aphanizomenon flos-aquae* on rat plasma lipids. *JANA*, 2, 59-65.

Laliberte, G., Lessard P., Noue, J., and Sylvestre, S. (1997).Effect of phosphorus addition onnutrient removal from wastewater with the cyanobacterium *Phormidiumbohneri*. *Bioresource Technology*, 59, 227–33.

Lapidot, M., Raveh, D., Sivan, A., Arad, S.M., and Shapira, M.(2002).Stable chloroplast transformation of the unicellular red alga *Porphyridium* species. *Plant Physiology*,129, 7-12.

Lavens, P., and Sorgeloos, P. (1996). Manual on the production and use of life food for aquaculture. *FAO Fisheries Technical Paper*, 361, 7–42.

Lee, K., and Lee, C.G. (2001).Effect of light/dark cycles on wastewater treatments by microalgae. *Biotechnology and Bioprocess Engineering*, 6, 194–199.

Lee, Y.K. (1997). Commercial production of microalgae in the Asia–Pacific rim. *Journal of Applied Phycology*, 9, 403–411.

Lembi, C.A., and Waaland, J.R.(1989).Algae and Human Affairs.Cambridge University Press.

León-Bañares, R., González-Ballester, D., Galván, A., and Fernández, A. (2004).Transgenic microalgae as green cell factories. *Trends in Biotechnology*,22, 45–52.

Li, W.I., Berman, F.W., Okino,T., Yokokawa, F., Shioiri, T., Gerwick, W.H., and Murray, T.F. (2001). Antillatoxin is a marine cyanobacterial toxin that potently activates voltage-gated sodium channels. *Proceedings of the National Academy of Sciences*, 98, 7599-7604.

Li, Y.Q., Horsman, M., Wang, B., Wu, N., andLan, C.Q. (2008). Effects of nitrogen sources on cell growth and lipid accumulation of green alga *Neochlorisoleoabundans*. *Applied Microbiologyand Biotechnology*, 81, 629–636.

Liang, Y., Beardall, J.,and Heraud, P. (2006).Changes in growth, chlorophyll fluorescence and fatty acid composition with culture age in batch cultures of *Phaeodactylum tricornutum* and *Chaetoceros muelleri* (Bacillariophycee). *Botanica Marina*, 49, 165-173.

Lilleheil, G., Andersen, R.A., Skulberg, O.M., and Alexander, J. (1997). Effects of a homoanatoxin-A-containing extract from *Oscillatoria formosa* (Cyanophyceae / Cyanibacteria) on neuromuscular transmission. *Toxicon*, 35, 1275-1289.

Lindberg,P., Park, S., and Melis, A. (2010). Engineering a platform for photosynthetic isoprene production in cyanobacteria, using *Synechocystis* as a model organism. *Metabolic Engineering*, 12, 70-79.

Lindblad, P., Christensson, K., Lindberg, P., Fedorov, A., Pinto, F., and Tsygankov, A. (2002). Photoproduction of H_2 by wildtype *Anabaena* sp. PCC 7120 and a hydrogen uptake deficient mutant: from laboratory experiments to outdoor culture. *International Journal of Hydrogen Energy*, 27, 1271-1281.

Liu, C.P., and Lin, L.P. (2001).Ultrastructural study and lipid formation of *Isochrysis* sp. CCMP1324. *Botanical Bulletin Academia Sinica*, 42, 207-214.

Lopes da Silva, T., Reis, A., Medeiros, R., Oliveira, A.C., and Gouveia, L. (2008). Oil production towards biofuel from autotrophic microalgae semicontinuous cultivations monitorized by flow citometry. *Applied Biochemistry and Biotechnology*, 159, 568-578.

Madamwar, D., Garg, N., and Shah, V. (2000).Cyanobacteria hydrogen production. *World Journal of Microbiology and Biotechnology*, 16, 757-767.

Maeda, K., Owada, M., Kimura, N., Omata, K., and Karube, I. (1995). CO_2 fixation from the flue gas on coal-fired thermal power plant by microalgae. *Energy Conversion and Management*, 36, 717–20.

Mahmood, N.A., and Carmichael, W.W. (1986). The pharmacology of anatoxin-a(s), a neurotoxin produced by the freshwater cyanobacterium *Anabaena flos-aquae* NRC 525-517. *Toxicon*, 24, 425-434.

Mahmood, N.A., and Carmichael, W.W. (1987).Anatoxin-a(s), an anti-cholinesterease from the cyanobacterium *Anabaena flos-aquae* NRC 525-5127. *Toxicon.* 25, 1221-1227.

Majmudar, G. (2007) Compositions of marine botanicals to provide nutrition to aging and environmentally damaged skin. US Patent 7, 303,753 B2.

Malakoff, D. (1997) Extinction on the high seas. *Science.* 277, 486-488.

Martínez, M.E., Sánchez, S., Jiménez, J.M., El-Yousfi, F., Munõz, L. (2000).Nitrogen andphosphorus removal from urban wastewater by the microalga *Scenedesmusobliquus*. *Bioresource Technology*, 73, 263–272.

Martins, R.F., Ramos, M.F., Herfindal, L., Sousa J.A., Skaerven, K., Vasconcelos, V.M. (2008). Antimicrobial and cytotoxic assessment of marine cyanobacteria – *Synechocystis* and *Synechococcus*. *Marine Drugs*, 6, 1-11.

Mason, C.P., Edwards, K.R., Carlson, R.E., Pignatello, J., Gleason, F.K., and Wood, J.M. (1982).Isolation of a chlorine containing antibiotic from a fresh water cyanobacterium. *Science*,215, 400-402.

Masukara, H., Nakamura, K., Mochimaru, M., and Sakurai, H. (2001).Photohydrogen production and nitrogenase activity in some heterocystous cyanobacteria. *Biohydrogen*, 2, 63-66.

Masukawa, H., Mochimaru, M., and Sakurai, H. (2002). Disruption of the uptake hydrogenase gene, but not of the bidirectional hydrogenase gene, leads to enhanced photobiological hydrogen production by the nitrogen-fixing cyanobacterium *Anabaena* sp. PCC7120. *Applied Microbiology and Biotecnology*, 58, 618-624.

Mata, T.M., Martins, A.A.,and Caetano, N.S. (2010). Microalgae for biodiesel production and other applications: a review. *Renewable and Sustainable Energy Reviews*, 14, 217–232.

Matern, U., Oberer, L., Falchetto, R.A., Erhard, M., Konig, W.A., Herdman, M., and Weckesser, J. (2001).Scyptolin A and B, cyclic depsipeptides from axenic cultures of *Scytonema hofmanni* PCC 7110. *Phytochemistry*, 58, 1087-1095.

Mathew, B., Sankaranarayanan, R., Nair, P.P., Varghese, C., Somanathan, T., Amma, B.P., Amma, N. S., and Nair, M.K. (1995).Evaluation of chemoprevention of oral cancer with *Spirulina fusiformis*, *Nutrition and Cancer*, 24, 197-202.

Matsumoto, H., Shioji, N., Hamasaki, A., and Ikuta, Y. (1996).Basic study on optimization of raceway-type algal cultivator. *Journal of Chemical Engineering of Japan*, 29, 541-543.

Mayfield, S.P., Franklin, S.E., and Lerner, R.A.(2003).Expression and assembly of a fully active antibody in algae. *Proceedings of the National Academy of SciencesUSA*,100, 438-442.

McKendry, P. (2002a). Energy production from biomass (part 2): conversion technologies. *Bioresource Technology*, 83, 47–54.

McKendry, P. (2002b). Energy production from biomass (part 3): gasification technologies. *Bioresource Technology*, 83, 55–63.

Melis, A., and Happe, T. (2001). Hydrogen production. Green algae as a source of energy. *Plant Physiology*, 127, 740-748.

Melis, A., Zhang, L., Forestier, M., Ghirardi, M.L., and Seibert, M. (2000).Sustained photobiological hydrogen gas production upon reversible inactivation of oxygen evolution in the green alga *Chlamydomonas reinhardtii*. *Plant Physiology*, 122, 127-136.

Melis, A. (2002). Green alga hydrogen production: progress, challenges and prospects. *International Journal of Hydrogen Energy*, 27, 1217–1228.

Melis, A. (2007). Photosynthetic H_2 metabolism in *Chlamydomonas reinhardtii* (unicellular green algae). *Planta*, 226, 1075–1086.

Meng, X., Yang, J., Xu, X., Zhang, L., Nie, Q., and Xian, M. (2009). Biodiesel production from oleaginous microorganisms. *Review Energy*, 34, 1-5.

Metting Jr, F.B.(1996). Biodiversity and application of microalgae. *Journal of Industrial Microbiology*, 17, 477-489.

Mian, P., Heilmann, J., Burgi, H.R., and Sticher, O. (2003). Biological screening of terrestrial and freshwater Cyanobacteria for antimicrobial activity, brine shrimp lethality and cytotoxicity. *Pharmaceutical Biology*, 41, 243-247.

Miao, X.L.,and Wu, Q.Y. (2004).High yield bio-oil production from fast pyrolysis by metabolic controlling of *Chlorellaprotothecoides*. *Journal of Biotechnology*, 110, 85–93.

Miao, X.L., and Wu, Q.Y. (2006).Biodiesel production from heterotrophic microalgal oil. *Bioresource Technology*, 97, 841–846.

Minoda, A., Sakagami, R., Yagisawa, F., Kuroiwa, T., and Tanaka, K.(2004).Improvement of culture conditions and evidence for nuclear transformation by homologous recombination in a red alga, *Cyanidioschyzon merolae* 10D. *Plant and Cell Physiology*, 45, 667-671.

Mishima, T., Murata, J., Toyoshima, M., Fujii, H., Nalajima, M., Hayashi, T., Kato, T., and Saiki, I. (1998). Inhibition of tumour invasion and metastasis by calcium spirulan (Ca-SP), a novel sulfated polysaccharide derived from a blue-green alga, *Spirulina platensis*. *Clinical and Experimental Metastasis*, 16, 541-550.

Miura, Y., Yamada, W., Hirata, K., Miyamoto, K., and Kiyohara, M. (1993).Stimulation of hydrogenproduction in algal cells grown under high CO_2 concentration and low temperature. *Applied Biochemistry and Biotechnology*, 39/40, 753-761.

Miyairi, S. (1995) CO_2 assimilation in a thermophilic cyanobacterium. *Energy Conversion and Management*, 36, 763-766.

Morowvat, M.H., Rasoul-Amini, S.,and Ghasemi, Y. (2010). *Chlamydomonas* as a "new" organism for biodiesel production. *Bioresource Technology*, 101, 2059–2062.

Mundt, S., Kreitlow, S., Nowotny, A., and Effmert, U. (2001).Biological and pharmacological investigation of selected cyanbacteria. *International Journal of Hygiene and Environmental Health*, 203, 327-334.

Mungo, F. (2005) A study into the prospects for marine biotechnology development in the United Kkingdom. *FMP Marine Biotechnology Group Report*, 2, 17-23.

Munoz, R., and Guieysse, B. (2006). Algal-bacterial processes for the treatment ofhazardous contaminants: a review. *Water Research*, 40, 2799-2815.

Murakami, M., Yamada, F., Nishide, T., Muranaka, T., Yamaguchi, N., and Takimoto, Y. (1998).The biological CO_2 fixation using *Chlorella* sp. with high capability in fixingCO_2. *Studies in Surface Science and Catalysis*, 114, 315–320.

Mussgnug, J.H. Klassen, V. Schlüter, A. Kruse, O. (2010). Microalgae as substrates for fermentative biogas production in a combined biorefinery concept. *Journal of Biotechnology*, 150, 51-56.

Murphy, V., Hughes, H., and McLoughlin, P. (2008).Comparative study of chromium biosorptionby red, green and brown seaweed biomass. *Chemosphere*, 70, 1128–1134.

Nagase, H., Eguchi, K., Yoshihara, K., Hirata, K., and Miyamoto, K. (1998) Improvement of microalgal NO_xremoval in bubble column and airlift reactors. *Journal of Fermentation and Bioengineering*, 86, 421-423.

Nagle, D.G., Paul, V.J., and Roberts, M.A. (1996).Ypaoamide, a new broadly acting feeding deterrent from the marine cyanobacterium *Lyngbyamajuscule*. *Tetrahedron Letters*, 37, 6263-6266.

Nakano, Y., Miyatake, K., Okuno, H., Hamazaki, K., Takenaka, S., Honami, N., Kiyota, M., Aiga, I., and Kondo, J. (1996) Growth of photosynthetic algae *Euglena* in high CO_2 conditions and its photosynthetic characteristics. *Acta Horticulturae*, 440, 49-54.

Noaman, N.H.,Fattah, A., Khaleafab, M., and Zaky, S.H.(2004).Factors affecting antimicrobial activity of *Synechococcusleopoliensis*. *Microbiological Research*, 159, 395-402.

Nunes, M.C., Raymundo, A.,and Sousa, I. (2006). Gelled vegetable desserts containing pea protein, κ-carrageenan and starch. *European Food Research and Technology*, 222, 622-628.

Olaizola, M. (2000).Commercial production of astaxanthin from *Haematococcus pluvialis* using 25,000-liter outdoor photobioreactors. *Journal of Applied Phycology*, 12, 499-506.

Olguín, E.J., Galicia, S., Mercado, G., and Perez, T. (2003). Annual productivity of *Spirulina*(*Arthrospira*) and nutrient removal in a pig wastewater recycle process undertropical conditions. *Journal of Applied Phycology*, 15, 249–57.

Ordög, V., Stirk, W.A., Lenobel, R., Bancirová, M., Strand, M.,and van Standen, J. (2004). Screening microalgae for some potentially useful agricultural and pharmaceutical secendary metabolites. *Journal of Applied Phycology*, 16, 309-314.

Oren, A. (2005). A hundred years of *Dunaliella* research: 1905-2005. *Saline Systems*, 1, 2.

Oswald, W.J.(1978). Engineering aspects of microalgae. In *Handbook of Microbiology* (Volume 2, pp. 519-552). Boca Raton, Florida: CRC Press.

Oswald, W.J.(1988). Large-scale algal culture systems (engineering aspects).In M.A. Borowitzka andL.J. Borowitzka (Eds.), *Micro-algal Biotechnology.*Cambridge University Press.

Oswald, W.J. (1991). Introduction to advanced integrated wastewater ponding systems. *Water Science and Technology*, 24, 1–7.

Oswald, W.J., and Gotaas, H.B.(1957).Photosynthesis in sewage treatment. *Transactions of the American Society of Civil Engineers*, 122, 73–105.

Oswald, W.J., Golueke, C.G., and Gee, H.K.(1959). Waste water reclamation through production of algae. Contribution 22, Water Resources Center, University of California, Berkley.

Oswald, W.J., and Golueke, C.G. (1960).Biological transformation of solar energy. *Advancesin Applied Microbiology*, 2, 223–262.

Oswald, W.J., Green, F.B., and Lundquist, T.J.(1994). Performance ofmethane fermentation pits in advanced integrated wastewaterpond systems. *Water Science and Technology*, 30, 287–295.

Ötles, S. and Pire, R. (2001). Fatty acid composition of *Chlorella* and *Spirulina* microalgae species. *Journal of AOAC International*, 84, 1708-1714.

Panda, D., Deluca, K., Williams, D., Jordan, M. A., and Wilson, L. (1998). Antiproliferative mechanism of action of cryptophycin-52: kinetic stabilization of microtubule dynamics by high-affinity binding to microtubule ends. *Cell Biology*, 95, 9313-9318.

Pandey, B.N. and Gupta, A.B. (1977).Antibiotic properties in *Chlorococcum humicolumn* (Naeg) Rabenh.(*Chlorophyceae*). *Phycologia,*16, 439 – 441.

Perales-Vela, H.V., Pena-Castro, J.M., andCanizares-Villanueva, R.O. (2006).Heavy metal detoxification in eukaryotic microalgae. *Chemosphere*, 64, 1-10.

Pesando, D., and Caram, B. (1984).Screening of marine algae from the French Mediterranean coast for antibacterial and antifungal activity. *Botanica Marina*, 27, 381-386.

Pesando, D. Gnassia-Barlli, M., and Gueho, E. (1979).Part I - Partial characterization of a specific antifungal substance isolated from marine diatom *Chactoceroslaunderi*. In H.A. Hoppe, T. Levring, and Y. Tanaka (Eds.), *Marine Algae in Pharamaceutical Science*(pp.447-459). Walter de Grayter.

Pinto, F.A.L., Troshina, O., and Lindblad, P. (2002).A brief look at three decades of research on cyanobacterial hydrogen evolution.*International Journal of Hydrogen Energy*,27, 1209-1215.

Piteira, M.F., Nunes, M.C., Raymundo, A., and Sousa, I. (2004). Effect of principal ingredients on quality of cookies with dietary fibre. In P.A. Williams and G.O. Phillips (Eds.), *Gums and Stabilisers for the Food Industry*(12th Edition, pp. 475-483). London: Royal Society of Chemistry.

Pittman, J.K., Dean, A.P., and Osundeko, O. (2011).The potential of sustainable algal biofuel production using wastewater resources. *Bioresource Technology*, 102, 17–25.

Posewitz, M.C., King, P.W., Smolinski, S.L., Zhang, L., Seibert, M., and Ghirardi, M.L.(2004). Discovery of two novel radical S-adenosylmethionine proteins required for the assembly of an active [Fe] hydrogenase. *Journalof Biological Chemistry*,279, 25711–25720.

Powell, R.C., Nevels, E.M.,and McDowell, M.E. (1961).Algae feeding in humans. *Journal of Nutrition*, 75, 7-12.

Prashantkumar, P., Angadi, S.B., and Vidyasagar, G.M. (2006).Antimicrobial of green and bluegreenalgae.*Indian Journal of Pharmaceutical Sciences*, 68, 647-648.

Pratt, R., Daniels, T.C., Eiler, J.B., Gunnison, J.B., and Kumler, W.D. (1944).Chlorellin, an antibacterial substance from *Chlorella*. *Science*, 99, 351-352.

Pulz, O. (2001). Photobioreactors: production systems for phototrophic microorganisms. *Applied Microbiology and Biotechnology*, 57, 287-293.

Pulz, O., and Gross, W. (2004).Valuable products from biotechnology of microalgae. *Applied Microbiology and Biotechnology*, 65, 635-648.

Qin, S., Sun, G.Q., Jiang, P., Zou, L.H., Wu, Y., and Tseng, C.K.(1999) Review of genetic engineering of *Laminaria japonica* (Laminariales, Phaeophyta) in China. *Hydrobiologia*,398-399, 469-472.

Radakovits, R., Jinkerson, R.E., Darzins, A., and Posewitz, M.C. (2010).Genetic engineering of algae for enhanced biofuel production.*Eukaryotic Cell*, 9, 486–501.

Radmer, R.J.,and Parker, B.C. (1994). Commercial applications of algae: opportunities and constraints. *Journal of Applied Phycology*, 6, 93-98.

Radmer, R.J. (1996). Algal diversity and commercial algal products. *BioScience*, 46, 263-270.

Rao, K.K., and Hall, D.O. (1996). Hydrogen production by cyanobacteria: potential, problems and prospects and prospects. *Journal of Marine Biotechnology*, 4, 10-15.

Raymundo, A.,Gouveia, L.,Batista, A.P.,Empis, J.,and Sousa, I. (2005). Fat mimetic capacity of *Chlorellavulgaris* biomass in oil-in-water food emulsions stabilised by pea protein.*Food Research International*, 38, 961-965.

Regulation (EC) N° 258/97 of the European Parliament and of the Council of 27 January 1997 concerning novel foods and food ingredients. *Official Journal of the European Comunities* n° L 043, 14/02/1997, pp. 1-6.

Reijnders, L. (2008). Do biofuels from microalgae beat biofuels from terrestrial plants? *Trends in Biotechnology*, 26, 349–350.

Richmond, A. (2004). *Handbook of Microalgal Culture: Biotechnology and Applied Phycology*. Blackwell Publishing Ltd.

Rittmann, B.E. (2008). Opportunities for renewable bioenergy using microorganisms. *Biotechnology and Bioengineering*, 100, 203–212.

Robles Centeno, P.O., Ballantine, D.L., and Gerwick, W.H. (1996). Dynamic of antibacterial activity in three species of caribbean marine algae as a function of habitat and life history. *Hydrobiologia*, 326/327, 457-462.

Robles Centeno, P.O., and Ballantine, D.L. (1999). Effect of culture conditions on production of antibiotically active metabolites by the marine alga *Spyridia filamentosa* (Ceramiaceae, *Rhodophya*). I. Light. *Journal of AppliedPhycology*, 11, 217-224.

Rodolfi, L., Zittelli, G.C., Bassi, N., Padovani, G., Biondi, N., Bonini, G., and Tredici, M.R. (2009). Microalgae for oil: strain selection, induction of lipid synthesis and outdoor mass cultivation in a low-cost photobioreactor. *Biotechnology and Bioengineering*, 102, 100–112.

Roessler, P.G., Bleibaum, J.L., Thompson, G.A., and Ohlrogge, J.G. (1994). Characteristics of the gene that encodes acetyl-CoA carboxylase in the diatom *Cyclotella cryptica*, *Annals of the New York Academy of Sciences*, 721, 250–256.

Romay, C., Ledon, N., and Gonzalez, R. (1999). Phycocyanin extract reduces leukotriene B4 levels in arachidonic acid-induced mouse-ear inflammation test. *Journal of Pharmacy and Pharmacology*, 51, 641-642.

Rosenberg, J.N., Oyler, G.A., Wilkinson, L., and Betenbaugh, M.J. (2008). A green light for engineered algae: redirecting metabolism to fuel biotechnology revolution. *Current Opinion in Biotechnology*, 19, 430-436.

Rovirosa, J., and San Martin, A. (1997). Antimicrobial activity of the brown alga *Stypopodium flabelliforme* constituents. *Fitoterapia*, 68, 473-475.

Salvador, N., Garrete, A.G., Lavelli, L., and Ribera, M.A. (2007). Antimicrobial activity of iberian macroalgae. *Scientia Marina*, 71, 101-113.

Sanford, J.C., Smith, F.D., and Russell, J.A. (1993). Optimizing the biolistic process for different biological applications. *Methods of Enzymology*, 217, 483–509.

Sayre, R.T., Wagner, R.E., Sirporanadulsil, S., and Farias, C. (2001). Transgenic algae for delivery antigens to animals. InternationalPatent Number W.O. 01/98335 A2.

Schenk, P., Thomas-Hall, S., Stephens, E., Marx, U., Mussgnug, J., Posten, C., Kruse, O., and Hankamer, B. (2008). Second generation biofuels: high-efficiency microalgae for biodiesel production. *Bioenergy Research*, 1, 20–43.

Schiedlmeier, B., Schmitt, R., Müller, W., Kirk, M.M., Gruber, H., Mages, W., and Kirk, D.L. (1994). Nuclear transformation of *Volvox carteri*. *Proceedings of the National Academy of Sciences USA*, 91, 5080-5084.

Schütz, K., Happe, T., Troshina, O., Lindblad, P., Leitão, E., Oliveira, P., and Tamagnini, P. (2004). Cyanobacterial H_2-production – a comparative analysis. *Planta*, 218, 350-359.

Schwartz, J., Shklar, G., and Reid, S. (1987). Regression of experimental hamster cancer by beta-carotene and algae extracts. *Journal of Oral and Maxillofacial Surgery*, 45, 510-515.

Schwartz, J., Shklar, G., Reid, S., and Trickler, D. (1988). Prevention of experimental oral cancer by extracts of *Spirulina - Dunaliella* algae. *Nutrition and Cancer*, 11, 127-134.

Schwartz, R.E., Hirsch, C.F., Sesin, D.F., Flor, J.E., Chartrain, M., Fromtling, R.E., Harris, G.H., Salvatore, M.J., Liesch, J.M., and Yudin, K. (1990).Pharmaceuticals from cultured algae. *Journal of Industrial Microbiology*. 5, 113-124.

Seckbach, J., Gross, H., and Nathan, M.B. (1971).Growth and photosynthesis of *Cyanidium caldarium* cultured under pure CO_2. *Israel Journal of Botany*,20, 84-90.

Sheehan, J., Dunahay, T., Benemann, J., and Roessler, P. (1998). A look back at the U.S. Department of Energy's Aquatic Species Program: biodiesel from algae. National Renewable Energy Laboratory. Report NREL/TP-580-24190.

Shi, J., Podola, B., and Melkonian, M. (2007).Removal of nitrogen and phosphorus fromwastewater using microalgae immobilized on twin layers: an experimentalstudy. *Journal of Applied Phycology*, 19, 417–423.

Shimamatsu, H. (2004). Mass production of *Spirulina*, an edible microalga. *Hydrobiologia*, 512, 39-44.

Shimizu, Y. (2003). Microalgal metabolites. *Current Opinion and Microbiology*, 6, 236-243.

Shimogawara, K., Fujiwara, S., Grossman, A., and Usuda, H. (1998).High-efficiency transformation of *Chlamydomonas reinhardtii* by electroporation. *Genetics*,148, 1821–1828.

Sialve, B., Bernet, N., and Bernard, O. (2009).Anaerobic digestion of microalgae as a necessary step to make microalgal biodiesel sustainable. *Biotechnology Advances*, 27, 409–416.

Siddhanta, A.K.,Mody, K.H., Ramavat, B.K., Chauhan, V.D., Garg, H.S., Goel, A.K., Doss, M.J., Srivastava, M.N., Patnaik, G.K., and Kamboj, V.P.(1997). Bioactivity of marine organisms: Part VIII-Screening of some marine flora of western coast of India. *Indian Journal of Experimental Biology*, 35, 638-643.

Singh, S., Kate, B.N., and Banerjee, U.C. (2005). Bioactive compounds from cyanobacteria and microalgae: an overview. *Critical Reviews in Biotechnology*. 25, 73-95.

Soltani, N., Khavar-Nejad, R.A., Tabatabaei-Yazdi, M., Shokravi, Sh.,and Fernández-Valiente, E. (2005).Screening of soil cyanobacteria for antifungal and antibacterial activity. *Pharmaceutical Biology*, 43, 455-459.

Spivak, C.E., Witkop, B.,and Albuquerque, E.X. (1980). Anatoxin-*a*: a novel, potent agonist at the nicotinic receptor. *Molecular Pharmacology*, 18, 384-294.

Spolaore, P., Joannis-Cassan, C., Duran, E., and Isambert, A. (2006). Commercial applications of microalgae. *Journal of Bioscience and Bioengineering*, 102, 87-96.

Stangenberg, M. (1968). Bacteriostatic effects of some algae and *Lemna Minor* extracts. *Hydrobiologia.*32, 88-96.

Steinbrenner, J., and Sandmann, G.(2006) Transformation of the green alga *Haematococcus pluvialis* with a phytoene desaturase for accelerated astaxanthin biosynthesis. *Applied and Environmental Microbiology* 72, 7477-7484.

Stephens, E., Ross, I.L., King, Z., Mussgnug, J.H., Kruse, O., Posten, C., Borowitzka, M.A., and Hankamer, B. (2010).An economic and technical evaluation of microalgal biofuels.*Nature Biotechnology*, 28, 126–128.

Stolz, P., and Obermayer, B. (2005). Manufacturing microalgae for skin care. *Cosmetics Toiletries*, 120, 99–106.

Subhadra, B., and Edwards, M. (2010). An integrated renewable energy park approach for algal biofuel production in United States. *Energy Policy*, 38, 4897-4902.

Sukumaran, P., and Thevanathan, R. (2011).Antibacterial properties of the green alga *Pithophoraoedogonia*(Mont.) Wittrock.*Report and Opinion*. 3, 53-60.

Sun, M., Qian, K., Su, N., Chang, H., Liu, J., and Shen, G.(2003). Foot-and-mouth disease virus VP1 protein fused with cholera toxin B subunit expressed in *Chlamydomonasreinhardtii* chloroplast. *Biotechnology Letters*,25, 1087-1092.

Sun, Y., Yang, Z.,Gao, X., Li, Q., Zhang, Q. and Xu, Z. (2005).Expression of foreign genes in *Dunaliella* by electroporation. *Molecular Biotechnology*,30, 185–192.

Sveshnikov, D., Sveshnikova, N., Rao, K., and Hall, D. (1997). Hydrogen metabolism of mutant forms of *Anabaenavariabilis* in continuous cultures and under nutritional stress. *FEMS Microbiology Letters*, 147, 297–301.

Takagi, M., Karseno, and Yoshida, T. (2006).Effect of salt concentration on intracellular accumulation of lipids and triacylglyceride in marine microalgae *Dunaliella* cells. *Journal of Bioscience and Bioengineering*, 101, 223–226.

Tamagnini, P., Leitão, E., Oliveira, P., Ferreira, D., Pinto, F., Harris, D., and Heidorn, T. (2007).Cyanobacterial hydrogenases.Diversity, regulation and applications.*FEMS Microbiology Review*, 31, 692-720.

Tan, C., Qin, S., Zhang, Q., Jiang, P., and Zhao, F.(2005).Establishment of a micro-particle bombardment transformation system for *Dunaliella salina*. *Journal of Microbiology*,43, 361-365.

Thillairajasekar, K., Duraipandiyan, V.,and Ignacimuthu, S. (2009).Antimicrobial activity of *Trichodesmium erythraeum* (Ehr) (microalga) from south east coast of Tamil Nadu, India. *International Journal of Integrative Biology*. 5, 167-170.

ten Lohuis, M.R., and Miller, D.J. (1998). Genetic transformation of dinoflagellates (*Amphidinium* and *Symbiodinium*): expression of GUS in microalgae using heterologous promoter constructs. *Plant Journal*,13, 427-435.

Teng, C., Qin, S., Liu, J., Yu, D., Liang, C., and Tseng, C.(2002). Transient expression of *lacZ* in bombarded unicellular green alga *Haematococcus pluvialis*. *Journal of Applied Phycology*,14, 495–500.

The National Academies (2008). Ecological impacts of climate change. Available at: http://dels.nas.edu/climatechange/index.shtml (last accessed on April, 2009).

Todd-Lorenz, R.,and Cysewski, G.R. (2000). Commercial potential for *Haematococcus* microalgae as a natural source of astaxanthin. *Trends in Biotechnology*, 18, 160-167.

Tokuda, H., Nishino, H., Shirahashi, H., Murakami, N., Nagatsu, A., and Sakakibra, J. (1996). Inhibition of 12-*O*-tetradecanoylphorbol-13-acetate promoted mouse skin papiloma bu digalactosyl diacylglycerols from the fresh water cyanobacterium *Phormidium tenue*. *Cancer Letters*, 104, 91-95.

Tredici, M.R. (2004). Mass production of microalgae: photobioreactors. InA. Richmond (Ed.),*Handbook of Microalgal Culture: Biotechnology and Applied Phycology* (pp. 178-214). Blackwell Publishing Ltd.

Ueno, Y., Kurano, N., and Miyachi, S.(1998). Ethanol production by dark fermentation in the marine green alga, *Chlorococcum littorale*.*Journal of Fermentation and Bioengineering*. 86, 38–43.

Vadiraja, B.B., Gaikwad, N.W.,and Madyastha, K.M. (1998). Hepatoprotective effect of *C*-phycocyanin: protection for carbon tetrachloride and R-(+)-pulegone-mediated hepatotoxicity in rats. *Biochemical and Biophysical Research Communications*, 249, 428-431.

Valencia, I., Ansorena, D.,and Astiasarán, I. (2007).Development of dry fermented sausages rich in docosahexaenoic acid from the microalgae *Schizochytrium* sp.: influence on nutritional properties, sensorial quality and oxidation stability. *Food Chemistry*, 104, 1087-109.

van Beilen, J.B. (2010). Why microalgal biofuels won't save the internal combustion machine. *Biofuels, Bioproducts and Biorefining*, 4, 41–52.

van den Hoek, C., Mann, D.G.,and Jahns, H.M. (1995).*Algae. An introduction to Phycology*.Cambridge University Press.

Vergara-Fernandez, A., Vargas, G., Alarcon, N., and Velasco, A. (2008).Evaluation of marine algae as a source of biogas in a two-stage anaerobic reactor system.*Biomass and Bioenergy*, 32, 338-344.

Walker, D.A. (2009). Biofuels, facts, fantasy, and feasibility.*Journal of Applied Phycology*, 21, 509–517.

Walter, C.S., and Mahesh, R. (2000).Antibacterial and antifungal activities of some marine diatoms in culture. *Indian Journal of Marine Sciences,* 29, 238-242.

Wang, B., Li, Y., Wu, N., Lan, C.Q. (2008).CO_2 bio-mitigation using microalgae.*Applied Microbiology and Biotechnology*, 79, 707–718.

Wang, Z., Pan, Y., Dong, T., Zhu, X., Kan, T., Yuan, L., Torimoto, Y., Sadakata, M., and Li, Q. (2007).Production of hydrogen from catalytic steam reforming of biooil using C12A7-O-based catalysts.*Applied Catalysis A*, 320, 24–34.

Wilde, E.W., and Benemann, J.R. (1993).Bioremoval of heavy-metals by the use ofmicroalgae. *Biotechnology Advances*, 11, 781-812.

Worms, I.A.M., Traber, J., Kistler, D., Sigg, L., and Slaveykova, V.I. (2010).Uptake of Cd(II) and Pb(II) by microalgae in presence of colloidal organic matter from wastewater treatment plant effluents. *Environmental Pollution*, 158, 369–374.

Wu, M., Okino, T., Nogle, L.M., Marquez, B.L., Williamson, R.T., and Sitchitta, N. (2000).Structure, synthesis and biological properties of kalkitoxin, a novel neurotoxin from the marine cyanobacterium *Lyngbya majuscula*. *Journal of the American Chemical Society*, 122, 12041-12042.

Xu, H., Miao, X., and Wu, Q. (2006).High quality biodiesel production from a microalga *Chlorella protothecoides* by heterotrophic growth in fermenters. *Journal of Biotechnology*. 126, 499-507.

Yamaguchi, K. (1997). Recent advances in microalgal bioscience in Japan, with special reference to utilization of biomass and metabolites: a review. *Journal of Applied Phycology,* 8, 487-502.

Yang, H., Lee, E.,and Kim, H. (1997).*Spirulina platensis* inhibits anaphylactic reaction. *Life Sciences*, 61, 1237-1244.

Yoshihara, K., Nagase, H., Eguchi, K., Hirata, K., and Miyamoto, K. (1996) Biological elimination of nitricoxide and carbon dioxide from flue gas by marine microalga NOA-113 cultivation in a longtubular photobioreactor. *Journal of Fermentation and Bioengineering*, 82, 351-354.

Yun, Y.S., Lee, S.B., Park, J.M., Lee, C.I., and Yang, J.W. (1997).Carbon dioxide fixation by algalcultivation using wastewater nutrients. *Journal of Chemical Technology and Biotechnology*,69, 451–455.

In: Microalgae: Biotechnology, Microbiology and Energy
Editor: Melanie N. Johnsen
ISBN 978-1-61324-625-2
© 2012 Nova Science Publishers, Inc.

Chapter 2

ASSESSING THE RENEWABILITY OF BIODIESEL FROM MICROALGAE VIA DIFFERENT TRANSESTERIFICATION PROCESSES

Ehiaze Ehimen, Zhifa Sun[*], and Gerry Carrington*
Physics Department, University of Otago,
Dunedin, New Zealand

ABSTRACT

Using process modelling tools, the conventional and in-situ transesterification processes for biodiesel production from microalgae biomass was modelled. The raw material and process energy requirements of the up-scaled process were obtained for the different transesterification processes, and a renewability assessment of the various schemes was carried out. The biomass cultivation and biodiesel production process renewability was assessed by comparing the minimum work required to restore the non-renewable resources degraded in the biomass and biodiesel production process with the useful work available from the main process products. If the maximum work obtained from the process products is larger than the restoration work, the process is considered as renewable. In a present day scenario (with the use of fossil fuel sources for the production of the process raw materials, such as for methanol and sulphuric acid production, and electricity), all the transesterification processes were shown to be non-renewable. The influence of the choice of the electricity generation scheme, raw material source and the type of heating fuels (including heating and drying technology) on the process renewability was also examined. The process renewability of the in-situ transesterification of microalgae lipids to biodiesel was found to significantly improve with the use of renewable electricity, reacting alcohols from biomass fermentation and heat pump technology to facilitate the biomass drying and process heating.

---

[*] Corresponding author. Tel: +64 3 4799420, E-mail:zhifa@physics.otago.ac.nz

Modelling and Assessment of the Transesterification Processes

Introduction

The potential for the use of microalgae biomass for biodiesel production has been examined in the literature (i.e. Nagle and Lemke, 1990; Miao and Wu, 2006; Chisti, 2007), with the advantages of this feedstock relative to common terrestrial oil sources (i.e. soybeans and rapeseed oils) highlighted.

With the cost of the biomass feedstock oil contributing to 60-75% of the final biodiesel cost (Mata et al., 2010), the relatively higher cost of microalgae oil (compared to traditional oil sources) is considered to be one of the factors potentially limiting its use (Chisti, 2007). The prohibitive cost associated with microalgae production and conversion was reported to be mainly tied with the process energy requirements (Chisti, 2007). The production economics of microalgae derived biofuels is however still being disputed, with some investigations indicating that microalgae oil is cost competitive with current conventional oil sources and petro-diesel (Huntley and Redalje, 2007).

To further advance the proposed use and competiveness of microalgae biomass for biodiesel production, Grobbelaar (2000) and Chisti (2007) put forward technological research aspects which could be investigated to support improvements in the energetics, economics and practicality of biofuel production from microalgae. These include enhancing the algal biology (i.e. at a genetic level), improved bioreactor engineering and optimisation of the biofuel production processes.

This chapter does not deal with the optimisation of the microalgae biomass or oil productivity or the photo-bioreactor technologies. The investigations in this study are based on the third approach above, and explore how the application of process modifications to the conventional oil extraction and transesterification process potentially influence the overall energetics of biodiesel production using microalgae biomass.

The production of biodiesel from microalgae has been largely investigated using the conventional route as demonstrated in Nagle and Lemke (1990) and Miao and Wu (2006). This involves the extraction of the lipids from the microalgae biomass followed by its conversion to alkyl esters and glycerol. Results of the methanolysis studies carried out by Nagle and Lemke (1990) indicated that the most important variable in the microalgae lipid conversion process was the type of catalyst used. The use of acid catalysts (HCl) was demonstrated to result in a higher fatty acid methyl ester (FAME) yield per microalgae lipid input compared to the use of alkaline catalyst (NaOH) under the same reaction conditions. This was attributed to the high characteristic free fatty acid (FFA) content of microalgae lipids (Miao and Wu, 2006), which would contribute to a reduction in the conversion efficiency with alkaline catalysts use. The use of inorganic acids as the reaction catalysts is largely considered for microalgae lipid transesterification, due to its insensitivity to the FFA content of this oil feedstock. This is advantageous since both the biodiesel producing transesterification and esterification reactions are facilitated by acidic catalysis.

The use of enzyme catalysed and supercritical transesterification methods as alternatives to the conventional method for the production of microalgae biodiesel have also been investigated (i.e. Li et al., 2007; Demirbaş, 2009).The supercritical method (based on the reacting alcohol properties) involves the use of high reaction temperatures and pressures i.e. > 240°C and >8 MPa respectively, without catalysts (Saka and Kusdiana, 2001). For the large scale production of microalgae biodiesel, it is proposed that cheap and energetically efficient conversion technologies are applied to overcome the disadvantageous costs associated with the microalgae biomass production. The relatively high costs of the lipase catalysis (Fukuda et al., 2001), and the prohibitive operational and energetic costs of the supercritical method (Marchetti and Errazu, 2008), when compared with the inorganic catalysed routes, limits the short-medium term commercialisation of those methods and will therefore not be examined further in this chapter. In addition, safety concerns about the use of supercritical transesterification would further hinder its use for microalgae biodiesel production (Saka, 2006). The use of inorganic catalysis for the transesterification process was solely considered for use in this study.

An alternative to the conventional lipid extraction and transesterification process, the 'in-situ' transesterification method (and its modifications), will be the main focus in the investigations as a potential tool for the improvement of the energetics of the microalgae biodiesel production in this study.

The in-situ process facilitates the conversion of the biomass oil to fatty acid alkyl esters (FAAE) directly from the oil bearing biomass, thereby eliminating the solvent extraction step required to obtain the oil feedstock, as in the conventional method. The application of the one step in-situ transesterification method for simplified biodiesel production has been discussed to potentially lead to the reduction in the energetic, raw material and economic requirements when compared to the use of the conventional oil extraction and transesterification process (Haas et al., 2007). This method may be especially advantageous for use with microalgae, since the extraction of microalgae lipids is usually accomplished by solvent extraction, rather than cheaper physical extraction methods (for example, expellers), as utilised in conventional oil crops (Johnson and Wen, 2009).

The influence of important reaction parameters (i.e. temperature, time, biomass moisture content and stirring) on FAME production from microalgae biomass using the in-situ transesterification process has been reported in Ehimen et al. (2010). The results of that study (Ehimen et al., 2010) show that increases in the reaction temperature and alcohol volume favoured the production of FAME. However, within the experimental conditions used, an increase in the reacting molar methanol to oil ratio > 315:1 did not show significant changes in the percentage mass conversion of the oil to FAME. This was with the use of dried *Chlorella* biomass (with an oil content of 0.276g/ g dry microalgae) as the feedstock (Ehimen et al., 2010).

The percentage mass content of the FAME in the biodiesel was shown to improve significantly with more process stirring for the in-situ process (Ehimen et al., 2010). Biomass drying was shown to play an important role, with an increase in the moisture content of the biomass resulting in significant reductions of the equilibrium FAME yields (g FAME/g biodiesel) (Johnson and Wen, 2009; Ehimen et al., 2010). Extensive biomass drying may therefore be needed prior to biodiesel production using the in-situ process. To further reduce the large reacting alcohol requirements for the in-situ process and improve the FAME yields, the use of ultrasound stirring, as well as the integration of co-solvents in the transesterification process has been demonstrated in the literature (i.e. Xu and Mi, 2011; Ehimen et al., 2011).

The aim of this study is to determine the potential energetic advantages associated with the use of the in-situ process (and its modifications i.e. the use of ultrasound agitation and the use of co-solvents) for microalgae biodiesel production compared with the conventional process. To achieve this goal, an adequate quantification technique is required to compare the transesterification processes. The renewability analysis in this chapter is based on an assessment that quantifies the net gains (or losses) obtained from the use of microalgae for biodiesel production.

With the aid of process modelling tools, a preliminary assessment and comparison of the renewability of microalgae derived biodiesel via the in-situ and conventional transesterification processes is carried out in this study. This involved an examination of the fossil fuel inputs into the microalgae biomass and fuel production process as well as the useful work obtainable from the fuel products from the conversion process. The simplified assessment used in this paper was also used to examine the influence of the choice of transesterification method (conventional and in-situ) on the biodiesel production process renewability. Schemes which could be potentially applied to improve the renewability of the use of microalgae for biodiesel were also highlighted.

Before any energetic evaluations can be made, the different microalgae biomass transesterification processes investigated were initially modelled in this chapter – "Process Modelling and Simulation".This initial investigation was carried out to: (i) develop a base case large scale design for the in-situ (including in-situ modifications) and conventional transesterification processes using microalgal biomass and identification of the net external process heating and cooling requirements (to be used as an indicator of the net process energy needs), and (ii) preliminarily compare the external energy requirement of the different transesterification processes.

The results obtained from the process model are used to obtain the energy and material requirements of the various transesterification processes. These requirements were then used in simplified assessment of the renewability of the different microalgae biodiesel production options and energy recovery routes.

PROCESS MODELLING AND SIMULATION

The Process Modelling Software

The process simulation software Aspen Plus® version 11.1.1 developed by Aspen technology Inc., Cambridge, Mass., USA, was used in the design of the reaction and separation units of the transesterification methods. With limited detailed knowledge on the comprehensive design of the process units i.e. distillation columns, the use of this commercial software for the process modelling was useful, because the software provided a simple means of designing and calculating the energy demands of the processes investigated. The commercial software was considered to be sufficient for the analysis in this paper since it is widely used in the literature where only a simplified modelling and comparison of biodiesel production systems is required i.e. Haas et al. (2006), Sheehan et al (1998), Myint and El-Hawagi (2009) and Kasteren and Nisworo (2007).

To ensure that the modelled processes described in this chapter reflect practical biodiesel production, empirical data on oil yields and fatty acid methyl esters (FAME) production from microalgae (*Chlorellasp*) obtained from the laboratory experiments (Ehimen et al., 2010; 2011) was used. In addition, assumptions made about the operation of downstream separation and purification processes for biodiesel production were based on current industrial practices as used in ASPEN assisted process models in the literature i.e. as in Sheehan et al. (1998) and Haas et al. (2006).

Preliminary Design Assumptions

A biodiesel production plant with a fixed biomass feedstock input of 1 t dry *Chlorella* biomass/ h was proposed. This input rate was selected to facilitate an even comparison of the transesterification routes evaluated in this study. The estimates of the energetic and process raw material requirements were consequently based on those required for the conversion of 1 t dry *Chlorella* biomass per hour.

Process Feedstock Choice and Specification

Microalgae (*Chlorella*) which is the lipid-bearing biomass feedstock for this study, is assumed to be dried as reported in Ehimen at al., (2010). The *Chlorella* oil has an oleic acid (C18:1) content of 61.8% (w/w), together with palmitic (4.37%), palmitoleic (0.44%), linolenic (19.94%), linoleic (12.22%) and eicosanoic (1.22%) acids (w/w) (Ehimen et al., 2010). The percentage of the free fatty acids (FFA) in the microalgae oil was 5.11% w/w (Ehimen et al., 2010). The esterification of these fatty acids, and the transesterification of the triglycerides which they form, results in FAME production. With oleic acid ($C_{18}H_{34}O_2$) as the principal fatty acid in the *Chlorella* oil, it and its corresponding triglyceride, triolein ($C_{57}H_{104}O_6$) were used as the model fatty acid and oil in this study. Since the thermal properties of the other fatty acids (and their triglycerides) are similar to that of oleic acid (Myint and El-Halwagi, 2009), it was expected that the results of the process models using oleic acid and triolein would be representative of the transesterification of the *Chlorella* lipids.

The decision steps outlined by Carlson (1996) were used for the selection of an appropriate method for the process modelling using the Aspen Plus® software, and for the estimation of any missing process parameters.

The assumptions guiding the process modelling for the in-situ transesterification process are elaborated in the sections:"The In-Situ Transesterification Process"and, the approaches used for the modified in-situ (using ultrasound agitation and co-solvents) and conventional process models are presented in the sections -"The Ultrasound Assisted In-Situ Transesterification Process" and"The Conventional Transesterification Process".

Unlike previous ASPEN assisted conventional biodiesel production models (i.e. Zhang et al., 2003; Haas et al., 2006; Kasteren and Nisworo, 2007) where the process models were based on the extracted oil, the conventional process modelled in this study uses the microalgae biomass as the starting point. The system boundaries employed here present a fair basis for comparing the different transesterification processes considered.

The *In-Situ* Transesterification Process

Transesterification Reaction, Acidic Catalyst and Reacting Alcohol

Sulphuric acid (H_2SO_4) and methanol (CH_3OH) were used as the process catalyst and reacting alcohol respectively.

The transesterification reaction was assumed to be facilitated using continuous reactors, with the representative triglyceride, triolein ($C_{57}H_{104}O_6$) converted to methyl oleate esters ($C_{19}H_{36}O_2$) with glycerol ($C_3H_5(OH)_3$) as the reaction by-products. Due to the presence of FFAs in the microalgae lipids, the conversion process in this study also accounted for the esterification reaction, with oleic acid as the representative FFA. The esterification reaction was assumed to occur in the reactors to completion simultaneously with the transesterification reaction.

The continuous transesterification reactors were modelled on the basis of the reaction stoichiometry and percentage mass conversion of the microalgae lipids to FAME using the results of the in-situ transesterification trials as in Ehimen et al. (2010).

From the results of the in-situ laboratory investigations, with mechanical stirring, no appreciable improvement in the oil to FAME conversion was observed for reacting molar ratios of alcohol to *Chlorella* oil greater than 315:1 (Ehimen et al., 2010). This was with the use of a H_2SO_4 acid concentration of 0.04 mol as the transesterification process catalyst and a reaction time of 8 h (Ehimen et al., 2010). It was decided that the process models in this chapter would be based on a fixed process reaction temperature of 60°C. Although the use of higher temperatures have been demonstrated to result in higher FAME conversions (Ehimen et al., 2010), the use of the 60°C temperature level was deemed adequate since this setting was reported to be typical for most commercial biodiesel production facilities (Sheehan et al., 1998).

The models used two reactors arranged in series as demonstrated by Haas et al. (2006) for the transesterification processes. For this process, the reactors were assumed to be agitated using mechanical stirrers which would ensure complete mixing and suspension of the reactor contents. The use of ultrasound agitation is later considered in the section "The Ultrasound Assisted In-Situ Transesterification Process".

The modelled transesterification reaction involved the use of a molar methanol to oil ratio of 315:1 which was shown to result in an 84% conversion of *Chlorella* oil to methyl esters in 4 h with the reaction temperature at 60°C (Ehimen et al., 2010).

The process flowchart of the mechanically stirred in-situ biodiesel production route using *Chlorella* biomass as feedstock is shown in Figure 1.

Fresh (and subsequently recycled) methanol (3115.4 kg/ h) in stream 1 and acidic catalyst, H_2SO (270 kg/ h) in stream 2 was pre-mixed (in MIXER) with the incoming dried *Chlorella* biomass (1000 kg/ h), all at ambient temperatures (20°C). The resulting mixture in stream 3 was heated in the heat exchanger HX-1 to meet the desired the reaction temperature of 60°C. The heated mixture (stream 3-1) was transferred to the temperature controlled continuously stirred transesterification reactors (REACTR-1 and 2). The simplified reactors were assumed to be properly insulated, so heat losses from the reactors were not considered further.

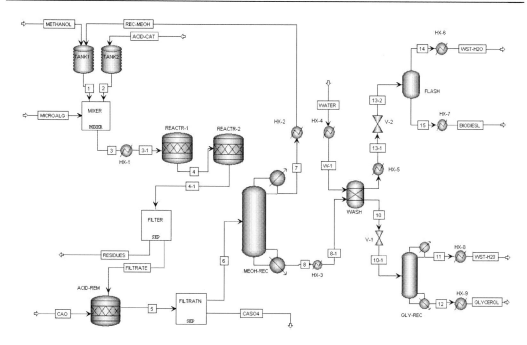

Figure 1. Flowsheet showing the process units and streams for the up-scaled in-situ transesterification of microalgae.

Using the percentage mass conversion of *Chlorella* lipids to FAME of 84%, and a biodiesel yield (glycerides and FAME) of 0.288 kg/kg *Chlorella* (Ehimen, 2011), this equates to a combined percentage mass FAME conversion of 97.4% obtained after the two reactors in series (i.e. an 84% conversion after the first reactor + a 13.4% conversion after the second).

Residues Filtration

After the transesterification reactors, the resulting stream (4-1) contains a mixture of the methyl esters, "unreacted" oil, biomass residues, glycerol, H_2SO_4 catalyst, excess methanol and the water resulting from the esterification process. This stream was subjected to a physical solid-liquid separation process (FILTER unit) to extract the solid biomass residues. This was considered to be composed of a simple belt filtration unit, using a cloth filter. The simple filtration model in ASPEN software was used for the design of the solid-liquid separation unit.

The solid filtration model for this process was based on empirical data from the laboratory investigations for the biomass filtration, post transesterification. This involved using a material mass balance of the component streams in the modelling of the separation unit (FILTER), with the assumption that 100% of the solid residues are recovered along with traces of the liquid streams entrained in the biomass. From the laboratory experiments, it was observed that there was a (1.6 ± 0.4) % increase in the weight of the vacuum filtered biomass residues owing to the liquids held in the solid biomass phase. This was estimated based on the residues weight before and after drying to a constant weight at 105°C in a force draft oven for 12 h. The mass flows of the different components in the residue stream was obtained by multiplying their mass flows in the inlet stream (4-1) by the factors shown in Table 1, which were obtained from experiments.

The solid biomass residues were separated as the stream – RESIDUES with the filtrate stream transferred to the acid catalyst removal step.

Table 1. Ratio of components in the residues after the filtration unit

Component	Fraction entrained in residues (kg/kg dry residues)
Methanol	1.4×10^{-2}
FAME	1.4×10^{-3}
H_2SO_4	1.3×10^{-4}
Glycerol	1.4×10^{-4}
Water	4.3×10^{-5}

Acid Catalyst Removal

In the reactor (ACID-REM), the H_2SO_4 fraction was assumed to be completely removed from the filtrate stream via a neutralisation reaction. This was facilitated by the introduction of calcium oxide (CaO) to the reactor, resulting in the formation of calcium sulphate ($CaSO_4$) and water (H_2O). The low costing CaO (compared to other alkali oxides) was selected for use in this model in line with previous work by Zhang et al. (2003), which considered its preferential use in commercial biodiesel production systems. The problem of using up the reacting H_2SO_4 however arises with the use of this method, rendering it unusable for recycling in the transesterification process. More research into the recycling of the acid catalyst is thus required.

The molar mass flow of the CaO utilised for the removal of the acid catalyst fraction was equivalent to the molar flow of the H_2SO_4 fraction in the incoming stream. After the neutralisation reaction, a solid-liquid separation process was carried out to facilitate the separation of the $CaSO_4$ from the liquid fraction of the stream. It must be mentioned that although the $CaSO_4$ production in the presence of water normally results in the formation of hydrates i.e. $CaSO_4.2H_2O$, this model only considered the anhydrous form, $CaSO_4$ as the filtered solid mass. This was carried out to allow for simplification of the process model as described in Zhang et al. (2003). The acid removal step could also have been considered to be carried out in the same step with the residues removal, this was however not carried out in this study to facilitate a close monitoring of the process streams and to minimise any potential model problems.

Methanol Recovery

Excess stoichiometric ratios of methanol to oil for the in-situ process were used to increase the percentage FAME conversion (w/w). The reacting alcohol level was also selected to ensure the complete immersion of the reacting microalgae biomass, thus ensuring full interaction between the reacting species. The in-situ transesterification model was designed to incorporate a methanol recovery unit for the excess reacting methanol. This was carried out to reduce raw materials cost and processes wastes, and to improve the separation of the transesterification products in the subsequent stages. The recovered methanol was recycled back to the reactor units where it was mixed with fresh methanol.

To model the methanol recovery unit, the rigorous vapour liquid fractionation model (RADFRAC®) in the ASPEN modelling software was used because of its reported suitability for two- and three- phase systems, as well as for narrow and wide boiling systems. This section will highlight only the important details used to model the distillation columns using the ASPEN software. This study used a simplified distillation set-up and does not to provide a comprehensive stage by stage analysis of the recovery columns, because the aim of this study is primarily to obtain the process energy requirement of the transesterification method investigated. The distillation column was designed to ensure the methanol mass fraction (w/w) in the recovered distillate stream (i.e. at the top of the distillation column) was >99%.

Two key column parameters; the reflux ratio (RR) and the number of stages (N) of the distillation column were used as inputs in the distillation model. The RR, which is the ratio of the liquid reflux from the condenser (stage 1) per unit quantity of the distillate product removed from the distillation column, has a significant influence on the energy requirement of the column. Increased RR values lead to a larger heating requirement with a corresponding reduction in the required number of distillation stages, N (Gilliland, 1940), and a reduction in the capital cost. However, increasing the column stages, N, reduced heating demands of the system only up to a point (Zhang et al., 2003). The column stages are also related to the required concentration of methanol at the top of the distillation column. The selection of a suitable RR and N for the distillation model is therefore important, since it provides a balance between operational and capital costs.

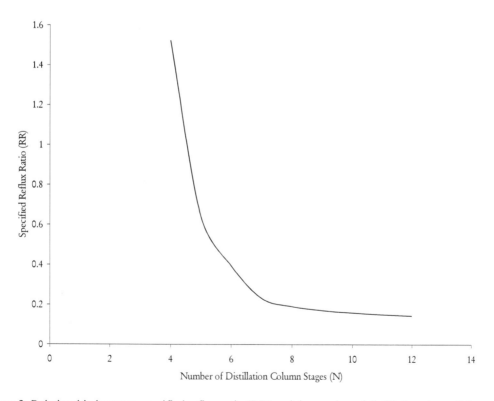

Figure 2. Relationship between specified reflux ratio (RR) and the number of distillation stages (N).

The estimation of these parameters was carried out using the simplified Gilliland empirical relation provided in the Aspen Plus® software. This was used to relate specified values of N to give RR, or specified values of RR to determine N. The RR was specified as shown in Figure 2 using the ASPEN Gillard relation, and its relationship with the number of theoretical distillation stages determined. An operational RR of 0.1 with N being 12 was selected for this study model since it was seen from Figure 2 that a further reduction of the RR (i.e. significant lower distillation heating demand) may not be achieved with the column stages > 12. The separation of the methanol component from the inlet stream (6) in the distillation column was then carried out. It was considered that the additional heat energy required for the distillation process was supplied via the reboiler unit of the distillation column MEOH-REC. The distillate stream (7) primarily containing methanol with 99% purity (on a mass fraction basis) was sent for re-use in the transesterification reactors. The heat exchangers (HX-2 and HX-3) were used to recover the useful heat energy available in streams 7 and 8. The recovered heat was considered for use for the heating requirements of other process units. The bottoms (stream 8) containing the biodiesel product was transferred to the water washing stage.

Water Washing

The bottoms product of the distillation column (stream 8) was subjected to an ester washing process using water to enhance the formation of two distinct and easily separable layers (methyl esters and glycerol layers). The water wash stage in this model (represented as WASH) was based on batch counter-current extraction column with the process specifications and mass flow balances as presented in Sheenan et al. (1998). This involves the use of water heated to 70°C (using HX-4) in the wash column where the separation of the stream contents was carried out. The mass flow rate of the process water input (WATER) for the wash process was equal to 20% of the mass flow of the FAME component in the incoming stream 8-1 (Sheenan et al., 1998). This was based on current biodiesel industrial practices (Sheenan et al., 1998). It was further assumed that a 100% recovery of the FAME and unreacted oil content in stream 8-1 was obtained after the wash. The methyl ester product and glycerides (stream 13-1) were accompanied by 10% of the wash water entrained as in Sheenan et al., 1998. The heat exchanger (HX-5) was used to heat the contents of the stream 13-1 to the temperature of 150°C, with which it can be separated in the FLASH unit.

Stream 13-1 was transferred to a flash unit where a further reduction of its water content was carried out to improve the fuel characteristics of the biodiesel product. The valve, V-2 was considered for use to reduce the stream pressure to 0.1 bar to ensure that the flash temperature (137°C) was kept lower than the decomposition temperature of the methyl esters. Stream 10 contained 100% of the glycerol fraction in stream 8-1, 90% of the wash water used in the abstraction process, and any left over methanol. This stream is transferred to the glycerol recovery step: the distillation column (GLY-REC). The valve V-1 was used to reduce the pressure of stream 10 to 0.7 bar to ensure that the reboiler temperatures were less than that of glycerol decomposition.

Methyl Ester Purification

For this study, the European biodiesel standard as specified in EN 14214 was used to specify further purification requirements of the biodiesel product. Based on the standard specification, it was decided that the biodiesel product, where possible, should contain a FAME mass fraction of ≥ 96.5 % (w/w). This was used to provide an appropriate means of comparing the different processes. The simple vacuum flash distillation process (FLASH) discussed above was used to upgrade Stream 13 to yield a FAME mass fraction of 97.5%

(w/w). This was higher than that of the specifications but was used since the same purification steps were used for the other investigated processes.

The fuel purification could alternatively be carried out via physical stripping methods i.e. as in Wang et al. (2009). This study however did not examine those alternative methods, instead considered only current industrial practices.

The condensed water vapour (WST-H2O) was treated as a process wastewater stream. The WST-H2O stream contained a mass flow of unreacted microalgae oil, water and FAME of 0.24, 10.37 and 0.68 kg/hr respectively. The treatment and recovery of the useful products in the waste stream were however not further examined. The heat exchangers (HX-6 and HX-7) were used to recover useful heat from the distillate and bottoms streams respectively.

Glycerol Recovery

The extent of the glycerol by-product purification depends on the proposed use of the glycerol product obtained after this step. It was decided in this study that the glycerol purification would be carried out to achieve a commercial grade purity i.e. $\geq 90\%$ (w/w), for sale as an industrial raw material (Zhang et al., 2003).

The RADFRAC® vacuum distillation routine of the Aspen Plus® software was used for modelling the GLY-REC distillation column. The theoretical column stages (N) and the reflux ratio (RR) inputs of 6 and 0.2 respectively were used to obtain a final glycerol product with a glycerol mass purity of 93% for the bottoms stream (Stream 12).It was considered that the reboiler unit was used to provide the distillation heating requirements. Vacuum distillation was used to ensure that the temperature of the bottom stream was kept below the decomposition temperature for glycerol (150°C) (Haas et al., 2006).The distillation was carried out with an absolute operating pressure of 70 kPa, which ensured that the temperature of the bottom stream from the column was 145.9°C.The distillate waste stream composed of a mass flow of methanol and water at 30.41 and 96.12 kg/h respectively was treated as a process waste stream. The heat exchangers (HX-8 and HX-9) were used to recover useful heat from the distillate and bottoms streams respectively.

The Ultrasound Assisted *In-Situ* Transesterification Process

The ultrasound assisted in-situ transesterification process was modelled out with the same process outline used for the in-situ model as in Figure 1, with mechanical stirrers in the reactors replaced with the use of ultrasonicators for the process agitation.

Process Inputs and Transesterification Reaction

A microalgae mass input of 1 t *Chlorella* biomass (MICROALG) was used for this process, with a reacting methanol mass flow of 1063.72 kg/h (stream 1). This provided the same reacting molar alcohol to oil ratio of 105: 1 used in the laboratory experiments, described in Ehimen et al. (2011).

The transesterification reaction was carried out using two continuous reactors in series as in section –"The In-SituTransesterification Process". Stream 3, containing the reaction feedstocks was heated to a temperature of 60°C and transferred to the reactors where this temperature was maintained. Unlike the mechanically stirred in-situ process, the reactor agitation was performed using low frequency ultrasound (24 kHz) connected to the transesterification reactors. After a transesterification time of 1 h, a 94% mass conversion of the microalgae lipids to FAME was assumed to be achieved (Ehimen et al., 2011). This corresponds to an overall conversion of 99.6% (w/w) after both reactors, with a total biodiesel yield of 0.295 kg/kg dried *Chlorella* biomass (i.e. as in Ehimen et al., 2011).

Downstream Processes and Biodiesel Upgrading

The *Chlorella* residues filtration and acid catalyst removal, post transesterification, were carried out as described in the sections: "*Residues Filtration*" and "*Acid Catalyst Removal*" respectively. The recovery of the excess methanol quantities was carried out to achieve a mass recovery (≥ 99%) of the methanol component in stream 6 (w/w). This was performed using a distillation column reflux ratio and theoretical column stages of 0.1 and 12 respectively. The washing and purification of the FAME and glycerol fractions were then modelled with specifications as in section "The In-Situ Transesterification Process". The final FAME mass purity in the BIODIESL stream was 98.9% (w/w). The final GLYCEROL stream contained a mass purity of 93% glycerol (w/w).

Integration of Co-Solvents in the *In-Situ* Transesterification Process

For the analysis presented in this chapter, diethyl ether was solely considered for modelling the co-solvents assisted in-situ transesterification process. Diethyl ether was selected because the FAME yield (g/g *Chlorella*) with the use of this solvent was significantly higher than that observed with the use of pentane (Ehimen et al., 2011). The modelling for the continuous co-solvent assisted in-situ transesterification process was carried out using the same process outline as the mechanically stirred in-situ transesterification method. A pre-mixing unit for the diethyl ether inputs and the reacting methanol was however incorporated as shown in Figure 3, with the process raw materials introduced at ambient temperatures (20°C).

The estimation of the diethyl ether component in the stream DIETET was based on the molar concentration of the reacting methanol. A reacting molar ratio of methanol to oil of 79:1 was selected. This was due to the fact that the highest percentage mass oil to FAME conversion was shown to be achieved using this alcohol level (Ehimen et al., 2011). The molar flow of the diethyl ether input (kmol/h) into the process was 0.128 times the molar flow of the reacting methanol (kmol/h). With a methanol mass flow of 781.3127 kg/h this relates to a diethyl ether mass input of 232.11 kg/h into the mixer (MIXSOL). The microalgae biomass inputs and acid catalysts were the same as the in-situ transesterification process in the section – "The In-Situ Transesterification Process".

The continuous transesterification process was assumed to be operated at a temperature of 60°C, with continuous stirring provided by the use of a mechanical stirring system. The transesterification reactors were operated at an elevated pressure of 300 kPa (absolute) to ensure that the reactor contents were in a liquid phase, because the temperature was higher than the normal boiling point of the diethyl ether co-solvent. After 4 h, a 70% mass conversion of the *Chlorella* lipids to FAME (w/w) was obtained (Ehimen et al., 2011), and a total percentage mass FAME conversion of 91% (w/w) was obtained after both reactors as estimated in the section "*Transesterification Reaction, Acidic Catalyst and Reacting Alcohol*".

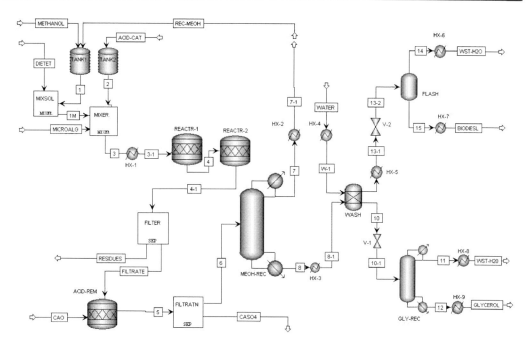

Figure 3. Flow sheet showing the process units and streams for the in-situ transesterification model for biodiesel production from microalgae with diethyl ether as a co-solvent.

The resulting stream (4-1) was subjected to the biomass residues filtration and acid catalyst removal steps as in section – "The In-Situ Transesterification Process". The methanol and diethyl ether components in stream 6 were recovered using distillation processes as earlier described. The specifications of the distillation column were for the recovery of 100% mass flow of the diethyl ether and ≥99% mass of the methanol component in the incoming stream 6. The biodiesel and glycerol were also recovered with schemes as described in section – "The In-Situ Transesterification Process". The final FAME purity in the BIODIESL stream was 91.6% (FAME mass fraction in the product), and the purity of the GLYCEROL stream was 93% glycerol (mass fraction of glycerol in the product stream).

The Use of Ultrasound and Co-Solvent for the *In-Situ* Transesterification Process

This modified in-situ transesterification process was carried out using low frequency ultrasound for the reaction agitation with diethyl ether as a process co-solvent. The molar ratio of reacting methanol to oil was 26:1 as described in Ehimen et al. (2011). The process model outline for the conversion of the microalgae biomass to biodiesel was similar to Figure 3, but with differences in the process inputs and stirring. The continuous transesterification reactors were assumed to be the same as in "The In-Situ Transesterification Process", with the process agitation provided by low frequency ultrasound (24 kHz). The reaction temperature was maintained at 60°C. With the use of diethyl ether, the percentage mass oil to FAME conversion was 96% (w/w) after a reaction time of 1 h after the two reactors in series, corresponding to a total biodiesel yield of 0.295 kg/kg dry *Chlorella* biomass (Ehimen et al.,

2011). The methanol and solvents recovery, biomass residues filtration and biodiesel and glycerol upgrading were then carried out using processes similar to those described in sections *"Residues Filtration"*–*"Glycerol Recovery"*.

The Conventional Transesterification Process

The process flowchart for the conventional transesterification route is shown in Figure 4.

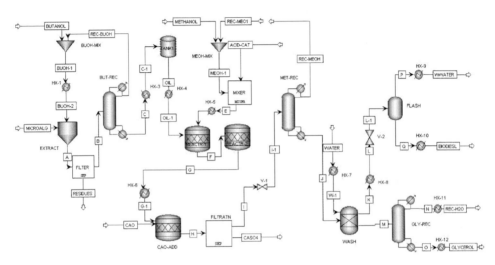

Figure 4. Flow sheet showing the process units and streams for the up-scaled conventional extraction and transesterification of microalgae oil.

Oil Extraction and Solvent Recovery

The industrial scaled conventional transesterification process in this study covered the extraction of lipids from the microalgae biomass, and the transesterification process.

As reported in Ehimen et al. (2009), the use of butanol and a chloroform-methanol mixture have been examined for the extraction of the microalgae (*Chlorella*) lipids. The results obtained showed similar oil contents for *Chlorella* biomass following extraction using both solvents. Only butanol was used as the solvent in this study modelling. The reason is that where energy recovery (as CH_4) from the residues post transesterification is proposed, butanol use is preferred, because the chloroform remaining in the *Chlorella* residues appeared to inhibit the CH_4 generation process (Ehimen et al., 2009).The choice of butanol as the process extraction solvent was also based on the fact that butanol could potentially be a renewable feedstock i.e. via acetone-butanol fermentation (Jones and Woods, 1986). Its use could further improve the overall 'greenness' of the biodiesel product by reducing fossil fuel derived process feedstocks. This is examined further in the section "Sensitivity Analysis of Ir".

Dry microalgae biomass (1000 kg/h) at ambient temperatures was transferred to an extraction reactor (EXTRACT) with 2000 kg/h of preheated butanol (from 20°C to 90°C) in stream BUOH-2, with continuous stirring throughout the extraction period. The heat exchanger HX-1 was used to heat the extraction solvent to the required extraction temperature of 90°C. The extraction unit was based on the results of the laboratory lipid extraction experiments, which showed that 0.275 kg of microalgae lipids were stripped per kg of the dried *Chlorella* biomass using butanol. The resulting solid-liquid mixture (stream A) was subjected to a filtration process (FILTER), as previously described, and the extracted microalgae residues collected. The filtrate (stream B) was sent to a distillation column to recover the butanol fraction.

The distillation column (BUT-REC) was used for the recovery of the 1-butanol solvent. Vacuum distillation was performed using the RADFRAC® model described earlier, with the RR and N at 0.2 and 6 respectively. A distillation column pressure of 10 kPa (absolute) was used to ensure that the bottom products were kept below 250°C, to prevent the decomposition of the triglycerides contents in the stream (Schawbo et al., 1988). Because butanol is a relatively expensive solvent[1], it was decided ≥99.9% of the mass butanol component of the incoming stream B (w/w) would be recycled. This was carried out to minimise process wastes. The recycled butanol (99.9 % mass purity, w/w) available in stream REC-BUOH was re-used in the extraction process.

The bottoms stream (C) containing the oil feedstocks were sent to a tank (TANK 1) for its use in the transesterification process. The heat exchanger HX-3 was used to recover some of the useful heat in stream C before the microalgae oil tank. The microalgae oil was then considered to be heated to the transesterification temperature of 60°C using HX-4. The use of only one heat exchanger between the bottoms of the distillation column (stream C) and the first of the continuous transesterification reactors (REACTR-1) could have been adequate for the recovery of the useful energy.

Transesterification Reaction

A 1:1 mass flow of H_2SO_4 to reacting oil was used as the reaction acidic catalyst. Prior to the transesterification reaction, fresh and recycled methanol were mixed in MEOH-MIX and mixed with H_2SO_4 (ACID-CAT) using MEOH-MIX. The resulting stream E was then heated to 60°C using HX-5. The microalgae oil feedstock (OIL) was also heated to 60°C (using HX-4) before it was transferred to the continuous transesterification reactors. The continuous acid catalysed process used two transesterification reactors in series (REACTR-1 and 2) as previously described. The reaction however involved the use of a molar reacting methanol to oil ratio of 50: 1. This was used since it was demonstrated to result in an 81% mass conversion of *Chlorella* oil to methyl esters in 4 h at the reaction temperature at 60°C (Ehimen, 2011). This corresponds to a 96.4% mass conversion of *Chlorella* oil to FAME (w/w) after the two transesterification reactors.

Methanol Recovery

After the transesterification process, the heat energy in stream G was recovered using HX-6, and acidic catalyst component in the mixed product stream (G) was removed using the reactor (CAO-ADD) as described in section: *Acid Catalyst Removal*. The methanol fraction in the filtrate stream (I) was subjected to a distillation process (using MET-REC) to recover the excess process methanol. Vacuum distillation with a RR and N of 0.1 and 8 respectively was used. The valve V-1 was used to reduce the pressures in stream I to the requirement of the distillation column. This involved fixing the column pressure at 50 kPa (absolute) to keep the

---

[1] Compared to the use of methanol and chloroform. The 2010 first quarter spot prices of methanol, chloroform and butanol were 200, 500 and 1813 US$/metric-t respectively (ICIS, 2010).

temperature of the distillation column bottoms product below the decomposition temperature of glycerol (150°C).

Biodiesel Purification

The separation of the biodiesel and glycerol fractions of the filtrate streams after the methanol recovery process was carried out using wash columns specifications similar to those in "The In-Situ Transesterification Process". The methyl ester rich stream (L) containing 93.0% (w/w) FAME, was subjected to a flash vacuum process. This resulted in a methyl ester stream (BIODIESL) contained 96.5% FAME (w/w).

Glycerol Purification

The glycerol upgrading process was performed using the ASPEN RADFRAC® routine with the conditions as described in section –*"Glycerol Recovery"*. This involved the use of a process RR and N of 1 and 6 respectively, and a column pressure of 70 kPa (absolute). A glycerol mass purity of 93.0% (w/w) in the stream GLYCEROL was obtained.

MODEL RESULTS AND ANALYSIS

This section presents the results of material and heating demands for the different conversion routes obtained from the process models. The external heat demand after process integration was used as a simple indicator of the process energy requirement. The process electrical power requirement for the operation of the pumps and valves in the process models were however not considered. This was due to the fact that it was considered that similar electrical requirements would be expected for the different process models. The electricity requirements for the process agitation are expected to differ, and its influence on the process energetics would be taken into account in the section "Renewability Assessment of the Different Transesterification Processes". This section deals mainly with the net heat energy requirements of the different transesterification processes. The energy requirement of the different transesterification processes was thus carried out on the basis of their annual minimum utility heating demand.

The models presented in this study are relatively basic with regard to the level of the details of the individual process units. The simulation was primarily aimed at estimating the energetic and raw material demand of the conversion routes. The models are not intended to replace detailed engineering analysis which would be required in the design and construction of biodiesel production plants.

The in-situ transesterification routes were seen to facilitate the microalgae-biodiesel conversion with the fewer processing steps compared with the conventional method, because the oil extraction step and the accompanying solvent recovery unit were eliminated. This agrees with the discussions by Haas et al. (2007). An increase in the biodiesel product can also be observed for the different in-situ transesterification processes over that of the conventional process. The reasons behind the increases of the biodiesel yield have been previously reported in Ehimen et al. (2011).

Process Integration and Energy Comparison

Before the heating demands of the different transesterification routes are estimated, the extent to which the process heating and cooling could be met internally was explored. This involved recovering heat energy from process streams which can be taken as heat sources or supplying heat energy to process streams which can be taken as heat sinks. External utilities (i.e. pressurised steam and cooling water) were then considered to meet the shortfall of the available heat sources and sinks to meet the process heat and cooling demand. The estimated minimum external utility demand was then used as a preliminary indicator of the energy requirements of the various transesterification processes. The stream mass flow rates and enthalpy changes of the process streams obtained using the ASPEN models were used to estimate the maximum recoverable process heat.

The heat integration method, pinch analysis as demonstrated by Linnhoff and Flowers (1978), Linnhoff and Vredeveld (1984) and Kemp (2007) was applied to achieve this goal. This technique is applied for minimising energy consumption in industrial processes. The maximum possible process heat recoverable was calculated using pinch analysis which involves the calculation of the thermodynamically accessible heat energy which could be consumed by the modelled process. The process stream data i.e. heat sinks and sink streams and mass flow rates, the composite curves used in the determination of the minimum energy consumption targets of the respective processes, and the theoretical heat energy savings are described in detail in Ehimen, 2011.

Results for the hot and cold utilities required (i.e. before and after the heat integration using the pinch analysis) for the various up-scaled in-situ and conventional processes obtained from Ehimen, 2011 are shown in Figure 5.

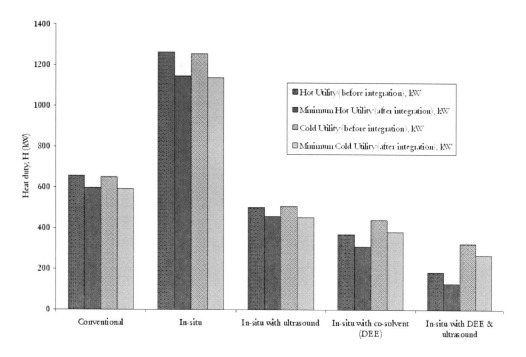

Figure 5. Heating and cooling utility requirements before and after heat integration (via pinch analysis).

An additional 548.3 and 547 kW of hot and cold utilities respectively were found to be required by the mechanically stirred in-situ process compared to the similarly sized conventional biodiesel production unit. This was based on the assumption that the same biomass inputs (1 t dry microalgae/ h) were used for the transesterification processes. The increased heat demand for the in-situ process was due to the energy required to recover (via distillation) the higher methanol volumes used in the in-situ transesterification method. A reduction in the utility heating requirement was obtained by the modifications applied to the mechanical stirred process. An 88.7% reduction of the process heating demand was obtained by the use of co-solvents and ultrasound respectively, compared with the mechanically agitated in-situ process. The transesterification process modifications appear to produce considerable reductions of the process heating demands due to the comparably reduced methanol quantities used in these processes. However, due to the differences in the mode of reactor agitation, the energy requirement for the process mixing must also be assessed to facilitate a proper comparison of the energy requirement for the different processes. This will be examined in "Renewability Assessment of the Different Transesterification Processes". In the following section, the process heating and raw material requirement estimated for the modelled industrial transesterification processes is further used for the renewability assessment.

The heat demands for the transesterification processes obtained in this study are intended to be used only as a first indicator of the process energy requirement. The cooling demands for the respective processes in this chapter were however not further explored since the process cooling requirement could be met using process water. This was since a minimum target temperature of 30°C was considered in this study for the streams requiring cooling.

RENEWABILITY ASSESSMENT OF THE DIFFERENT TRANSESTERIFICATION PROCESSES

The renewability analysis in this study is based on an assessment that quantifies the net gains (or losses) obtained from the use of microalgae for biodiesel production. The renewability evaluation in this chapter was based on the process energy availability, or exergy, obtained using both the first and second laws of thermodynamics. The exergy (B) is the work that a system can produce if it is brought to physical, thermal and chemical equilibrium with its environment in a reversible manner. This corresponds to the maximum work extractable from a given system in a steady flow process (Ayres et al., 2006). The main objective of this chapter was to use the exergy based approach for the acquisition of a suitable indicator which would be used to quantify the renewability of different biodiesel production processes.

This renewability indicator used in this paper would be initially presented in the section – "Quantification of the Renewability Indicator". The system boundaries for the renewability evaluation were then defined, and the renewability of the biodiesel production processes evaluated.

QUANTIFICATION OF THE RENEWABILITY INDICATOR

A process or resource can be regarded as renewable when regeneration mechanisms exist to return the resource to its original state i.e. a cyclic transformation is facilitated. The production of biodiesel from microalgae, and its subsequent combustion in diesel engines, could be considered as a truly renewable resource, when the ideal process cycle shown in Figure 6 is feasible.

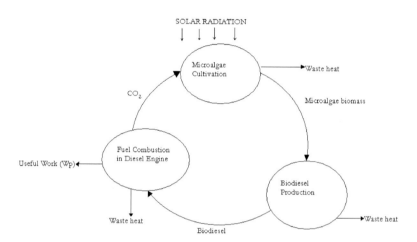

Figure 6. The ideal microalgae-biodiesel cycle.

In the ideal process, solar energy is harnessed by the thermochemical cycle which involves the following step-wise carbon transformations (Figure 6):

i. Microalgae biomass is naturally cultivated using water, atmospheric CO_2, nutrients, and the photosynthetic process driven solely by solar energy.
ii. The lipid fraction of the microalgae biomass is obtained and converted to biodiesel without the need for fossil fuels.
iii. The biodiesel produced is combusted in diesel engines to produce useful work (Wp), with CO_2 and water as the process by-products.

For the ideal process, it is assumed that all of the chemical reactants and by-products are recyclable, with only low quality heat rejected by the microalgae-biodiesel cycle to the environment. Furthermore, all of the CO_2 released from the fuel combustion process and the organic process wastes after the biodiesel production process are assumed to be recycled into the microalgae production process. The renewability of the microalgae biodiesel production process could however be altered when non-renewable resources are consumed in the biofuel processing units, or in the restoration of the environment to its initial state. Resources classified as non-renewable resources (NRRs), are those which are consumed at a faster rate than they can be regenerated. The work required for the restoration of the environment to its initial state is greater than that produced from these resources (Berthiaume et al., 2001).Fossil fuels i.e. petroleum and coal are regarded as NRRs. The direct use of these fossil fuels and their by-products (i.e. fertilisers) in the microalgae biodiesel production process although increasing the microalgae biomass and biodiesel productivities, could potentially reduce the overall renewability of the conversion process.

The NRRs consumed during the biomass and the fuel production process must be accounted for to establish the renewability of the biodiesel production process. The process exergy consumption approach as described by Berthiaume et al., (2001) and Szargut et al., (1987) was used in the renewability assessments in this study. This paper therefore compares the maximum work (Wp) obtainable from the final products of the transesterification process with the minimum work (Wr) required for restoration of the non-renewable resources consumed in the different production processes to their initial environmental state (Berthiaume et al., 2001).

The renewability indicator (Ir) used in this study was obtained as defined by Berthiaume et al. (2001) using Eq. 1:

$$I_r = (W_p - W_r)/W_p \tag{1}$$

The process renewability can be evaluated on the basis of the Ir estimated for the different processes, i.e.:

- $I_r = 1$, is indicative of a fully renewable system, i.e. with $W_r = 0$, as for the ideal microalgae-biodiesel system.
- $0 < I_r < 1$, for a partially renewable system.
- $I_r = 0$, suggests a system in which the work produced by the biofuel and the restoration work required are equal. In such cases, the non-renewable resources could be considered for use directly instead of the produced biofuel.
- $I_r < 0$, for a process which consumes more restoration work than it produces. Such processes are considered as non-renewable, and are less efficient than similar processes where the non-renewable resources are used directly for the intended biofuel application.

This renewability indicator was applied to assess the microalgae transesterification processes examined and to compare the different processes studied.

THE SYSTEM BOUNDARIES USED IN THIS STUDY

As thermodynamically required, the boundaries for the microalgae biodiesel production systems were selected in this study to start with the cultivation and harvesting of the microalgae biomass and end after the biodiesel and glycerol upgrading process.

The system inputs include microalgae cultures, CO_2, nutrients, electricity, fossil fuels and process chemicals. The system outputs are biodiesel, glycerol, calcium sulphate ($CaSO_4$), microalgae biomass residues, and process heat and material wastes. Figure 7 shows the boundaries of the system and the various material and energy flows considered in this chapter.

The material and energy inputs for the transesterification processes were considered on the basis of the requirements for the conversion of 1 t dry microalgae biomass/h. With the process assumed to be in steady state, and the energy input for the recovery of the excess methanol and solvents accounted for, only the additional raw materials added to make up the required reacting quantities were used as process inputs (at ambient temperature and pressure).

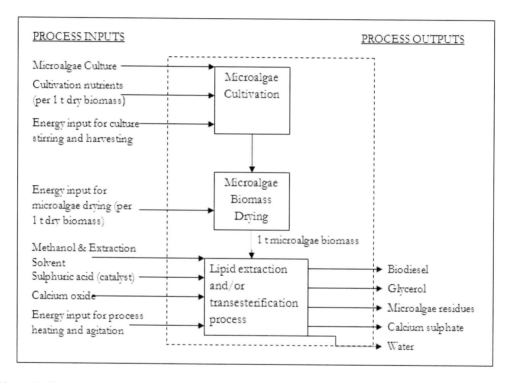

Figure 7. The system boundary (represented by the dashed lines) and material and energy flows considered in this study.

ESTIMATION OF THE PROCESS WR

Microalgae Biomass Production and Harvesting

The minimum work inputs into the microalgae production and harvesting stages are still contentious, as shown in Chisti, 2008 and Reijnders, 2008. With the biomas production process Wr depend on the specific microalgae species cultivated, the type of bioreactor used, the microalgae culture and the climatic conditions available. The differences in the literature reports are mainly due to the developmental nature of microalgae technology aimed for cheap biofuel production.

With the assumption of the use of *Chlorella* biomass with a lipid content of 27.6 % (w/w, on a dry biomass basis) and an assumed biomass productivity of 20 g dry microalgae/m^2/day, the cultivation and harvesting systems as presented in (Ehimen et al, 2010; Goldman, 1980) was considered to be suited for use in this chapter.

The mean values for the major elements found in microalgae (expressed in percentage of ash free dry weight) are reported to be: Carbon (C), 52%; Hydrogen (H), 8.0%; Oxygen (O), 31.0%; Nitrogen (N), 8.0%; and Phosphorus (P), 1.0% (Benefield and Randall,1980). This elemental composition was similar to those obtained for the *Chlorella* samples in Ehimen et al (2010) which was used as the basis for most of the study assumptions. These nutrients must be adequately supplied to the microalgae cultures to facilitate biomass synthesis and maintain productivity levels (Becker, 1994). Atmospheric CO_2 was considered to meet the culture carbon requirement, with the biomass H and O fractions obtained from the culture water. Only the major nutrients, N and P were considered in this chapter as process fertilisers for the biomass cultivation step. Ammonia (NH_3) and urea ($CO[NH_2]_2$) are commonly used as N sources in microalgae cultivation systems (Becker, 1994). Only the use of NH_3 was considered in this paper since it is relatively cheaper and is reported to be easier to handle than urea for application in microalgae cultivation systems (Neenan et al, 1986). Triple super-phosphate (TSP) was used as the reference P fertilizer due to its widespread use in agricultural applications (Patzek, 2004). The areal NH_3 and TSP requirements for the microalgae biomass cultivation process was same as presented in Ehimen (2010).

Mixing of the microalgae cultures, pumping of the water and microalgae cultures, as well as the harvesting and concentration of the algal biomass (i.e. via centrifugation) are some of the processes which require electrical input for their operation (Becker, 1994; Becker and Venkatraraman, 1980; Neenan et al, 1986). The process electricity demand presented in Chisti (2008a) was used in this chapter to assess the biomass production electricity requirement. The electrical energy requirement for biomass cultivation and harvesting was reported to be 8.77 and 0.30 MJ/kg microalgae oil respectively (Chisti, 2008). This conservative electrical energy demand was 9 times less than that demonstrated in Becker and Venkatraraman (1980) for similar raceway cultivation systems, and is taken to represent the process electrical requirement in a present day scenario. With an estimated extracted lipid content of 0.276g lipids/g *Chlorella* biomass (Ehimen et al., 2010) this relates to an electrical energy input of 2448.9 MJ/t dry *Chlorella* biomass produced for the microalgae cultivation and harvesting processes.

The renewability assessment carried out was initially based on electricity generated from typical fossil fuel generation systems. An analysis of the influence of electricity source (i.e. the use of hydro-electricity) on the overall process renewability was later conducted under the sensitivity analysis section in this chapter.

Microalgae Biomass Drying

Although detailed reports on the energy requirement for the microalgae drying was not found in the literature, this processing step has been indicated to potentially contribute a significant energy input (Chen et al, 2009). The energetic demand for the microalgae biomass drying was thus included in this study. This involved a simple assessment of the use of two drying methods (a) a conventional hot air dryer and (b) a heat pump dryer for the reduction of the moisture content of the post harvested biomass to levels suited for the transesterification process. Post-harvest microalgae moisture content of 72.5% (w/w, on the basis of the dried biomass) as described in Ehimen et al. (2010) was used in this study.

The specific moisture extraction rate (SMER) which is the rate of water extracted from the product per unit energy consumed in the drying process (kg/kWh) was used to assess the effectiveness of the energy used in the biomass drying processes. During the drying process, the SMER value reduces with time as the moisture removal from the biomass becomes more difficult, due to hygroscopic retention of moisture in the product. This effect can be observed for wood drying using dehumidifiers as demonstrated in Sun et al., (1999), where the reported process SMERs clearly decreases towards the end of the drying process. Although factors such as the thickness of biomass layers, the kiln air state and drying time could influence the SMER of the considered drying process, only reported average SMERs were used for the preliminary estimation of the biomass drying energy requirement. A maximum SMER value of 1.55 kg moisture/kWh, based on the latent heat of water vaporisation at 100°C (Raghavan et al. 2005) was used to estimate the conventional dryer energy demand. This was used to represent best-case scenarios for conventional dryers (Raghavan et al. 2005). For the removal of 725 kg of moisture from the post harvested microalgae, this relates to an energy input of 467.74 kWh for the recovery of 1t dry biomass.

For the biomass drying using a heat pump dryer, a SMER of 4 kg moisture/kWh electricity (Carrington, 2010) was considered in this study.

For the biomass drying stage, it was initially considered that fossil fuels combustion was used to generate the energy consumed in the drying process. The use of heat pump dryers for the biomass drying and its influence on the renewability of the considered processes is subsequently explored later in the sensitivity analysis section -"Sensitivity Analysis of Ir". In a fossil based scenario, it is initially assumed that diesel combustion is used to meet this energy requirement. This was carried out with an overly optimistic thermal efficiency of 100% considered for the combustion process, even though best operating efficiencies of ≈ 80% is usually seen for direct fired heating system (Carrington, 1978). The analysis presented in section – "Estimation of the Process Wr" was initially based on the use of separate heat and power (SHP) installations (i.e. electrical generators and/or onsite boilers) for providing the process electrical and heating requirement. The use of combined heat and power (CHP) systems as an efficient approach for power and thermal energy generation from a single fuel source has been discussed widely in the literature (i.e. Pilavachi, 2000; Graus et al., 2007). The use of CHP systems for the process electricity and heat generation and its influence on the process renewability was however not considered in this study due to the considered use of a 100% operating efficiency for heating fuel combustion in the SHP installations.

With the same biomass cultivation and drying processes considered to be applied for the production of a unit of microalgae biomass, the same energy inputs for the biomass production, harvesting and drying steps were used for all the different transesterification processes.

Table 2. Minimum restoration work required for the NRRs depleted in the in-situ transesterification of 1 t *Chlorella* biomass (mechanical stirring)

Process unit	NRRs	Unit	Quantity of NRRs	CNEx (MJ/unit)	Wr (MJ)
Microalgae biomass production					
	Ammonia	kg	80.00	30.9[2]	2472.00
	P_2O_5	kg	20.00	7.52[3]	150.4
	Electricity associated with cultivation and harvesting	MJ	2448.90	4.17[4]	10211.91
	Microalgae drying (diesel)	kg	37.93	53.20	2017.88
Biodiesel production					
	Methanol	kg	104.2	73.08[5]	7614.94
	H_2SO_4	kg	270.00	9.10[6]	2457.00
	CaO	kg	154.37	10.05[7]	1551.42
	Reactor stirring (electricity)	MJ	0.82	4.17	3.42
	Process heating (diesel)	kg	92.86	53.20[8]	4940.15
			Total Wr (MJ)		31419.28

Microalgae Biodiesel Production

The microalgae lipids are extracted and transesterified (conventional transesterification) or transesterified directly (in-situ transesterification) from the dried biomass. This involves the use of methanol as the reaction alcohol, and H_2SO_4 as the process catalyst. To estimate the process renewability of biodiesel production via the considered transesterification processes, the minimum restoration work (Wr) required for the restoration of the non-renewable resources consumed in the different transesterification processes was initially estimated. The maximum useful work available from the process products (Wp) was then determined. The Wp and Wr were then used to estimate the renewability indicator (Ir) in section "Renewability Evaluation of the Different Transesterification Processes".

---

[2] CNEx values based on ammonia gas production via steam reforming of natural gas (Szargut & Morris, 1987).
[3] CNEx value for P_2O_5 obtained from Berthiaume et al. (2001) based on the production energy values demonstrated in Wittmus et al. (1975).
[4] Typical CNEx value for electricity generated from fossil based fuels (Szargut & Morris, 1987).
[5] Methanol produced from crude oil (Szargut & Morris, 1987).
[6] CNEx of sulphuric acid based on production via the application of the Frasch process and sulphur combustion using sulphur from the ground as the process feedstock (Szargut et al, 1988).
[7] CNEx of calcium oxide based on CaO production via the mining, crushing and calcining of limestone as the process feedstock (Szargut et al, 1988).
[8] Typical diesel fuel from crude oil CNEx value (Szargut & Morris, 1987).

Mechanically Stirred *In-Situ* Transesterification

Cumulative Exergy Consumed for Production of NRRs Consumed in the Mechanically Stirred In-Situ Process

Using the process material and minimum heat demands, the cumulative net exergy (CNEx) consumed for the production of the NRRs which are used in the mechanically stirred in-situ process is shown in Table 2. The process inputs (i.e. methanol, catalyst and CaO and outputs i.e. biodiesel, glycerol and $CaSO_4$) were based on the equivalent requirements for the in-situ transesterification of 1 t dried microalgae biomass as obtained from the process models as detailed in "Modelling and Assessment of the Transesterification Processes".

For the stirring of the in-situ transesterification reactors, a mixing electrical energy requirement of 1.42 MJ /t biodiesel produced (Sorguven and Özilgen, 2010), was used in this study. A total hourly mass flow of biodiesel of 0.28 t was shown for the mechanically stirred in-situ transesterification process after the conversion of 1 t dry *Chlorella*. The stirring electrical energy requirement for the two in-situ reactors therefore corresponds to 0.80 MJ for the stirring of the two transesterification reactors (i.e. 1.42 MJ/t× 0.28 t × 2 reactors).

It was also initially considered that the process heat demands would be supplied via diesel fuel combustion in boilers. A maximum useful work (exergy) for the diesel heating fuel of 44.4 MJ/kg (Szargut and Morris, 1987) was used for in this analysis. The quantity of heating fuel (kg/t dry microalgae biomass consumed) required to meet the minimum heat demand (after process integration) of the different transesterification processes was estimated with the assumption of a 100% conversion of the fuel to heat.

The influence of the use of other common boiler fuels i.e. coal, natural gas and wood pellets on the process renewability is subsequently examined in section"Sensitivity Analysis of Ir". Furthermore, the use of heat pump heating (using electricity) to meet the process heating demand was also explored in that section.

The cumulative exergy values for the production of the NRRs consumed for the process transesterification raw materials, heating fuel and electricity requirements(CNEx), considered to be representative of a present day scenario, were largely obtained from Szargut and Morris (1987).

This analysis used the CNEx values from Szargut and Morris (1987), since they has been widely used in recent literature (i.e. Berthiaume et al, 2001; Yang et al, 2009; Sorguven and Özilgen, 2010) to determine the minimum net exergy cost for the NRRs production in a present day production scenario.

Useful Work Available (Wp) from the Products of the Mechanical Stirred In-Situ Transesterification Process

The exergy content of the biodiesel and glycerol product was used to account for the useful work (Wp) obtainable from the transesterification process. The exergy content of the FAME and oil were obtained from the heat of combustion of the extracted products. The sum of the minimum work available from the transesterification products was used to estimate the process Wp. The minimum useful work of the FAME and unreacted oil content of the biodiesel product were treated separately to quantify the work available from the transesterification product. The total useful work (Wp) available from the products derived from the in-situ transesterification (with mechanical agitation) of 1 t dry microalgae is shown in Table 3.

Table 3. Useful work available from the in-situ transesterification products (per 1 t microalgae biomass converted/ h) using a reacting ratio of methanol to oil of 315:1 (mechanical agitation)

Product	Exergy (MJ/kg)	Product quantity (kg)	Wp (MJ)
FAME	40.70	281.88	11472.52
Microalgae oil	40.15	6.76	271.41
Glycerol	18.52[9]	27.69	512.82
CaSO$_4$	0.06[9]	374.73	22.48
		Total Wp (MJ)	12278.71

In-Situ Transesterification with Ultrasound Agitation

Net Exergy Consumed for Restoration of NRRs Used in the Ultrasound Agitated In-Situ Transesterification of 1 t Microalgae Biomass to Its Initial State

For the low frequency ultrasonication of the in-situ mixture, the assessment in this section was based on the process electrical energy estimates obtained from the industry(Tyrrell, 2010).

Table 4. Minimum restoration work required for the NRRs depleted in the ultrasound agitated in-situ transesterification of 1 t microalgae biomass with a molar methanol to oil ratio of 105:1

Process unit	NRRs	Unit	Quantity of NRRs	CNEx (MJ/unit)	Wr (MJ)
Wr for microalgae production					14852.19
Biodiesel production					
	Methanol	kg	56.64	73.08	4139.25
	H$_2$SO$_4$	kg	279.00	9.10	2538.9
	CaO	kg	159.52	10.05	1603.18
	Reactor stirring (electricity)	MJ	28.80	4.17	120.10
	Process heating (diesel)	kg	37.20	53.20	1979.04
			Total Wr (MJ)		25232.66

Two 4 kW, 20 kHz ultrasonicator attached to the reactors were considered for the estimation of the agitation electrical energy requirement in this section. This was due to the fact that this commercially available ultrasonicator size was reported to be suited for biodiesel reactors with a flow rate of 1-3 m^3 transesterification mixture/h (Tyrrell, 2010). This ultrasonicator size was also reported to be adequate for microalgae cell disruption for

---

[9] Ayres & Ayres (1999).

processes with a biomass flow rate of 0.1-0.8 m^3 microalgae/h (Tyrrell, 2010). Thus, the use of the considered ultrasonicator size seems adequate for the mass flow rates involved for the in-situ transesterification of 1 t of microalgae biomass/ h using a molar methanol to oil ratio of 105:1.

The minimum work required to restore the NRRs depleted in the ultrasound agitated in-situ transesterification of 1 t *Chlorella* biomass is given in Table 4. The excess methanol was recovered using distillation methods and considered for re-use in the transesterification process.

Useful Work Available (Wp) from the Products of the Ultrasound Agitated In-Situ Transesterification of 1 t Chlorella Biomass

The mass flow of the main products following the application of the ultrasound assisted in-situ transesterification process for the conversion of 1 t *Chlorella* was used to estimate the Wp obtainable from the process. The sum of the work available from the products (Wp) for the considered transesterification process is shown in Table 5.

Table 5. Useful work available (Wp) from the ultrasound agitated in-situ transesterification products (for 1 t microalgae biomass converted/ h) using a reacting ratio of methanol to oil of 105:1

Product	Exergy (MJ/kg)	Product quantity (kg)	Wp (MJ)
FAME	40.70	295.58	12030.11
Microalgae oil	40.15	1.00	40.15
Glycerol	18.52	29.01	537.26
CaSO$_4$	0.06	387.22	23.23
		Total Wp (MJ)	12630.75

In-Situ Transesterification with Diethyl Ether Used as a Process Co-Solvent

Net Exergy Cost for the Production of NRRs Used in the Mechanically Agitated In-Situ Transesterification Process Using Diethyl Ether

The cumulative exergy consumed for the production of the materials and energy inputs of the mechanically stirred in-situ transesterification process (1 t microalgae biomass/ h) incorporating the use of diethyl ether (DEE) as a co-solvent are shown in Table 6. The energetic requirement for the stirring of the transesterification reactors was assumed to be same as for the mechanically stirred in-situ process in the section "Mechanically Stirred In-Situ Transesterification". Since the process was considered to be in steady state, and with the diethyl ether reactant completely recovered from the transesterification product stream as described earlier in the section "Integration of Co-Solvents in the In-Situ Transesterification Process", the diethyl ether mass inputs were not considered in the accounting of the NRRs consumed in the process. This was used since the heat requirements for the co-solvent recovery was included in the process energy assessment.

Table 6. Restoration work required to restore the NRRs consumed in the mechanically stirred in-situ transesterification of 1 t microalgae biomass with diethyl ether (DEE) as a co-solvent to its initial state

Process unit	NRRs	Unit	Quantity of NRRs	CNEx (MJ/unit)	Wr (MJ)
Wr for microalgae biomass production					14852.19
Biodiesel production					
	Methanol	kg	46.59	73.08	3404.80
	H2SO4	kg	270.00	9.1	2457.00
	CaO	kg	154.37	10.05	1551.42
	Reactor stirring (electricity)	MJ	0.82	4.17	3.42
	Process heating (diesel)	kg	18.82	53.2	1001.22
			Total Wr (MJ)		23270.05

Useful Work Available (Wp) from the Products of the Ultrasound Agitated In-Situ Transesterification Process with Diethyl Ether

The sum of the useful work available from the main products of the mechanically stirred in-situ transesterification of 1t *Chlorella* biomass with a reacting methanol to oil ratio of 79:1 using DEE as a co-solvent is given in Table 7.

Table 7. Useful work (Wp) available from the main in-situ transesterification products (for 1 t microalgae biomass converted/ h) using a reacting ratio of methanol to oil of 79:1 and diethyl ether as co-solvent

Product	Exergy (MJ/kg)	Product quantity (kg)	Wp (MJ)
FAME	40.70	264.24	10754.57
Microalgae oil	40.15	23.77	954.37
Glycerol	18.52	25.86	478.93
CaSO$_4$	0.06	374.73	22.48
		Total Wp (MJ)	12210.34

In-Situ **Transesterification of Microalgae with Co-Solvent (Diethyl Ether) and Ultrasound Agitation**

Net Exergy Cost for Production of NRRs Used in the Mechanically Agitated In-Situ Transesterification Process Using Diethyl Ether

The minimum work required to restore the degraded NRRs consumed in the in-situ transesterification of *Chlorella* biomass involving the use of DEE as co-solvent with ultrasonic agitation to its initial state is presented in Table 8.

Table 8. Restoration work required for the NRRS degraded in the ultrasonically agitated in-situ transesterification of 1 t microalgae biomass with diethyl ether as a co-solvent

Process unit	NRRs	Unit	Quantity of NRRs	CNEx (MJ/unit)	Wr (MJ)
Wr for microalgae biomass production					14852.19
Biodiesel production					
	Methanol	kg	36.54	73.08	2670.34
	H_2SO_4	kg	279	9.1	2538.90
	CaO	kg	159.52	10.05	1603.18
	Reactor stirring (electricity)	MJ	28.8	4.17	120.10
	Process heating (diesel)	kg	10.46	53.2	556.47
			Total Wr (MJ)		22341.18

Useful Work Available (Wp) from the Products of the Ultrasound Agitated In-Situ Transesterification of 1 t Chlorella Biomass

Similarly, the sum of the product chemical exergies for the transesterification of 1 t *Chlorella* biomass with a molar reacting ratio of methanol to oil of 26:1 is given in Table 9.

Table 9. Useful work available (Wp) from the transesterification products (per 1 t microalgae biomass) using ultrasound agitation and diethyl ether as co-solvent

Product	Exergy (MJ/kg)	Product quantity (kg)	Wp (MJ)
FAME	40.70	284.73	11588.51
Microalgae oil	40.15	10.83	434.82
Glycerol	18.52	27.95	517.63
$CaSO_4$	0.06	387.22	23.23
		Total Wp (MJ)	12564.19

Conventional Extraction and Transesterification of the Microalgae Biomass Lipids

Cumulative Exergy Consumed for the Production of the Resources Used in the Extraction and Transesterification of 1 t Dry Chlorella Biomass

For the conventional process, the material inputs included the extraction solvents required for the initial stripping of the oil feedstocks from the microalgae. The conventional transesterification process was assumed to be conducted with a reacting methanol to oil ratio of 50:1 as described in "Conventional Transesterification Process". The process heat input was based on the minimum hot utility requirement after the application of process integration techniques. The conventional transesterification process was also assessed on the basis of the exergy consumed for the production of the resources to facilitate the conversion of 1 t dry microalgae biomass.

The minimum restoration work required for the NRRs consumed in the extraction and transesterification of 1 t of dried *Chlorella* biomass to its initial environmental state is given in Table 10.

Table 10. Minimum work required to restore the NRRs depleted in the mechanically stirred conventional transesterification of 1 t microalgae biomass to its initial state

Process unit	NRRs	Unit	Quantity of NRRs	CNEx (MJ/unit)	Wr (MJ)
Total Wr for microalgae production					14852.19
Biodiesel production					
	Methanol	kg	30.02	73.08	2193.86
	H_2SO_4	kg	260.00	9.10	2366.00
	Butanol	kg	2.00	60.13	120.26
	CaO	kg	148.67	10.05	1494.13
	Process stirring (electricity)	MJ	1.23	4.17	5.13
	Process heating (diesel)	kg	48.40	53.20	2574.88
			Total Wr (MJ)		23606.45

For the estimation of the cumulative exergy consumed for the butanol production, it was initially considered that the process butanol was derived from crude oil. This section used an exergy cost of 60.13 MJ/kg butanol produced[10]. This estimation is similar to crude oil derived ethanol as put forward by Sorguven and Özilgen (2010). The influence of the use of butanol derived from fermentation processes on the transesterification process renewability would be examined later in the sensitivity analysis section (Sensitivity Analysis of Ir).

Useful Work Available (Wp) from the Products of the Mechanical Stirred Conventional In-Situ Transesterification of 1 t Chlorella Biomass

The sum of the useful work (Wp) available from the main products of the conventional lipid extraction and transesterification process is presented in Table 11.

Table 11. Useful work (Wp) available from the conventional lipid extraction and transesterification products (for 1 t microalgae biomass converted) using a reacting molar methanol to oil ratio of 50:1

Product	Exergy (MJ/kg)	Product quantity (kg)	Wp (MJ)
FAME	40.70	239.85	9761.89
Microalgae oil	40.15	8.17	328.03
Glycerol	18.52	23.52	435.59
$CaSO_4$	0.06	360.90	21.65
		Total Wp	10547.16

---

[10] Butanol was initially considered to be synthesised via the acid catalysed hydration of propylene derived from the autothermic cracking of crude oil.

Renewability Evaluation of the Different Transesterification Processes

Using the renewability indicator (Ir) presented in Eq. 1, Figure 8 compares the calculated Ir for the different transesterification processes assessed in sections "Mechanically Stirred In-Situ Transesterification" – "Conventional Extraction and Transesterification of the Microalgae Biomass Lipids". It must be noted that here the processes are considered to have their process heating supplied via diesel combustion and electricity using fossil fuel based sources. The renewability assessment provided here is thus aimed at representing that of a present day production scenario.

From Figure 8, it can be seen that for all the processes examined, the estimated Ir was < 0. This means that more work was required for the restoration of the process to an equilibrium environmental state than was produced from the main products of the transesterification process. Moreover, this is the most favourable assessment, because the potential work outputs are the maximum possible and the work inputs are the minimum possible. The application of the process modifications considered in this study to the in-situ transesterification process was observed to improve its renewability. The modified processes i.e. involving the use of low frequency ultrasound and co-solvents were seen to be more renewable than the use of the conventional method for microalgae FAME production under the process conditions of a present day scenario.

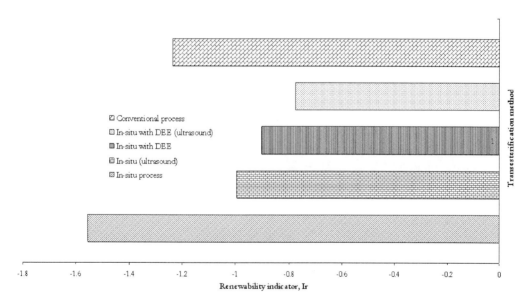

Figure 8. Renewability indicators for the different transesterification processes in a fossil fuel based scenario (per 1 t microalgae biomass converted).

The use and recovery of large reacting alcohol quantities for the in-situ transesterification process appear to negatively affect the overall renewability of this method. The use of a reacting molar ratio of methanol to oil of 315:1 for the mechanically stirred in-situ process (including the heat requirement for the alcohol recovery) was shown to contribute ≈40% of the overall Wr calculated for this process. The in-situ method involving the use of ultrasound and co-solvent aided in reducing the Wr, thus improving the transesterification process renewability. However that method was also seen to exhibit negative renewability (Ir) values mainly due to the source of the resources employed for the process electricity and raw materials. It was initially considered in this study that the resources consumed in the microalgae production and conversion was derived from fossil fuel based sources (i.e. methanol from crude oil). This assumption was based on a present day scenario where most of the industrial inputs are considered to be derived directly or indirectly from such sources. The influence of the source of these raw materials on the process renewability was further explored in the section below. Changes in the electricity generation scheme and the choice of the fuel source for the process heat generation were also examined in the section below since they could also influence the process Ir. The overall influence of these factors on the renewability of the different microalgae production and transesterification processes is presented in the following section.

Sensitivity Analysis of Ir

Electricity Generation Scheme

The cumulative exergy consumed for the production of the process electricity inputs used in the renewability assessment in sections "Mechanically Stirred In-Situ Transesterification" - "Conventional Extraction and Transesterification of the Microalgae Biomass Lipids", was based on typical values for fossil fuels generated electricity of 4.17 MJ/MJ electricity as put forward by Szargut and Morris (1987). Compared to other process inputs, the electricity requirement for the different biomass production and transesterification processes (from fossil fuel sources) were seen to contribute the most to the overall process Wr (≈30-46% of the total Wr). In scenarios where electricity is generated from alternative sources, the exergy consumed for the production of the process electricity would differ. In locations with hydro-generated electricity, the minimum restoration work required for the electricity production is 0.006 MJ/MJ of generated hydro-electricity (Berthiaume et al., 2001).The cumulative exergy cost of 4.17 MJ/MJ electricity associated with fossil fuel based electricity generation was thus replaced with 0.006 MJ/MJ and Tables 2, 4, 6, 8 and 10 re-evaluated. The Wr and Ir of the different processes were then recalculated. Assuming the biomass production and transesterification electricity requirements were met using hydro-electricity, the changes in the calculated Ir are given in Table 12. DEE in Table 12 refers to the diethyl ether co-solvent.

With the use of hydroelectricity, an improvement in the process renewability was observed for all processes compared with the use of fossil fuel generated electricity. Compared to the Wr estimated earlier for the fossil fuel based generated electricity driven processes, percentage reductions in the total Wr of 32.50, 40.95, 43.90, 46.25 and 43.28 % were obtained for the in-situ (mechanical stirred), in-situ (ultrasonicated), in-situ (with diethyl ether), ultrasound agitated in-situ with diethyl ether, and conventional processes respectively using hydro-electricity. With the use of hydro-generated electricity, an improvement in the renewability (compared with fossil fuel based electricity) for all the considered processes was seen, as shown in the Ir column in Table 12. This shows that it is essential to use a renewable electricity source to make the biomass production and transesterification process renewable, here, only the ultrasound agitated in-situ process with diethyl ether was deemed renewable. The use of other sources i.e. wind or solar for electricity generation were not further considered, since it was expected that similar CNEx values as that seen for hydro-electricity would be obtained using these electricity sources for the considered processes.

Table 12. Process renewability with the use of hydro-electricity

Process	Wr	Wp	Ir
In-situ (mechanical stirring)	21218.65	12279.23	-0.73
In-situ (ultrasound)	14915.60	12630.75	-0.18
In-situ with DEE (mechanical stirring)	13069.49	12210.34	-0.07
In-situ with DEE (ultrasound)	12024.09	12564.20	0.05
Conventional process (mechanical stirring)	13404.16	10547.16	-0.27

Source of Process Raw Materials (Inputs)

The evaluation in the sections: "Mechanically Stirred In-Situ Transesterification" – "Conventional Extraction and Transesterification of the Microalgae Biomass Lipids", was based on the assumption that the process inputs were derived from fossil based sources. Under that scenario, the cumulative exergy consumed for the production of the process raw materials was seen to significantly influence the process renewability. For example, the exergy consumed for the methanol used for the different transesterification process was seen to contribute 10.18-25.93% of the total estimated Wr. The influence of using alternatively sourced raw material inputs for the transesterification processes was considered in this section, here, the use of reacting alcohols from biomass fermentation was examined.

The large scale production and use of ethanol via biomass fermentation has been widely detailed in the literature (i.e. Lin and Tanaka, 2006; Hattori and Morita, 2010). The use of ethanol as a reacting alcohol in the transesterification process has also been demonstrated in the literature (i.e. Al-Widyan and Al-Shyoukh, 2002). The use of ethanol as the reacting transesterification alcohol is considered in this section in order to investigate the influence of fermentation sourced alcohols on the process renewabilities. The cumulative exergy cost (CNEx) required for the production of ethanol via the fermentation of sugars of 26.40 MJ/kg ethanol (Sorguven and Özilgen, 2010) was used in this analysis. The same assumption of using biomass derived ethanol was also applied to the conventional process to facilitate an even comparison with the in-situ processes.

Butanol was used as the extraction solvent in this study because it can be produced in large quantities via fermentation with various biomass materials i.e. starch, sugars and non-cellulosic substrates (including algae biomass) as feedstocks (Jones and Woods, 1986). Instead of the use of butanol derived from propylene, as previously evaluated in "Conventional Extraction and Transesterification of the Microalgae Biomass Lipids", the use of butanol derived from fermentation processes on the conventional extraction and transesterification process was examined. The process butanol solvent was assumed to be produced using the acetone-butanol-ethanol (ABE) fermentation process. A conservative exergy cost of 26.40 MJ/kg butanol produced was considered for use in this section. This was based on cumulative exergy consumed for the production of fermentation derived alcohols, as presented by Sorguven and Özilgen, 2010.

The minimum restoration work required for the production of the reacting alcohol, as presented in Tables 2, 4, 6, 8 and 10 were replaced with those described in this section. The Ir of the different processes was then recalculated. This evaluation also considered that the process electricity requirement was met using hydro-generated electricity. The influence of the use of the alcohols obtained from biomass fermentation on the Wr of the processes and their corresponding Ir is shown in Figure 9. The sum of the useful work (Wp) available from the main products of the different transesterification processes is also resented in that figure.

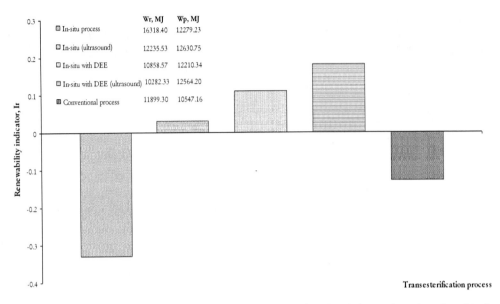

Figure 9. Influence of the use of fermentation alcohols and hydro-electricity on the microalgae biodiesel renewability. DEE refers to the reactions using diethyl ether as a co-solvent.

With the use of fermentation derived alcohols and hydro-electricity for the microalgae production and its conversion to biodiesel, it was observed that all the modified in-situ transesterification processes (i.e. using ultrasound agitation and/or co-solvents) could be deemed renewable. The ultrasound agitated in-situ process using diethyl ether as a co-solvent was however seen to be superior to all the considered transesterification processes with regards to the process renewability. The mechanically stirred in-situ transesterification process involving the use of reacting methanol to oil molar ratios of 315:1 was shown to still be non-renewable with the use of hydro-electricity and fermentation alcohols for the conversion process. The overall process heat demands (for the biomass drying and recovery of the stoichiometrically excess methanol) were seen to contribute significantly to the minimum restoration work calculated for the process. With the use of diesel fuel to meet the process heating needs, the exergy cost associated with the process heating contributed 42.64% of the total process Wr.

The influence of the choice of the process heating fuels was thus carried out in the section: "*Choice of Process Heating Fuels or Technology*" to explore if any improvement in the process renewability can be achieved. The use electrically driven heat pumps to meet the heat requirement of the different processes was also considered in that section.

Choice of Process Heating Fuels or Technology

The use of petro-diesel fuels for the operation of boilers was initially considered to meet the minimum process heating requirement. The influence of the choice of the boiler fuels on the process renewabilities was carried out in this section. This section investigates the effect of the use of the common fossil fuel derived industrial boiler fuels: coal and natural gas on the estimated Ir. The use of wood pellets to meet the process heating demands was also explored.

It was also considered that the process heat is supplied using heat pumps as have been well demonstrated for the heating of water and other fluids (i.e. Carrington and Knopp, 1983; Anderson et al., 1985). The influence of coal and natural gas on the process renewability was investigated due to their widespread use for industrial heat generation. The useful work available using coal and natural gas of 29.00 and 50.75 MJ/kg respectively as presented in Szargut and Morris (1987) were used in this section. The quantity of heating fuel (kg/t microalgae biomass consumed) required to meet the minimum heat demand (after process integration) of the different transesterification processes was estimated with the assumption of a 100% conversion of the fuel to heat. Although this assumption is not realistic, it was used to represent the most optimistic conversion scenarios with the use of such fuel sources in this study. The cumulative exergy cost associated with the production of the heating fuels of 30.44 and 57.87 MJ/kg consumed was used for coal and natural gas respectively (Szargut and Morris, 1987).

The average useful work available (Wp) value of 20.0 MJ/kg for dry softwood (Bilgen et al., 2004) was used to represent the exergy content of the wood pellets. A conservative cumulative exergy cost (CNEx) of 6.4 MJ/kg wood pellets was used for the process renewability evaluation. This CNEx value associated with the wood pellet production was presented in Patzek and Pimentel (2005) and Theka and Obernbergera (2004). This estimated value incorporates the exergy costs for the production of the NRRs consumed in the biomass cultivation, wood felling and collection, the wood chipping and drying process. The wood drying process was considered to be performed using best available technology tube bundle dryers with wood residues used as the process heating fuel (Theka and Obernbergera, 2004).

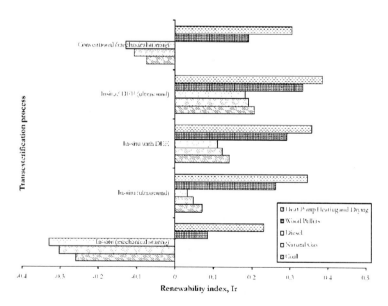

Figure 10. Influence of choice of heating fuel on the renewability of the different microalgae biodiesel processes.

For the heat pump process heating, a conservative thermal conversion efficiency of 40% (Carrington, 2010) was considered for use to estimate the electricity inputs with the use of this method. The process was then assumed to be driven using hydro-electricity. Furthermore, as earlier highlighted, the use of drying technologies such as isothermal heat pump drying

(Catton et al., 2010), which could reduce the estimated exergy cost for the restoration of the NRRs degraded in the biomass drying process was also considered in this section. This was carried out using a SMER of 4 kg moisture/ kWh (Carrington, 2010) as previously mentioned in the section -"Microalgae Biomass Drying".

The estimated Ir with the use of the different process heating fuels (including biomass drying) is shown in Figure 10.

The process renewabilities presented in Figure 10 were estimated on the basis of the use of hydro-generated electricity and fermentation derived reacting alcohol for the biomass and biofuel production process. The electricity and heat generation was assumed to be carried out using separate heat and power (SHP) systems with a 100% operating efficiency. The heating fuels considered in this section was used to meet both the heat demands of the biomass drying process, as well as that of the process heating.

The results in Figure 10 show that the type of heating fuel or technology used to meet the heat requirement of the biomass drying and process heating significantly affects the renewability of the process. The use of heat pumps for the biomass drying and process heating (with a maximum heating temperature of 120°C (Carrington, 2010)), was seen to result in the best improvements for the renewability indicators (Ir) of all the transesterification processes compared to the Irs obtained using heat sources. Wood pellets combustion as a heat source was shown to reduce the total Wr of all the considered transesterification processes by 8.1-16.0% with the process operated with hydro-electricity, compared with the use of fossil fuels for process heating. The modified ultrasound agitated in-situ process using diethyl ether was seen to be the most renewable of the considered transesterification processes. The mechanically stirred in-situ process which was previously considered as non-renewable with fossil based fuels was seen to be renewable with the use of heat pump drying and heating, as well as using wood pellets for the process heating.

Although the use of heat pump heating and drying is not as commercially applied compared to the conventional boiler and drying systems, the use of this technology holds a lot of promise for the long term viability of biodiesel production from microalgae.

Comparing the process Ir of the different fossil fuel derived heating fuels, the use of coal was observed to lead to increased renewabilities when compared to diesel fuel and natural gas. This was attributed to the current ease of coal extraction and processing compared to the other fossil fuels.

CONCLUSION

To facilitate a comparison of the different transesterification processes in this study, the investigated processes were modelled and their renewabilities evaluated.

The transesterification processes were modelled using commercial chemical engineering modelling software, ASPEN Plus®. This involved the use of a fixed microalgae input of 1 t-dry/h and reaction conditions similar to the laboratory experiments. The process models were used to estimate the energy and raw material requirement of the different transesterification processes.

The process renewability was adjudged by comparing the minimum restoration work consumed for the production of the non-renewable resources depleted by the conversion process with the maximum work obtained from the main process products. The results obtained showed that in a present day fossil fuel based scenario, all the processes were deemed to be non-renewable. Here, the ultrasound agitated in-situ transesterification process with diethyl ether as co-solvent was shown to be less non-renewable than the other conversion processes.

Generally, the applied modifications to the in-situ transesterification process (i.e. use of low frequency ultrasound and co-solvent) were demonstrated to exhibit significant improvements in process renewability compared to the mechanically stirred in-situ process. The results of this study show promise with the use of the modified in-situ processes for microalgae biodiesel production.

Altering the electricity generation scheme from a fossil based source to the use of 100% hydro-electricity (or wind electricity) for the biomass production and transesterification methods led to an improvement in the process renewability for all the investigated routes. All of the processes considered were shown to be non-renewable here, with the exception of the ultrasound agitated stirred in-situ process using diethyl ether as co-solvent. A significant improvement in the process renewability (with all the considered modified in-situ processes exhibiting positive Ir values) was obtained with the use of alcohols sourced from biomass fermentation. The use of heat pumps to supply the process heat and facilitate the biomass drying was shown to increase the process renewability compared to the use of commonly used fossil heating fuels and wood pellets. The mechanically stirred in-situ process was shown to be renewable with the use of fermentation derived alcohols for the transesterification process, with the process heat and biomass drying supplied using heat pumps and wood pellets. Although not covered in this study, the use of combined heat and power (CHP) could potentially improve the renewability of the considered processes, since less NRRs would be consumed for the process electricity and heat generation.

To improve the renewability of the biomass production and transesterification processes, energy generation from the post transesterified microalgae residues i.e. Ehimen et al. (2010, 2011) could be examined. The use of the digestion process for CH_4 recovery and waste treatment could potentially provide a significant increase in the useful work available from the process products due to the CH_4 produced. Furthermore, recycling the nutrient rich digestate after the anaerobic digestion process into the microalgae cultivation unit is another scheme which could be used to further increase the overall renewability of microalgae biodiesel production using the in-situ process. The cultivation nutrients could also be supplied using the anaerobically digested manure or organic wastes as have been demonstrated by Levine et al. (2010) for *Neochloris oleoabundans* cultivation aimed for biodiesel production. In that case, the digestate would be considered as the major nutrient supply without the need for the use of in-organic fertilisers as have been reported in this chapter. Post-treated organic industrial and municipal wastes could therefore be considered for use for the microalgae cultivation step.

REFERENCES

Al-Widyan, M.I., Al-Shyoukh, A.O. 2002.*Bioresour. Technol.* 85, 253-256.
Anderson, J.A., Bradford, R.A., Carrington, C.G. 1985. *Int. J. Energ.* Res. 9, 65-89.

Ayres, R.U., Ayres, L.W. 1999. Accounting for Resources, 2. The Life Cycle of Materials. Edward Elgar. Cheltenham, UK.

Ayres, R.U., Ayres, L.W., Masini A. 2006. In: Sustainable Metals Management. Eds. A. von Gleich, R.U. Ayres, S. Göß ling-Reisemann. Springer Publishing, Dordrecht, the Netherlands. pp 141-194.

Becker, E.W. 1994. Microalgae: Biotechnology and Microbiology. Cambridge Studies in Biotechnology. Cambridge University Press, Cambridge.

Becker, E.W., Venkataraman, L.V. 1980.In: Algae Biomass. Eds. G.Shelef, C.J.Soeder. Elsevier/North-Holland Biomedical Press. pp 35-50.

Benefield, L.D.,Randall, C.W. 1980. Biological Process Design for Wastewater Treatment. Prentice-Hall Inc, Englewood Cliffs, New Jersey.

Berthiaume, R., Bouchard, C. 1999. *Trans. Can. Soc. Mech. Eng.* 23: 187-196.

Carlson, E.C. 1996. *Chem. Eng. Prog.* 92, 35-46.

Carrington, C.G. 1978. Heating with a Heat Pump. *Ind. Shellman* 41, 19-21.

Carrington, C.G. 2010. *Personal Communication.* 18 October 2010.

Carrington, C.G., Knopp T.C. 1983. *Int. J. Energ. Res.* 7, 255-267.

Catton, W., Carrington, C.G., Sun, Z.F. 2010.*Int. J. Energ. Res*. DOI: 10.1002/ er.1704.

Chen, P., Min, M., Chen, Y., Wang, L., Li, Y., Chen, Q., Wang, C., Wan, Y., Wang, X., Cheng, Y., Deng, S., Hennessy, K., Lin, X., Liu, Y., Wang, Y., Martinez, B., Ruan, R. 2009. *Int. J. Agric. Biolog. Eng.* 2, 1-30.

Chisti, Y. 2007. *Biotechnol.Adv.* 25, 294-306.

Chisti, Y. 2008. *Trends. Biotechnol.* 26, 351-352.

Demirbaş, A. 2009.*Energ. Sources* A. 31, 163-168.

Ehimen, E.A. 2011. PhD Thesis. University of Otago, Dunedin, New Zealand.

Ehimen, E.A. 2010. *Energ. Sources A*. 32: 1111-1120.

Ehimen, E.A., Connaughton, S., Sun, Z., Carrington, C.G. 2009. *GCB Bioenerg.* 1, 371-381.

Ehimen, E.A., Zhifa, Z.F., Carrington, C.G. 2010. *Fuel*89, 677-684.

Ehimen, E.A., Zhifa, Z.F., Carrington, C.G. 2011 (in press). *Procedia Environ. Sci.*

Fukuda, H., Kondo, A., Noda, H. 2001. *J. Biosci. Bioeng.* 92, 405-416.

Gilliland, E.R. 1940. *Ind. Eng. Chem.* 32, 1220-1223.

Goldman, J.C. 1980. In: Algae Biomass Eds. G.Shelef, C.J.Soeder. Elsevier/North-Holland Biomedical Press. pp 343 – 359.

Graus, W.H.J., Voogt, M., Worrell, E. 2007. *Energ. Policy* 35, 3936-3951.

Haas, M.J., McAloon, A.J., Yee, W.C., Foglia, T.A. 2006. *Bioresour. Technol*. 97, 671-678.

Haas, M.J., Scott, K.M., Foglia, T.A., Marmer, W.N. 2007. *J. Am. Oil. Chem. Soc*. 84, 963-970.

Hattori, T., Morita, S. 2010. *Plant Prod. Sci.* 13, 221-234.

Huntley, M.E., Redalje, D.G. 2007. *Mitigation and Adaptation Strategies for Global Change* 12, 573-608.

ICIS. 2010.http://www.icispricing.com. Accessed on 29-08-2010.

Johnson, M.B., Wen, Z. 2009. *Energ. Fuels*. 23, 5179-5183.

Jones, D.T., Woods, D.R. 1986. *Microbiological Reviews* 50, 484-524.

Kasteren van, J.M.N, Nisworo, A.P. 2007.*Resour. Conserv. Recycl.* 50, 442-458.

Kemp I.C. 2007. Pinch Analysis and Process Integration. A Users guide on Process Integration for the Efficient Use of Energy. 2nd Edition. Butterworth-Heinemann, Oxford, UK. 391 pp.

Levine, R.B., Constanza-Robinson, M.S., Spatafora, G.A. 2010. Biomass.Bioenerg. DOI: 10.1016/j.biombioe.2010.08.035

Li, X., Xu, H., Wu Q. 2007.*Biotechnol. Bioeng.* 98, 764-771.

Lin, Y., Tanaka, S. 2006. *Appl. Microbiol. Biotechnol.* 69, 627-642.

Linnhoff, B., Flower, J.K. 1978. *AiChE J.* 24, 633-642.

Linnhoff, B., Vredeveld, D.R. 1984. *Chem. Engr. Prog.* 80, 30-40.

Marchetti, J.M., Errazu, A.F. 2008. *Energ. Convers. Management* 49, 2160-2164.

Mata, T.M., Martins, A.A., Caetano, C.S. 2010. *Renew. Sustain. Energ. Rev.* 14, 217-232.

Miao, X., Wu, Q. 2006.*Bioresour. Technol.* 97, 841-846.

Myint, L.L., El-Halwagi, M.M. 2009. Clean Technol. Environ. *Policy* 11, 263-276.

Nagle, N., Lemke, P. 1990. *Appl. Biochem.Biotechnol.* 24/25, 355–361.

Neenan, B., Feinberg, D., Hill, A., McIntosh, R., Kerry, K. 1986. Fuels from Microalgae: Technology Status, Potential and Research Requirements. Solar Energy Research Institute. Golden, Colorado. Task No. 4513.20.

Patzek, T.W., Pimentel D. 2005.*Critical Rev. Plant Sci.* 24, 327-364.

Pilavachi, P.A. 2000. *Appl. Thermal Eng.* 20, 1421-1429.

Raghavan, G.S.V., Rennie, T.J., Sunjka, P.S., Orsat, V., Phaphuangwittayakul, W., Terdtoon, P. 2005. *Brazilian J. Chem. Eng.* 22, 195-201.

Reijnders, L. 2008. *Trends. Biotechnol.* 26, 349-350.

Saka, S. 2006. NEDO Biodiesel Production Process by Supercitical Methanol Technologies.The 2nd Joint International Conference on "Sustainable Energy and Environment (SEE 2006)".21-23 November 2006, Bangkok, Thailand. Paper C-043(O).

Saka, S., Kusdiana, D. 2001. Fuel 80, 225-251.

Schwabo, A.W., Dykstrab, G.J., Selkeo, E., Sorenson, S.C., Pryde, E.H. 1988.*J. Am. Oil.Chem. Soc.* 65, 1781-1786.

Sheehan, J., Camobreco, V., Duffield, J., Graboski, M., Shapouri, H. 1998. Life cycle analysis of Biodiesel and Petroleum Diesel for Use in an Urban Bus.U.S. Department of Energy, Office of Fuels Development. NREL/SR-580-24089. Golden, Colorado, US.

Sorguven, E., Özilgen, M. 2010. *Renew. Energ.* 35, 1956-1966.

Sun, Z.F., Carrington, C.G., Bannister, P. 1999. *Drying Technol* 17, 731-743.

Szargut, J., Morris, D.R. 1987. *Energ. Res.* 11, 245-261.

Szargut, J., Morris, D.R., Steward, F.R. 1988. Exergy Analysis of Thermal, Chemical, and Metallurgical Processes. Hemisphere Publishing Corporation. New York, US.

Theka, G., Obernbergera, I. 2004. *Biomass. Bioenerg.* 27, 671-693.

Tyrrell, J. 2010. Personal communication with Dr. John Tyrrell, DKSH NZ, Auckland.

Wang, Y., Wang, X., Liu, Y., Ou, S., Tan, Y., Tang, S. 2009. *Fuel Process. Technol.* 90, 422-427.

Wittmus, H., Olson, L., Lane, D. 1975. *J. Soil. Water. Conserv.* 3, 72-75.

Xu, R., Mi, Y., 2011. J. Am. Oil. Chem. Soc. DOI: 10.1007/s11746-010-1653-3.

Yang, Q., Chen, B., Ji, X., He, Y.F., Chen, G.Q. 2009.*Comm. Nonlinear. Sci. Num. Simul.* 14, 2450-2461.

Zhang, Y., Dube M.A., McLean, D.D., Kates, M. 2003. *Bioresour. Technol.* 89, 1-16.

Chapter 3

TOXICITY AND REMOVAL OF ORGANIC POLLUTANTS BY MICROALGAE: A REVIEW

Lin Ke[1,2], Yuk Shan Wong[3] and Nora F. Y. Tam[2,4]*

[1]College of Environmental Science and Engineering,
South China University of Technology, Guangzhou,
Guangdong 510006, PR China
[2]Department of Biology and Chemistry,
City University of Hong Kong, Tat Chee Avenue, Kowloon Tong,
Kowloon, Hong Kong SAR, PR China
[3]Department of Biology, The Hong Kong University of Science and Technology,
Clear Water Bay, Kowloon, Hong Kong SAR, PR China
[4]State Key Laboratory on Marine Pollution,
City University of Hong Kong, Tat Chee Avenue, Kowloon Tong,
Kowloon, Hong Kong SAR, PR China

ABSTRACT

The ubiquity and persistence of organic pollutants in the aquatic environment are of potential risk to aquatic habitats and human health due to their highly toxic, mutagenic and carcinogenic properties. Bioremediation, a cost-effective technology to remove organic pollutants from contaminated water bodies, involves a number of biological processes, including accumulation, transformation and degradation, mediated mainly by microorganisms such as bacteria, fungi and microalgae. Eukaryotic microalgae are dominant primary producers and play a central role in the fixation and turnover of carbon and other nutrient elements. However, their role in the remediation of aquatic contaminants and the relative importance of the processes involved are much less understood, as compared to bacteria and fungi. Further, most of the studies on the novel remediation technology using microalgae have concentrated on metals and nutrients and much less on organic pollutants. Screening of tolerant species is a crucial step in bioremediation, and the understanding of the response and adaptive changes in

---

* Corresponding author: N.F.Y. Tam, Tel.: +852-3442 7793; fax: +852-3442 0522; E-mail: bhntam@cityu.edu.hk

microalgal cells to toxicity induced by organic pollutants is equally important as it serves as a scientific basis of remediation practices. However, there has been very little published information on the toxicity, resistance and adaptations of microalgae to persistent organic pollutants (POPs). This paper reviews the recent research on the sensitivity, tolerance and adaptations of microalgae to the toxicity of various POPs, including organochlorine pesticides (OCPs), polychlorinated biphenyls (PCBs), polycyclic aromatic hydrocarbons (PAHs), polybrominated diphenyl ethers (PBDEs) and some other emerging environmental endocrine disrupting compounds. The review focuses on the physiological and biochemical changes in microalgae and their relationships with tolerance. The mechanisms and factors affecting the capacity of microalgal cells to remove POPs, as well as the feasibility, limitations and future research directions of employing microalgae in POPs remediation, are also addressed.

1. INTRODUCTION

The release of a large number of organic pollutants, in particular persistent organic pollutants (POPs), into the environment is the consequence of the rapid growth of chemical and agrochemical industries, since early 1900s. POPs, mainly of anthropogenic origin, are of increasing public concern due to their persistent nature in the environment, their bioaccumulation potential and their adverse impact on humans. The twelve Stockholm Convention POPs, notoriously known as the "dirty twelve", include nine organochlorine pesticides (OCPs) of aldrin, toxaphene, chlordane, dieldrin, endrin, heptachlor, dichlordiphenyltrichloroethanes (DDT), mirex and hexachlorobenzene (HCB) and three industrial chemical products or by-products of polychlorinated biphenyls (PCBs), dibenzo-*p*-dioxins and dibenzo-*p*-furans (UNEP-Chemicals, 2004). In addition to the "dirty twelve", there are many thousands of other potentially harmful POPs in the environment, some of which, such as carcinogenic polycyclic aromatic hydrocarbons (PAHs), brominated flame-retardants (BFRs) and certain perfluorinated chemicals (PFCs) have spawned a great deal of research interest (Jones and de Voogt, 1999; Muir and Howard, 2006; Yogui and Sericano, 2009). PAHs are accidental and unintentional by-products generated from various industrial activities and incomplete combustion processes, while BFRs and PFCs are contaminants of emerging concern. In the context of this review paper, the scope of POPs is not limited to the "dirty twelve", but otherwise refer to all organic pollutants that "possess toxic characteristics in a broad sense, are persistent, bioaccumulate, are prone to long-range transboundary atmospheric transport and deposition and are likely to cause significant adverse human health or environmental effects near to and distant from their sources" (Ballschmiter et al., 2002). Special focus is given to OCPs, PCBs, PAHs, PBDEs and some environmental endocrine disrupting compounds (EDCs), such as nonylphenol (NP) and estrogens.

POPs are substances generally characterized by low water solubility and high lipophilicity, favouring partitioning into organic matter and lipids of organisms (Jones and de Voogt, 1999). Their environmental persistence and semi-volatile nature also suggest the potential for long-distance atmospheric transport of POPs before deposition occurs. Because of their known or suspected adverse impacts on humans, the use of many OCPs, such as DDT, chlordane, heptachlor and pentachlorophenol, has been banned or severely restricted in most developed countries. However, their use has continued in developing countries due to its cost-effectiveness and broad-spectrum activities (Hussein et al., 1994; Colborn and Smolen,

1996). It was reported that the cumulative global usage (in metric tons) is 450,000 for toxaphene, 1,500,000 for DDT, 550,000 for hexachlorocyclohexane (HCH) and 720,000 for lindane (Voldner and Li, 1995).

The widespread use of OCPs, coupled with the rising industrial and agricultural activities over the last century, has led to tremendous input of POPs into marine and freshwater ecosystems through various routes of entry, including leakage, effluents, wastewater discharges, atmospheric deposition and terrestrial runoff. It was estimated that oil spills contribute 170,000 metric tons of PAHs to aquatic environments each year (Eisler, 1987), and in China, the riverine input of total PBDEs from the Pearl River Delta to the coastal ocean is about two metric tons per year (Guan et al., 2007). Although with low water solubility, the high affinity of POPs for organic particulates may exert significant adverse effects on aquatic biota through bioaccumulation and trophic transfer in aquatic food webs (Lovell et al., 2002; Kelly et al., 2007).

Microalgae, including cyanobacteria, are the principal primary producers of aquatic ecosystems (Falkowski and Raven, 1997), which form the basis of the food chain for more than two thirds of the world's biomass and are responsible for approximately half of the global photosynthetic activity (Day et al., 1999). Any change in the proliferation of microalgae could induce a global alteration of the equilibrium (Rioboo et al., 2009). Their ubiquitous distribution, their central role in the fixation and turnover of carbon and other nutrient elements and recognition of their heterotrophic abilities have aroused an increasing interest on microalgae among researchers in the last two decades (Borde et al., 2003). Microalgae are known to serve as the raw material for the production of high-value chemicals or biofuels (Costa and de Morais, 2010). In addition, they are of high application potential in the removal of nutrients and hazardous pollutants, such as PAHs, phenolics, organic solvents and heavy metals, from contaminated aquifers (Hosetti and Frost, 1998; Lei et al., 2003; Muñoz and Guieysse, 2006).

There are several biological processes, such as accumulation, transformation and degradation, involved in microalgae-mediated remediation but relatively few published materials are available as compared to bacteria and fungi. Further, most of the studies and review articles on microalgal bioremediation have concentrated on metals and nutrients, with much less attention given to POPs. In bioremediation, screening of tolerant species is the first crucial step and requires a sufficient understanding of the sensitivity, resistance and adaptations of microalgae to POP-induced toxicity.

Such research however, is scattered and shows a lack of systematic summary in and of the literature. This paper attempts to fill these knowledge gaps by providing a systematic review of recent research on (1) sensitivity, tolerance and adaptations of microalgae to the toxicity of a number of important POPs, including OCPs, PCBs, PBDEs, PAHs, NP and oestrogens. The action mode of toxicity, the physiological and biochemical responses and their connections to algal sensitivity/tolerance are the focuses; (2) the removal mechanisms, such as bioconcentration and biotransformation, as well as the factors affecting the removal capacity of microalgae, aiming to evaluate the feasibility of using microalgae in bioremediation of POPs and (3) the limitations of employing microalgae in toxicity bioassays and remediation of POPs. Finally, some thoughts on future research directions in this field are discussed.

2. TOXICITY OF PERSISTENT ORGANIC POLLUTANTS (POPs) TO MICROALGAE

Organochlorine Pesticides (OCPs)

OCPs, such as aldrin, toxaphene, chlordane, dieldrin, endrin, heptachlor, DDT, mirex and HCB (i.e., the nine pesticides in the initial list of the Stockholm Convention on POPs), were extensively used on a worldwide basis after World War II in agriculture and for vector disease control. Although these pesticides met with considerable success, their negative impacts on the environment and human health are also evident and have been of great concern for decades owing to their chronic toxicity, persistency and susceptibility to biomagnification.

The mechanism of action of OCPs is primarily the disruption of the nervous system's function in target organisms or insects (DeLorenzo et al., 2001). Although OCPs are not specifically designed to affect microalgae, they may also be toxic to non-target species despite of the possibly different modes of action, compared to that for target species (Lal and Saxena, 1982). In aquatic microorganisms, pesticides have been shown to interfere with respiration, photosynthesis and biosynthetic reactions, as well as cell growth, division and molecular composition (DeLorenzo et al., 2001). Some algal toxicity results from OCPs studies are summarized in the following.

Among all OCPs, DDT is the most notorious of these chemicals, and thus has received the most extensive research attention since 1960s. DDT has been widely used to control vectors of anthropod disease and agricultural pests (Laws, 1981). DDT and its metabolites (e.g., DDD and DDE) inhibit cell growth and the photosynthetic system in microalgae through the inhibition of electron transport system (Powers et al., 1979; Samson and Popovic, 1988). It has been observed that DDT interacts with the photosynthetic electron transport chain in vascular plants at two sites: in the oxidation of photosystem (PS II) and the electron transport chain (Akbar and Rogers, 1985). DDT was also suggested to be the cause of reduction in fluorescence parameters in microalgae (Chung et al., 2007b). Various different responses were observed among marine and freshwater microalgae to the toxicity of DDT. Lee et al. (1976) found that concentrations of DDT below 10 mg L^{-1} inhibited photosynthesis in marine microalgae. At concentrations between 3.6 and 36 mg L^{-1} p,p'-DDT, the photosynthetic CO_2 fixation in the green microalga *Selenastrum capricornutum* was also inhibited. *Chlorella* sp. was affected by less than 0.3 mg L^{-1} DDT. The blue-green microalga *Anabaena* sp. was found to be sensitive to DDT at 1 mg L^{-1} (Lal and Lal, 1988). The toxicity of DDT and its metabolites caused the alteration in the species composition of marine microalgae (Mosser et al., 1972b). Megharaj et al. (1999) have reported for the first time that DDT and its metabolites altered the microalgal species composition in the soil ecosystem, and the cyanobacteria present in the untreated soil were sensitive to DDT and its metabolites while the unicellular green microalgae belonged to the Division of Chlorophyceae became dominant in the soils receiving high doses of these compounds.

Diverse responses in microalgae to other OCPs have been reported. Aldrin and dieldrin at high concentrations (100 mg L^{-1}) lowered the ATP levels in microalgae but had no effect on their population densities (Clegg and Koevenig, 1974). Aldrin, dieldrin and endrin at low concentrations (up to 1 mg L^{-1}) had no significant effect on respiration of green and blue-green microalgae (Vance and Drummond, 1969). Different pesticides are different in their toxicity to microalgae. Mirex had no effect on marine microalgae, including *Chlorococcum* sp., *Dunaliella tertiolecta* and *Chlamydomonas* sp. at 0.2 μg L^{-1} (Hollister et al., 1975), and 100 mg L^{-1} of both mirex and methoxychlor (each at 50 mg L^{-1}) only slightly inhibited their growth (19 and 17% less than control, respectively) (Kricher et al., 1975). Chlordane was reported to pose toxic effects on the photosynthesis of estuarine phytoplankton at 1 mg L^{-1}, while concentrations ranging from 0.1 to 100 μg L^{-1} significantly stimulated the growth of the green microalga *Scenedesmus quadricauda* (Glooschenko et al., 1979). Glooschenko and Lott (1977) also reported that the growth of a soil microalga *Chlamydomonas* sp. was stimulated by chlordane in concentrations from 0.1 to 50 μg L^{-1}, while chlordane at 100 μg L^{-1} was inhibitory to its cell division. Marine microalgae have been considered to have generally greater tolerance to aldrin than the freshwater forms (Hill and Wright, 1978). The toxicity testing results of OCPs using microalgae are summarized in Table 1.

Polychlorinated Biphenyls (PCBs)

The toxic effects of PCBs on microalgae have been extensively studied since the early 1970s. Previous results have shown that PCBs inhibited the growth and photosynthesis of microalgae (Moore and Harriss, 1972; Mosser et al., 1972a; Harding, 1976), as well as altered the size and species composition of phytoplanktons under laboratory and field conditions (Mosser et al., 1972b; Fisher and Wurster, 1973; Fisher et al., 1974). Mayer et al. (1998) found that the toxicity of PCBs to the unicellular green microalga *S. capricornutum* was related to external aqueous concentrations as well as to the internal algal bound PCB concentrations. Estimates of EC_{50} values for three PCB congeners based on the measured aqueous concentrations ranged within a factor of 17 (14-241 nmol L^{-1}), while the respective range of TCC_{50} (toxic cell concentrations) based on the internal toxicant concentrations was 6.7-14.3 mmol kg^{-1} wet weight, indicating that changing the basis from external to internal concentrations reduced the range by almost one order of magnitude. The authors concluded that the estimated EC_{50} values were mainly due to the differences in bioconcentration behaviour rather than the different intrinsic toxicities, and PCBs exerted their acute toxicity by a relatively non-specific mode of action. PCBs were also found to be toxic to marine picoplanktons, as evidenced by a significant decrease of abundance, biomass and cell size of autotrophic and heterotrophic picoplanktons and of bacterial secondary production (Caroppo et al., 2006). Table 2 summarizes the toxic effect of PCBs on microalgae.

Polycyclic Aromatic Hydrocarbons (PAHs)

The possible toxic effects of PAHs on photosynthetic organisms, such as the inhibition of photosynthesis and respiration, are non-specific (Huang et al., 1997; Tripuranthakam et al., 1999). Zbigniew (1987) found that *S. quadricauda* exposed to various levels of PAHs led to significant decreases in the cell density and chlorophyll a content.

Table 1. Toxicity of organochlorine pesticides (OCPs) to microalgae
(DDT, dichlordiphenyltrichloroethanes; DDE, dichlorodiphenyldichloroethylene; DDD, dichlorodiphenyldichloroethane; EC_{50} and EC_{10}, concentrations of toxicant result in 50% (EC_{50}) or 10% (EC_{10}) reductions of endpoint relative to control at a given exposure time; F_v/F_m, variable fluorescence to maximal fluorescence ratio; OD_{650}, optical density at 650 nm; G6PDH, glucose-6-phorsphate-dehydrogenase; AP, acid phosphatise; -, not available)

Compound	Species name	Endpoint	Exposure time	Value (µg L^{-1} or µg kg^{-1})	References
DDTs					
DDT	Selenastrum capricornutum	EC_{50}, F_v/F_m	96 hr	>500,000	Chung et al., 2007a
		EC_{10}, F_v/F_m	96 hr	4,970-56,950	
		EC_{50}, OD_{650}	96 hr	215,060-276,430	
		EC_{10}, OD_{650}	96 hr	520-3,830	
		EC_{50}, Chlorophyll a	96 hr	127,790-379,450	
		EC_{10}, Chlorophyll a	96 hr	820-5,870	
	Chlorococcum hypnosporum	EC_{50}, F_v/F_m	96 hr	>1,000,000	
		EC_{10}, F_v/F_m	96 hr	>1,000,000	
		EC_{50}, OD_{650}	96 hr	135,410-196,980	
		EC_{10}, OD_{650}	96 hr	14,960-26,240	
		EC_{50}, Chlorophyll a	96 hr	553,570-823,080	
		EC_{10}, Chlorophyll a	96 hr	26,850-39,210	
	Chlorococcum meneghini	EC_{50}, F_v/F_m	96 hr	>700,000	
		EC_{10}, F_v/F_m	96 hr	33,750-157,220	
		EC_{50}, OD_{650}	96 hr	45,910-79,800	
		EC_{10}, OD_{650}	96 hr	2,360-3,970	
		EC_{50}, Chlorophyll a	96 hr	48,510-60,790	
		EC_{10}, Chlorophyll a	96 hr	2,560-2,700	

Table 1. (Continued)

Compound	Species name	Endpoint	Exposure time	Value (µg L^{-1} or µg kg^{-1})	References
	Marine microalgae	Inhibition of photosynthesis	-	<10	Lee et al., 1976
p,p'-DDT	*Anabaena* sp.	Growth inhibition	-	1,000	Lal and Lal, 1978
	S. capricornutum	Inhibition of photosynthesis	-	3.6-36	Lee et al., 1976
DDD		EC$_{50}$, F$_v$/F$_m$	96 hr	>500,000	Chung et al., 2007a
		EC$_{10}$, F$_v$/F$_m$	96 hr	520-57,470	
		EC$_{50}$, OD$_{650}$	96 hr	292,230-366,400	
		EC$_{10}$, OD$_{650}$	96 hr	620-1,100	
		EC$_{50}$, Chlorophyll a	96 hr	200,690-492,000	
		EC$_{10}$, Chlorophyll a	96 hr	790-22,570	
	C. hypnosporum	EC$_{50}$, F$_v$/F$_m$	96 hr	>1,000,000	
		EC$_{10}$, F$_v$/F$_m$	96 hr	>1,000,000	
		EC$_{50}$, OD$_{650}$	96 hr	444,770-466,480	
		EC$_{10}$, OD$_{650}$	96 hr	109,820-125,620	
		EC$_{50}$, Chlorophyll a	96 hr	440,650-607,790	
		EC$_{10}$, Chlorophyll a	96 hr	56,020-57,760	
	C. meneghini	EC$_{50}$, F$_v$/F$_m$	96 hr	>700,000	
		EC$_{10}$, F$_v$/F$_m$	96 hr	28,010-49,490	
		EC$_{50}$, OD$_{650}$	96 hr	138,010-256,450	
		EC$_{10}$, OD$_{650}$	96 hr	11,470-15,230	
		EC$_{50}$, Chlorophyll a	96 hr	109,840-179,020	
		EC$_{10}$, Chlorophyll a	96 hr	13,380-15,320	

Table 1. (Continued)

Compound	Species name	Endpoint	Exposure time	Value (µg L^{-1} or µg kg^{-1})	References
	Scenedesmus quadricauda	25% biomass reduction	10 d	100	Stadnyk et al., 1971
		51% biomass reduction	10 d	1,000	
DDE	*S. capricornutum*	EC$_{50}$, F$_v$/F$_m$	96 hr	>1,000,000	Chung et al., 2007a
		EC$_{10}$, F$_v$/F$_m$	96 hr	1,950-78,330	
		EC$_{50}$, OD$_{650}$	96 hr	379,150-575,740	
		EC$_{10}$, OD$_{650}$	96 hr	490-1,470	
		EC$_{50}$, Chlorophyll a	96 hr	>1,000,000	
		EC$_{10}$, Chlorophyll a	96 hr	840-523,030	
	C. hypnosporum	EC$_{50}$, F$_v$/F$_m$	96 hr	>1,000,000	
		EC$_{10}$, F$_v$/F$_m$	96 hr	44,320-54,170	
		EC$_{50}$, OD$_{650}$	96 hr	246,870-246,060	
		EC$_{10}$, OD$_{650}$	96 hr	41,660-54,720	
		EC$_{50}$, Chlorophyll a	96 hr	688,890-724,190	
		EC$_{10}$, Chlorophyll a	96 hr	71,670-78,240	
	C. meneghini	EC$_{50}$, F$_v$/F$_m$	96 hr	>300,000	
		EC$_{10}$, F$_v$/F$_m$	96 hr	70-1,010	
		EC$_{50}$, OD$_{650}$	96 hr	9,980-34,750	
		EC$_{10}$, OD$_{650}$	96 hr	800-1,730	
		EC$_{50}$, Chlorophyll a	96 hr	8,420-9,290	
		EC$_{10}$, Chlorophyll a	96 hr	1,920-2,230	
DDT, DDD, DDE	*Thalassiosira pseudonana*	Growth rate inhibition		500-1,000	Mosser et al., 1974
Mirex	*Chlorococcum* sp.	No growth inhibition	168 hr	0.2	Hollister et al., 1975
	Dunaliella tertiolecta	No growth inhibition	168 hr	0.2	
	Chlamydomonas sp.	No growth inhibition	168 hr	0.2	

Table 1. (Continued)

Compound	Species name	Endpoint	Exposure time	Value ($\mu g\ L^{-1}$ or $\mu g\ kg^{-1}$)	References
	Nitzschia sp.	No growth inhibition	168 hr	0.2	Hollister et al., 1975
	Thalassiosira pseudonana	No growth inhibition	168 hr	0.2	
	Porphyridium cruentum	No growth inhibition	168 hr	0.2	Kricher et al., 1975
		19% growth reduction		100	
Dieldrin	S. quadricauda	22% biomass reduction	10 d	100	Stadnyk et al., 1971
Aldrin, dieldrin, endrin	Green and bluegreen microalgae	No respiration inhibition		1,000	Vance and Drummond, 1969
Aldrin, dieldrin	Microalgae	No effect on population density		100,000	Clegg and Koevenig, 1974
		ATP content reduction		100,000	
Hexachloro-benzene	S. capricornutum	EC_{50}, G6PDH	48 hr	1,000	Kong et al., 1998
		EC_{50}, AP	48 hr	4,500	
		Dry weight reduction	48 hr	3,000	
		Protein reduction	48 hr	3,000	
Chlorodane	S. quadricauda	Respiration stimulation	24 hr	0.1-100	Glooschenko et al., 1979
		Growth stimulation	12 hr	1-100	Glooschenko and Lott, 1977

**Table 2. Toxicity of polychlorinated biphenyls (PCBs) to microalgae
(NOEC, no observed effect concentration; MEC, concentration caused maximum effect;
LC_{50}, concentration of toxicant results in 50% mortality of a sample population at a given
exposure time; SOD, superoxide dismutase; APx, ascorbate peroxidise;
ECx (x=10, 50, 90), concentrations of toxicant result in x% reductions of endpoint
relative to control at a given exposure time; OR, outside measured range)**

Compound	Species name	Endpoint	Exposure time	Value ($\mu g\ L^{-1}$)	References
Aroclor 1242	*Cylindrotheca closterium*	Growth, RNA and chlorophyll synthesis inhibition	14 d	100	Keil et al., 1971
	Chlamydomonas reinhardtii	Initial growth inhibition; recovered after 22 d	Up to 22 d	0.2-20	Morgan, 1972
Aroclor 1248	*Chlamydomonas* sp.	Growth inhibition	28 d	11-111	Christensen and Zielski, 1980
Aroclor 1254	*Lingulodinium polyedrum*	NOEC, growth	48 hr	100	Leitão et al., 2003
		NOEC, growth	96 hr	25	
		MEC, growth	48 hr	300	
		MEC, growth	96 hr	250	
		LC_{50}	48 hr	144	
		LC_{50}	96 hr	122	
		Induction of SOD, APx and peridinin	48 hr	120	
		No change in MDA content	48 hr	120	
	Thalassiosira pseudonana	Growth inhibition	7 d	25-100	Mosser et al., 1972
		No growth inhibition	7 d	10	
	Skeletonema costatum	Growth inhibition	7 d	10	
	Dunaliella tertiolecta	No growth inhibition	7 d	1,000	
	Euglena gracilis	No growth inhibition	7 d	100	
	C. reinhardtii	Slight growth inhibition	7 d	100	
PCB 31	*Selenastrum capricornutum*	EC_{10}, growth	48 hr	29.1	Mayer et al., 1998
		EC_{50}, growth	48 hr	62.1	
		EC_{90}, growth	48 hr	99.9	
PCB 48		EC_{10}, growth	48 hr	53.1	
		EC_{50}, growth	48 hr	OR	
		EC_{90}, growth	48 hr	OR	
		EC_{10}, growth	70 hr	23.9	
		EC_{50}, growth	70 hr	39.1	
		EC_{90}, growth	70 hr	54.0	
PCB 105		EC_{10}, growth	48 hr	1.8	
		EC_{50}, growth	48 hr	4.6	
		EC_{90}, growth	48 hr	8.2	

Naphthalene (Nap) reduced the cell density of *S. capricornutum* (Gaur, 1988). Petersen and Dahllof (2007) showed that exposure of arctic sediment to 30 µg g^{-1} pyrene (Pyr), a 4-ring PAH, severely reduced the benthic microalgal biomass and their beneficial functions as reflected by a decrease in ^{14}C-incorporation and uptake of ammonium, nitrate and silicate, which indirectly affected the bacterial community in terms of increased oxygen consumption and DNA degradation; but this Pyr-contaminated sediment did not have any toxic effect on benthic microalgae when exposed to UV radiation. Some other studies, however, have reported that the decomposition of PAHs due to photolysis and biodegradation, as well as the accumulation of PAH metabolites could modify the toxicity of PAHs to aquatic organisms (Miller et al., 1988; O'Brien, 1991). The half-lives for photolysis of PAHs in water ranged from 35 min for anthracene (Ant) to 20 hr for benzo[a]pyrene (BaP), and the susceptibility of PAHs to photolysis may reduce the bioavailability and thus the toxicity of some water-soluble PAHs to microalgae or other aquatic organisms (Sims and Overcash, 1983; Djomo et al., 2004). However, it is generally accepted the toxicity of PAHs to plants is enhanced by light, especially UV radiation (Huang et al., 1993; Ren et al., 1994; Arfsten et al., 1996). The toxic effect of a few PAHs, such as BaP and Ant, on microalgae has been found to be photo-enhanced (Cody et al., 1984; Schoeny et al., 1988; Gala and Giesy, 1994; Warshawsky et al., 1995). Such phototoxicity might be due to the production of quinines under light (Warshawsky et al., 1995). Different microalgal species have been shown to respond differently to PAHs. Schoeny et al. (1988) found that the growth of three species of *Ankistrodesmus braunii*, *Scenedesmus acutus* and *S. capricornutum* were inhibited under exposure to BaP while the other four species of *Anabaena flosaquae*, *Chlamydomonas reinhardtii*, *Euglena gracilis* and *Ochromonas malhamensis* were not affected. Similarly, the species-specific toxicity was found for *N*-heterocyclic PAH acridine, and the EC$_{50}$ values obtained from seven microalgal species ranged from 0.08 mg L^{-1} for *Nitzschia sigma* to 0.78 mg L^{-1} for *Chlamydomonas eugametos* and *Navicula salinarum* (Dijkman et al., 1997). Such species variability in the response to PAH toxicity may be a reflection of the morphological, sub-cellular and physiological differences, including cell size, shape, cell wall and membrane composition and growth rates, among different microalgal species (Tam et al., 1998). A summary on the microalgal toxicity due to PAHs is given in Table 3.

The toxicity of PAHs to microalgae appears to be highly associated with their intrinsic physicochemical properties, including water solubility, octanol-water portioning coefficient (K_{ow}), reactivity potential of hydroxylated radicals with PAHs and the Homo-Lumo (highest occupied molecular orbital-lowest unoccupied molecular orbital) gap (Mekenyan et al., 1995). Djomo et al. (2004) reported that the toxic effect of PAHs having low K_{ow} values and low coefficients of volatilization, including Nap, phenanthrene (Phe) and Ant, on microalgal photosynthetic activity was less than that of BaP, which has the highest K_{ow} value. The observed discrepancy in EC$_{50}$ values of Ant (1 mg L^{-1}) and Phe (50 mg L^{-1}), which have identical K_{ow} values (that is, 4.45), may be due to their differences in susceptibility to photo-oxidation; Ant is about three times higher than Phe in terms of photo-oxiation capacity (Djomo et al., 2004). The reactivity potential of PAHs with hydroxylated radicals, which is favoured by the excitation of nitrate ions (Zepp et al., 1987), may also play an important role in the difference in toxicity among different PAH compounds. Previous studies have shown that the more the PAH with reactive hydroxylated radicals, the higher the PAH's toxicity on microalgae.

Table 3. Toxicity of polycyclic aromatic hydrocarbons (PAHs) to microalgae (EC_{50} and EC_{10}, concentrations of toxicant result in 50% (EC_{50}) or 10% (EC_{10}) reductions of endpoint relative to control at a given exposure time; F_v/F_m, variable fluorescence to maximal fluorescence ratio; OD_{650}, optical density at 650 nm; LC_{50}, the concentration of toxicant that results in 50% mortality of a sample population at a given exposure time; NOEC, no observed effect concentration)

Compound	Species name	Endpoint	Exposure time (hr)	Value ($\mu g\ L^{-1}$ or $\mu g\ kg^{-1}$)	References
Naphthalene	Selenastrum capricornutum	EC_{50}, F_v/F_m	96	>1,000,000	Chung et al., 2007a
		EC_{10}, F_v/F_m	96	710-556,100	
		EC_{50}, OD_{650}	96	366,580-470,470	
		EC_{10}, OD_{650}	96	3,350-56,880	
		EC_{50}, Chlorophyll a	96	672,780-760,690	
		EC_{10}, Chlorophyll a	96	17,030-98,110	
		LC_{50}	4	2,960	Millemann et al., 1984
	Scenedesmus subspicatus	EC_{50}, growth	168	54,320-102,900	Djomo et al., 2004
		EC_{10}	168	7,090-9,730	
		NOEC	168	3,960-5,170	
	Nitzschia palea	LC_{50}	4	2,820	Millemann et al., 1984
	Chlorella vulgaris	LC_{50}	48	33,000	Kauss and Hutchinson, 1975
	Chlorococcum hypnosporum	EC_{50}, F_v/F_m	96	>1,000,000	Chung et al., 2007a
		EC_{10}, F_v/F_m	96	89,170-360,000	
		EC_{50}, OD_{650}	96	460,720-488,840	
		EC_{10}, OD_{650}	96	42,970-83,480	
		EC_{50}, Chlorophyll a	96	403,390-439,070	
		EC_{10}, Chlorophyll a	96	84,460-88,740	
	Chlorococcum meneghini	EC_{50}, F_v/F_m	96	>1,000,000	
		EC_{10}, F_v/F_m	96	>1,000,000	
		EC_{50}, OD_{650}	96	494,910-556,940	
		EC_{10}, OD_{650}	96	20,890-35,090	
		EC_{50}, Chlorophyll a	96	454,430-474,620	
		EC_{10}, Chlorophyll a	96	112,550-165,320	
Acenaphthene	S. capricornutum	LC_{50}	96	520	USEPA, 1978

Compound	Species name	Endpoint	Exposure time (hr)	Value (µg L^{-1} or µg kg^{-1})	References
Phenanthrene		EC$_{50}$, F$_v$/F$_m$	96	>500,000	Chung et al., 2007a
		EC$_{10}$, F$_v$/F$_m$	96	2,490-437,300	
		EC$_{50}$, OD$_{650}$	96	7,360-32,320	
		EC$_{10}$, OD$_{650}$	96	330-630	
		EC$_{50}$, Chlorophyll a	96	4,910-8,020	
		EC$_{10}$, Chlorophyll a	96	380-1,440	
		LC$_{50}$	4	940	Millemann et al., 1984
	S. subspicatus	EC$_{50}$, growth	168	34,880-72,500	Djomo et al., 2004
		EC$_{10}$	168	4,200-5,730	
		NOEC	168	2,280-2,970	
	N. palea	LC$_{50}$	4	870	Millemann et al., 1984
	C. hypnosporum	EC$_{50}$, F$_v$/F$_m$	96	>500,000	Chung et al., 2007a
		EC$_{10}$, F$_v$/F$_m$	96	8,350-15,710	
		EC$_{50}$, OD$_{650}$	96	127,590-253,710	
		EC$_{10}$, OD$_{650}$	96	2,830-3,140	
		EC$_{50}$, Chlorophyll a	96	85,590-290,010	
		EC$_{10}$, Chlorophyll a	96	2,930-3,810	
	C. meneghini	EC$_{50}$, F$_v$/F$_m$	96	>500,000	
		EC$_{10}$, F$_v$/F$_m$	96	3,830-20,930	
		EC$_{50}$, OD$_{650}$	96	20,650-38,010	
		EC$_{10}$, OD$_{650}$	96	670-3,810	
		EC$_{50}$, Chlorophyll a	96	8,570-9,270	
		EC$_{10}$, Chlorophyll a	96	2,140-2,530	
Anthracene	S. subspicatus	EC$_{50}$, growth	168	740-1,500	Djomo et al., 2004
		EC$_{10}$	168	5-14	
		NOEC	168	2-4	
Fluoranthene	S. capricornutum	LC$_{50}$	96	54,400	USEPA, 1978
Pyrene		EC$_{50}$, F$_v$/F$_m$	96	>500,000	Chung et al., 2007a
		EC$_{10}$, F$_v$/F$_m$	96	960-413,800	
		EC$_{50}$, OD$_{650}$	96	331,460-430,700	
		EC$_{10}$, OD$_{650}$	96	540-760	

Table 3. (Continued)

Compound	Species name	Endpoint	Exposure time (hr)	Value ($\mu g \ L^{-1}$ or $\mu g \ kg^{-1}$)	References
		EC_{50}, Chlorophyll a	96	>500,000	
		EC_{10}, Chlorophyll a	96	510-2,760	
	S. subspicatus	EC_{50}, growth	168	16,980-20,700	Djomo et al., 2004
		EC_{10}	168	2,110-2,750	
		NOEC	168	1,240-1,600	
	C. hypnosporum	EC_{50}, F_v/F_m	96	>1,000,000	Chung et al., 2007a
		EC_{10}, F_v/F_m	96	49,860-150,000	
		EC_{50}, OD_{650}	96	284,170-400,220	
		EC_{10}, OD_{650}	96	31,610-40,860	
		EC_{50}, Chlorophyll a	96	178,910-711,670	
		EC_{10}, Chlorophyll a	96	27,910-30,620	
	C. meneghini	EC_{50}, F_v/F_m	96	>1,000,000	
		EC_{10}, F_v/F_m	96	>1,000,000	
		EC_{50}, OD_{650}	96	252,750-595,290	
		EC_{10}, OD_{650}	96	14,380-28,910	
		EC_{50}, Chlorophyll a	96	900,860-999,770	
		EC_{10}, Chlorophyll a	96	28,360-45,600	
Benzo[a]pyrene	S. subspicatus	EC_{50}, growth	168	1,280-1,720	Djomo et al., 2004
		EC_{10}	168	20-40	
		NOEC	168	9-11	
Acridine	S. capricornutum	LC_{50}	96	900	Blaylock et al., 1985

Vysotskaya and Bortun (1984) revealed that the reactivity potential of hydroxylated radicals with PAHs followed a descending order of benzo[a]anthracene (BaA) > BaP > Pyr > Ant > Phe, which is similar to that of PAH toxicity, BaP > Pyr > Ant > Phe > Nap, reported by Djomo et al. (2004). A measure of the energy stabilization of toxicants in the form of the Homo-Lumo gap was also found to provide a useful index to explain the persistence, light absorption and photo-induced toxicity of PAHs (Mekenyan et al., 1994, 1995). Their experimental results showed that the photo-induced toxicity of PAHs was consistent within the Homo-Lumo gap "window" of 7.2-70.4 eV (e.g., 6.8 eV for BaP, 7.2 eV for Pyr, 7.2 eV for Ant, 8.2 eV for Phe and 10.1 eV for Nap), with a reversed trend to the ranking of toxicity. Djomo et al. (2004) stated that a compound characterized by less energy required for maintaining a stable structure was more toxic. Some PAH compounds are even more toxic to microalgae than OCPs, such as DDTs. Chung et al. (2007a) developed a 4-day solid-phase microalgal bioassay to compare the toxic effects of DDTs (DDT, DDD and DDE) and PAHs (Nap, Phe and Pyr) on three species of microalgae, *S. capricornutum*, *Chlorococcum hypnosporum* and *Chlorococcum meneghini*, using different endpoint measurements. The bioassay results showed that both PAHs and DDTs were toxic to microalgae in the 4-day exposure test with Phe being the most toxic compound, and fluorescence emission by microalgae was less sensitive than cell density and chlorophyll a concentration. Among these three species, *S. capricornutum* was the most sensitive species for PAHs (EC_{50} of 9.4 mg kg^{-1} for Phe), while *C. meneghini* was the most sensitive one for DDTs (EC_{50} of 20.0 mg kg^{-1} for DDE), and the microalgal tests were more sensitive to PAHs and DDTs toxicities than the US EPA standard seed germination/root elongation test using *Lolium perenne*.

Polybrominated Diphenyl Ethers (PBDEs)

PBDEs are flame-retardants used in a variety of products, such as electronic equipment, household appliances, cables, furniture and textiles (De Wit, 2002). PBDEs are emerging contaminants found ubiquitous in air, water, sediments and biota (Hites, 2004). The global presence of PBDEs has raised considerable concern because they are highly hydrophobic, biodegradation-resistant and susceptible to bioaccumulation (De Wit, 2002). PBDEs have been reported to cause a variety of adverse biological effects on fish and mammals, but the effects on microalgae remain largely unknown (Kallqvist et al., 2006). Limited toxicity data showed that PBDEs were not toxic to microalgae; however, the bioaccumulation of PBDEs in microalgae would cause significant toxic effects on organisms at higher trophic levels. Evandri et al. (2009) showed that 2,2',4,4',5-brominated diphenyl ether (BDE-99) was not toxic to *Raphidocelis subcapitata* at a concentration up to 100 mM after 24 hours of exposure but was toxic to *Daphnia magna* with decreases in survival and impairment of reproduction, probably because this animal fed on *R. subcapitata* that accumulated BDE-99. Kallqvist et al. (2006) also observed that 2,4,2',4'-tetrabromodiphenyl ether (BDE 47) caused growth inhibition in a marine diatom *Skeletonema costatum* at concentrations > 6.6 µg L^{-1} (the no-observed-effect concentration [NOEC]), which was much lower than the realistic environmental concentrations in sewage (ng L^{-1}) and in natural water (pg L^{-1}) (North, 2004; Oros et al., 2005).

Other Emerging Endocrine Disrupting Compounds (EDCs)

EDCs are substances capable of interfering with the reproductive processes in wildlife and humans (Palace et al., 2001; Hunt et al., 2003; Younglai et al., 2005). EDCs have been found in aquatic environments at unacceptable levels of µg L^{-1} and may be a health risk to ecosystems and humans (Kolpin et al., 2002). However, the potential impact of these compounds on non-target organisms that have no endocrine system is largely unknown (Daughton and Ternes, 1999). Microalgae, on one hand have been found to be able to bioconcentrate EDCs, such as natural and synthetic oestrogens and NP, which are very likely to be transferred to higher trophic levels (Lai et al., 2002; Correa-Reyes et al., 2007). On the other hand, EDCs may pose negative effects on microalgae (Escher et al., 2005; Pavlić et al.,

2005). NPs are metabolic products from the microbial degradation of alkylphenol polyethoxylate, a class of non-ionic surfactant used worldwide (Ying et al., 2002; Vazquez-Duhalt et al., 2006). NPs are ubiquitous in the environment with significant concentrations found in air, water, soils and sediments. The responses of microalgae to NP exposure have been reported to be dose and species specific. *Microcystis aeruginosa* was shown to be able to resist high concentrations of NP (1-2 mg L^{-1}), and NP at low concentrations (0.02-0.5 mg L^{-1}) even stimulated its growth and toxin production (Wang and Xie, 2007). Gao et al. (2011) found that *Chlorella vulgaris* exhibited more efficient and rapid responses against NP-induced oxidative stress than *S. capricornutum*. The chlorophyll content of *C. vulgaris* was reduced and the photosynthesis-related gene transcription was inhibited after 24- to 48-hour exposure to NP, with the lowest transcript levels of *psa*B, *psb*A and *rbc*L decreased to only 18.5%, 7% and 4% of the control, respectively (Qian et al., 2010). NP is not the only chemical in question; other EDCs also cause alternations of the growth, cellular content and antioxidant responses in microalgae. Yang et al. (2002) observed an increase of the lipid content in a marine diatom *S. costatum* when exposed to 6.0 mg L^{-1} of 2,4-dichlorophenol. Oestrogens at environmental concentrations had no effect on the growth of *N. incerta*, but the antioxidant responses were induced by NP, bisphenol A (BPA) and 17-α-ethinylestradiol (EE2) (Liu et al., 2010).

Li et al. (2009) reported that BPA changed the growth rate of a diatom *Stephanodiscus hantzschii* at concentrations higher than 3 mg L^{-1}, and chlorophyll a content was inhibited in a similar way as the cell density. Table 4 is a summary of research on the toxicity of EDCs to microalgae.

Combined Toxic Effects of Pollutants

When pollutants are applied in combination, their combined toxicity would be exhibited in one of the three modes of interactions, synergism, addition or antagonism. Baścik-Remisiewicz et al. (2010) found that the combination of cadmium (Cd) and Ant at low concentrations, or at EC_{10} for Cd and Ant, inhibited the growth rate more significantly than when the toxins applied individually, and the interaction type between Cd and Ant was synergistic, but such interaction shifted to antagonistic at high concentrations or at EC_{50}. The impact of the addition of both PCB compound (Aroclor 1260) and heavy metals (zinc (Zn) and lead (Pb)) on picophytoplanktons, in terms of abundance, biomass and bacterial carbon production, was less rapid and less negative than when PCB was applied individually, indicating a possible antagonistic effect between metals and PCBs (Caroppo et al., 2006). Antagonistic and additive effects for mixed Ant, Cd and chloridazone (a herbicide) were the most frequently observed interaction modes on the growth of three *Desmodesmus* species. The mixed toxicants had no effect on the profiles of superoxide dismutase (SOD) izoforms, and chloroplasts might be the main target site where the interaction effects of these toxicants occurred (Zbigniew and Wojciech, 2006). The same research team further observed a multi-fold increase in the SOD activity in these microalgae in response to the combined action of Ant and Cd, suggesting a possible synergistic effect between Ant and Cd (Wojciech and Zbigniew, 2010).

Table 4. Toxicity of endocrine disrupting compounds (EDCs) to microalgae (NP, nonylphenol; BPA, bisphenol A; EE2, 17α-ethynylestradiol; E2, estradiol; LC$_{50}$, concentration of toxicant results in 50% mortality of a sample population at a given exposure time; EC$_{50}$ and EC$_{10}$, concentrations of toxicant result in 50% (EC$_{50}$) or 10% (EC$_{10}$) reductions of endpoint relative to control at a given exposure time; NOEC, no observed effect concentration)

Compound	Species name	Endpoint	Exposure time	Value (µg L^{-1})	References
NP	*Chlamydomonas segnis*	LC$_{50}$	24 hr	1500	Weinberger, 1984
		Photosynthesis inhibition	24 hr	500-750	
	Selenastrum capricornutum	EC$_{50}$	96 hr	410	Ward and Boeri, 1990b
		NOEC	96 hr	92	Naylor, 1995; Weeks et al., 1996
		EC$_{50}$, growth	24 hr	1050	Gao and Tam, 2011
		EC$_{50}$, growth	72 hr	80-3,300	Graff et al., 2003
	Skeletonema costatum	EC$_{50}$, growth	96 hr	27	Ward and Boeri, 1990a; Naylor, 1995; Weeks et al., 1996
		NOEC	96 hr	10	
	Scenedesmus subspicalus	EC$_{50}$, growth	72 hr	1300	Hüls, 1996
		EC$_{10}$, growth	72 hr	500	
		EC$_{50}$, growth	72 hr	870-980	Hense et al., 2003
		EC$_{10}$, growth	72 hr	370-550	
	Chamydomoras reinhardii	Membrane disruption	72 hr	500-700	Weinberger and Rea, 1981
	Chlorella pyrenoidosa	LC$_{50}$	24 hr	1500	
	Chlorella vulgaris	EC$_{50}$, growth	24 hr	1300	Qian et al., 2011
	Microcystis aeruginosa	EC$_{50}$, growth	12 d	670-2960	Wang et al., 2007
	Navicula incerta	EC$_{50}$, growth	96 hr	200	Liu et al., 2010
	Microalgae	Cell division, dry wt	24 hr	2500	Weinberger et al., 1987
	Green algae	NOEC, biomass	96 hr	694	Brooke, 1993a
		Growth	72 hr	750	Moody et al., 1983
BPA	*Stephanodiscus hantzschii*	EC$_{50}$, growth	96 hr	8650	Li et al., 2009
	S. capricornutum	EC$_{50}$, growth	96 hr	2700	Alexander et al., 1988
		EC$_{50}$, growth	96 hr	2500	Stephenson, 1983
	S. costatum	EC$_{50}$, growth	96 hr	1000	Alexander et al., 1988
	N. incerta	EC$_{50}$, growth	96 hr	3730	Liu et al., 2010
EE2		EC$_{50}$, growth	96 hr	3210	
E2		EC$_{50}$, growth	96 hr	>1000	

3. RESISTANCE OF MICROALGAE TO POPs

Detoxification

Once taken up into the cells, organic pollutants are subject to endogenous enzymatic biotransformation and elimination processes, which are so called detoxification (Torres et al., 2008). The biotransformation of organic pollutants in photosynthetic organisms can be divided into three phases (Sandermann, 1994; Komoβa et al., 1995; Thies and Grimme, 1996; Thies et al., 1996). In Phase I, functional groups, such as hydroxyl, amino and thiol groups, are introduced into organic compounds (transformation) catalyzed by microsomal monooxygenase (MO) enzymes or mixed-function oxidases (MFO), including cytochrome P-450 (Cyt P450), cytochrome b5 (Cyt b5) and NADPH cytochrome P450 reductase (P450R) (Thies et al., 1996; Pflugmacher and Sandermann, 1998a; Barque et al., 2002). These more hydrophilic, or more water-soluble, metabolites are reacted with large and often polar compounds like glucoside, glutathione and amino acid in Phase II (conjugation). The covalent addition of these large, polar compounds to xenobiotic compounds facilitates their excretion (Warshawsky et al., 1990; Pflugmacher and Sandermann, 1998b). In photosynthetic organisms, Phase III is characterized by compartmentalization of the conjugated xenobiotic metabolites in the cell wall fraction or in the vacuole (Avery et al., 1995; Alivé et al., 2003; Jabusch and Swackhamer, 2004).

Antioxidant Responses

Photosynthetic organisms, including plants and algae have evolved different antioxidant defence systems to combat oxidative damage due to the excess accumulation of reactive oxygen species (ROS) produced from cellular electron transport activities during photosynthesis and respiration (Asada and Takahashi, 1987). Membrane lipid peroxidation is one of the adverse effects of oxidative stress on plant cells, and the malondialdehyde (MDA) concentration is frequently used to quantify the level of lipid peroxidation (Halliwell and Gutteridge, 1989; De Zwart et al., 1999). In response to the oxidative stress, a large battery of antioxidant enzymes, including SOD, peroxidase (POD) and catalase (CAT) in plants could be induced to scavenge ROS. SOD acts as the first line of antioxidant defence against the toxic effect of elevated levels of ROS by catalyzing the conversion of superoxide ($O_2^{\cdot-}$) to hydrogen peroxide (H_2O_2) and oxygen (O_2), and H_2O_2 can be further decomposed by CAT and POD (Parida et al., 2004). In addition to antioxidant enzymes, a number of compounds with high-reducing potentials, such as glutathione (GSH) and ascorbic acid, also serve as antioxidants to scavenge ROS (Law et al., 1983; Smirnoff, 1996). GSH is the major water-soluble antioxidant in plant and algal cells, which can either directly reduce most of the ROS (Noctor et al., 1998) or act as a cofactor or substrate in enzymatic reactions involved in controlling cellular ROS levels (Okamoto et al., 2001; Pinto et al., 2003). These enzymatic and non-enzymatic defence systems have been used as non-specific markers of a general metabolic shift in response to exposure to a toxicant (Lytle and Lytle, 2001). Various antioxidant responses, including SOD activity and GSH content, in microalgal cells were found induced by stress of single-type pollutants, such as metals or PAHs (Rijstenbil et al.,

1994; Tripathi and Gaur, 2004; Lei et al., 2006), or by combinations of different types of contaminants (Zbigniew and Wojciech, 2006; Wojciech and Zbigniew, 2010). When compared to heavy metals, microalgal antioxidant response data on organic pollutants, singly or combined, are very scarce. Geoffroy et al. (2002) found that the antioxidant enzymes, CAT, ascorbate peroxidase (APx), glutathione reductase (GR) and glutathione-S-transferase (GST) in a microalga *S. obliquus* were significantly stimulated by a commonly used herbicide, oxyfluorfen, while the stimulation of antioxidant responses by diuron (another commonly used herbicide) was only observed with GR and GST, with no significant changes in the other two enzymes. The antioxidant responses in both *C. vulgaris* and *S. capricornutum* were induced by NP, but the former species was more effective in scavenging NP-induced ROS as it had higher GSH content and its CAT and POD responses were more sensitive, with a greater extent of responses within a short period of time, i.e., a shorter cause-effect time than the latter microalga (Gao and Tam, 2011). Some of the antioxidant parameters may have a connection to the process of PAH metabolism and may play an important role in microalgal tolerance/adaptations to PAH-induced toxicity. Lei et al. (2006) reported that when exposed to pyrene (Pyr), the GST activity in *Scenedesmus platydiscus* and *S. capricornutum* was significantly increased but such an increase was not observed in *S. quadricauda* and *C. vulgaris*, which were either lacking in ability to metabolize Pyr (*C. vulgaris*) or Pyr-sensitive (*S. quadricauda*). Some antioxidants, including activities of GPx, SOD and CAT in all Pyr-treated species, did not show much difference to those in the control samples. The authors suggested that Pyr-enhanced GSH metabolism might be important in Pyr biotransformation. When free *S. capricornutum* was exposed to mixed heavy metals and PAHs, the antioxidant responses were induced, but the induction was exposure time and heavy metal dose dependent, with GSH content being the most sensitive, followed by SOD and POD activity was the least sensitive parameter (Wang et al., 2011). Studies on the antioxidant response in microalgae to POPs and other environmental stressors are summarized in Table 4.

REMOVAL OF POPS BY MICROALGAE

Bioconcentration

There are a number of microbial processes involved in the removal of POPs, including biosorption (adsorption onto the cell surface and absorption into the cells) and cellular transformation, degradation and storage. The concentrations of POPs accumulated in microalgal cells can be several orders of magnitude higher than in the surrounding medium (Harding and Phillips, 1978; Swackhamer and Skoglund, 1991). This process is simple, passive diffusion and is referred to as bioconcentration (Schwarzenbach et al., 2003). Microalgae are effective in bioconcentrating hydrophobic POPs and the rate of bioconcentration is compound and species dependent (Table 5). The bioconcentration of POPs in microalgae is of fundamental importance as many POPs, such as PAHs, PCBs, DDT and PBDEs show a high potential for biomagnification in food webs and can cause harm to top predators, particularly humans and wildlife (Munoz et al., 1996; D'Adamo et al., 1997; Borga et al., 2005; Kelly et al., 2007; Evandri et al., 2009).

Table 5. Antioxidant responses in microalgae to different environmental stresses (SOD, superoxide dismutase; CAT, catalase; GSH, glutathione; MDA, malondialdehyde; POD, peroxidase; APx, ascorbate peroxidase; GPx, glutathione peroxidase; GST, glutathione-S-transferase; GR, glutathione reductase; S: Stimulation; I: Inhibition; N: No change)

Stress type	Species name	Antioxidant	Response	References
UV	*Chlamydomonas nivalis*	Phenolic compounds	S	Duval et al., 2000
UV-B	*Platymonas subcordiformis*	SOD, CAT, GSH	I	Zhang et al., 2005
		MDA	S	
UV-B	*Chlamydomonas* sp.	SOD, POD, CAT	S	Wang et al., 2009
Irradiation	*Cladophora glomerata*	CAT, APx	S	Choo et al., 2004
	Enteromorpha ahlneriana	SOD	N	
Cu	*Scenedesmus bijugatus*	GPx, GST	S	Nagalakshmi & Prasad, 2001
		GR, GSH	I	
Cu	*Scenedesmus vacuolatus*	SOD, CAT, GSH, MDA	S	Sabatini et al., 2009
	Chlorella kessleri	SOD, CAT, GSH, MDA	I	
Cu	*Chlorella vulgaris*	MDA	S	Mallick, 2004
Tributyltin chloride	*Cladophora* sp.	GST	S	Pflugmacher et al., 2000
Microcystins	*Scenedesmus quadricauda*	GST, GPx, MDA	S	Mohamed, 2008
		GSH	I	
Pyrene	*Selenastrum capricornutum*	GPx, SOD, CAT, MDA	N	Lei et al., 2006
		GST	S	
Arochlor 1254	*Lingulodinium polyedrum*	SOD, APx, peridinin	S	Leitão et al., 2003
		β-carotene	N	
Nonylphenol	*Microcystis aeruginosa*	SOD, GST, GSH	S	Wang & Xie, 2007
	C. vulgaris, S. capricornutum	GSH, CAT, POD	S	Gao & Tam, 2011
PAHs and heavy metals	*S. capricornutum*	GSH, SOD, POD	S	Wang et al., 2011

Many reports demonstrate that the bioconcentration of a compound in microalgae is positively associated with the octanol-water partition coefficient (K_{ow}) (Casserly et al., 1983; Swackhamer and Skoglund, 1993). The variability in bioconcentration of POPs among different microalgal species may be due to the differences in their lipid contents (Canton et al., 1977). The possession of different metabolic enzymatic systems in different microalgal species was reported to be a possible reason of species-specific removal or bioconcentration of BaP (Kirso and Irha, 1998).

Table 6. Bioconcentration of persistent organic pollutants (POPs) in microalgae (BCF, bioconcentration factor; OCPs, organochlorine pesticides; DDT, dichlordiphenyltrichloroethanes; PAHs, polycyclic aromatic hydrocarbons; PCBs, polychlorinated biphenyls; -, not available)

Compound	Species name	Concentration in medium (µg L^{-1})	Log(BCF)	References
OCPs				
DDT	Microcystis sp.	1,000	2.4	Vance & Drummond, 1969
	Anabaena sp.	1,000	2.4	
	Scenedesmus sp.	1,000	2.1	
	Oedogonium sp.	1,000	2.3	
	Amphidinium carteri	Ambient	4.9	Cox, 1972
		0.7	4.0	Rice & Sikka, 1973
	Syracosphaera sp.	Ambient	4.4	Cox, 1972
	Anacystis nidulans	1,000	2.9	Gregory et al., 1969
	Scenedesmus obliquus	1,000	2.8	
	Euglena gracilis	1,000	2.0	
	Skeletonema costatum	0.7	4.6	Rice & Sikka, 1973
	Cyclotella nana	0.7	4.8	
	Tetraselmis chuii	0.7	3.8	
	Ankistrodesmus amalloides	0.72	2.8	Neudorf & Kban, 1975
	Cylindrotheca closterium	100	2.3	Keil & Priester, 1969
	Paramecium bursaria	1,000	2.4	Gregory et al., 1969
	Selenastrum capricornutum	6	4.6-5.5	Halling-Sørensen et al., 2000
	Chlorella vulgaris	10	2.6	Kikuchi et al., 1984
	Nitzschia closterium	10	4.9	Kikuchi et al., 1984
Dieldrin	Microcystis sp.	1,000	2.1-2.4	Vance & Drummond, 1969
	Anabaena sp.	1,000	2.1-2.4	
	S. obliquus	1-20	3.1	Reinert, 1972
	Scenedesmus sp.	1,000	2.1-2.4	Vance & Drummond, 1969

Table 6. (Continued)

Compound	Species name	Concentration in medium (µg L^{-1})	Log(BCF)	References
	Oedogonium sp.	1,000	2.1-2.4	
	A. amalloides	0.72	2.5	Neudorf & Kban, 1975
	Zalerion maritimum	10,000-100,000	3.3	Sguros & Quevedo, 1978
Chlordane	Scenedesmus quadricauda	0.1-100	3.8-4.2	Glooschenko et al., 1979
	A. amalloides	0.72	3.7	Moore et al., 1977
Endrin	S. quadricauda	1,000	2.2	Vance & Drummond, 1969
Hexachlorobenzene	Scenedesmus sp.	80	1.6-2.2	Koelmans & Sánchez Jiménez, 1994
PAHs				
Naphthalene	S. capricornutum	-	4.1	Casserly et al., 1983
Phenanthrene		1,200	4.2-4.5	Halling-Sørensen et al., 2000
Phenanthrene		-	4.4	Casserly et al., 1983
Pyrene		-	4.6	
Benzo[a]pyrene	C. vulgaris	5	3.2-3.4	Janikowska & Wardas, 2002
		50	3.1-3.4	
		500	3.2-3.5	
PCBs				
PCB-31	S. capricornutum	310	4.5-5.3	Halling-Sørensen et al., 2000
PCB-47	Nannochloropsis oculata	50	2.97	Wang et al., 1998
		500	2.71	
PCB-49	S. capricornutum	16	4.3-5.3	Halling-Sørensen et al., 2000
PCB-105		9	5.0-5.7	
PCB-153		1	4.5-5.5	

Many abiotic and biotic factors that affect microalgal physiology and physical property of the compound also play a role in the variability of bioconcentration or bioavailability of POPs. For instance, a change in nitrogen status was reported to increase the lipid content and affected the bioconcentration of hydrophobic organic compounds in *S. capricornutum* (Halling-Sørensen et al., 2000). Excretion of microalgal exudates or dissolved organic carbon

(DOC) would increase the POPs content in the aqueous phase and lower the partitioning or bioavailability of POPs to microalgal cells, leading to an underestimation of the role of the microalgae in bioconcentration (Sijm et al., 1995; Koelmans et al., 1999; Gerofke et al., 2005).

Biotransformation and Biodegradation

This section focuses on the species variability in biotransformation, biodegradation and the detoxification process of POPs, especially OCPs, PAHs and EDCs, in freshwater and marine microalgae. The microalgae-mediated biotransformation and biodegradation of POPs, such as PAHs and OPCs, have been studied and reviewed in the past few decades (e.g., Lal and Saxena, 1982; Cerniglia, 1992; Semple et al., 1999). Juhasz and Naidu (2000) summarized the 19 microalgal species capable of degrading Nap, Phe, Pyr and BaP. Cerniglia et al. (1980) showed that 18 microalgal species from different taxonomic groups, including cyanobacteria, diatoms, green, red and brown microalgae, were able to metabolize Nap. The degradability of PAHs varies among different species. *S. capricornutum*, as well as *A. flosaquae, A. braunii, C. reinhardtii, E. gracilis, O. malhamensis* and *S. acutus*, also metabolized BaP to different extents (Schoeny et al., 1988). *Agmenellum quadruplicatum* PR-6 degraded about 2.4% of the Phe added to the medium (Narro et al., 1992). Lei et al. (2002) reported that six species, namely *Chlamydomonas* sp., *C. miniata, S. platydiscus, S. quadricauda, S. capricornutum* and *Synechocystis* sp., degraded 34 to 100% of Pyr (0.1 mg L^{-1}) in seven days. Many metabolic reactions were involved in PAH degradation by eukaryotic microalgae and some pathways were similar to those in prokaryotic bacteria. For instance, *S. capricornutum* was found to metabolize BaP into four different *cis*-dihydrodiols of BaP, indicating that this microalga, similar to bacteria, degraded BaP through the dioxygenase pathway (Juhasz and Naidu, 2000). A cyanobacterium, *A. quadruplicatum* PR-6, transformed Phe through the monooxygenase pathway and produced Phe *trans* -9,10-dihydrodiol as intermediates (Narro et al., 1992). Another cyanobacterium, *Oscillatoria* sp., with photoautotrophic growth oxidized 2-ring Nap to 1-hydroxynaphthanene and naphthalene *cis*-1,2-dihydrodiol (Cerniglia et al., 1980). These degradation studies based on PAHs reflect that microalgae could metabolize PAHs through both monooxygenase and dioxygenase pathways and the latter one is generally considered as bacterial mediated biotransformation.

In addition to PAHs, the important metabolic reactions involved in microalgal biotransformation and degradation of OCPs include reductive dechlorination, dehydrochlorination, oxidation and isomerization (reviewed by Lal and Saxena, 1982). The biotransformation of emerging EDCs by microalgae has attracted increasing research interests in recent decades. Lai et al. (2002) reported that estradiol valerate was hydrolyzed to estradiol and then to estrone within three hours of incubation and 50% of the estradiol was further metabolized to an unknown product under light, while other parent and intermediate compounds, such as estrone, hydroxyestrone, estriol and ethinylestradiol were relatively stable in the culture of *C. vulgaris*. The capability of biotransformation of EE2 by 11 microalgae was compared, and *S. capricornutum, S. quadricauda, S. vacuolatus* and *A. braunii* were found to be able to biotransform the substrate and the metabolites were different from the other 7 species (Greca et al., 2008). Previous studies on the removal and biotransformation of POPs by microalgae are summarized in Table 7.

Table 7. Biotransformation and removal of persistent organic pollutants (POPs) by microalgae (DDT, dichlordiphenyltrichloroethanes; DDE, dichlorodiphenyldichloroethylene; DDD, dichlorodiphenyldichloroethane; PCBs, polychlorinated biphenyls; PAHs, polycyclic aromatic hydrocarbons; Phe, phenanthrene; Fla, Fluoranthene; Pyr, pyrene; EDCs, endocrine disrupting compounds; NP, nonylphenol; BPA, bisphenol A; EE2, 17α-ethynylestradiol; E2, estradiol; dw, dry weight; -, not available; R, removal from medium; BT, biotransformation; BA, bioaccumulation)

Compound	Species name	Initial cell density	Concentration in medium (µg L^{-1})	Exposure time	R (%)	BT (%)	BA (%)	References
DDTs								
DDT	Ankistrodesmus amalloides	3x10^4 cells mL^{-1}	0.72	30 d	-	DDE: 3.5; DDD: 0.8	-	Neudorf & Khan, 1975
	Cyclotella nana	17 mg dw L^{-1}	0.7	2 hr	62.8	-	-	Rice & Sikka, 1973
		8 mg dw L^{-1}	0.7	2 hr	48.8	-	-	
	Isochrysis galbana	39 mg dw L^{-1}	0.7	2 hr	50.0	-	-	
		19 mg dw L^{-1}	0.7	2 hr	42.0	-	-	
	Olisthodiscus luteus	108 mg dw L^{-1}	0.7	2 hr	49.7	-	-	
		54 mg dw L^{-1}	0.7	2 hr	38.0	-	-	
	Amphidinium carteri	66 mg dw L^{-1}	0.7	2 hr	28.4	-	-	
		33 mg dw L^{-1}	0.7	2 hr	23.7	-	-	
	Tetrasalmis chuii	106 mg dw L^{-1}	0.7	2 hr	55.4	-	-	
		53 mg dw L^{-1}	0.7	2 hr	33.3	-	-	
	Skeletonema costatum	29 mg dw L^{-1}	0.7	2 hr	93.8	-	-	
		15 mg dw L^{-1}	0.7	2 hr	55.9	-	-	
	Dunaliella tertiolecta	10^4 cells mL^{-1}	80	14 d	97.8	-	-	Bowes, 1972
	C. nana	10^4 cells mL^{-1}	80	14 d	96.6	-	-	
	Thalassiosira fluviatilis	10^4 cells mL^{-1}	80	12 d	94.4	-	-	
	A. carteri	10^4 cells mL^{-1}	80	14 d	93.2	-	-	

Compound	Species name	Initial cell density	Concentration in medium (μg L^{-1})	Exposure time	R (%)	BT (%)	BA (%)	References
p,p'-DDE	Chlorella pyrenoidosa	-	-	-	82.0	-	-	Sodergren, 1971
Dieldrin	Cyclotella cryptica	-	0.1	1 hr	85-90	-	-	Werner & Morschel, 1978
	Chlamydomonas spp.	-	0.1	27 hr	~60	-	-	
	Nitzschia sp.	-	0.1	1 hr	85-90	-	-	
	C. nana	20 mg dw L^{-1}	1.7	2 hr	13	-	-	Rice & Sikka, 1972
	I. galbana	20 mg dw L^{-1}	1.7	2 hr	15.5	-	-	
	O. luteus	20 mg dw L^{-1}	1.7	2 hr	13	-	-	
	A. carteri	20 mg dw L^{-1}	1.7	2 hr	2.3	-	-	
	T. chuii	20 mg dw L^{-1}	1.7	2 hr	16	-	-	
	S. costatum	20 mg dw L^{-1}	1.7	2 hr	42	-	-	
PCBs								
Clophen A50	C. pyrenoidosa	-	-	-	88	-	-	Sodergren, 1971
PCB mix	Scenedesmus acutus	-	40.5	7-25 d	45-100	-	-	Véber et al., 1980
	Chlamydomonas geitleri	-	40.5	6-7 d	64-100	-	-	
PAHs								
Phe	Skeletonema costatum	~10^5 cells mL^{-1}	950	168 hr	60	16	44	Hong et al., 2008
	Nitzschia sp.	~2.5x10^3 cells mL^{-1}	300	168 hr	95	55	40	
Fla	S. costatum	~10^5 cells mL^{-1}	170	168 hr	50	8	42	

Table 7. (Continued)

Compound	Species name	Initial cell density	Concentration in medium (μg L^{-1})	Exposure time	R (%)	BT (%)	BA (%)	References
	Nitzschia sp.	~2.5x10^3 cells mL^{-1}	100	168 hr	90	2	88	Lei et al., 2007
	Chlorella vulgaris	34.5 mg dw L^{-1}	1,000	24-168 hr	65.4	46.7	18.7	
	Scenedesmus platydiscus	34.5 mg dw L^{-1}	1,000	24-168 hr	80.2	47.6	32.6	
	S. quadricauda	34.5 mg dw L^{-1}	1,000	24-168 hr	79.9	55.5	24.4	
	Selenastrum capricornutum	34.5 mg dw L^{-1}	1,000	24-168 hr	88.9	80.9	8	
Pyr	C. vulgaris	34.5 mg dw L^{-1}	1,000	24-168 hr	71.5	41.3	30.2	
	S. platydiscus	34.5 mg dw L^{-1}	1,000	24-168 hr	76.1	48.5	27.6	
	S. quadricauda	34.5 mg dw L^{-1}	1,000	24-168 hr	75	35.2	39.8	
	S. capricornutum	34.5 mg dw L^{-1}	1,000	24-168 hr	83.2	73.5	9.7	
Phe+Fla	S. costatum	~10^5 cells mL^{-1}	170 each	168 hr	Phe: 50; Fla: 47	Phe: 38; Fla: 10	Phe: 12; Fla: 37	Hong et al., 2008
	Nitzschia sp.	~2.5x10^3 cells mL^{-1}	100 each	168 hr	Phe: 90; Fla: 90	Phe: 40; Fla: 2	Phe: 50; Fla: 88	
Fla+Pyr	C. vulgaris	34.5 mg dw L^{-1}	1,000 each	24-168 hr	Fla: 76.1; Pyr: 74.8	Fla: 55.2; Pyr: 48.1	Fla: 20.9; Pyr: 26.7	Lei et al., 2007
	S. platydiscus	34.5 mg dw L^{-1}	1,000 each	24-168 hr	Fla: 87.8; Pyr: 85.3	Fla: 54; Pyr: 51.5	Fla: 33.8; Pyr: 33.7	
	S. quadricauda	34.5 mg dw L^{-1}	1,000 each	24-168 hr	Fla: 87.2; Pyr: 81.1	Fla: 64.4; Pyr: 46.7	Fla: 22.8; Pyr: 34.8	

Table 7. (Continued)

Compound	Species name	Initial cell density	Concentration in medium (μg L^{-1})	Exposure time	R (%)	BT (%)	BA (%)	References
	S. capricornutum	34.5 mg dw L^{-1}	1,000 each	24-168 hr	Fla: 90.8; Pyr: 89.1	Fla: 79.6; Pyr: 76.1	Fla: 11.2; Pyr: 13.1	
Compound	Species name	Initial cell density	Concentration in medium (μg L^{-1})	Exposure time	R (%)	BT (%)	BA (%)	References
PAH mix		10^7 cells mL^{-1}	Phe: 76,600; Fla: 20,900; Pyr: 8,300	96 hr	Phe: 88.2; Fla, Pyr: 100	Phe: 69.7; Fla, Pyr: 100	Phe: 18.5; Fla, Pyr: 0	Chan et al., 2006
		10^7 cells mL^{-1}	Phe: 8,400; Fla: 9,500; Pyr: 9,700	72 hr	Phe, Fla, Pyr: 100	Phe: 85.5; Fla, Pyr: 100	Phe: 14.5; Fla, Pyr: 100	
EDCs								
NP	*Navicula incerta*	10^5 cells mL^{-1}	1-1,000	96 hr	-	3.96-25.5	-	Liu et al., 2010
	C. vulgaris	1 mg L^{-1} chl. a	1,000	24 hr	86	47	39	Gao & Tam, 2011
		1 mg L^{-1} chl. a	1,000	96 hr	94.1	61.4	32.7	
		1 mg L^{-1} chl. a	2,000	24 hr	89.6	32.5	57.1	
		1 mg L^{-1} chl. a	2,000	96 hr	98	85.4	12.6	
	S. capricornutum	1 mg L^{-1} chl. a	1,000	24 hr	86.3	24.5	61.8	
		1 mg L^{-1} chl. a	1,000	96 hr	97	85.1	11.9	
		1 mg L^{-1} chl. a	2,000	24 hr	94.6	12.3	82.3	
		1 mg L^{-1} chl. a	2,000	96 hr	97	72.8	24.2	
BPA	*N. incerta*	10^5 cells mL^{-1}	1-1,000	96 hr	-	3.7-37.8	-	Liu et al., 2010
BPA	*Stephanodiscus hantzschii*	1.3x10^4 cells mL^{-1}	10-9,000	16 d	26-99	-	-	Li et al., 2009
EE2	*N. incerta*	10^5 cells mL^{-1}	1-1,000	96 hr	-	4.7-31.3	-	Liu et al., 2010
E2	*N. incerta*	10^5 cells mL^{-1}	1-1,000	96 hr	-	51.8-58.3	-	

There have been extensive studies on the transformation or degradation of POPs by microalgae; however, most were conducted in laboratory conditions with a single species and a single contaminant. Research on using mixed microbial species is limited. The algal-algal or algal-bacterial interactions, as well as the presence of other types of pollutants, in the removal of POPs are basically unknown. Ke et al. (2010) recently found that mixed heavy metals (Cd, Zn, copper and nickel) enhanced the ability of *S. capricornutum* in removing fluorene (Flu) and Phe (low molecular weight PAHs) but had no effect on fluoranthrene (Fla), Pyr and BaP (high molecular weight PAHs) in the medium contaminated with this PAH mixture, indicating different contaminants would interact and affect the removal ability of microalgae. In nature, POPs do not exist in isolation but among a variety of pollutants of different kinds. The combined effect of mixed pollutants on the performance of microalgae in the degradation and removal of POPs is largely unknown and should be further evaluated. As a novel technology, the application of microalgae in the bioremediation of POP-contaminated water remains at its infant stage as many of the influencing factors affecting the removal, particularly under field conditions, and the scale-up problems are still unknown. Future research should focus on the removal of mixed contaminants under field conditions, the interactions between different microalgal species and the possible use of mixed species.

Conclusion

The ecological importance of microalgae in ecosystems has been fully recognized. The difference in the adaptation of different microalgae to the changing environment suggests that the sensitivity or tolerance of microalgae to pollutants would be species-specific. Considering this intrinsic property, microalgae not only have been widely used in toxicity bioassays, but also have a promising potential in the bioremediation of contaminated aquifers and wastewater.

Use of Microalgae in Toxicity Assays

One of the challenges to use microalgae in the assessment of POPs toxicity, as suggested by Torres et al. (2008), may be that the low levels of pollutants readily present in individual cells may not be sufficient to reduce growth of microalgae, whereas biomagnification through the food web may cause drastic impacts on organisms at higher trophic levels. The selection of sensitive biomarkers, therefore, appears to be of equal importance to the selection of appropriate microalgal species. It is generally accepted that the species that will be used in toxicity bioassays should be sensitive to toxicant exposure. Compared to toxicity data using animals, the phytotoxicity data are very limited and part of them are derived from non-sensitive species and, perhaps, from non-sensitive endpoints or response measurements (Lytle and Lytle, 2001). Antioxidant responses in microalgae have been considered as a sensitive biomarker and have received increasing research attention. However, the database in this regard is very limit, especially for organic pollutants. Research also showed that the sensitive species identified as a potential bioindicator by the traditional screening test might not be the species having the most sensitive antioxidant biomarkers, as the metabolic activity of a species would affect its sensitivity to the toxicant (Tang et al., 1998; Lei et al., 2003). Therefore, the selection criteria of microalgal species should be re-evaluated. The feasibility of using antioxidant responses as biomarkers is in need of further research.

Another challenge is the difficulty of applying free microalgal cells in risk assessments in field conditions or in natural contaminated environments. A possible alternative is to use immobilized cells, entrapping the microalgae in a matrix. However, the effects of the matrix in protecting the cells and the possible changes in biological and biochemical parameters when the cells are entrapped are not clear due to limited available data, which also require in-depth studies.

Bioremediation Using Microalgae

In the bioremediation perspective, the biotransformation and biodegradation ability a microalga possesses is one of the most crucial criteria in the screening process, since the ultimate goal of the bioremediation of POPs is to breakdown (or ideally mineralize) the complex compound molecules into simple, harmless forms, such as H_2O and CO_2. However, this goal is far from being achieved. An important drawback is that microalgae normally could not completely degrade POPs to harmless forms; instead, transformation or partial degradation is the dominant process. A promising manipulation is to use mixed species, such as a combination of bacteria and microalgae. Microalgae are responsible for oxygen supply and the initial degradation, while bacteria are responsible for further degradation and mineralization. Although studies using combined microalgae and bacteria have been reported, none of them have specifically targeted POPs, and this merits further investigations.

Another knowledge gap is that the combined effects of different kinds of pollutants on the degradation and biotransformation processes in microalgae are largely unknown. Ke et al. (2010) suggested that mixed pollutants, such as heavy metals and PAHs, would stimulate the degradation of low molecular weight PAHs, which may be attacked by the excess cellular accumulation of ROS under heavy metal stress. This study provides some evidence that microalgae have the potential to remove different types of pollutants and they are also feasible to treat the combined wastewater from different origins. However, more research on the interaction between PAHs and other POPs must be carried out to better understand the underlined mechanisms involved in the removal process. It is also essential to evaluate the toxicities of metabolites produced during the degradation and transformation processes, many of which have been shown to be more toxic than their parent compounds.

ACKNOWLEDGMENTS

The work described in this paper was supported by a research grant supported from Shenzhen Science, Technology and Information Bureau, Shenzhen Government (Project No. [2008]121).

REFERENCES

Akbar, S. and Rogers, L. J. (1985). Effects of DDT on photosynthetic electron flow in *Secale* species. *Phytochemistry*, 24, 2785-2789.

Alexander, H. C., Dill, D. C., Smith, L. W., Guiney, P. D. and Dorn, P. (1988). Bisphenol a: Acute aquatic toxicity. *Environmental Toxicology*, 7, 19-26.

Alivés, C., Torres-Márquez, M. E., Mendoza-Cozátl, D. and Moreno-Sanchéz, R. (2005). Time-course development of the Cd^{2+} hyper-accumulating phenotype in (*Euglena gracilis*). *Archives of Microbiology*, 184, 83-92.

Arfsten, D. P., Schaeffer, D. J. and Mulveny, D. C. (1987). The effects of near ultraviolet radiation on the toxic effects of polycyclic aromatic hydrocarbons in animals and plants: A review. *Ecotoxicology and Environmental Safety*, 33, 1-24.

Asada, K. and Takahashi, M. (1987). Production and scavenging of active oxygen in photosynthesis. In Kyle, D. J., Osmond, C. B. and Arntzen, C. J. (Eds.), Photoinhibition (pp. 227-287). Elsevier, Amsterdam.

Avery, S. V., Codd, G. A. and Gadd, G. M. (1995). Characterization of caesium transport in the microalga (*Chlorella salina*). *Biochemical Society Transactions*, 23, 468S.

Ballschmiter, K., Hackenberg, R., Jarman, W. M. and Looser, R. (2004). Man-made chemicals found in remote areas of the world: The experimental definition for POPs. *Environmental Science and Pollution Research*, 9, 274-288.

Barque, J. P., Abahamid, A., Flinois, J. P., Baune, P. and Bonaly, J. (2002). Constitutive overexpression of immunoidentical forms of PCP-induced (*Euglena gracilis*) CYP-450. *Biochemical and Biophysical Research Communications*, 298, 277-281.

Baścik-Remisiewicz, A., Aksmann, A., Żak, A., Kowalska, M. and Tukaj, Z. (2010). Toxicity of cadmium, anthracene, and their mixture to *Desmodesmus subspicatus* estimated by algal growth-inhibition ISO standard test. *Archives of Environmental Contamination and Toxicology* (published online).

Blaylock, B. G., Frank, M. L. and McCarthy, J. F. (1985). Comparative toxicity of copper and acridine to fish, *Daphnia* and algae. *Environmental Toxicology and Chemistry*, 4, 63-71.

Borde, X., Guieysse, B., Delgado, O., Muñoz, R., Hatti-Kaul, R., Nugier-Chauvin, C., Patin, H. and Mattiasson, B. (2003). Synergistic relationships in algal-bacterial microcosms for the treatment of aromatic pollutants. *Bioresource Technology*, 86, 293-300.

Borgå, K., Wolkers, H., Skaare, J. U., Hop, H., Muir, D. C. G. and Gabrielsen, G. W. (2005). Bioaccumulation of PCBs in Arctic seabirds: influence of dietary exposure and congener biotransformation. *Environmental Pollution*, 134, 397-409.

Brooke, L. T. (1993). Acute and Chronic Toxicity of Nonylphenol to Ten Species of Aquatic Organisms. Report to the W.S. Environmental Protection Agency, Duluth, MN (contract no. 68-C1-0034). Lake Superior Research Institute, University of Wisconsin-Superior, Superior, Wisconsin (cited in Servos et al., 1999).

Canton, J. H., van Esch, G. J., Greve, P. A. and van Hellemond, A. B. A. M. (1977). Accumulation and elimination of α-hexachlorocyclohexane (α-HCH) by the marine algae *Chlamydomonas* and *Dunaliella*. *Water Research*, 11, 111-115.

Caroppo, C., Stabili, L., Aresta, M., Corinaldesi, C. and Danovaro, R. (2006). Impact of heavy metals and PCBs on marine picoplankton, *Environmental Toxicology*, 21, 541-551.

Casserly, D. M., Davis, E. M., Downs, T. D. and Guthrie, R. K. (1983). Sorption of organics by *Selenastrum capricornutum*. *Water Research* 17, 1591-1594.

Cerniglia, C. E. (1992). Biodegradation of polycyclic aromatic hydrocarbons. *Biodegradation*, 3, 351-368.

Christensen, E. R. and Zielski, P. A. (1980). Toxicity of arsenic and PCB to a green alga (*Chlamydomonas*). *Bulletin of Environmental Contamination and Toxicology*, 25, 43-48.

Chung, M. K., Hu, R., Wong, M. H. and Cheung, K. C. (2007a). Comparative toxicity of hydrophobic contaminants to microalgae and higher plants. *Ecotoxicology*, 16, 393-402.

Chung, M. K., Hu, R., Cheung, K. C. and Wong, M. H. (2007b). Screening of PAHs and DDTs in sand and acrisols soil by a rapid solid-phase microalgal bioassay. *Ecotoxicology*, 16, 429-438.

Clegg, T. J. and Koevenig, J. L. (1974). The effect of four chlorinated hydrocarbon pesticides and one organophosphate pesticide on ATP levels in three species of photosynthetic freshwater algae. *Botanical Gazette*, 135, 369-372.

Cody, T. E., Radike, M. J. and Warshawsky, D. (1984). The phytotoxicity of benzo[a]pyrene in the green alga *Selenastrum capricornutum*. *Environmental Research*, 35, 122-132.

Colborn, T. and Smolen, M. J. (1996). An epidemiological analysis of persistent organochlorine contaminants in cetaceans. *Reviews of Environmental Contamination and Toxicology*, 146, 91-172.

Correa-Reyes, G., Viana, M. T., Marquez-Rocha, F. J., Licea, A. F., Ponce, E. and Vazquez-Duhalt, R. (2007). Nonylphenol algal bioaccumulation and its effect through the trophic chain. *Chemosphere*, 68, 662-670.

Costa, J. A. V. and de Moraisa, M. G. (2010). The role of biochemical engineering in the production of biofuels from microalgae. *Bioresource Technology*, 102, 2-9.

D'Adamo, S. Trotta, P. P. and Sansone, G. (1997). Bioaccumulation and biomagnifications of polycyclic aromatic hydrocarbons in aquatic organisms. *Marine Chemistry*, 56, 45-49.

Daughton, C. G. and Ternes, T. A. (1999). Pharmaceuticals and personal care products in the environment: agents of subtle change? *Environmental Health Perspectives*, 107, 907-938.

Day, J. G., Benson, E. E. and Fleck, R. A. (1999). In vitro culture and conservation of microalgae: Applications for aquaculture, biotechnology and environmental research. In Vitro Cellular and Developmental Biology-Plant, 35, 127-136.

DeLorenzo, M. E., Scott, G. I. and Ross, P. E. (2001). Toxicity of pesticides to aquatic microorganisms: A review. *Environmental Toxicology and Chemistry*, 20, 84-98.

Della Greca, M., Pinto, G., Pistillo, P., Pollio, A., Previtera, L. and Temussi, F. (2008). Biotransformation of ethinylestradiol by microalgae. *Chemosphere*, 70, 2047-2053

De Zwart, L. L. Meerman, J. H. N., Commandeur, J. N. M. and Vermeulen, N. P. E. (1999). Biomarkers of free radical damage: applications in experimental animals and in humans. *Free Radical Biology and Medicine*, 26, 202-226.

De Wit, C. A. (2002). An overview of brominated flame retardants in the environment. *Chemosphere,* 46, 583-624.

Dijkman, N. A., van Vlaardingen, P. L. A. and Admiraal, W. A. (1997). Biological variation in sensitivity to N-heterocyclic PAHs; effects of acridine on seven species of micro-algae. *Environmental Pollution*, 95, 121-126.

Djomo, J. E., Dauta, A., Ferrier, V., Narbonne, J. F., Monkiedje, A., Njine, T. and Garrigues, P. (2004). Toxic effects of some major polyaromatic hydrocarbons found in crude oil and aquatic sediments on *Scenedesmus subspicatus*. *Water Research*, 38, 1817-1821.

Eisler, R. (1987). Polycyclic aromatic hydrocarbon hazards to fish, wildlife, and invertebrates: A synoptic review. U.S. Fish and Wildlife Service Biological Report 85(1.11).

Escher, B. I., Bramaz, N., Eggen, R. I. L. and Richter, M. (2005). *In vitro* assessment of modes of toxic action of pharmaceuticals in aquatic life. *Environmental Science and Technology*, 39, 3090-3100.

Evandri, M. G., Costa, L.G. and Bolle, P. (2003). Evaluation of brominated diphenyl ether-99 toxicity with *Raphidocelis subcapitata* and *Daphnia magna*. *Environmental Toxicology and Chemistry*, 22, 2167-2172.

Falkowski, P. G. and Raven, J. A. (1997). Aquatic Photosynthesis. Blackwell, Oxford.

Fisher, N. S. and Wurster, C. F. (1973). Individual and combined effects of temperature and polychlorinated biphenyls on the growth of three species of phytoplankton. Environmental Pollution, 5, 205-212.

Fisher, N. S., Carpenter, E. J., Remsen, C. C. and Wurster, C. F. (1974). Effects of PCB on interspecific competition in natural and gnotobiotic phytoplankton communities in continuous and batch cultures. *Microbial Ecology*, 1, 39-50.

Gala, W. R. and Giesy, J. P. (1992). Photo-induced toxicity of anthracene to the green alga, *Selenastrum capricornutum*. *Archives of Environmental Contamination and Toxicology*, 23, 316-323.

Gao, Q. and Tam, N. F. Y. (2011). Growth, photosynthesis and antioxidant responses of two microalgal species, *Chlorella vulgaris* and *Selenastrum capricornutum*, to nonylphenol stress. *Chemosphere*, 82, 346-354.

Gaur, J. P. (1988). Toxicity of some oil constituents to *Selenastrum capricornutum*. *Environmental Chemistry*, 16, 617-620.

Geoffroy, L., Teisseire, H., Couderchet, M. and Vernet, G. (2002). Effect of oxyfluorfen and diuron alone and in mixture on antioxidative enzymes of *Scenedesmus obliquus*. *Pesticide Biochemistry and Physiology*, 72, 178-185.

Gerofke, A., Kamp, P. and McLachlan, M. S. (2005). Bioconcentration of persistent organic pollutants in four species of marine phytoplankton. *Environmental Toxcology and Chemistry*, 24, 2908-2917.

Glooschenko, V. and Lott, J. N. A. (1977). Effects of chlordane on green-algae *Scenedesmus quadricauda* and *Chlamydomonas* sp. Canadian Journal of Botany, 55, 2866-2872.

Glooschenko, V., Holdrinet, M., Lott, J. N. A. and Frank, R. (1979). Bioconcentration of Chlordane by the Green Alga *Scenedesmus quadricauda*. *Bulletin of Environmental Contamination and Toxicology*, 21, 515-520.

Graff, L., Isnard, P., Cellier, P., Bastide, J., Cambon, J. P., Narbonne, J. F., Budzinski, H. and Vasseur, P. (2003). Toxicity of chemicals to microalgae in river and in standard waters. *Environmental Toxicology and Chemistry*, 22, 1368-1379.

Guan, Y. F., Wang, J. Z., Ni, H. G., Luo, X. J., Mai, B. X. and Zeng, E. Y. (2007). Riverine inputs of polybrominated diphenyl ethers from the Pearl River Delta (China) to the coastal ocean. *Environmental Science and Technology*, 41, 6007-6013.

Halling-Sørensena, B., Nyholm, N., Kusk, K. O. and Jacobsson, E. (2000). Influence of nitrogen status on the bioconcentration of hydrophobic organic compounds to *Selenastrum capricornutum*. *Ecotoxicology and Environmental Safety*, 45, 33-42.

Halliwell, B. and Gutteridge, J. M. C. (1989). Free Radicals in Biology and Medicine (2nd edition). Clarendon Press, Oxford.

Harding Jr., L. W. and Phillips Jr., J. H. (1978). Polychlorinated biphenyl (PCB) effects on marine phytoplankton photosynthesis and cell division. *Marine Biology*, 49, 93-101.

Hense, B. A., Jüttner, I., Welzl, G., Severin, G. F., Pfister, G., Behechti, A. and Schramm, K. W. (2003). Effects of 4-nonylphenol on phytoplankton and periphyton in aquatic microcosms. *Environmental Toxicology and Chemistry*, 22, 2727-2732.

Hill, I. R. and Wright, S. J. L. (1978). Pesticide Microbiology. Academic, London, UK.

Hites, R. A. (2004). Polybrominated diphenyl ethers in the environment and in people: a meta-analysis of concentrations. *Environmental Science and Technology*, 38, 945-956.

Hollister, T. A., Gerald, E., Walsh, G. E. and Forester, J. (1975). Mirex and marine unicellular algae: Accumulation, population growth and oxygen evolution. Bulletin of *Environmental Contamination and Toxicology*, 14, 753-759.

Hosetti, B. and Frost, S. (1998). A review of the control of biological waste treatment in stabilization ponds. *Critical Reviews in Environmental Science and Technology*, 28, 193-218.

Huang, X. D., Dixon, D. G. and Greenberg, B. M. (1993). Impacts of UV radiation and photomodification on the toxicity of PAHs to the higher plant *Lemna gibba* (Duckweed). *Environmental Toxicology and Chemistry*, 12, 1067-1077.

Huang, X. D., McConkey, B. J., Babu, T. S. and Greenberg, B. M. (1997). Mechanisms of photoinduced toxicity of photomodified anthracene to plants: Inhibition of photosynthesis in the aquatic higher plant *Lemna gibba* (duckweed). *Environmental Toxicology*, 16, 1707-1715.

Hüls, A. G. (1996). Determination of the effects of nonylphenol on the growth of *Scenedesmus subspicatus*. 86.91.SAG (algal growth inhibition test according to UBA Feb. 1984), Report AW-185 (cited in Servos et al., 1999).

Hunt, P. A., Koehler, K. E., Susiarjo, M., Hodges, C. A., Ilagan, A., Voigt, R. C., Thomas, S., Thomas, B. F. and Hassold, T. J. (2003). Bisphenol A exposure causes meiotic aneuploidy in the female mouse. *Current Biology*, 13, 546-553.

Jabusch, T. W. and Swackhamer, D. L. (2004). Subcellular accumulation of polychlorinated biphenyls in the green alga (*Chlamydomonas reinhardtii*). *Environmental Toxicology and Chemistry*, 23, 2823-2830.

Jones, K. C. and de Voogt, P. (1999). Persistent organic pollutants (POPs): state of the science. *Environmental Pollution*, 100, 209-221.

Juhasz, A. L. and Naidu, R. (2000). Bioremediation of high molecular weight polycyclic aromatic hydrocarbons: a review of the microbial degradation of benzo[a]pyrene. *International Biodeterioration and Biodegradation*, 45, 57-88.

Källqvist, T., Grung, M. and Tollefsen, K. E. (2006). Chronic toxicity of 2,4,2',4'-tetrabromodiphenyl ether on the marine alga *Skeletonema costatum* and the crustacean *Daphnia magna*. *Environmental Toxicology*, 25, 1657-1662.

Kauss, P. and Hutchinson, T. C. (1975). The effects of water-soluble petroleum components on the growth of *Chlorella vulgaris* Beijerinck. *Environmental Pollution*, 9, 157-174.

Ke, L., Luo, L., Wang, P., Luan, T. and Tam, N. F. Y. (2010). Effects of metals on biosorption and biodegradation of mixed polycyclic aromatic hydrocarbons by a freshwater green alga *Selenastrum capricornutum*. *Bioresource Technology*, 101, 6950-6961.

Keil, J. E., Priester, L. E. and Sandifer, S. H. (1971). Polychlorinated biphenyl (Arochlor 1242): Effects of uptake on growth, nucleic acids, and chlorophyll of a marine diatom. *Bulletin of Environmental Contamination and Toxicology*, 6, 156-159.

Kelly, B. C., Ikonomou, M. G., Blair, J. D., Morin, A. E. and Gobas, F. A. P. C. (2007). Food web-specific biomagnification of persistent organic pollutants. *Science*, 317, 236-239.

Kelly, B. C., Ikonomou, M. G., Blair, J. D. and Gobas, F. A. (2008). Bioaccumulation behaviour of polybrominated diphenyl ethers (PBDEs) in a Canadian Arctic marine food web. *The Science of the Total Environment*, 401, 60-72.

Kirso, U. and Irha, N. (1998). Role of algae in fate of carcinogenic polycyclic aromatic hydrocarbons in the aquatic environment. *Ecotoxicology and Environmental Safety*, 41, 83-89.

Koelmans, A. A., van der Woude, H., Hattink, J. and Niesten, D. J. M. (1999). Long-term bioconcentration kinetics of hydrophobic chemicals in *Selenastrum capricornutum* and *Microcystis aeruginosa*. *Environmental Toxicology*, 18, 1164-1172.

Kolpin, D. W., Furlong, E. T., Meyer, M. T., Thurman, E. M., Zaugg, S. D., Barber, L. B. and Buxton, H. T. (2002). Pharmaceuticals, hormones, and other organic wastewater contaminants in U.S. streams 1999-2000-a national reconnaissance. *Environmental Science and Technology*, 36, 1202-1211.

Komoβa, D., Langebartels, C. and Sandermann, H. (1995). Metabolic processes for organic chemicals in plants. In Trapp, S. and McFarlane, J. C. (Eds.), *Plant Contamination* (pp. 69-103). CRC Press, Boca Raton, FL.

Kong, F., Hu, W. and Liu, Y. (1998). Molecular structure and biochemical toxicity of four halogeno-benzenes on the unicellular green alga *Selenastrum capricornutum*. *Environmental and Experimental Botany*, 40, 105-111.

Kricher, J. C., Urey, J. C. and Hawes, M. L. (1975). The effects of mirex and methoxychlor on the growth and productivity of *Chlorella pyrenoidosa*. *Bulletin of Environmental Contamination and Toxicology*, 14, 617-620.

Lai, K. M., Scrimshaw, M. D. and Lester, J. N. (2002). Biotransformation and bioconcentration of steroid estrogens by *Chlorella vulgaris*. *Applied and Environmental Microbiology*, 68, 859-864.

Lal, R. and Saxena, D. M. (1982). Accumulation, metabolism, and effects of organochlorine insecticides on microorganisms. *Microbiological Reviews*, 46, 95-127.

Lal, R. and Lal, S. (1988). Pesticides and Nitrogen Cycle, Vol. 3. CRC Press, Boca Raton, FL.

Law, M. Y., Charles, S. A. and Halliwell, B. (1983). Glutathione and ascorbic acid in spinach (*Spinacia oleracea*) chloroplast. The effect of hydrogen peroxide and paraquat. *Biochemical Journal*, 210, 899-903.

Laws, E. A. (1981). Aquatic Pollution. John Wiley and Sons, New York.

Lee. S. S., Fang, S. C. and Freed, V. H. (1976). Effect of DDT on photosynthesis of *Selenastrum capricornutum*. *Pesticide Biochemistry and Physiology*, 6, 46-51.

Lei, A. P., Wong, Y. S. and Tam, N. F. Y. (2002). Removal of pyrene by different microalgal species. *Water Science and Technology*, 46, 196-201.

Lei, A. P., Wong, Y. S. and Tam, N. F. Y. (2003). Pyrene-induced changes of glutathione-S-transferase activities in different microalgal species. *Chemosphere*, 50, 293-301.

Lei, A., Hu, Z., Wong, Y. S. and Tam, N. F. Y. (2006). Antioxidant responses of microalgal species to pyrene. *Journal of Applied Phycology*, 18, 67-78.

Lei, A. P., Hu, Z. L., Wong, Y. S. and Tam, N. F. Y. (2007). Removal of fluoranthene and pyrene by different microalgal species. *Bioresource Technology*, 98, 273-280.

Leitão, M. A., Cardozo, K. H. M., Pinto, E. and Colepicolo, P. (2003). PCB-induced oxidative stress in the unicellular marine dinoflagellate *Lingulodinium polyedrum*. *Archives of Environmental Contamination and Toxicology*, 45, 59-65.

Li, R., Chen, G. Z., Tam, N. F. Y., Luan, T. G., Shin, P. K. S., Cheung, S. G. and Liu, Y. (2009). Toxicity of bisphenol A and its bioaccumulation and removal by a marine microalga *Stephanodiscus hantzschii*. *Ecotoxicology and Environmental Safety*, 72, 321-328.

Liu, Y., Guan, Y., Gao, Q., Tam, N. F. Y. and Zhu, W. (2010). Cellular responses, biodegradation and bioaccumulation of endocrine disrupting chemicals in marine diatom *Navicula incerta*. *Chemosphere*, 80, 592-599.

Lovell, C.R., Eriksen, N.T. and Lewitus, A. J. (2002). Resistance of the marine diatom *Thalassiosira* sp. to toxicity of phenolic compounds. *Marine Ecological Progress Series*, 229, 11-18.

Lytle, J. S. and Lytle, T. F. (2001). Use of plants for toxicity assessment of estuarine ecosystems. *Environmental Toxicology and Chemistry*, 20, 68-83.

Mayer, P., Halling-Sørensen, B., Sijm, D. T. H. M. and Nyholm, N. (1998). Toxic cell concentrations of three polychlorinated biphenyl congeners in the green alga *Selenastrum capricornutum*. *Environmental Toxicology*, 17, 1848-1851.

Megharaj, M., Boul, H. L. and Thiele, J. H. (1999). Effect of DDT and its metabolites on soil algae and enzymatic activity. *Biology and Fertility of Soils*, 29, 130-134.

Mekenyan, O. G., Ankley, G. T., Veith, G. D. and Call, D. J. (1994). QSARs for photoinduced: acute lethality of polycyclic aromatic hydrocarbons to *Daphnia magna*. *Chemosphere*, 28, 567-582.

Mekenyan, O. G., Ankley, G. T., Veith, G. D. and Call, D. J. (1995). QSARs for photoinduced toxicity of aromatic compounds. *SAR and QSAR in Environmental Research*, 4, 139-145.

Miller, R. M., Singer, G. M., Rosen, J. D. and Bartha, R. (1988). Photolysis primes biodegradation of benzo[a]pyrene. *Applied and Environmental Microbiology*, 54, 1724-1730.

Millemann, R. E., Birge, W. J., Black, J. A., Cushman, R. M., Daniels, K. L., Franco, P. J., Giddings, J. M., McCarthy, J. F. and Stewart, A. J. (1984). Comparative acute toxicity to aquatic organisms of components of coal-derived synthetic fuels. *Transactions of the American Fisheries Society*, 113, 74-85.

Moore, S. A., Jr and Harriss R. C. (1974). Differential sensitivity to PCB by phytoplankton. *Marine Pollution Bulletin*, 5, 174-176.

Moody, R. P., Weinberger, P. and Greenhalgh, R. (1983). Algal fluorometric determination of the potential phytotoxicity of environmental pollutants. In Nriagu, J. O. (Ed.), Aquatic Toxicology (pp. 503-512). John Wiley and Sons, New York (cited in Servos et al., 1999).

Morgan, J. R. (1972). Effects of Aroclor 1242 (a polychlorinated biphenyl) and DDT on cultures of an alga, protozoan, dephnid, ostracod and guppy. *Bulletin of Environmental Contamination and Toxicology*, 5, 226-230.

Mosser, J. L., Fisher, N. S., Teng, T. C. and Wurster, C. F. (1972a). Polychlorinated biphenyls: Toxicity to certain phytoplankters. *Science*, 175, 191-192.

Mosser, J. L., Fisher, N. S. and Wurster, C. F. (1972b). Polychlorinated biphenyls and DDT alter species composition in mixed cultures of algae. *Science*, 176, 533-535.

Mosser, J. L., Teng, T. C., Walther, W. G. and Wurster, C. F. (1974). Interactions of PCBs, DDT and DDE in a marine diatom. *Bulletin of Environmental Contamination and Toxicology*, 12, 665-668.

Muir, D. G. and Howard, P. (2006). Are there other persistent organic pollutants? A challenge for environmental chemists. *Environmental Science and Technology*, 40, 7157-7166.

Munoz, M. J., Ramos, C. and Tarazona, J. V. (1996). Bioaccumulation and toxicity of hexachlorobenzene in *Chlorella vulgaris* and *Daphnia magna*. *Aquatic Toxicology*, 35, 211-220.

Muñoz, R. and Guieysse, B. (2006). Algal–bacterial processes for the treatment of hazardous contaminants: A review. *Water Research*, 40, 2799-2815.

Narro, M. L., Cerniglia, C. E., van Baalen, C. and Gibson, D. T. (1992). Metabolism of phenanthrene by the cyanobacterium *Agmenellum quadruplicatum* PR-6. *Applied and Environmental Microbiology*, 58, 1352-1359.

Naylor, C. G. (1995). Environmental fate and safety of nonylphenol ethoxylates. *American Association of Textile Chemists and Colourists*, 27, 29-33.

Noctor, G. and Foyer, C. H. (1998). Ascorbate and glutathione: keeping active oxygen under control. *Annual Review of Plant Physiology and Plant Molecular Biology*, 49, 249-279.

North, K. D. (2004). Tracking polybrominated diphenyl ether releases in a wastewater treatment plant effluent, Palo Alto, California. *Environmental Science and Technology*, 38, 4484-4488.

O'Brien, P. J. (1991). Molecular mechanisms of quinone cytotoxicity. *Chemico-Biological Interactions*, 80, 1-41.

Okamoto, O. K., Asano, C. S., Aidar, E. and Colepicolo, P. (1996). Effects of cadmium on growth and superoxide dismutase activity of the marine microalga *Tetraselmis gracilis* (Prasinophyceae). *Journal of Phycology*, 32, 74-79.

Oros, D. R., Hoover, D., Rodigari, F., Crane, D. and Sericano, J. (2005). Levels and distribution of polybrominated diphenyl ethers in water, surface sediments, and bivalves from the San Francisco Estuary. *Environmental Science and Technology*, 39, 33-41.

Palace, V. P., Wautier, K., Evans, R. E., Baron, C. L., Werner, J., Ranson, C., Klaverkamp, J. F. and Kidd, K. (2001). Effects of 17-β estradiol exposure on metallothionein and fat soluble antioxidant vitamins in juvenile lake trout (*Salvelinus namaycush*). *Bulletin of Environmental Contamination and Toxicology*, 66, 591-596.

Parida, A. K., Das, A. B. and Mohanty, P. (2004). Defense potentials to NaCl in a mangrove, *Bruguiera parviflora*: Differential changes of isoforms of some antioxidative enzymes. *Journal of Plant Physiology*, 161, 531-542.

Pavlíc, Z., Vidakovíc-Cifrek, Z. and Puntaríc, D. (2005). Toxicity of surfactants to green microalgae *Pseudokirchneriella subcapitata* and *Scenedesmus subspicatus* and to marine diatoms *Phaeodactylum tricornutum* and *Skeletonema costatum*. *Chemosphere*, 61, 1061-1068.

Petersen, D. G. and Dahllöf, I. (2007). Combined effects of pyrene and UV-light on algae and bacteria in an arctic sediment. *Ecotoxicology*, 16, 371-377.

Pflugmacher, S. and Sandermann Jr., H. (1998a). Cytochrome P450 monooxygenases for faty acids and xenobiotics in marine macroalgae. *Plant Physiology*, 117, 123–128.

Pflugmacher, S. and Sandermann Jr., H. (1998b). Taxonomic distribution of plant glucosyltransferase acting on xenobiotics. *Phytochemistry*, 49, 507-511.

Pinto, E., Sigaud-kutner, T. C. S., Leitão, M. A. S., Okamoto, O. K., Morse, D. and Colepicolo, P. (2003). Heavy metal-induced oxidative stress in algae. *Journal of Phycology*, 39, 1008-1018.

Powers, C. D., Wurster, C. F. and Rowland, R. G. (1979). DDE inhibition of marine algal cell division and photosynthesis per cell. *Pesticide Biochemistry and Physiology*, 10, 306-312.

Qian, H., Pan, X., Shi, S., Yu, S., Jiang, H., Lin, Z. and Fu, Z. (2011). Effect of nonylphenol on response of physiology and photosynthesis-related gene transcription of *Chlorella vulgaris*. Environmental Monitoring and Assessment (published online).

Ren, L., Huang, X. D., McConkey, B. J., Dixon, D. G. and Greenberg, B. M. (1994). Photoinduced toxicity to three polycyclic aromatic hydrocarbons (fluoranthene, pyrene, and naphthalene) to the duckweed *Lemna Gibba* L. G-3. *Ecotoxicology and Environmental Safety*, 28, 160-171.

Rijstenbil, J. W., Derksen, J. W. M., Gerringa, L. J. A., Poortvliet, T. C. W., Sandee, A., Berg, M., Drie, J. and Wijnholds, J. A. (1994). Oxidative stress induced by copper: defense and damage in the marine planktonic diatom *Ditylum brightwellii*, grown in continuous cultures with high and low zinc levels. *Marine Biology*, 119, 583-590.

Rioboo, C., O'Connor, J. E., Prado, R., Herrero, C. and Ci, Á. (2009). Cell proliferation alterations in *Chlorella* cells under stress conditions. *Aquatic Toxicology*, 94, 229-237.

Samson, G. and Popovic, R. (1988). Use of algal fluorescence for determination of phytotoxicity of heavy metals and pesticides as environmental pollutants. *Ecotoxicology and Environmental Safety*, 16, 272-278.

Sandermann, H. (1994). Invited review: higher plant metabolism of xenobiotics: the "green liver" concept. *Pharmacogenetics*, 4, 225-241.

Schoeny, R., Cody, T., Warshawsky, D. and Radike, M. (1988). Metabolism of mutagenic polycyclic aromatic hydrocarbons by photosynthetic algal species. *Mutation Research*, 197, 289-302.

Schwarzenbach, R. P., Gschend, P. M. and Imboden, D. M. (2003). Environmental Organic Chemistry. (2nd edition). John Wiley and Sons, Hoboken, NJ.

Semple, K. T., Cain, R. B. and Schmidt, S. (1999). Biodegradation of aromatic compounds by microalgae. *FEMS Microbiology Letters*, 170, 291-300.

Servos, M. R. (1999). Review of the aquatic toxicity, estrogenic responses and bioaccumulation of alkylphenols and alkylphenol polyethoxylates. *Water Quality Research Journal of Canada*, 34, 123-177.

Sijm, D. T. H. M., Middelkoop, J. and Vrisekoop, K. (1995). Algal density dependent bioconcentration factors of hydrophobic chemicals. *Chemosphere*, 31, 4001-4012.

Sims, R. C. and Overcash, M. R. (1983). Fate of polynuclear aromatic compounds (PNAs) in soil-plant systems. *Residue Reviews*, 88, 1-68.

Smirnoff, N. (1996). The function and metabolism of ascorbic acid in plants. *Annals of Botany*, 78, 661-669.

Stadnyk, L., Campbell, R. C. and Join, B. T. (1971). Pesticide effect on grow and ^{14}C-assimilation in a fresh water alga. *Bulletin of Environmental Contamination and Toxicology*, 6, 1-8.

Stephenson, R. R. (1983). Diphenylal propane: acute toxicity to *Daphnia magna* and *Selenastrum capricornutum*. Group Research Report. Shell Research Limited, Sittingbourne Research Centre, Sittingbourne, Kent, England.

Swackhamer, D. L. and Skoglund, R. S. (1993). Bioaccumulation of PCBs by algae: Kinetics versus equilibrium. *Environmental Toxicology and Chemistry*, 12, 831-838.

Tam, N. F. Y., Wong, Y. S. and Craig, G. S. (1998). Removal of copper by free and immobilized microalga, Chlorella vulgaris. In Wong, Y. S. and Tam, N. F. Y. (Eds.), Wastewater Treatment with Algae (pp. 17-36). Springer-Verlag, Berlin, Germany.

Thies, F., Backhaus, T., Bossman, B. and Grimme, L. H. (1996). Xenobiotic biotransformation in unicellular green algae. *Plant Physiology*, 112, 361-370.

Torres, M. A., Barros, M. P., Campos, S. C., Pinto, S. C., Raiamani, S., Sayre, R. T. and Colepicolo, P. (2008). Biochemical biomarkers in algae and marine pollution: a review. *Ecotoxicology and Environmental Safety*, 71, 1-15.

Tripathi, B. N. and Gaur, J. P. (2004). Relationship between copper- and zinc-induced oxidative stress and proline accumulation in *Scenedesmus* sp. *Planta*, 219, 397-404.

Tripuranthakam, S., Duxbury, C. L., Babu, T. S. and Greenberg, B. M. (1999). Development of a mitochondrial respiratory electron transport bioindicator for assessment of hydrocarbon toxicity. In Henshel, D., Black, M. C., Harris, M. C., (Eds.), Environmental Toxicology and Risk Assessment, Vol. 8. Special Technical Publication (pp. 350-361). American Society for Testing and Materials, Philadelphia, PA.

UNEP-Chemicals (2004). Global Inventory of Persistent Organic Pollutant Laboratories (1st edition). United Nations Environment Programme, Geneva, Switzerland.

US Environmental Protection Agency (USEPA) (1978). In-depth Studies on Health and Environmental Impacts of Selected Water Pollutants. U.S. *Environmental Protection Agency Contract* No. 68-01-4646.

Vance, B. D. and Drummond, W. (1969). Biological concentration of pesticides by algae. *Journal of Phycology*, 61, 360-362.

Vazquez-Duhalt, R., Marquez-Rocha, F., Ponce, E., Licea, A. F. and Viana, M. T. (2005). Nonylphenol, an integrated vision of a pollutant. Scientific review. *Applied Ecology and Environmental Research*, 4, 1-25.

Voldner, E. C. and Li, Y. F. (1995). Global usage of selected persistent organochlorines. *The Science of the Total Environment*, 160/161, 201-210.

Vysotskaya, N. A. and Bortun, L. N. (1984). The mechanism of radiation-induced homolytic substitution of condensed aromatic hydrocarbons in aqueous solutions. *Radiation Physics and Chemistry*, 23, 731-738.

Wang, J. and Xie, P. (2007). Antioxidant enzyme activities of *Microcystis aeruginosa* in response to nonylphenols and degradation of nonylphenols by *M. aeruginosa*. *Environmental Geochemistry and Health*, 29, 375-383.

Wang, P., Ke, L., Luo, L., Luan, T., Wong, M. H. and Tam, N. F. Y. (2011). Combined effects of co-contaminated polycyclic aromatic hydrocarbons and heavy metals on biochemical responses in free and immobilized *Selenastrum capricornutum*. (In preparation).

Ward, T. J. and Boeri, R. L. (1990a). Acute static toxicity of nonylphenol to the marine alga (*Skeletonema costattun*). Prepared for the Chemical Manufactures Association by Resource Analysts. Study No. 8970-CMA (cited in Servos et al., 1999).

Ward, T. J. and Boeri, R. L. (1990b). Acute static toxicity of nonylphenol to the freshwater alga (*Selenastrum capricornutum*). Prepared for the Chemical Manufactures Association by Resource Analysts. Study No. 8969-CMA (cited in Servos et al., 1999).

Warshawsky, D., Cody, T., Radike, M., Reilman, R., Schumann, B., LaDow, K. and Schneider, J. (1995). Biotransformation of benzo[a]pyrene and other polycyclic aromatic hydrocarbons and heterocyclic analogs by several green algae and other algal species under gold and white light. *Chemico-Biological Interactions*, 97, 131-148.

Weeks, J. A., Adams, W. J., Guiney, P. D., Hall, J. F. and Naylor, C. G. (1996). Risk assessment of nonylphenols and its ethoxylates in U.S. river water and sediment. Proceedings of the 4th CESIO World Surfactants Congress, 3, 276-291 (cited in Servos et al., 1999).

Weinberger, P. (1984). Screening for Potential Ecotoxicology of Solvents, Surfactants and Detergents: Final Contract Report. DSS file 24SU, KN107-2-4304 (cited in Servos et al., 1999).

Weinberger, P., DeChacin, C. and M. Czuba, M. (1987). Effects of nonylphenol, a pesticide surfactant, on some metabolic processes of *Chlamydomonas segnis*. *Canadian Journal of Botany*, 65, 696-702.

Weinberger, R. and Rea, M. (1981). Nonylphenol: a perturbant additive to an aquatic ecosystem. In Bermington et al. (Eds.), Proceedings of the 7th Annual Aquatic Toxicity Workshop (pp. 371-380), Canadian Technical Report on Fish and Aquatic Sciences (cited in Servos et al., 1999).

Wojciech, P. and Zbigniew, T. (2010). The combined effect of anthracene and cadmium on photosynthetic activity of three *Desmodesmus* (Chlorophyta) species. *Ecotoxicology and Environmental Safety*, 73, 1207-1213.

Yang, S., Wu, R. S. S. and Kong, R. Y. C. (2002). Physiological and cytological responses of the marine diatom *Skeletonema costatum* to 2,4-dichlorophenol. *Aquatic Toxicology*, 60, 33-41.

Ying, G. G., Williams, B. and Kookana, R. (2002). Environmental fate of alkylphenols and alkylphenol ethoxylates. A review. *Environmental International*, 28, 215-226.

Yogui, G. T. and Sericano, J. L. (2009). Polybrominated diphenyl ether flame retardants in the U.S. marine environment: A review. *Environment International*, 35, 655-666.

Younglai, E. V., Holloway, A. C. and Foster, W. G. (2005). Environmental and occupational factors affecting fertility and IVF success. Human Reproduction Update, 11, 43-57.

Zbigniew, T. (1987). The effects of crude and fuel oils on the growth, chlorophyll a content and dry matter production of a green alga *Scenedesmus quadricauda* (Turp.) bréb. *Environmental Pollution*, 47, 9-24.

Zbigniew, T. and Wojciech, P. (2006). Individual and combined effect of anthracene, cadmium, and chloridazone on growth and activity of SOD izoformes in three *Scenedesmus* species. *Ecotoxicology and Environmental Safety*, 65, 323-331.

Zepp, R. G., Hoigné, J. and Bader, H. (1987). Nitrate-induced photooxidation of trace organic-chemical in water. *Environmental Science and Technology* 21, 443-450.

In: Microalgae: Biotechnology, Microbiology and Energy
Editor: Melanie N. Johnsen
ISBN 978-1-61324-625-2
© 2012 Nova Science Publishers, Inc.

Chapter 4

MICROALGAL ENGINEERING: THE METABOLIC PRODUCTS AND THE BIOPROCESS

Jorge Alberto Vieira Costa*, *Michele Greque de Morais and Michele da Rosa Andrade*

Laboratory of Biochemical Engineering,
College of Chemistry and Food Engineering,
Federal University of Rio Grande (FURG), Brazil

ABSTRACT

Many analyses have been carried out about the future possibility of exhausting the planet's resources and its ability to sustain its inhabitants. The use of microorganisms and their metabolic products by humans is one of the most significant fields of biotechnology activities. Microalgae are descendants of the first photosynthetic life forms. More than 3,500 million years ago the oxygen atmosphere was made up by cyanobacteria and other forms of life could evolve. Since then, microalgae have contributed to regulating the planet's biosphere. The use of solar energy through photosynthesis in microalgae cultivation is a clean, efficient and low cost process, because the sun's energy is virtually free and unlimited. The biomass of microalgae and its products are employed in many fields: feed and food additives in agriculture, fertilizers, in the food industry, pharmaceutics, perfume making, medicine, in biosurfactants, biofuels and in science. This wide use is due to its fast growth, non-toxicity, assimilability (85–95%), high protein content (60–70%), well-balanced amino acid composition, richness in vitamins, minerals, fatty acids, biopigments, biopolymers, and the fact that is has a great variety of biologically active agents in appreciable amounts. Microalgae are considered to be efficient immunopotentiators and have anticarcinogenic and antiviral effects. Producing biocompounds from microalgae has the additional advantage of simultaneously fixing large amounts of carbon dioxide. The use of these alternative sources reduces costs and can generate carbon credits. Microalgae can be grown on land that is unsuitable for agriculture and farming, or on inhospitable land such as deserts, using brackish water and/or wastes from the desalination process. The composition and rates of photosynthesis and growth of these organisms are strongly dependent on growth conditions.

---

* author for correspondence: jorge@pq.cnpq.br.

Manipulating these conditions can result in higher yield metabolites that are of interest. At the end of the process, according to the characteristics of the microalgal biomass obtained, it can be used to produce different compounds. The objective of this study was to present traditional and advanced bioproducts obtained from microalgal biomass and to describe the characteristics of the cultivation process.

Keywords: bioproducts, bioprocess, microalgae.

1. INTRODUCTION

Microalgae are a group of prokaryotic and eukaryotic photosynthetic organisms, of which more than 50,000 species are currently known, including mainly unicellular organisms that can form filaments or aggregates. The photosynthesis of cyanobacteria was responsible for the formation of the Earth's atmosphere more than 3,500 million years ago.

Microalgae cultures became the focus of research after 1948 in the United States, Germany and Japan. The commercial cultivation of microalgae started around 1960 for *Chlorella* and 1970 for *Spirulina* in Mexico, both of which were used as food supplements. *Dunaliella salina* was used to produce β-carotene in Australia in 1986, *Haematococcus pluvialis* was used for astaxanthin production and other species were used for aquaculture. Afterwards, the cultivation of microalgae spread throughout the world, to countries such as Israel, the United States and India (Borowitzka, 1999). Since then, microalgae have become attractive sources of bioproducts such as carotenoids and, in recent years, their use has expanded to include biofuels and environmental agents in the treatment of wastewater and CO_2 fixation.

The use of microalgae has already been studied in space missions of the European Space Agency for oxygen production and processing of human waste in aircraft (Godia et al., 2002). In the ocean, microalgae are responsible for the fixation of more than 50% of the planet's CO_2 emissions (Takahashi, 2004). Microalgae have been the object of advances in genetic engineering, and studies of gene sequencing and gene modification have been carried out with microalgae of the genus *Chlamydomonas* (Morowvat et al. 2010; *Rupprecht*, 2009, Bañares et al., 2004) with the aim of maximizing the production of metabolites, and with *Anabaena* for the production of bio-insecticides (Boussiba and Zaritsky, 2004).

The high rates of growth, along with the possibility of manipulation of culture conditions and genetic changes have led to the large number of studies to obtain microalgal biomass and bioproducts that are currently used in many fields, such as feed and food additives in agriculture, fertilizers, in the food industry, in pharmaceuticals, in the production of perfumes, medicines, biosurfactants, biofuels and in science.

The kinetics of growth and the cellular composition of microalgae depend on the conditions that the microalgae are exposed to. Manipulating these conditions allows the maximization of production of compounds that are of interest. Several studies have assessed the reaction of microalgae to external conditions and the effect on growth and biomass composition. This feature, combined with their high rates of growth, means that studies with microalgae are easier when compared with studies of higher organisms, where the results take longer to obtain.

2. PHOTOBIOREACTORS TO PRODUCTION PROCESSES

The bioreactors used for microalgae biomass production can be divided into open and closed types. Each has its inherent characteristics regarding the building material, the operation and the control variables, which leads to differences in the cost of installation and operation.

Open reactors are mainly raceway and circular types, built out of concrete, PVC and fiberglass. They are about 50 cm deep, which allows the incidence of sunlight on the cultures and they can be up to 5,000 m^2 in area (Belay, 1997). Open tanks can be covered, according to the site characteristics and the biomass's intended use. The covers are made out of transparent and flexible plastic, which may protect against UV rays. The agitation in open reactors is accomplished by rotating blades in raceway reactors and by rotational arms with a fixed axis in a circular tank.

Open reactors are the most commonly used for commercial microalgae cultivation, because of the low cost of installation and operation and use of sunlight for photosynthesis. However it is difficult to control the temperature of the liquid medium, evaporation of water, concentration of dissolved gases and contamination by other microorganisms. Temperature and rainfall are the main environmental factors that cultures are exposed to in open tanks, which limit their use to certain areas or seasons.

The low temperature in some locations may reduce the productivity of microalgal biomass. The selection of species that are resistant to regional conditions where the cultivation will be carried out is one alternative of minimizing this problem. In the extreme south of Brazil (a sub-tropical climate zone) a strain of *Spirulina* has been isolated from Mangueira Lagoon, which is used for pilot scale cultivation of biomass for institutional alimentation, and it has an adequate productivity during all seasons (Morais et al., 2009).

Closed reactors are constructed of transparent material, allowing light to penetrate. These reactors can be constructed from glass, plastic or acrylic, in the form of tubes that are vertical (Morais and Costa, 2007), inclined (Ugwu et al., 2005), slabs (Sierra et al., 2008) or helical (Travieso et al., 2001). Among the photobioreactors, the use of bubbles columns or air-lift is widely used for the cultivation of microalgae due to the low shear stress and ease of gas exchange (Malea et al., 2006). Two air-lift systems have been tested for semicontinuous cultivation of *Haematococcus pluvialis*, one of which is a flat panel airlift photobioreactor and the other a conventional cylindrical airlift photobioreactor (Issarapayup et al., 2009).

The cultures can be illuminated by artificial light or sunlight (Travieso et al., 2001). Sunlight is used both for photosynthesis and for heating the liquid medium in cold months (Masojídek et al., 2003). The disadvantages of this type of reactor are that they need to be cooled in places with high humidity and irradiance, they need agitation that will provide uniform light and to avoid the deposition of cells. However, productivity in closed photobioreactors is usually greater than in open tanks. *Haematococcus pluvialis* is a promising source of astaxanthin, but it has low rates of growth, and is susceptible to contamination, therefore closed reactors are recommended for its cultivation, in order to control variables (Kaewpintong et al., 2006; Kim et al. 2006; Suh et al., 2006). *Porphyridium cruentum* is another microalga with these characteristics, and cultures in closed reactors are recommended. Even so, due to the lower installation and operation cost, several studies have assessed the growth characteristics of these microalgae in open reactors (Fuentes et al., 1999).

The use of immobilized cells is reported for various microalgae such as *Chlorella vulgaris* (Lau et al. 1998); *Anabaena, Chlorella* (Rai and Mallick, 1992) and *Botryococcus* (Bailliez et al., 1986). Microalgae cells are usually immobilized by adsorption and entrapment in gel. For gel entrapment, alginate is widely used, and adsorption supports include polyurethane foam and coral stone. In addition to the challenges of using immobilized cells (deviations from ideal behaviour in the reactor, diffusional effects, cell viability and stability of the support), photoautotrophic cultivation of microalgae still requires light, and thus a transparent support and operation with low cell density (Malik, 2002; Vilchez et al., 1997). Another complicating factor for cultivation of immobilized microalgae is the fact that the support is attacked by the chelating compounds of the culture media, such as EDTA (Lee and Shen, 2004). *Anabaena variabilis* was immobilized by Markov et al. (1995) for CO_2 fixation and hydrogen production in hydrophilic cellulosic hollow fibers.

The largest field of immobilized microalgae application is wastewater treatment, where the microalgae are widely used for the absorption of nitrogen, phosphorus and heavy metals (Shi et al., 2007). One of the biggest challenges in the use of microalgae for wastewater treatment is the separation of cells in the treated effluent. In this case, the immobilization avoids problems with energy costs, separation or loss of cells (Malik, 2002), as reported for *Scenedesmus bicellularis* (Kaya and Picard, 1996) and *Chlorella* (Hernandez et al., 2006).

3. OPERATION OF CULTURES

The cultivation of microalgae biomass is carried out in batch, fed batch, semicontinuous or continuous operation. The variation in the nutrient concentration and the physicochemical conditions of the culture medium, which occurs in batch cultures, exposes the biomass to different conditions as changes in light with the increase in cell concentration and the concentration of nutrients that are consumed over time. However, batch cultures are less susceptible to contamination by other organisms since they are under aseptic conditions. The time and costs associated with downtime for cleaning, loading and unloading of the reactor make the use of this type of operation virtually impossible for commercial cultivation.

Repeated batch culture, or semi continuous mode, in which successive batches of biomass are harvested and the medium is replaced (Radmann et al., 2007, Otero et al., 1998) overcomes some of these problems; the cells can stay in the reactor for longer than with simple batch cultivation, and the nutrient concentrations can be controlled.

The semicontinuous cultivation of microalgae is reported by Otero et al., (1998) for *Phaeodactylum tricornutum*; Fábregas et al. (1998) for *Porphyridium cruentum,* Fábregas et al. (1995) for *Dunaliella tertiolecta*, Voltolina et al., (2005) for *Scenedesmus obliquus,* Morais et al. (2009); Masojídek et al. (2003); and Travieso et al. (2001) for *Spirulina platensis*. The main parameters that influence the kinetics in semicontinuous cultivation are the cut-off concentration and replacement rate of fresh medium, both of which are the subject of several studies (Radmann et al., 2007; Reinehr and Costa, 2006; Fábregas et al., 1996).

For continuous cultivation of microalgae, installations with more equipment and controls are required than in the other methods of cultivation, but the maintenance of constant conditions and cells with high rates of growth means this system is often used. The continuous cultivation of *Spirulina platensis* was studied by Vernerey et al. (2001) and Morist et al. (2001); *Isochrysis galbana* was studied by Otero et al. (1997); *Porphyridium cruentum* by Fuentes et al. (1999); and *Porphyridium purpureum* by Baquerisse et al. (1999).

4. NUTRIENTS FOR CULTIVATION

Although microalgae are photosynthetic organisms, essentially photoautotrophic, mixotrophic cultivation is also used for biomass production. In this type of cultivation, autotrophic metabolism (inorganic carbon source) occurs simultaneously with heterotrophic metabolisms (organic carbon source) (Andrade and Costa, 2007; Chen and Zhang, 1997).

Culture media chemically defined for the main species of microalgae have been studied for many years in the literature (Watanabe, 1960; Zarrouk, 1966; Ripka et al., 1979). They are mainly inorganic media, which consist of combinations of salts (carbonates and bicarbonates) and they are used as a source of carbon. CO_2 can also be used as a source of carbon for microalgae growth by injecting it into cultures of natural sources such as the air, or concentrated sources such as combustion gases. Some microalgae require organic nutrients such as vitamins (Watanabe, 1960).

The cost of nutrients is an important factor in microalgae cultivation, and therefore alternative sources have been sought to reduce spending. In the commercial cultivation of microalgae, sources of nutrients available in nature are used, with complex composition, such as sea water, industrial effluents, and fertilizers, among others. On the other hand, this search for alternative nutrients favours the use of microalgae in various processes that can generate mainly environmental benefits. These processes include the use of liquid and gas effluents as sources of nutrients, resulting in the treatment of these effluents and in environmental benefits combined with the biomass production.

Studies of alternative sources of nutrients include natural waters of alkaline lakes, such as water from Mangueira Lagoon in the extreme south of Brazil that is used for the cultivation of native *Spirulina* (Costa et al., 2002); brackish water or waste water from desalination, industrial, agro-industrial and municipal government wastewater, as sources of phosphorus and nitrogen; combustion gases from thermoelectric power plants as a source of CO_2 (Benemann, 1997; Maeda et al., 1995, Sakai et al., 1995). For heterotrophic or mixotrophic growth, glucose, acetate, glycerol or other sources of nutrients that are residues from the sugar industry are used (Olguín et al., 1995), including molasses (Andrade and Costa, 2007). In accordance with the requirements, two or more sources of these nutrients can be combined, complementing each other, or synthetic components of the culture media can be added (Morais et al., 2009).

5. THE USE OF CO_2 AS A CARBON SOURCE FOR MICROALGAE

The photoautotrophic nature of microalgae enables CO_2 from burning fuel to be injected into cultures as a carbon source for biomass growth. Microalgae contribute through these processes, to the mitigation of the greenhouse effect, mainly in power plants, where large quantities of CO_2 are emitted by burning coal. The Department of Energy in the US estimated in 2000 that 7% of all CO_2 emissions on the planet originate from thermal power plants (USDOE, 2000).

The use of microalgae for CO_2 fixation of the combustion gases from coal is interesting from an operational standpoint, because the gas stream, which contains about 12 to 15% CO_2 can be piped to the cultures and the gas can be used before being released to the atmosphere, where CO_2 concentration is about 0.038% (Takahashi, 2004).

The use of microalgae for CO_2 fixation is an efficient use for the gas, because the biomass generated can be used as raw material for biofuel production (Benemann, 1997) or used for direct burning alone or together with coal in power plants. It is estimated that 1,000 hectares of microalgae can process about 210,000 tons/year of CO_2 (Kadam, 2002)

Studies on the use of microalgae for CO_2 fixation focus on the tolerance of microalgae to the other gases in the combustion gases (especially nitrogen and sulfur oxides), on the increased absorption of gas in liquid media, on the resistance of cells to high concentrations of CO_2 and on the composition of the produced biomass.

Monoraphidium minutum resisted concentrations of up to 200 ppm of SOx and 150 ppm of NOx when grown in simulated combustion gas, with 13.6% CO_2, 0.015% of NO, 0.02% of SO_2, and 5% of O_2 (Brown, 1966). *Botryococcus braunii*, *Chlorella vulgaris* and *Scenedesmus sp.* were grown with 10% CO_2 which resulted in yields of 217.50 mg $L^{-1}.d^{-1}$ for the biomass of *Scenedesmus*, with 9% lipids, and yields of 26.55 mg.$L^{-1}.d^{-1}$ for *B. braunii*, with 21% lipids. *Chlorella vulgaris* grown with different CO_2 concentrations (0,10, 20 and 30% v/v) after 4 days of culture had a greater cell concentration at the largest gas concentrations (Jeong et al., 2003), which indicates that the cell adapts to the presence of CO_2.

Chlorella isolated from hot springs in Japan withstood temperatures higher than 42 °C with 40% CO_2 (Sakai et al., 1995). *Scenedesmus obliquus* and *Chlorella kessleri* have been isolated in southern Brazil in a thermal power plant (Morais and Costa, 2007).

By injecting combustion gas from liquefied petroleum gas (0.3 vvm), the lipid concentration in the biomass of *Scenedesmus* microalgae increased 1.9 times and in the biomass of *Botryococcus* it increased 3.7 times (Yoo et al., 2010). *Chlorella sp.* cultivated with 2%, 5%, 10% and 15% CO_2 in a current of 0.25 vvm, had higher specific growth rates with increasing cell density which was inoculated in the bioreactor, which indicates that highly concentrated cultures of microalgae enable the use of gases with higher CO_2 concentrations (Chiu et al., 2008).

6. WASTEWATER TREATMENT AND BIOREMEDIATION AS A NUTRIENT SOURCE FOR MICROALGAE

The use of microalgae for the tertiary treatment of wastewater and bioremediation combines the growth of biomass with environmental benefits through the removal of organic and inorganic nutrients, and heavy metals. The applications mainly include the removal of carbon, nitrogen, phosphorus, sulfur compounds, cyanide, chromium and zinc from industrial effluents or contaminated environments. The treatment by microalgae can be combined with CO_2 sequestration, resulting in a double environmental benefit, which can be even further maximized if the biomass is collected and used as biofuel (Muñoz and Guieysse, 2006).

Scenedesmus, *Spirulina* and *Chlorella* are the most commonly genera of microalgae used to remove nitrogen and phosphorus from effluent, followed by *Nannochloris*, *Botryococcus* and *Phormidium* (Órpez et al., 2009).

The use of microalgae not only results in consumption of nutrients from the effluent, it also provides oxygen through photosynthesis that is required by the bacteria that aerobically degrade the nutrients. This reduces the need for agitation and aeration.

There are several challenges when it comes to using microalgae to treat wastewater. These include: the tolerance of microalgae to toxic chemicals in the effluent, such as phenols and heavy metals; the area of land that is required; the separation of the biomass from the effluent; the control of species present (some species may be concentrated if they are not removed during harvesting by chemical flocculation); and the different optimum conditions for treating wastewater and achieving high yields of microalgae (for efficient treatment, the nutrients must be exhausted, while for the highest yields the concentration of nutrients must be kept constant). The use of immobilized cells has been proposed (Mallick, 2002) to manage some of these problems.

Microalgae are used to remove heavy metals (Cr, Hg, Cd, Ag, Pb and Sn) from effluents and contaminated environments, and have the advantage of removing metals that are present in trace concentrations in large volumes of liquid. Other methods such as electrodialysis, evaporation, precipitation, and reverse osmosis are economically unviable (Rangsayatorna et al., 2002). Other metals such as Zn, Cu and Ni, which at low concentrations can be micronutrients, have toxic effects at high concentrations and should be removed from wastewater and watercourses. The cells can be used as free or immobilized cells (Mallick, 2002).

Metals can be removed by adsorption on the surface of microalgae cells without being metabolized (biosorption), metabolized within the cell (bioaccumulation), or chemically transformed by the cells. Biosorption is a non-specific process and depends on the interaction between cell surface molecules (usually proteins, carbohydrates and lipids) with the metals. The removal of a metal by bioaccumulation is a specific process that depends on the cell identification of the metal as a nutrient or toxic substance, and the extent of removal depends on this (Chojnacka et al., 2004).

According to their nature and concentration, metals can be toxic to algae, a fact that has prompted many studies regarding the kinetics of consumption of these components, the tolerance to the toxic effects of metals, the kinetics of cell growth and the composition of biomass grown in the presence of metals.

Scenedesmus obliquus and *Scenedesmus quadricauda* were grown at concentrations up to 8.0 ppm of zinc. Between 0.5 and 8.0 ppm, the increase in the concentration of metal increased its adsorption but decreased the dry weight of cells, chlorophyll a and b content, carotenoids and amino acids (Omar, 2002).

Spirulina platensis was studied for the biosorption of cadmium by Rangsayatorna et al. (2002). The maximum capacity of metal adsorption was 98.04 $mg.g^{-1}$ of biomass and the higher the concentration of cadmium, the more damage to cells, which caused changes in the thylakoid membranes and cell lysis. The maximum capacity of cadmium adsorption for the microalgae *Tetraselmis suecica* was 40.22 $mg.g^{-1}$, with higher capacity of intracellular removal when compared with surface removal (Rama et al., 2010). Mixotrophic cultivation of *Spirulina sp.* 0.25 gL^{-1} was studied by Chojnacka et al. (2005), for the removal of metals from the real effluent of a copper smelter and refinery, which had various metals and high concentrations of mercury, cadmium and ammoniacal nitrogen. No nutrient was added to the effluent and its composition was sufficient for the growth of microalgae; the *Spirulina* concentrated the metals in its cells with concentration factors ranging between 80 and 4,250.

The ability to selectively remove and concentrate the metals enables them to be retrieved from the microalgal biomass by processes such as incineration or elution. This alternative, although incipient, is a field worthy of investigation for the recovery of metals, especially those of high value (Mallick, 2002).

7. Recovery of Biomass

The recovery of the biomass from the liquid medium is the most costly operation in the production of microalgae, because it involves the separation of solids from a small diameter from a low concentration liquid. Low concentrations of solids in the cultivation of microalgae, compared to other processes are required in order to provide appropriate growth conditions, such as a low concentration of toxic compounds, and to allow the incidence of light.

Because the conditions that provide energy savings in the recovery are the opposite of those conditions that give the best yield, the recovery of biomass from diluted media is a broad field of study. Several studies have focused on increasing yields and concentrations in the liquid medium, on reducing the energy costs of recovery and on the overall cost of production. Chen and Zhang (1997) investigated the advantages of mixotrophic cultivation of *Spirulina platensis* and obtained cell concentrations of up to 10 gL^{-1}. However, several studies focus on the concentration to be maintained in the bioreactor to optimize productivity, according to the incidence of light. Vonshak et al. (1982) reported that 0.40 to 0.50 gL^{-1} is the optimal concentration to be maintained in *Spirulina* cultivation for maximum photosynthetic efficiency; however, this value depends on the height of the culture (Richmond et al., 2003).

The operations to recover biomass of microalgae from the culture medium include filtration, centrifugation, flocculation, flotation or gravitational decantation once agitation has ceased. These operations can be used alone or in combination, such as flocculation and centrifugation. The culture medium can be recycled to the reactor, to reutilize the nutrients.

The use of chemical flocculants, such as aluminum sulfate $Al_2(SO_4)_3$, ferric sulfate $(Fe_2(SO_4)_3)$ and ferric chloride $(FeCl_3)$ is acknowledged to be the cheapest method of collection; however, it has the disadvantage of accumulating metals in the medium and can contaminate the biomass (Grima et al., 2003).

Flotation is caused by the interruption of aeration and agitation of the medium, resulting in an agglomeration of cells that can be collected. This method avoids the use of chemical agents, and is similar to gravitational decantation. Filamentous microalgae such as *Spirulina* and *Anabaena* are easier to harvest, due to the sizes of filaments, when compared with other unicellular microalgae such as *Dunaliella*, *Chlorella*, *Scenedesmus*, allowing lower centrifugation speeds and lower loss of load in filters. Techniques for recovery of smaller cells include microfiltration and ultrafiltration (Brennan and Owende, 2010).

The harvesting of microalgae is an important area of research for biomass production, and the search for efficient and automated methods of harvesting microalgae biomass is imperative to reduce production costs. About 40 tons of *Spirulina* are produced annually on the banks of Kossorom lake, and are recovered by filtration in wicker baskets and the biomass is dried by placing it on the ground under the sun (Abdulqader et al., 2000).

For drying, which removes the residual water in the biomass, techniques such as spray-drier, drum-drying, sun-drying or lyophilisation are used, according to the final product to be obtained. The most used technique is spray-drying, for products with high added value, but this can cause degradation of the biomass's pigments (Grima et al., 2003).

8. MICROALGAL BIOMASS IN NUTRITION AND AS A FEED

The field of intentional application of microalgae with the longest record is food. Some authors report evidence that the microalga *Nostoc* was used in Asia 2000 years ago, and that the genus *Spirulina* was used by the Aztec civilization in Lake Texcoco, Mexico about 700 years ago. In Africa, there is evidence of *Spirulina* use in Lake Chad from the same period, where even today the biomass is produced using simple techniques and consumed by people living by the lake (Abdulqader et al., 2000). Microalgal biomass is used as food due to its high protein content, which can be up to 74% in *Spirulina* (Cohen, 1997) with a balanced amino acid composition, high digestibility, absence of toxic compounds, and the presence of vitamins, minerals, polyunsaturated fatty acids and pigments with antioxidant properties. The biomass of some microalgae, such as *Spirulina* and *Chlorella* are GRAS certified (Generally Recognized As Safe) by the FDA (Food and Drug Administration). Studies with other microalgae for food include the microalgae *Porphyridium cruentum* (Fuentes et al., 2000).

The microalgae *Spirulina*, *Chlorella* and *Scenedesmus*, along with fungi and bacteria are sources of single cell protein (SCP), defined as proteins extracted from microbial biomass grown for this purpose (Anupama and Ravindra, 2000). The proteins of microalgae generally have a higher nutritional value than vegetables, such as wheat or rice, but lower than proteins of animal origin such as milk, meat and egg. Currently, microalgal biomass is eaten whole, added to food supplements in capsule form or as a component in the formulation of foods such as biscuits (Morais et al., 2006). In animal nutrition, microalgae are used in aquaculture, mainly for fish and shrimp, including *Tetraselmis sp., Spirulina sp., Chaetoceros, Chlorella, Isochrysis, Pavlova, Phaeodactylum, Nannochloropsis, Skeletonema and Thalassiosira* (Chuntapa et al. 2006; Borowitzka, 1997; Yamaguchi, 1997). It is also used for pigs, chickens and flamingos (Grinstead et al., 2000). Microalgal biomass is used for animal feed because of its nutritional properties, as well as for the coloration it provides to aquatic animals, especially salmon, carp, trout and oysters. The microalgae *Haematococcus* (producer of *astaxanthin*) and *Spirulina* (a source of carotenoids) are widely used for this purpose (Zhang et al., 2009; Domínguez et al., 2005; Lorenz and Cysewski, 2000).

9. THERAPEUTIC EFFECTS ATTRIBUTED TO MICROALGAE

The use of microalgal biomass as food produces therapeutic benefits in addition to the nutritional benefits. Most of the studies in the literature focus on the microalgae *Chlorella* and *Spirulina*, the most consumed microalgae in the world as foodstuffs.

Due to the presence of γ-linolenic acid, *Spirulina* and *Chlorella* (Colla, 2008) have a hypocholesterolemic effect. In hypercholesterolemic patients, γ-linolenic acid is 170 times more effective than linoleic acid, which is commonly found in plants. γ-linolenic acid also acts against the symptoms of premenstrual tension, multiple sclerosis and Parkinson's disease (Cohen, 1997).

Some studies report that *Spirulina* stimulates the repair mechanisms of DNA that has been damaged by radiation effects, attributing this to the microalgae's polysaccharides. Some of these also have immunostimulating and regulatory activities, (Qishen, 1989), such as b-1,3-glucan from *Chlorella* (Iwamoto, 2004).

Anticancer properties of *Spirulina* are attributed, in part, to the antioxidant properties of carotenoids, such as β-carotene, the precursor of vitamin A in mammals. A study by Semba et al. (1994) showed that the transmission of the HIV virus from an infected mother to her child is strongly dependent on vitamin A deficiency, so the availability of foods that have high levels of this vitamin is important in populations with high rates of HIV infection, such as those in parts of Africa. With HIV/AIDS, *Spirulina* has also been shown to reduce the rate at which the virus replicates, to reduce the viral load and to boost the immune system of HIV-positive patients. This effect was attributed to the polyanions that regulate homeostasis in the microalgae in alkaline environments (Belay, 2004; Teas et al., 2004).

Cyanovirin-N, a compound that prevents virus replication *in vitro* was isolated from an aqueous extract of the cyanobacterium *Nostoc ellipsosporum* (Boyd et al. 1996; Gustafson et al. (1996) cited by Schaeffer and Krylov, 2000). Another *in vivo* study carried out by Hirahashi et al. (2002) demonstrated that the aqueous extract of the microalgae activated the human immune system.

Pernicious anemia, caused by vitamin B_{12} deficiency, can be prevented by diets containing *Spirulina*, due to its content of this vitamin. Anemias caused by iron deficiency, which mainly occur in pregnant women and children are another example of problems that *Spirulina* can help resolve.

Spirulina also has beneficial effects against child malnutrition as reported by Fox (1993), who described a case of severely malnourished children who recovered with the administration of 1g of *Spirulina* daily for 3 months. Degbey et al. (2004) evaluated the daily administration of 10 grams of *Spirulina* in malnourished children between 6 and 24 months of age, and found that within in seven days their clinical symptoms had improved.

10. MICROALGAL BIOPRODUCTS

In addition to the health benefits that microalgae provide through the use of biomass in foodstuffs, biologically active compounds can be obtained from the biomass, such as antioxidants, polyunsaturated fatty acids (Omega-3 and Omega-6), phycocyanin (phycobili-proteins) phycoerythrin, astaxanthin, beta carotene and antibiotics, among others.

For use in food, biomass is the product itself, but it can be partitioned to obtain an intracellular or extracellular byproduct. The cellular localization of the metabolite defines the downstream processes that are required to obtain the final product. When biomass is the actual product, it is generally sufficient to recover it from the liquid medium, to wash it (to remove salts) and to dry it. In the case of intracellular bioproducts such as pigments, lipids, biopolymer granules, among others, the downstream processes include the operations of breaking down the cell, separation of the product and its purification (Grima et al., 2003). In the case of products that are excreted in the medium, such as polysaccharides (De Philippis and Vincenzini, 1998), there is no need to break down the cells, only to recover the product from the liquid and purify it. In such cases, the biomass can be reused.

Methods that dispense with the destruction of the cells to obtain the intracellular products have been studied. The extraction of beta-carotene from *Dunaliella salina* without disintegration of the cells was studied by Hejazi and Wijffels (2004). They cultivated the microalgae in a flat panel, two-phase bioreactor with a biocompatible organic solvent continuously recirculated through the aqueous phase for extraction of the product. Because the biomass remained in the reactor, there was no need for growth phases, recovery or breaking down of the cells to extract the bioproduct.

To break down the cells, grinding, heat and pressure (autoclaves) are used as well as acids or bases, ultrasound, freezing and thawing.

Solvent extraction is widely used for metabolites such as lipids, mainly fatty acids, and pigments such as astaxanthin and beta-carotene. The presence of a cell wall in eukaryotic microalgae, such as *Chlorella* and *Scenedesmus*, influences the extraction operation, because it hinders the contact of solvent with the desired solute.

10.1. Carotenoids

Microalgae are an important source of carotenoids which have antioxidant and colorant properties for the food, cosmetics and animal feed industries. Carotenoids have effects that are anticarcinogenic, anti-aging, that strengthen the immune system, are anti-inflammatory, and that protect against Parkinson's and Alzheimer's. Natural carotenoids compete with synthetic ones that are produced at lower prices, however, the trend of increased demand for natural products coupled with their effects on living organisms that consume them, means they are gaining in popularity.

H. pluvialis is the most common microalgal source of astaxanthin. Its cells contain between 1 to 5% of the pigment (Lee and Soh, 1991). The microalgae are grown in open tanks or photobioreactors, and a two-stage process can be used, with optimal conditions for growth and production of the pigment that is a secondary metabolite. For the accumulation of astaxanthin to occur, the culture must receive an excess of light and a lack of nutrients.

The use of flashing light increased the production of astaxanthin per photon by 400% when compared with pigment produced under continuous light (Kim et al, 2006).

Astaxanthin is intracellular, but during processing and storage it is exposed and is susceptible to oxidation, especially with light and temperature, and it is therefore a challenge to maintain the stability of the highly unsaturated pigment during production (Gouveia and Empis, 2003). One solution that has been studied to protect it from oxidation was encapsulating it with chitosan (Kittikaiwan et al., 2007).

Although *Haematococcus* is the main source of this pigment, there are also reports of astaxanthin being produced by *Chlorella zofingiensis* (Ip et al., 2004).

Dunaliella sp. (especially *D. salina*) is a source of beta-carotene, containing about 4% of the pigment in its biomass; sometimes this content is reported to be as high as 14% (Spolaore et al., 2006).

As with *Haematococcus*, two-stage cultivation results in increased productivity of beta-carotene with *Dunaliella*. The first stage has an excess of nitrogen and low salinity, and the second stage has limited nitrogen and increased salinity (Amotz, 1995). An increase in the concentration of carotenoids associated with increased salinity, light and temperature was observed in species of *Dunaliella* by Borowitzka and Borowitzka (1988). Due to the conditions of salinity (2.0-4.0 M of NaCl), the cultures can be grown in open bioreactors, which minimizes the risk of contamination (Amotz, 2004).

Other carotenoids from microalgae include lutein, produced by *Muriellopsis sp.* and *Chlorella vulgaris* (Li et al., 2001; Shi et al., 1999) and the botryoxanthins produced by *Botryococcus braunii* (Okada et al., 1996).

10.2. Phycobiliproteins

Along with carotenoids and chlorophyll, phycobiliproteins are other pigments that absorb sunlight for the photosynthesis of microalgae. Commercially, phycobiliproteins are used as colorants in the food industry, especially in candy and ice cream, in the cosmetics industry with lipsticks, and in the pharmaceutical industry as biochemical diagnostics. Several studies have reported their beneficial effects such as anticarcinogenic and immunomodulating activity, mainly attributed to their antioxidant properties (Benedetti et al., 2004; Estrada et al. 2001; Bhat and Madyastha, 2000).

The phycobiliproteins include phycocyanin and phycoerythrin. The main microalgal sources of phycobiliproteins are the two genera *Porphyridium* (Singh et al., 2009) and *Spirulina* (Silveira et al., 2007). Because phycocyanin is a photosynthetic pigment, strictly heterotrophic cultivation of microalgae is not suitable for production of the colorant. Thus, phycocyanin production was studied in mixotrophic cultivation of *Spirulina platensis* (Chen and Zhang, 1997) and produced 10.24 $g.L^{-1}$ of biomass and 795 $mg.L^{-1}$ of phycocyanin. These values are about 5 and 3 times greater, respectively, than those obtained in autotrophic cultures of microalgae produced under the same conditions.

The mixotrophic cultivation of *Spirulina* was also studied by Chen et al. (1986). The presence of glucose and acetate increased the growth and production of phycocyanin, which was 322 mg/l in cultures containing glucose, with a specific growth rate of 0.62 d^{-1} and cell concentration (2.66 $g.L^{-1}$). The values obtained in cultures with acetate as the carbon source, were 246 $mg.L^{-1}$, 0.52 d^{-1} and 1.81 $g.L^{-1}$, respectively of *Spirulina* biomass.

Fuentes et al., 2000, studied *Porphyridium cruentum* cultures and collected the biomass at different growth stages and with different residence times in the reactor. The highest quantity of pigments found was the phycobiliproteins (20.20 $mg.g^{-1}$ of phycoerythrin and 2.62 $mg.g^{-1}$ of phycocyanin) and the smallest quantity was the carotenoids (1.02 $mg.g^{-1}$).

The extraction and purification of phycobiliproteins is a vast field of study (Minkova et al., 2003). Silveira et al. (2007) optimized conditions for extraction of phycocyanin from *Spirulina* using a factorial design and response surface techniques, and found that water was the best solvent for pigment extraction, with a biomass-solvent ratio of 0.08 $g.mL^{-1}$, for 4h at 25 °C. Patil et al. (2008), studied the separation and fractionation of C-Phycocyanin and allophycocyanin from *Spirulina platensis* using aqueous two phase extraction (PEG and salt), a non-chromatographic method. They obtained an increase in purity when using multiple extractions and the combination of two phase extraction with membrane process (ultrafiltration).

11. POLYUNSATURATED FATTY ACIDS - PUFAS

There is a great demand for polyunsaturated fatty acids, especially the ω3 and ω6 families such as: eicosapentaenoic acid (EPA) 20: ω3, 6, 9, 12, 15; gamma-linolenic acid (GLA) 18:3 ω6, 9, 12; docosahexaenoic acid (DHA) 22:6 ω3, 6, 9, 12, 15, 18; and arachidonic acid (AA) 20:4 ω6, 9, 12, 15. This demand is due to their effect in combating chronic inflammation such as rheumatism, skin diseases, poor circulation, arteriosclerosis, and hypercholesterolemia (Ward and Singh, 2005). Comparing eukaryotic microalgae with cyanobacteria the former generally contains a higher concentration of polyunsaturated fatty acids (Guschina and Harwood, 2006)

Traditional sources of PUFAS include fish that feed on microalgae, and oils extracted from these fish, whose residual taste and ease of oxidation limits their application in foods. The main microalgae used for production of PUFAs are of the genera *Nanochloropsis*, *Phaeodactylum*, *Porphyridium* (Fuentes et al., 2000), *Spirulina* (Sajilata et al., 2008), *Isochrysis galbana*, *Nitzschia*, *Crypthecodinium* (DHA) and *Schizochytrium* (Guschina and Harwood, 2006).

The content and composition of lipids in microalgae varies according to the species, with the stages of growth and with the growing conditions. Conditions that increase the production of lipids include nitrogen deficiency, excess light and high salinity. Among these, the effect of nitrogen limitation is the most widely studied for stimulating the synthesis of fatty acids in microalgae.

The total lipid content of *Haematococcus pluvialis* was 15.61% under optimal conditions and increased to 34.85% when the cells were exposed to continuous light without nitrogen limitation. Under nitrogen limitation, the total lipid content was 32.99% dw, with the fatty acid profile similar in both conditions, containing palmitic, stearic, oleic, linoleic, linolenic and linolelaidic acids (Damiani et al., 2010).

The lipid content and fatty acid composition of *Schizochytrium limacinum* was studied at temperatures from 16 to 37 °C and salinities of 0 to 3.6% (16, 23, 30, 37 °C). The best growth and lipid concentration occurred at 16-30 °C and 0.9-3.6% salinity. The growth and lipid concentration were reduced at 37 °C and zero salinity (Zhu et al., 2007).

Olguín et al., (2001) cultivated *Spirulina sp.* with seawater supplemented with effluent from the anaerobic process of treatment of waste from pigs and they studied the effects of two light intensities (66 and 144 µmol.m^{-2}.s^{-1}) and nitrogen deficiency. The biomass obtained from the complex medium had 28.6% lipids when exposed to lower light intensities and 18% under the higher light intensity.

In Zarrouk medium, these results were 8.0 and 6.4%, respectively. Light intensity only affected the palmitoleic acid content (C16:1). The concentration of linolenic acid (C18:3) was influenced by the composition of the culture medium, and was higher in Zarrouk medium (31.20%) than in complex medium (28.13%).

12. The Use of Microalgae in Agriculture

Microalgae are used as fertilizer or soil conditioners because of their ability to fix nitrogen, produce extracellular polysaccharides (De Philippis and Vincenzini, 1998) and provide nutrients, improving the absorption capacity of water and soil composition (Riley, 2002; Issa et al., 2007). These organisms are natural components of soils, either alone or in colonies, and can be added to improve soil characteristics. Microalgae can grow in symbiosis with plants or microalgal biomass can be used integrally as biofertilizer or after the extraction of biocompounds (Grimston et al., 2001).

Studies were carried regarding the screening of cyanobacteria to improve N fertility and structural stability of degraded soils, and evaluate their effectiveness in semiarid soils of South Africa. Out of 97 isolated cyanobacteria, three strains of Nostoc were heterocystous, with appreciable nitrogenase activity and the ability to produce exocellular polysaccharides. The strains that had one of these properties were deficient in the other. The ability of these to influence the dry matter (DM) of maize and soil C and N contents was tested in nitrogen-poor soil. The authors found that indigenous cyanobacteria strains screened for greater N_2-fixing ability have the potential to improve the productivity of N-poor soils in semiarid regions in South Africa (Maqubela et al., 2010).

Domracheva et al. (2010) studied the fungicidal effect of the cyanobacteria *Nostoc Linckia*, *N. commune*, and *Microchaeta tenera* on the pathogen *Fusarium*. Films of cyanobacteria *N. paludosum*, *N. linckia*, and *Microchaeta tenera* on the lawns of *F. oxysporum*, *F. nivale* and *F. culmorum* resulted in growth delay, drying, and lysis of the fungal mycelium (Domracheva et al., 2010). The anti-*Fusarium* activity of cyanobacteria in soil was also observed in the simulation experiment without plants. The introduction of *F. culmorum* spores into the soil, and addition of cyanobacterial cultures resulted in suppression of the fungus and a significant decrease in mycelial length compared to the experiment without cyanobacteria.

13. Biopolymers Extract to Biomass

The polyhydroxyalkanoates (PHAs) are natural polyesters, and consist of units of hydroxyalkanoic acids that have similar properties to petrochemical plastics, and are also biodegradable. They are produced as a reserve of carbon and energy and are accumulated within the cells of various microorganisms which are under special conditions of growth (Sudesh et al. 2000).

Polyhydroxyalkanoates are a class of polymers, the most studied of which are poly-3-hydroxybutyrate (PHB) and poly(3-hydroxybutyrate-co-3-hydroxyvalerate) (PHB-HV). The composition of the lateral chain or of the R radical, and the n value determine the identity of the monomer unit (Figure 1).

During the synthesis of PHB, two molecules of acetyl-CoA are coupled to form acetoacetyl-CoA in a condensation reaction catalyzed by the enzyme 3-β-cetotiolase. The enzyme 3-β-cetotiolase competes for acetyl-CoA with several other metabolic pathways, including the formation of acetate, citrate and fatty acid synthesis. The product is reduced to (R)-3-hydroxybutyryl-CoA in a reaction catalyzed by the enzyme NADPH-dependent acetoacetyl reductase. High concentrations of NADH and NADPH inhibit the citrate synthase enzyme, which is responsible for the entry of acetyl-CoA in TCA (tricarboxylic acids cycle), which means the acetyl-CoA remains available for the 3-β-cetotiolase. PHB is synthesized by the polymerization of molecules of (R)-3-hydroxybutyryl-CoA by the enzyme PHA synthase (Khana and Srivastava, 2005).

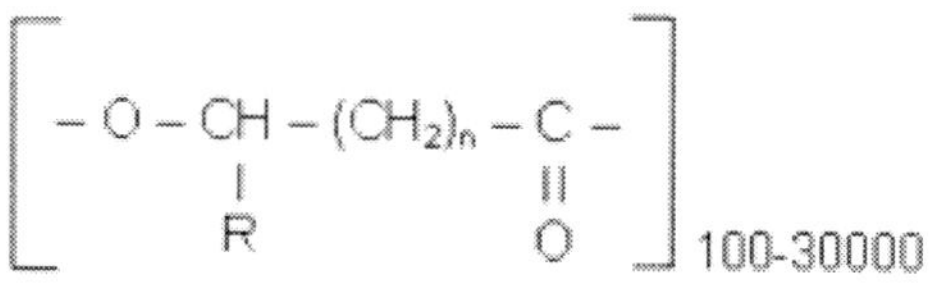

$$\left[-O-CH-(CH_2)_n-C- \atop \quad\quad R \quad\quad\quad\quad \|\atop \quad\quad\quad\quad\quad\quad O \right]_{100-30000}$$

Figure 1. Structure of polyhydroxyalkanoates (Lee, 1996).

The biosynthesis of PHB-HV requires a precursor such as propionic acid, itaconate, valeric acid, acetic acid and oleic acid. The synthesis of PHB-HV consists of the formation of propionyl-CoA from propionic acid and free CoA in a reaction catalyzed by the enzyme acyl-CoA synthetase.

The enzyme β-cetotiolase catalyzes not only the condensation of two molecules of acetyl CoA, but also the condensation of one molecule of propionyl CoA and one acetyl CoA molecule, forming the skeleton with 5 carbons and 3 keto-valeryl-CoA. The enzyme 3-ketoacyl-CoA reductase NADPH-dependent is not specific and may catalyze the reduction of 3-keto-valeryl-CoA to D-3-hydroxy-valeryl-CoA. The units D-3-hydroxy-valeryl-CoA also serve as substrate for PHA synthase enzyme activity, although this enzyme has a much higher specific activity with units D-3-hydroxy-butyryl-CoA.

To reduce the cost of biopolymer production, microalgal strains with high productivity are required, as well as the development of strategies for cultivation and use of low cost carbon sources. Several studies have shown that the synthesis of PHB is directly linked to the cultivation and nutritional conditions that the culture is exposed to. The growth phase, light-dark cycles, temperature, pH, nutrient limitation (phosphorus and nitrogen), excess nutrients (carbon), mixotrophic cultures, chemotrophic cultures and limitations of gas exchange (oxygen) have been investigated in order to increase the amount of PHB in the cells (Sharma et al. 2007).

Nishioka et al. (2001) obtained 55% of PHB with *Synechococcus sp.* MA19 under limited phosphate conditions. Sharma et al. (2007), cultivated *Nostoc muscorum* with 0.17% sodium acetate, 5 mg.L^{-1} K$_2$HPO$_4$, 0.16% glucose and 95h in the dark and obtained 47.4% of PHB. Experiments that were carried out with *Synechocystis* sp. PCC 6803 in non-continuous cultivation had a maximum accumulation of PHB in the stationary phase of growth (4.5% w/w), compared to the lag phase (1.8%) and the logarithmic phase (2.9%) (Panda et al. 2006).

Panda et al. (2006) obtained 11.2% of PHB with the microalga *Synechocystis* sp. PCC 6803 supplemented with 0.4% sodium acetate; however, the interaction of nitrogen deficiency with 0.4% acetate provided an accumulation of 14.6% of PHB. The addition of 0.5% sodium acetate in the cultivation of *Spirulina platensis* significantly increased the intracellular content of PHB (more than 10%). The presence of sodium acetate in the medium increases the concentration of acetyl CoA, which is the intermediate compound in the synthesis of PHB (Jau et al. 2005).

The use of microalgae is an environmentally correct method. Besides helping to reduce the greenhouse effect by biofixing CO_2 and treating waste, the biomass that is formed may be used for the development of biopolymers.

14. BIOFUELS

Increased energy demand and the need to reduce emissions of greenhouse gases motivate the growing number of studies on new energy sources that can be produced and consumed without increasing the concentration of atmospheric CO_2. Microalgae convert solar energy and CO_2 from natural sources or from fossil fuels into salvageable biomass, including the production of biofuels (Demirbas, 2010). Microalgae were first investigated as sources of biofuels in 1980 and since then there has been a growing number of studies and industrial facilities that exploit this capability (*Rupprecht*, 2009).

An important area of debate is the production of ethanol from starchy materials that are also foods such as sugar cane and maize, another is the production of biodiesel by oil crops used in food, such as soya and corn. Microalgae contribute to the solution of this controversy between biofuel and food (Zhang et al. 2010), because they require less land and do not need potable water for their production. They can grow in saline water or even in the desert, and do not compete with agricultural land and do not have harvest periods. In addition, microalgae can be genetically manipulated to increase the productivity of a compound of interest for biofuel production.

Microalgae fit into the concept of biorefineries (Demirbas, 2009; Mussgnug et al., 2010), that integrate biomass conversion processes to produce biofuels, energy and chemicals, similar to oil refineries. Biofuels produced from microalgae can be in liquid form such as ethanol, gas form such as bio-methane, or solid, such as the biomass itself.

14.1. Biomethane from Microalgae

Biomethane is the main fuel component in biogas, originating from the anaerobic digestion of organic matter, in concentrations ranging from 30 to 80% (v/v) of gas. Other gas components are carbon dioxide (CO_2) at about 30 to 45% (v/v), traces of hydrogen sulfide (H_2S) and water vapor. Small amounts of H_2, CO and volatile compounds such as amines and organic acids may be present (Kapdi et al., 2005).

Methane can be biologically produced by anaerobic digestion of organic matter, and is a product of the treatment of municipal and industrial wastes. In comparison to other biofuels, biomethane has the advantage that it is produced through biological processing of the entire organic fraction of biomass, unlike bioethanol and biodiesel, whose substrates are carbohydrates and lipids from the biomass, respectively. Thus, anaerobic digestion, compared to other processes of biofuel production, is operationally simpler and eliminates the drying and chemical pretreatment of raw material, using less energy and generating less waste (Mussgnug et al., 2010).

The use of algal biomass as a substrate for anaerobic digestion was studied by Demuynck and Nyns (1984), and microalgal biomass by Hernandez and Cordova (1993), Samson and

LeDuy (1986) and Keenan (1977). However, there are hardly any current records of microalgal biomass use for the production of biomethane (Mussgnug et al., 2010). Among the species studied for the production of biomethane are *Spirulina* (Costa et al., 2008), *Chlorella*, *Chlamydomonas* and *Dunaliella* (Mussgnug et al., 2010).

In the anaerobic digestion of microalgae, greater decompositions and efficiencies are accomplished than in the digestion of other materials, which can be mainly attributed to the low ash content in microalgal biomass and due to the absence of the cellulosic wall in cyanobacteria (Tomaselli, 1997) and organic compounds that degrade with difficulty such as lignin, aromatics, organochlorines, among others, present in other raw materials (Kusçu and Sponza, 2006; Angelidaki and Ahring, 2000).

The residual biomass from the production of other biofuels, such as the defatted biomass to produce biodiesel, or the fermented biomass for ethanol production, can be used as a substrate for the production of biomethane. After the production of biomethane, the residual biomass can be used as fertilizer.

Some of the problems with the use of microalgae as a raw material for biomethane are: the high cost of biomass; increasing the efficiency of converting biomass to biomethane; and the resistance of anaerobic bacteria to high concentrations of ammonia in the digestion of biomass with high protein content such as *Spirulina* (Costa et al., 2008). Furthermore, the physicochemical and operational parameters of the anaerobic process, such as temperature, pH, retention time and organic load are currently being studied.

The costs of biomass production can be reduced with the development of technologies for the use of microalgae in CO_2 sequestration, from which large quantities of biomass can be generated, and by using microalgal biomass produced from wastewater treatment.

14.2. Biodiesel

Biodiesel is a renewable, non-toxic, biodegradable and CO_2-neutral energy source. This biofuel has characteristics similar to normal diesel so it can be used directly in diesel-burning engines with less emission of CO or SO_x (Xin et al., 2011).

In recent years, it has become a hot topic for the exploitation of renewable and environment-friendly energy forms. Conventional biodiesel mainly comes from soybean and vegetables oils, palm oil, sunflower oil, rapeseed oil as well as restaurant waste oil (Huang et al., 2010).

Compared with conventional oil crops, microalgae are more attractive as feed-stock for biodiesel production, due to their high photosynthesis efficiency, lipid content, and the fact that the cultivation of microalgae does not require as much land as terraneous plants. Furthermore, biodiesel produced from microalgae does not compete with the production of food and other products derived from crops (Xin et al., 2011, Huang et al., 2010).

Many microalgal species can be induced to accumulate substantial lipid content. Although the mean lipid content varies from 1 to 70%, some species may contain as much as 90% (w/w_{DW}) under certain conditions. The synthesis pathways of triglycerides in microalgae may consist of the following three steps: the formation of acetyl coenzyme A in the cytoplasm, the elongation and desaturation of the carbon chain of fatty acids, and the biosynthesis of triglycerides in the microalgae. The fatty acid profile of the microalgal cell is also relevant, because the heating power of the resulting biodiesel is dependent upon its composition. Most of lipids are saturated and unsaturated fatty acids containing 12-22 carbon atoms, often of the ω3 and ω6 types (Amaro et al., 2011).

The structure of the biodiesel determines whether the biofuel is a feasible substitute for conventional energy.

The production cost is generally high for biodiesel (Huang et al. 2010). Microalgae have many advantages over other forms as they can be used to produce biodiesel at low cost. This is due to the fact that they use solar energy, simultaneously fix CO_2, and do not need extra organic carbon (unlike biological nitrification-denitrification). They discharge oxygenated effluents into water bodies, avoid the problem of sludge handling, and harvested microalgal biomass has a high economic potential (for feedstock, fertilizers, biogas, biofuels, and so on) (Xin et al., 2011).

Different nutritional and processing factors, cultivation conditions and growth phases will affect the fatty acid composition of microalgae, such as nitrogen deficiency and salt stress induced by the accumulation of C18:1 in all species, and of C20:5 to a lesser extent in *B. braunii* (Mata et al., 2010).

The intrinsic ability to produce large quantities of lipids and oil depends on the species and strain rather than the genus. Moreover, the lipid content increases when microalgae cells are subjected to unfavorable culture conditions, such as high salinity, nitrogen starvation, and high light intensity. In addition, the lipid composition in microalgae also depends on the age of the culture and the different life-cycle stages (Damiani et al., 2010).

14.3. Ethanol Produced from Microalgae

Ethanol is the most widely used liquid biofuel in the world, especially in the United States and Brazil, where sugar cane and corn are mainly used as sources of sugars for fermentation (first generation ethanol). Factors that contribute to the search for alternative sources for ethanol production include, besides the use of land and water, the respiratory problems that sugarcane burning causes in humans (Uriarte et al., 2009).

Currently a large number of studies have addressed the use of lignocellulose to produce second generation ethanol which is a vast field of international research (Sims et al., 2010). Although the research has not been exhausted, third-generation ethanol, produced from microalgal biomass is becoming another alternative (Subhadra and Edwards, 2010 Gressel, 2008).

Microalgae are interesting substrates for the production of ethanol, because in addition to their kinetic characteristics of growth and the conditions in which they grow, the concentration of carbohydrates in their biomass can be increased. The level of carbohydrates, and lipids, in the biomass of microalgae can be increased by nitrogen (Rigano et al., 1998) and phosphorus limitation (Dismukes et al. 2008).

Lighting is another factor that interferes with the carbohydrate content in the microalgal biomass. Fuentes et al. 1999 found that during the light period, *Porphyridium cruentum* stored energy as carbohydrates, and protein synthesis was low, while at night, the stored carbohydrates were consumed.

14.4. Hydrogen

Some microalgae have the ability to produce hydrogen, which is absent in other photosynthetic organisms, including higher plants (Rupprecht, 2009).

Hydrogen has attracted the attention of researchers as a source of energy because the product of its combustion is water, and there are no CO_2 emissions. Hydrogen can be produced by microalgae in a single stage using a direct or indirect biophotolysis reaction.

In direct water biophotolysis, there is simultaneous production of O_2 and H_2 by the action of hydrogenase. The enzymatic reaction of hydrogenase is inhibited by the O_2 that is produced, and the production of H_2 ceases quickly (Ghirardi et al., 1997), which is the main barrier for H_2 production by direct biophotolysis.

For large scale use, the O_2 that is generated needs to be removed by inert gases, or using substances that absorb the O_2, such as sodium dithionite (Kojima and Lin, 2004). In indirect biophotolysis, O_2 and H_2 are produced in different stages: CO_2 is fixed as carbohydrates through photosynthesis, and these are fermented into H_2. Although light is required, it is the most appropriate method for H_2 production by microalgae (Benemann, 2004). In both cases of biophotolysis, the system needs to be closed to collect the H_2 gas.

Another method of obtaining hydrogen from microalgal biomass is anaerobic digestion. As H_2 is an intermediary in the reaction sequence for methane production by anaerobic digestion, the process conditions can be manipulated to increase production of this compound. In this case, the microalgal biomass is used as substrate for anaerobic digestion, without biological activity (Costa et al., 2008). One of the advantages of producing hydrogen by anaerobic digestion is the continuous production of hydrogen, and it does not require light (Zhang et al., 2007).

The microalga that has been most studied for biofuel production is the genus *Chlamydomonas* (Fouchard et al., 2008; Nami, 2006), and there are also studies on *Chlorella* (Kojima and Lin, 2004) and *Anabaena* (Markov et al., 1995).

15. HYDROCARBONS

Long linear hydrocarbons similar to those present in petroleum can be obtained from the microalgal biomass of the genus *Botryococcus*, especially *Botryococcus braunii*. This colony-forming microalga produces linear and mono diunsaturated hydrocarbons C29 to C31 and penta-unsaturated branched chains of up to C34 in concentrations of up to 86% of the dry weight of biomass (Ji et al., 2010, Banerjee et al., 2002). To maximize the levels of these hydrocarbons, the microalgae should be exposed to stressful conditions, and the composition of these hydrocarbons depends on cultivation conditions.

These conditions include light, temperature, salinity and concentration of nutrients, especially nitrogen. Tran et al. (2009) optimized the medium for biomass and lipids production, regarding the content of sodium carbonate, potassium phosphate, calcium chloride, magnesium sulphate, ferric citrate, and sodium nitrate. They found that the components with the greatest effect were potassium phosphate and magnesium sulphate. The optimum concentrations of potassium phosphate and magnesium sulphate were 0.06 and 0.09 $g.L^{-1}$, respectively, for growth and 0.083 and 0.1 $g.L^{-1}$, respectively, for lipid production.

Botryococcus was grown at different salinities (NaCl 17 mM to 85 mM). The biomass yields increased with higher concentration of sodium chloride and maximum biomass was achieved in 17 mM and 34 mM salinity. The hydrocarbon content varied in the range of 12-28% in different salinities and maximum hydrocarbon content was observed in 51 mM and 68 mM of salinity (Rao et al., 2007).

To circumvent the high cost of *Botryococcus* biomass production, the culture has been studied in effluents (Órpez et al., 2009; An et al., 2003) and in combustion gas (Yang, et al. 2004; Murakami and Ikenouchi, 1997). Studies on the production of hydrocarbons from *Botrycoccus* include the extraction of hydrocarbons (Kita et al., 2010).

CONCLUSION

Microalgae have been present since the formation of the Earth's atmosphere. Later they were used as food, and more recently as sources of bioproducts. Currently, in addition to all these functions, they are important environmental agents and a promising sources of biofuels. The increasing world population, the need for food, the demand for bioactive substances, global warming and energy shortage problems are global in scale. It would be too bold and unfounded to suggest that microalgae can solve all these problems, because no single solution is capable of such a thing, but the microalgae are important agents in this scenario.

Lowering production costs, finding alternative sources of nutrients, minimizing costs of downstream processes, increasing productivity, developing high-efficiency bioreactors, discovering the exact relationship between biomass composition and physicochemical and environmental impacts, and resolving the problem of incidence of light to obtain highly concentrated crops are all challenges that still need to be overcome if we are to fully utilize the potential of these microorganisms.

REFERENCES

Abdulqader, G; Barsanti, L; Tredici, MR. Harvest of *Arthrospira platensis* from Lake Kossorom (Chad) and its household usage among the Kanembu. *Journal of Applied Phycology,* 2000, 12, 493–498.

Amaro, HM; Guedes, AC; Malcata, FX. Advances and perspectives in using microalgae to produce biodiesel. *Applied Energy*. 2011, IN PRESS.

Amotz, BA. Industrial Production of Microalgal Cell-mass and Secondary Products: Major Industrial Species. In: Richmond A, editor. *Handbook of microalgal culture: biotechnology and applied phycology*. Oxford: Blackwell Publishing; 2004; 255.

Amotz, BA. New mode of Dunaliella biotechnology: two-phase growth for b-carotene production. *Journal of Applied Phycology*, 1995, 7, 65–68.

Andrade, MR; Costa, JAV. Mixotrophic cultivation of microalga *Spirulina platensis* using molasses as organic substrate. *Aquaculture*, 2007, 264, 130-134.

Angelidaki, I; Ahring, BK. Methods for increasing the biogas potential from the recalcitrant organic matter contained in manure. *Water Science and Technology*, 2000, 41, 189–194.

Anupama and Ravindra, P. Value-added food: single cell protein. *Biotechnology Advances*, 2000, 18, 459-479.

Bailliez, C; Largeau, C; Berkaloff, C; Casadevall. E. Immobilization of *Botryococcus braunii* in alginate: influence on chlorophyll content, photosynthetic activity and degeneration during batch cultures. *Applied Microbiology and Biotechnology*, 1986, 23, 361–366.

Bañares, LR; Ballester, GD; Galváan, A; Fernández, E. Transgenic microalgae as green cell-factories. *Trends in Biotechnology*, 2004, 22, 45–52.

Banerjee, A; Sharma, R; Chisti, Y. Banerjee, UC. *Botryococcus braunii*: A renewable source of Baquerisse, D; Nouals, S; Isambert, A; Santos, PF; Durand, G. Modelling of a continuous pilot photobioreactor for microalgae production. *Journal of Biotechnology*, 1999, 70, 335–342.

Belay, A; Mass culture of *Spirulina* outdoors: The Earthrise farms experience. In: Vonshak A, editor. *Spirulina platensis (Arthrospira) Physiology, cell-biology and biotechnology*. London: Taylor and Francis; 1997; 131.

Benedetti, S; Benvenuti, F; Pagliarani, S; Francogli, S; Scoglio, S; Canestrari, F. Antioxidant properties of a novel phycocyanin extract from the blue-green alga *Aphanizomenon* flos-aquae. *Life Sciences*, 2004, 75, 2353–2362.

Benemann, JR. CO_2 mitigation with microalgae systems. *Energy Conversion and Management*, 1997, 38, S475–S479.

Benemann, JR. Hydrogen and Methane Production by Microalgae. In: Richmond A, editor. *Handbook of microalgal culture: biotechnology and applied phycology*. Oxford: Blackwell Publishing; 2004; 403.

Bhat, BV; Madyastha, KM. C-Phycocyanin: A Potent peroxyl radical scavenger in vivo and in vitro. *Biochemical and Biophysical Research Communications*, 2000, 275, 20–25.

Borowitzka, M; Borowitzka, L. *Dunaliella* In: Borowitzka, LJ, Borowitzka, MA. editors. *Microalgal biotechnology*. Cambridge: Cambridge University Press; 1988; Cambridge University Press, 1988; 27-58.

Borowitzka, MA. Commercial production of microalgae: ponds, tanks, tubes and fermenters. *Journal of Biotechnology*, 1999, 70, 313–321.

Borowitzka, MA. Microalgae for aquaculture: opportunities and constraints. *Journal of Applied Phycology*, 1997, 9, 393–401.

Boussiba, S; Zaritsky, A. N_2-fixing cyanobacteria as a gene delivery system for expressing mosquitocidal toxins of *Bacillus thuringiensis* subsp. Israelensis. In: Richmond A, editor. *Handbook of microalgal culture: biotechnology and applied phycology*. Oxford: Blackwell Publishing; 2004; 525.

Boyd, MR; Gustafson, K; McMahon, J; Shoemaker, R. Discovery of cyanovirin-N, a novel HIV-inactivating protein from *Nostoc ellipsosporum* that targets viral gp120. *International Conference on AIDS*, 1996, 11, 71.

Brennan, L; Owende, P. Biofuels from microalgae—A review of technologies for production, processing, and extractions of biofuels and co-products. *Renewable and Sustainable Energy Reviews*, 2010, 14, 557–577.

Brown, LM.. Uptake of carbon dioxide from flue gas by microalgae. *Energy Conversion and Management,* 1996, 37, 6-8, 1363-1367.

Chen, F; Zhang, Y. High cell density mixotrophic culture of *Spirulina platensis* on glucose for phycocyanin production using a fed-batch system. *Enzyme and Microbial Technology*, 1997, 20, 221-224.

Chen, F; Zhang, Y; Guo, S. Growth and phycocyanin formation of *Spirulina platensis* in photoheterotrophic culture. *Biotechnology Letters*, 1986, 18, 5. 603-608.

Chiu, SY; Kao, CY; Chen, CH; Kuan, TC; Ong, SC; Lin, CS. Reduction of CO_2 by a high-density culture of *Chlorella* sp. in a semicontinuous photobioreactor. *Bioresource Technology*, 2008, 99 3389–3396.

Chojnacka, K; Chojnacki, A; Górecka H. Biosorption of Cr^{3+}, Cd^{2+} and Cu^{2+} ions by blue-green algae *Spirulina* sp.: kinetics, equilibrium and the mechanism of the process. *Chemosphere*, 2005, 59(1), 75–84.

Chuntapa, B; Powtongsook, S; Menasveta. P. Water quality control using *Spirulina platensis* in shrimp culture tanks. *Aquaculture*, 2003, 220, 355–366.

Cohen, Z. The chemicals of *Spirulina*. In: Vonshak A, editor. *Spirulina platensis (Arthrospira) Physiology, cell-biology and biotechnology*. London: Taylor and Francis, 1997, 175.

Colla, LM; Baisch, ALM; Costa, JAV. *Spirulina platensis* effects on the levels of total cholesterol, HDL and triacylglycerols in rabbits fed with a hypercholesterolemic diet. *Brazilian Archives of Biology and Technology*, 2008, 51(2), 405-411.

Costa, JAV; Santana, FB; Andrade, MR; Lima, MB; Franck, DT. Microalga biomass and biomethane production in the south of Brazil. *Biotechnology Letters*, 2008, 136S, S402–S403.

Damiani, MC; Popovich, A; Constenla, D; Leonardi, PI. Lipid analysis in *Haematococcus pluvialis* to assess its potential use as a biodiesel feedstock. *Bioresource Technology*, 2010, 101, 3801–3807.

De Philippis, R; Vincenzini, M. Exocellular polysaccharides from cyanobacteria and their possible applications. *FEMS Microbiology Reviews*, 1998, 22, 151-175.

Degbey, H; Hamadou, B; Oumarou, H. Evaluation of effectiveness of the supplementation in *Spirulina* of the usual alimentation of severe malnourished children. In: *International Symposium: CSSD Cyanobacteria for Health, Science and Development*, 2004, 12.

Demirbas, A. Use of algae as biofuel sources. *Energy Conversion and Management*, 2010, 51, 2738–2749.

Demuynck, M; Nyns, E. Biogas plants in Europe. *Journal of International Solar Energy*, 1984, 2(6), 477–85, 1984.

Dismukes, GC; Carrieri, D; Bennette, N; Ananyev, GM; Posewitz, MC. Aquatic phototrophs: efficient alternatives to land-based crops for biofuels. *Current Opinion in Biotechnology* 2008, 19(3), 235–40.

Domínguez, A; Ferreira, M; Coutinho, P; Fábregas, J; Otero, A. Delivery of astaxanthin from *Haematococcus pluvialis* to the aquaculture food chain. *Aquaculture*, 2005, 250, 424–430.

Domracheva, LI; Shirokikh, IG; Fokina, AI; Anti-*Fusarium* activity of cyanobacteria and actinomycetes in soil and rhizosphere. *Microbiology*, 2010, 79(6), 871–876.

Estrada, JEP; Bescós, PB; Fresno, AMV. Antioxidant activity of different fractions of *Spirulina platensis* protean extract. *Il Farmaco* 2001, 56, 497–500.

Fábregas, J; García, D; Morales, E; Dominguez, A; Otero, A. Renewal rate os semicontinuous cultures of the microalga *Porphyridium cruentum* modifies phycoerythrin, exopolysaccharide and fatty acid productivity. *Journal of Fermentation an Bioengineering*, 1998, 86(5), 477-481.

Fábregas, J; Patiño, M; Morales, ED. Cordero, B; Otero, A. Optimal renewal rate and nutrient concentration for the production of the marine microalga *Phaeodactylum tricornutum* in semicontinuous cultures. *Applied and Environmental Microbiology*, 1996, 62(1), 266–268.

Fábregas, J; Patiño, M; Vega, BOA; Tobar, JL; Otero, A. Renewal rate and nutrient concentration as tools to modify productivity and biochemical composition of cyclostat cultures of the marine microalga *Dunaliella tertiolecta*, *Applied Microbiology and Biotechnology*, 1995, 44, 287–292.

Fouchard, S; Pruvost, J; Degrenne, B; Legrand, J. Investigation of H_2 production using the green microalga *Chlamydomonas reinhardtii* in a fully controlled photobioreactor fitted with on-line gas analysis. *International Journal of Hydrogen Energy*, 2008, 33, 3302–3310.

Fox, RD. Health benefits of *Spirulina* and proposal for a nutrition test on children suffering from kwashiorkor and marasmus. In: Doumengue F; Chastel, DH; Toulemont A, Eds. *Spiruline* algue de vie. Bulletin de l'Institut Oceanographique Monaco, Musee Oceanographique, 1993, 12, 179-185.

Fuentes, MMR; Sánchez, GJL; Sevilla, FJM.; Fernández, AFG; Pérez, SJA; Grima, EM. Outdoor continuous culture of Porphyridium cruentum in a tubular photobioreactor: quantitative analysis of the daily cyclic variation of culture parameters. *Journal of Biotechnology*, 1999, 70, 271–288.

Fuentes, RMM; Fernández, GGA; Pérez, JAS, Guerrero, JLG. Biomass nutrient profiles of the microalga *Porphyridium cruentum*. *Food Chemistry*, 2000, 70, 345-353.

Ghirardi, ML; Togasaki, RK; Seibert, M. Oxygen-sensitivity of algal H_2 production. *Applied Biochemistry and Biotechnology*, 1997, 63, 141–51.

Godia, F; Albiol, J; Montesinos, JL; Perez, J; Creus, N; Cabello, F; Mengual, X; Montras, A; Lasseur, C. MELISSA: a loop of interconnected bioreactors to develop life support in Space. *Journal of Biotechnology*, 2002, 99, 319-330.

Gouveia, L; Empis, J. Relative stabilities of microalgal carotenoids in microalgal extracts, biomass and fish feed: effect of storage conditions. *Innovative Food Science and Emerging Technologies*, 2003, 4, 227–233.

Gressel, J. Transgenics are imperative for biofuel crops. *Plant Science*, 2008, 174, 246–263.

Grima, EM; Belarbi, EH; Fernández, FGA; Medina AR; Chisti, Y. Recovery of microalgal biomass and metabolites: process options and economics. *Biotechnology Advances*, 2003, 20 491–515.

Grimston, MC; Karakoussis, V; Fouquet, R; Vorst, RVD; Pearson, P; Leach, M. The European and global potential of carbon dioxide sequestration in tackling climate change. *Climate Policy*, 2001, 1 (2), 155–171.

Grinstead, GS; Tokach, MD; Dritz, SS; Goodband, RD; Nelssen, JL. Effects of *Spirulina platensis* on growth performance of weanling pigs. *Animal Feed Science and Technology*, 2000, 83, 237-247.

Gurbuz, F; Ciftci, H; Akcil, A. Biodegradation of cyanide containing effluents by *Scenedesmus obliquus*. *Journal of Hazardous Materials*, 2009, 162, 74–79.

Guschina, IA; Harwood, JL. Lipids and lipid metabolism in eukaryotic algae. *Progress in Lipid Research*, 2006, 45, 160–186.

Hejazi MA; Wijffels, RH. Milking of microalgae. *Trends in Biotechnology*, 2004, 22(4), 189–94.

Hernández, EPS; Córdoba, LT. Anaerobic digestion of *Chlorella vulgaris* for energy production. *Resources, Conservation and Recycling*, 1993, 9, 127–132.

Hernandez, J; Bashan, LE; Bashan, Y. Starvation enhances phosphorus removal from wastewater by the microalga *Chlorella* spp. co-immobilized with *Azospirillum brasilense*. *Enzyme and Microbial Technology*, 2006, 38, 190–198.

Hirahashi, T; Matsumoto, M; Hazeki, K; Saeki, Y; Ui, M; Seya, T. Activation of the human innate system by *Spirulina*: augmentation of interferon production and NK cytotoxicity by oral administration of hot water extract of *Spirulina platensis*. *International Immunopharmacology,* 2002, 2, 423–434.

Huang, G; Chen, F; Wei, D; Zhang, X, Chen, G. Biodiesel production by microalgal biotechnology. *Applied Energy*, 2010, 87, 38–46.

Hydrocarbons and other chemicals. *Critical Reviews in Biotechnology*, 2002, 22(3), 245-279.

Ip, PF; Wong, KH; Chen, F. Enhanced production of astaxanthin by the green microalga *Chlorella zofingiensis* in mixotrophic culture, *Process Biochemistry*, 2004, 39, 1761–1766.

Issa, OM; Défarge, C; Bissonnais, YL; Marin, B; Duval, O; et al. Effects of the inoculation of cyanobacteria on the microstructure and the structural stability of a tropical soil. *Plant Soil*, 2007, 290(1,2), 209–219.

Issarapayup, K; Powtongsook, S; Pavasant, P. Flat panel airlift photobioreactors for cultivation of vegetative cells of microalga *Haematococcus pluvialis*. *Journal of Biotechnology*, 2009, 142, 227–232.

Iwamoto, H. Industrial production of microalgal cell-mass and secondary products – Major industrial species. In: Richmond A, editor. *Handbook of microalgal culture: biotechnology and applied phycology*. Oxford: Blackwell Publishing; 2004; 255.

Jau, M; Yew, S; Toh, PSY; Chong, ASC; CHU, W; Phang, S; Najimudin, N; Sudesh, K. Biosynthesis and mobilization of poly(3-hydroxybutyrate) [P(3HB)] by *Spirulina platensis*. *Int J Biol Macromol.* 2005, 36, 144 – 151.

Jeong, ML; Gillis, JM; Hwang, JY. Carbon dioxide mitigation by microalgal photosynthesis. *Bulletin of the Korean Chemical Society*. 2003, 24, 12, 1763-1766.

Ji, L; Yan, K; Meng, F; Zhao, M. The oleaginous *Botryococcus* from the triassic yanchang formation in Ordos Basin, Northwestern China: morphology and its paleoenvironmental significance. *Journal of Asian Earth Sciences*, 2010, 38, 175–185.

Kadam, KL. Environmental implications of power generation via coal-microalgae cofiring. *Energy*, 2002, 27, 905-922.

Kaewpintong, K; Shotipruk, A; Powtongsook, S. Photoautotrophic high-density cultivation of vegetative cells of *Haematococcus pluvialis* in airlift bioreactor. *Bioresource Technology*, 2006, 98, 288–295.

Kapdi, SS; Vijay, VK; Rajesh, SK; Prasad, R. Biogas scrubbing, compression and storage: perspective and prospectus in Indian context. *Renewable Energy*, 2005, 30, 1195–1202.

Kaya, VM; Picard, G. Stability of chitosan gel as entrapment matrix of viable *Scenedesmus bicellularis* cells immobilized on screens for tertiary treatment of wastewater, *Bioresource Technology*, 1996, 56, 147–55.

Keenan, JD. Bioconversion of solar energy to methane. *Energy*, 1977, 2, 365-373.

Khanna, S; Srivastava, AK. Recents advances in microbial polyhydroxyalkanoates. *Process Biochem.* 2005, 40, 607 – 619.

Kim, Z.H.; Kim, SH; Lee, HS; Lee, CG. Enhanced production of astaxanthin by flashing light using *Haematococcus pluvialis*. *Enzyme and Microbial Technology*, 2006, 39, 414–419.

Kita, K; Okada, S; Sekino, H; Imou, K; Yokoyama, S; Amano, T. Thermal pre-treatment of wet microalgae harvest forefficient hydrocarbon recovery. *Applied Energy*, 2010, 87, 2420–2423.

Kita, K; Okada, S; Sekino, H; Imou, K; Yokoyama, S; Amano, T. Thermal pre-treatment of wet microalgae harvest for efficient hydrocarbon recovery. *Applied Energy*, 2010, 87, 2420–2423.

Kittikaiwan, P; Powthongsook, S; Pavasant P; Shotipruk, A. Encapsulation of *Haematococcus pluvialis* using chitosan for astaxanthin stability enhancement. *Carbohydrate Polymers*, 2007, 70, 378–385.

Kojima, E; Lin, B; Effect of partial shading on photoproduction of hydrogen by *Chlorella*. *Journal of Bioscience and Bioengineering*, 2004, 97(5) 317–321.

Kusçu, OS; Sponza, DT. Treatment efficiencies of a sequential anaerobic baffled reactor (ABR)/completely stirred tank reactor (CSTR) system at increasing p-nitrophenol and COD loading rates. *Process Biochemistry*, 2006, 41, 1484–1492.

Lau, PS; Tam, NFY; Wong, YS. Effect of carrageenan immobilization on the physiological activities of *Chlorella vulgaris*. *Bioresource Technology,* 1998, 63, 115–121.

Lee, YK; Shen, H. Basic Culturing Techniques. In: Richmond A, editor. *Handbook of microalgal culture: biotechnology and applied phycology*. Oxford: Blackwell Publishing; 2004; 40.

Lee, EY; Jendrossek, D; Schirmer, A; Choi, Y; Steinbüchel, A. Biosynthesis of copolyesters consisting of 3-hydroxybutyric acid and medium-chain-lenght 3-hydroxyalkanoic acids from 1,3 butanediol or from 3-hydroxybutyrate by *Pseudomonas* sp. A33. *Appl Microbiol Biotechnol*. 1995, 42, 901 – 909.

Lee, YK; Soh, CW. Accumulation of astaxanthin in *Haeamatococcus lacustris*. *Journal of Phycology*, 1991, 27, 575–577.

Li, HB; Chen, F; Zhang, TY; Yang, FQ; Xu, GQ. Preparative isolation and purification of lutein from the microalga *Chlorella vulgaris* by high-speed counter-current chromatography. *Jornal of Chromatography,* 2001, 905, 151–55.

Lorenz, TR; Cysewski, GR. Commercial potential for *Haematococcus* microalgae as a natural source of astaxanthin. *Trends in Biotechnology*, 2000, 18, 160–167.

Maeda, K; Owada, M; Kimura, N; Omata, K; Karube, I. CO_2 fixation from the flue gas on coal-fired thermal power plant by microalgae. *Energy Conversion and Management*, 1995, 36(6-9), 717-720.

Malea, GMC; Del Rio, E; Casas, JL; Acien, FG; Fernandez, JM; Rivas, J; Guerrero, MG; Molina, E. Comparative analysis of the outdoor culture of *Haematococcus pluvialis* in tubular and bubble column photobioreactors. *Journal of Biotechnology*, 2006, 123, 329–342.

Mallick, N. Biotechnological potential of immobilized algae for wastewater N, P and metal removal: A review. *BioMetals*, 2002, 15, 377–390.

Maqubela, MP; Mnkeni, PNS; Muchaonyerwa, P, et al. Effects of cyanobacteria strains selected for their bioconditioning and biofertilization potential on maize dry matter and soil nitrogen status in a South African soil. *Soil Science and Plant Nutrition*, 2010, 56 (4), 552-559.

Markov, SA; Bazin, MJ; Hall, DO. Hydrogen photoproduction and carbon dioxide uptake by immobilized *Anabaena variabilis* in a hollow-fiber photobioreactor. *Enzyme and Microbial Technology*, 1995, 17, 306-310.

Masojídek, J; Papácek, S; Sergejevová, M; Jirka, V; Cervený, J; Kunc, J; Korecko, J; K; Verbovikova, O; Kopecký, J; Stys, D; Torzillo, G. A closed solar photobioreactor for cultivation of microalgae under supra-high irradiance: basic design and performance. *Journal of Applied Phycology*, 2003, 15, 239–248.

Mata, TM; Martins, AA; Caetano, NS; Microalgae for biodiesel production and other applications: A review. *Renewable and Sustainable Energy Reviews*, 2010, 14(1), 217-232.

Minkova, KM; Tchernov, AA; Tchorbadjieva, MI; Fournadjieva, ST; Antova, RE; Busheva, MC. Purification of C-phycocyanin from *Spirulina* (*Arthrospira*) *fusiformis*. *Journal of Biotechnology*, 2003, 102, 55-59.

Morais, MG; Costa, JAV. Biofixation of carbon dioxide by *Spirulina* sp. and *Scenedesmus obliquus* cultivated in a three-stage serial tubular photobioreactor. *Journal of Biotechnology*, 2007, 129, 439–445.

Morais, MG; Miranda, MZ; Costa, JAV. Biscoitos de chocolate enriquecidos com *Spirulina platensis*: caracteristicas fisico-quimica, sensorial e digestibilidade. *Alimentos e Nutrição*, 2006, 17, 333-340.

Morais, MG; Radmann, EM; Andrade, MR; Teixeira, GG; Brusch, LRF; Costa, JAV. Pilot scale semicontinuous production of *Spirulina biomass* in southern Brazil. *Aquaculture*, 2009, 294, 60–64.

Morist, A; Montesinos, JL; Cusidó, JA; Gòdia, F. Recovery and treatment of *Spirulina platensis* cells cultured in a continuous photobioreactor to be used as food. *Process Biochemistry*, 2001, 37, 535-547.

Morowvat, MH; Amini, SR; Ghasemi, Y. Chlamydomonas as a "new" organism for biodiesel production. *Bioresource Technology*, 2010, 101, 2059–2062.

Muñoz, R; Guieysse, B. Algal–bacterial processes for the treatment of hazardous contaminants: A review. *Water Research*, 2006, 40, 2799 – 2815.

Mussgnug, JH; Klassen, V; Schlüter, A; Kruse, O. Microalgae as substrates for fermentative biogas production in a combined biorefinery concept. *Journal of Biotechnology*, 2010, 150, 51–56.

Nami, HK. A unique feature of hydrogen recovery in endogenous starch to alcohol fermentation of the marine microalga, *Chlamydomonas perigranulata*. *Applied Biochemistry and Biotechnology*, 2006, 131, 808–828.

Nishioka, N; Nakai, K; Miyake, M; Asada, Y; Taya, M. Production of poly-β-hydroxybutyrate by thermophilic cyanobacterium Synechoccocus sp. MA19 under phospate-limited conditions. *Biotechnol Lett.* 2001, 23, 1095 – 1099.

Okada, S; Matsuda, H; Murakami, M; Yamaguchi, K. Botryoxanthin A, A member of a new class of carotenoids from the green microalga *Botryococcus braunii* Berkeley. *Tetrahedron Letters*, 1996, 37(7), 1065-1068.

Olguín, E; Doelle, H; Mercado, G. Resource recovery through recycling of sugar processing by-products and residuals resources. *Conservation and Recycling*, 1995, 15, 85 – 94.

Olguín, EJ; Galicia, S; Guerrero, OA; Hernández, E. The effect of low light flux and nitrogen deficiency on the chemical composition of *Spirulina* sp. (*Arthrospira*) grown on digested pig waste. *Bioresource Technology*, 2001, 77, 19-24.

Omar, HH. Bioremoval of zinc ions by *Scenedesmus obliquus* and *Scenedesmus quadricauda* and its efect on growth and metabolism. *International Biodeterioration and Biodegradation*, 2002, 50, 95–100.

Órpez, R; Martínez, ME; Hodaifa, G; Yousfib, FE; Jbari, N; Sánchez, S. Growth of the microalga *Botryococcus braunii* in secondarily treated sewage. *Desalination*, 2009, 246, 625–630.

Otero, A; Domínguez, A; Lamela, T; García, D; Fábregas, J. Steady-states of semicontinuous cultures of a marine diatom: effect of saturating nutrient concentrations. *Journal of Experimental Marine Biology and Ecology.* 1998, 227, 23–34.

Otero, A; García, D; Morales, ED; Arán, J; Fábregas, J. Manipulation of the biochemical composition of the eicosapentaenoic acid-rich microalga *Isochrysis galbana* in semicontinuous cultures. *Biotechnology and Applied Biochemistry*, 1997, 26, 171-177.

Panda, B; Jain, P; Sharma, L; Mallick, N. Optmization of cultural and nutritional conditions for accumulation of poly-β-hydroxybutyrate in Synechocystis sp. PCC 6803. Bioresour Technol. 2006, 97, 1296 – 130.

Patil, G.; Chethana, S; Madhusudhan, MC; Raghavarao, KSMS. Fractionation and purification of the phycobiliproteins from *Spirulina platensis*. *Bioresource Technology*, 2008, 99, 7393–7396.

Qishen, PK. Radioprotective effect of extract from *Spirulina* in mouse bone narrow cells studied by using the micronucleus test. *Toxicology Letters,* 1989, 48, 165–169.

Radmann, EM; Reinehr, CO; Costa, JAV. Optimization of the repeated batch cultivation of microalga *Spirulina platensis* in open raceway ponds. *Aquaculture*, 2007, 265, 118-126.

Rai, LC; Mallick, N. Removal and assessment of toxicity of Cu and Fe to *Anabaena doliolum* and *Chlorella vulgaris* using free and immobilized cells. *World Journal of Microbiology and Biotechnology*, 1992, 8, 110–114.

Rama, MP; Torres, E; Suárez, C; Herrero, C; Abalde, J. Sorption isotherm studies of Cd(II) ions using living cells of the marine microalga *Tetraselmis suecica* (Kylin) Butch. *Journal of Environmental Management*, 2010, 91, 2045-2050.

Rangsayatorn, N; Upatham, ES; Kruatrachue, M; Pokethitiyook, P; Lanza GR. Phytoremediation potential of *Spirulina* (*Arthrospira*) *platensis*: biosorption and toxicity studies of cadmium. *Environmental Pollution*, 2002, 119, 45–53.

Reinehr, CO; Costa, JAV. Repeated batch cultivation of the microalga *Spirulina platensis*. *World Journal of Microbiology and Biotechnology*, 2006, 22, 937–943.

Richmond, A; Wu, ZC; Zarmi, Y. Efficient use of strong light for high photosynthetic productivity: interrelationships between the optical path, the optimal population density and cell-growth inhibition. *Biomolecular Engineering*, 2003, 20, 229-236.

Rigano, VDM; Vona, V; Esporito, S; Carillo, P; Carfagna, S; Rigano, C. The physiologican significance of light and dark NH_4^+ metabolism in *Chlorella sorokiniana*. *Phytochemistry*, 1998, 47, 177-181.

Riley, H;. Effects of algal fibre and perlite on physical properties of various soils and on potato nutrition and quality on a gravelly loam soil in southern Norway. *Acta Agriculturae Scandinavica, Section B - Plant Soil Science*, 2002, 52(2–3), 86–95.

Rippka, R; Deruelles, J; Waterbury, J; Herdman, M; Stanier, R. Generic assignments, strain histories and properties of pure cultures of cyanobacteria. *Journal of General Microbiology*, 1979, 111, 1-61.

Rupprecht, J. From systems biology to fuel—*Chlamydomonas reinhardtii* as a model for a systems biology approach to improve biohydrogen production. *Journal of Biotechnology*, 2009, 142, 10–20.

Sajilata, MG; Singhal, RS; Kamat, MY. Fractionation of lipids and purification of c-linolenic acid (GLA) from *Spirulina platensis*. *Food Chemistry*, 2008, 109, 580–586.

Sakai, N; Sakamoto, Y; Kishimoto, N; Chihara, M; Karube, I. *Chlorella* strains from hot springs tolerant to high temperature and high CO_2. *Energy Conversion and Management*, 1995, 36(6-9), 693–696.

Samson, R; Leduy, A. Biogas production from anaerobic digestion of *Spirulina maxima* algal biomass. *Biotechnology and Bioengineering*, 1986, 24, 1919–1924.

Schaeffer, DJ; Krylov, VS. Anti-HIV activity of extracts and compounds from algae and cyanobacteria. *Ecotoxicology and Environmental Safety*, 2000, 45, 208-227.

Semba, RD; Miotti, PG; Chiphangwi, JD; Saah, AJ; Canner, JK; Dallabetta, GA; Hoover, DR. Maternal vitamin A deficiency and mother-to-child transmition of HIV-1. *The Lancet*. 1994, 343(8913), 1593-1597.

Sharma, L; Singh, AK; Panda, B., Mallick, N. Process optimization for poly-β-hydroxybutyrate production in a nitrogen fixing cyanobacterium, Nostoc muscorum using response surface methodology. Bioresour Technol. 2007, 98, 987 – 993.

Shi, J; Podola, B; Melkonian, M. Removal of nitrogen and phosphorus from wastewater using microalgae immobilized on twin layers: an experimental study. *Journal of Applied Phycology* 2007, 19, 417–423.

Shi, XM; Liu, HJ; Zhang, XW; Chen, F. Production of biomass and lutein by *Chlorella protothecoides* at various glucose concentrations in heterotrophic cultures. *Process Biochemistry*, 1999, 34, 341–347.

Sierra, E; Acién, FG; Fernández, JM; García, JL; González, C; Molina, E. Characterization of a flat plate photobioreactor for the production of microalgae. *Chemical Engineering Journal*, 2008, 138, 136–147.

Silveira, ST; Burkert, JFM; Costa, JAV; Burkert, CAV; Kalil, S.J. Optimization of phycocyanin extraction from *Spirulina platensis* using factorial design. *Bioresource Technology*, 2007, 98, 1629–1634.

Sims, REH; Mabee, W; Saddler, JN; Taylor, M. An overview of second generation biofuel technologies. *Bioresource Technology*, 2010, 101, 1570–1580.

Singh, NK; Parmar, A; Madamwar, D. Optimization of medium components for increased production of C-phycocyanin from *Phormidium ceylanicum* and its purification by single step process. *Bioresource Technology*, 2009, 100, 1663–1669.

Spolaore, P; Cassan, CJ; Duran, E; Isambert, A. Commercial applications of microalgae. *Journal of Bioscience and Bioengineering*, 2006, 101(2), 87–96.

Subhadra, B; Edwards, M; An integrated renewable energy park approach for algal biofuel production in United States. *Energy Policy*, 2010, 38, 4897–4902.

Sudesh, K; Abe, H; Doi, Y. Synthesis, structure and properties of polyhydroxyalkanoates: biological polyesters. *Progress in Polymer Science*, 2000, 25, 1503-1555.

Suh, IS; Joo, HN; Lee, CG A novel double-layered photobioreactor for simultaneous Haematococcus pluvialis cell growth and astaxanthin accumulation. *Journal of Biotechnology*, 2006, 125, 540–546.

Takahashi, T. The fate of industrial carbon dioxide. *Science*, 2004, 305(5682), 352-353.

Teas, J.; Hebert, JR; Fitton, JH; Zimba, PV. Algae – a poor man's HAART? *Medical Hypotheses*, 2004, 62, 507-510.

Tokosoglu, O; Unal, MK. Biomass nutrient profiles of three microalgae: *Spirulina platensis, Chlorella vulgaris* and *Isochrisis galbana. Journal of Food Science*, 2003, 68(4), 1144–1148.

Tomaselli, L. Morphology, ultrastructure and taxonomy of *Arthrospira* (*Spirulina*). In: Vonshak A. editor *Spirulina platensis (Arthrospira) Physiology, cell-biology and biotechnology*. London: Taylor and Francis; 1997; 1.

Tran, HL; Kwon, JS; Lee, CG. Optimization for the growth and the lipid productivity of *Botryococcus braunii* LB572. *Journal of Bioscience and Bioengineering*, 2009, 108, S41–S56.

Travieso, L; Hall, DO; Rao, KK; Benítez, F; Sánchez, E; Borja, R. A helical tubular photobioreactor producing *Spirulina* in a semicontinuous mode. *International Biodeterioration and Biodegradation*, 2001, 47, 151–155.

Ugwu, CU; Ogbonna, JC; Tanaka, H. Characterization of light utilization and biomass yields of *Chlorella sorokiniana* in inclined outdoor tubular photobioreactors equipped with static mixers. *Process Biochemistry,* 2005, 40, 3406–3411.

Uriarte, M; Yackulic, CB; Cooper, T; Flynn, D; Cortes, M; Crk, T; Cullman, G; McGinty, M; Sircely, J. Expansion of sugarcane production in São Paulo, Brazil: Implications for fire occurrence and respiratory health. *Agriculture, Ecosystems and Environment*, 2009, 132, 48–56.

USDOE. Carbon dioxide emissions from the generation of electric power in the United States. CO_2 emissions report. Washington, DC: US Department of Energy (USDOE), July 2000.

Vernerey, A; Albiol, J; Lasseur, C; Godia, F. Scale-up and design of a pilot-plant photobioreactor for the continuous culture of *Spirulina platensis. Biotechnology Progresses*, 2001, 17, 431–438.

Vilchez, C; Garbayo, I; Lobato, MV; Vega, JM. Microalgae-mediated chemicals production and wastes removal. *Enzyme and Microbial Technology*, 1997, 20, 562-572.

Voltolina, D; Villa, HG; Correa, G. Nitrogen removal and recycling by *Scenedesmus obliquus* in semicontinuous cultures using artificial wastewater and a simulated light and temperature cycle *Bioresource Technology*, 2005, 96(3), 359-362.

Vonshak, A; Abeliovich, A; Boussiba, S; Arad, S; Richmond, A. Production of *Spirulina* biomass: effects of environmental factors and population density. *Biomass,* 1982, 2, 175–185.

Ward, OP; Singh, A. Omega-3/6 fatty acids: Alternative sources of production. *Process Biochemistry*, 2005, 40, 3627–3652.

Watanabe, A. List of algal strains in collection at the institute of applied microbiology university of Tokyo. *Journal of Genetic and Applied Microbiology*, 1960, 6, 4.

Xin, L; Hu H; Gan K; Yang J. Growth and nutrient removal properties of a freshwater microalga Scenedesmus sp. LX1 under different kinds of nitrogen sources. Ecological Engineering. 2010, 36, 379–381.

Yamaguchi, K; Recent advances in microalgal bioscience in Japan, with special reference to utilization of biomass and metabolites: a review. *Journal of Applied Phycology*, 1997, 8, 487–502.

Yang, S; Wang, J; Cong, W; Cai, Z; Ouyang, F. Effects of bisulfite and sulfite on the microalga *Botryococcus braunii*. *Enzyme and Microbial Technology*, 2004, 35, 46–50.

Yoo, C; Jun, SY; Lee, JY, Ahn, CY; Oh, HM. Selection of microalgae for lipid production under high levels carbon dioxide. *Bioresource Technology*, 2010, 101, S71–S74.

Zarrouk, C. 1966. Contribution a l'étude d'une cyanophycée. influence de dívers facteurs physiques et chimiques sur la croissance et photosynthese de *Spirulina maxima* geitler, Ph.D. Thesis, University of Paris.

Zhang, BY; Geng, YH; Li, ZK; Hu, HJ; Li, YG. Production of astaxanthin from *Haematococcus* in open pond by two-stage growth one-step process. *Aquaculture*, 2009, 295, 275–281.

Zhang, Z; Lohr, L; Escalante, C; Wetzstein, M. Food versus fuel: What do prices tell us? *Energy Policy*, 2010, 38, 445–451.

Zhang, ZP; Tay, JH; Show, KY; Yan, R; Liang, DT; Lee, DJ; Jiang, WJ. Biohydrogen production in a granular activated carbon anaerobic fluidized bed reactor. *International Journal of Hydrogen Energy*, 2007, 32, 185 – 191.

Zhu, L; Zhang, X; Ji, L; Song, X; Kuang, C. Changes of lipid content and fatty acid composition of *Schizochytrium limacinum* in response to different temperatures and salinities. *Process Biochemistry*, 2007, 42, 210–214.

Chapter 5

HYDROTHERMAL CARBONIZATION OF MICROALGAE AND OTHER LOW CELLULOSIC BIOMASS MATERIALS

Steven M. Heilmann*, Marc G. von Keitz and Kenneth J. Valentas
BioTechnology Institute, University of Minnesota

ABSTRACT

Hydrothermal carbonization is a process in which biomass is heated in water under pressure at temperatures below 250 °C to create a char product. A significant advantage of the process is that water is not removed from the char by evaporation but by filtration, providing a favorable energy input to output ratio.

With higher plants, the chemistry derives primarily from lignin, hemicellulose and cellulose components. Cellulose is the most recalcitrant of these and requires relatively high temperatures and long reaction periods for conversion into a highly carbonized char product.

Our approach has been to examine biomass materials having relatively low cellulose contents, applying reaction temperatures generally below 225 °C, and for reaction periods of less than 2 hours. These conditions are believed to be conducive to continuous processing and to increase the carbon content in the char primarily via a dehydration mechanism, rather than by loss of carbon dioxide. Substrates that have been examined in this manner included microalgae, cyanobacteria, and fermentation residues such as distiller's grains, brewer's grains and others.

It was determined during the course of these investigations that fatty acids created by hydrolysis of lipids during the process do not chemically contribute to char formation but are adsorbed onto the char and can be recovered by solvent extraction in high yield. Therefore, the process provides fatty acid, char, and aqueous filtrate products, all of which have utility.

---

[*] Email: heilm001@umn.edu.

INTRODUCTION

Problems associated with maintaining so-called "advanced" lifestyles based on fossil fuels seem almost insurmountable. Unbalanced use of these resources by western nations and an ever increasing global population desiring these lifestyles have resulted in an insatiable appetite for fossil fuels for transportation; electrical power generation; and as feedstocks for petrochemicals and refined products, plastics, and fertilizers. Aside from rapid depletion of fossil fuel resources, when petroleum, coal and natural gas are obtained from subterranean locations and burned, new carbon dioxide is formed and released into the atmosphere. This provides an ever increasing greenhouse gas shroud that traps thermal energy around the earth, resulting in global warming, significantly altered climates, and rising water levels in the world's oceans. All these issues, if left unchecked, will result in serious ecological changes, massive population redistribution, and economic and national security instabilities that will confront present and future generations.

Utilization of biomass and biomass-derived products to replace fossil carbon materials has the potential to partially mitigate the release of "new" carbon dioxide into the atmosphere. Biomass is present in great abundance around the world and its combustion is generally regarded as a "carbon neutral" event in that only the carbon sequestered by the plant is released as carbon dioxide back into the atmosphere.

Among the many challenges associated with biomass conversion into practical fuel products is that thermochemical treatment is often required to increase the energy content and density of the fuel product. Water is ubiquitous in nature and is present in high concentrations in most biomass materials, and thermal energy is required to remove water from the biomass for its conversion into a higher fuel density product. Removal of water requires considerable energy (1100 BTU/lb) and, more often than not, a situation is created in which the thermal energy obtained on combustion of the biomass product is less than the energy required for its formation.

Hydrothermal processes are thermochemical processes that actually require water for effective conversion into the desired energy product, and none of the processes removes water from products by evaporation. As indicated in Figure 1, the most thermally severe operations are employed in *gasifications* conducted at up to 700 °C, or 350 – 380 °C under catalytic conditions. Products are gases collected by distillation (with heat recycle) and include hydrogen, methane, carbon monoxide and, predominantly in terms of mass yields, carbon dioxide. The predominance of the latter, of course, diminishes the yield of the fuel as well as its energy content unless it is removed from the fuel gases at significant additional cost. *Liquefaction* is an intermediate hydrothermal operation in terms of temperature, and the desired products are hydrophobic oils that can phase separate and be isolated from the aqueous suspending medium. However, very complicated mixtures of organic compounds are created and conversion into conventional liquid transportation fuels is a significant challenge. Hydrothermal *carbonization* (HTC) is the least thermally severe of the various processes, and a char product is formed in moderate yield that can be isolated and "de-watered" by filtration.

Hydrothermal Methods

	Carbonization		Liquefaction	Gasification	
	Microalgae	Lignocellulosics		With catalyst	Without catalyst
Reaction medium	Liquid water	Liquid water	Liquid to near supercritical water	Near supercritical water	Above supercritical water
Temp. (°C)	170-210	170-250	250-350	350-380	600-700
Pressure (barometric)	10-20	10-20	50-200	180-300	250-300
Catalyst	None	Citric acid or $FeSO_4$	K_2CO_3 or KOH	Ru and Ni	None
Reaction time (h)	0.5-2.0	4-16	0.25	<1	0.25
Main products	Char and filtrate	Char	Phenol-rich oily liquid	CH_4 and CO_2	H_2, CH_4 and CO_2
Product separation	Filtration	Filtration	Phase separation	Gas collection	Gas collection

Figure 1. Summary of Hydrothermal Methods.

HTC of lignocellulosic biomass materials has been known since the early 20th century [1], and the area has been recently reviewed [2]. With lignocellulosic materials, hemicellulose and even lignin to a considerable extent are solubilized and can undergo carbonization, i.e., a process in which the carbon to oxygen ratio is increased, in the hot aqueous medium. In a report examining the HTC of wheat straw [3], for example, essentially all of the hemicellulose was degraded to the extent that it was completely soluble in water and up to 62% of the lignin was reduced to water-dispersible aggregates at 195 °C; cellulose at that temperature was only moderately affected under the conditions, exhibiting a slight (11%) reduction in molecular weight. A more thermally severe condition of 230 °C for 4 hours was required to obtain a char from pure cellulose in 33% yield [4].

Examination of non-lignocellulosic biomass substrates has been limited to water-soluble substrates such as sugars for preparing porous carbon materials [5] and metal/carbon-based composites [6]. Our work with microalgae [7] is believed to be the first report in the technical literature of HTC conducted with an insoluble, non-lignocellulosic biomass substrate. This research has been expanded by investigation of other low cellulose-containing substrates, e.g., distiller's grains and other fermentation residues.

MICROALGAE

Microalgae are excellent candidates for HTC. They are among the most efficient plants in terms of photosynthetic activity, literally being capable of metabolizing sufficient carbon dioxide to double their biomass in just a few hours [8]. Additionally, several varieties grow well in salt water, thus potentially utilizing the oceans of the world as "open ponds" for production, minimizing predation by bacteria, and not competing with food products for arable land [9]. Several varieties also contain significant quantities of lipids, even exceeding half the mass of the microorganism in several instances [10]. Furthermore, the fatty acids forming the lipids present in many strains of microalgae are generally in the $C_{12} - C_{18}$ range that when converted into hydrocarbon liquid fuels have adequate energy densities and low melting points required in aviation fuels. Though the present microalgae manufacturing industry is small at about 5000 tons of microalgae per year [11] employed primarily as sources of nutraceuticals, microalgae are currently being considered as viable alternatives to fossil fuels for liquid transportation fuels, and a budding algal oil industry posits increased availability and reduced cost of microalgae in the future.

From a practical processing perspective, microalgae are well suited for HTC. Since cellulose is not a major component of microalgae, HTC reactions can be conducted under relatively mild conditions, reducing the process energy requirement. The overall chemistry that takes place will derive from proteins, lipids, and carbohydrates and should produce distinctly different products than from lignocellulosic materials. Furthermore, the inherent small size of microalgae requires that no mechanical energy be expended for breaking down the physical size of the material, and the high surface to volume ratios of these very small microorganisms suggest that reaction temperatures could be achieved relatively rapidly, possibly enabling continuous processing operations.

Investigation of microalgae as a class of substrates for HTC began with *Chlamydomonas reinhardtii* that was grown in house. The growth process provided 15 – 20 grams of microalga on a dry weight basis from photoautotrophic growth supplied by fluorescent light rings to four 20 L carboys sparged with a gentle flow carbon dioxide (5% in air) over the course of 4-5 days [7]. The microalga was harvested and concentrated to 10% solids by centrifugation. Freeze-dried algae were utilized in experiments to minimize weighing and mass transfer errors often encountered with slurries that may not be homogeneous.

Investigations began utilizing a relatively thick-walled 71 mL unstirred stainless steel reactor that was capable of maintaining reaction pressures up to 3000 psi. Heat was applied using an induction heating system that achieved the desired reaction temperature in about 7 minutes and was accurate to +/- 3 °C. With this system, however, a significant lack of reproducibility of results was observed such that reliable conclusions based on the data obtained could not be made. The small batch size (typically 3 grams of microalga in 57 mL of water), change in starting material from one small lot to another, the unstirred condition, and relatively imprecise temperature control with the system were factors thought to contribute to the unacceptable variation observed in %C values and char yields.

Accordingly, a larger 450 mL stirred stainless steel reactor of thinner wall construction was purchased. With the proper volume and thermal expansion safety factors considered, a total reaction volume of 300 mL could process 15 – 30 grams of microalga at 5 to 10% solids which seemed to bracket a reasonable concentration range, at least for *Chlamydomonas reinhardtii*.

Furthermore, a larger water cooled coil for the 450 mL unit was obtained for the induction heating system that provided better temperature control of +/− 1 °C, and a reaction temperature of 200 °C could be achieved in 12 minutes. Another advantage of this heating method was that, when the unit was not actually in "heat" mode, the temperature surrounding the reactor was ambient. Therefore, any exotherms that may accompany these reactions, as have been observed with lignocellulosic materials [12], could be more easily controlled compared to a heating mantle arrangement in which the reactor would be warmed by the mantle even when not actively heating.

With this larger unit, *Chlamydomonas reinhardtii* and several other green microalgae and cyanobacteria were examined under a variety of conditions. At this point, each system was examined using either citric or oxalic acid as a buffer/catalyst. An acidic pH and addition of citric acid had been shown to be effective with lignocellulosic materials [12], and it had also been reported that alkaline pH tends to favor carbonization by loss of carbon dioxide [13]. This latter aspect was especially undesirable. Processes that generate carbon dioxide are disadvantageous for several reasons: 1) they contribute to global warming, 2) loss of carbon in carbon dioxide is counter-productive as a carbonization mechanism, and 3) creation of gaseous products results in elevated reaction pressures and increased cost and complexity of reaction equipment. It was an objective of our efforts to carbonize microalgae by loss of water rather than by loss of carbon dioxide.

The results obtained with the various green and blue green (cyanobacteria) microalgae are reported in Figure 2. Elemental analyses are reported for both starting microalgae and corresponding chars obtained in the process, and comparison of %C values of the two (in bold print in the following table) provides a quick determination of the relative effectiveness of the reaction conditions in achieving useful levels of carbonization.

Of the cyanobacteria examined in Figure 2, degrees of carbonization achieved varied depending on the microalgal substrate. With *Aphanizomenon* and *Spirulina* species, %C values increased ca. 14% in chars compared to starting algae. *Synechocystis* was better, increasing almost 20%. With all the cyanobacteria, however, char yields ranged from 16-27% and were distinctly lower than yields obtained with green microalgae that ranged from 32-39%.

This may be attributed to a more extensive dissolution of cyanobacteria under the hydrothermal conditions due to lack of a cell wall. Though yields were higher, the green microalgae also gave mixed results as substrates in the process. *Chlorella* and *Nannochloropsis* species exhibited %C level increases of ca. 14% in chars, while *Chlamydomonas* increased 21%. The *Scenedesmus/Chlorella* blend provided the lowest level of carbonization at ca. 10%.

The result with *Scenedesmus* was perhaps not surprising as this microalga possesses a cell wall consisting of a crosslinked polyether material that was probably not affected under the reaction conditions [14].

Chlamydomonas, on the other hand, possesses a cell wall consisting primarily of protein [15] that is softened and rendered more permeable or actually broken down under the severe hydrolytic conditions of the process.

Alga	Conditions [% solids, temp. oC, Citric (CA) or oxalic (OA) acid]	Char % Yield	%C	%H	%N
CYANOBACTERIA					
Aphanizomenon flosaquae[a]	*freeze-dried alga*	--	48.1	7.4	11.5
Aphanizomenon flosaquae	*(5%,213,2h,2%CA)*	16	62.7	8.5	7.1
Synechocystis sp.[b]	*freeze-dried alga*	--	48.2	7.3	9.1
Synechocystis sp.	*(5%,213,2h,2%CA)*	18	67.3	9.1	5.5
Spirulina[c]	*freeze-dried alga*	--	44.4	6.2	7.3
Spirulina	*(15%,213,3h,2%OA)*	27	58.4	6.7	6.6
GREEN MICROALGAE					
Chlamydomonas reinhardtii	*freeze-dried alga*	--	51.6	7.9	9.8
Chlamydomonas reinhardtii	*(7.5%,203,2h,2%OA)*	39	72.7	9.7	9.8
Scenedesmus/ Chlorella blend[d]	*freeze-dried alga*	--	43.3	6.7	7.2
Scenedesmus/ Chlorella blend	*(7.5%,203,3h,2%OA)*	39	53.8	7	3.5
Chlorella sp.[e]	*freeze-dried alga*	--	50.8	7.2	10.1
Chlorella sp.	*(10%,200,3h,2% OA)*	34	64.3	8.1	7
Chlorella sp.[f]	*freeze-dried alga*	--	51.5	7.1	10.1
Chlorella sp.	*(20%,200,2h,2%CA)*	32	66.2	8	7.3
Nannochloropsis sp.[g]	*freeze-dried alga*	--	49.4	7.2	7.4
Nannochloropsis	*(20%,200,2h,2%OA)*	37	62.6	8	4.2

[a] = *Aphanizomenon flosaquae* was obtained from Klamath Algae Products, Inc. (Klamath Lake, OR).
[b] = *Synechocystis sp.* was grown in house [7].
[c] = *Spirulina* was purchased from a local health food store.
[d] = This material was obtained from a Minnesota lake and was examined microscopically and determined to consist primarily of *Scenedesmus* and *Chlorella* microalgae.
[e] = *Chlorella sp.* was purchased from a local health food store.
[f] = *Chlorella sp.* received from Biocentric Algae (Santa Ana, CA).
[g] = *Nannochloropsis sp.* was obtained from Brine Shrimp Direct (Ogden, UT).

Figure 2. Hydrothermal carbonization of various green and blue green microalgae.

EVALUATION OF REACTION CONDITIONS

A significant issue, even with the experiments with *Chlamydomonas reinhardtii* prepared in house, was that multiple lots of the microalga were evaluated singly or in combination in multiple experiments designed to examine one reaction variable. With the various lots of algae grown, factors such as slight changes in temperature, illumination, pH control, mineral nutrients, carbon dioxide supply, nitrogen or phosphorous depletion, osmotic pressure, and population density during the growth process, all greatly influence the chemical composition even within same genus and species of microalgae [16]. Therefore, it was difficult to ensure that the actual starting material was chemically the same in multiple experiments. Other commercial sources of green microalgae were investigated, and *Dunaliella salina*, marketed as a source of β-carotene, was available from a source in China. This green microalga is quite different than *Chlamydomonas reinhardtii*. *D. salina* is grown in salt water and essentially has no cell wall, with osmotic pressures within and outside the cell membrane being compensated by production of glycerol [17]. This microalga has also displayed the capability of producing high levels of lipids under appropriate conditions [10]. Five kilograms of the material were purchased as a spray-dried material for extensive experimentation, and, because of the relatively large quantity at our disposal, replicate experiments could be conducted to determine the degree of reproducibility achievable with HTC processing. The spray-dried solid contained 5.8% moisture and 23.2% solubles when extracted with tepid water. The level of solubles seemed very high, although a sizeable portion was probably derived from the salt water growth medium. The freeze-drying process may also have caused glycerol to be extracted on rehydration and contributed to the very high solubles level. It was decided to evaluate the microalga as received, however, since additional processing would add significantly to the cost of any conversion process. Another factor was that as various species of microalgae become available from the algal oil industry, they will undoubtedly not be pristine monocultures but will likely include other invading microorganisms and contaminants, whether grown in open ponds or photobioreactors.

AGITATION

The influence of stirring as a reaction variable was initially considered using *Dunaliella salina* examined at 15% solids in distilled water containing 3% citric acid for 1.25 h at 200 °C. When the system was examined without stirring, a solid mass or clump of char was obtained that possessed a somewhat hard shell.

When the same reaction was conducted with stirring (88 rpm), some coating of solid char was observed on the reactor walls but most of the material was suspended in the aqueous medium and could be isolated by filtration as a free flowing solid. Both materials were obtained in the same yield (39%). However, since the agitation appeared to create a more homogenous product and was probably better suited to scale up operations, stirring was employed in all subsequent experiments.

Effects of Freeze-Drying

Due to the relatively small quantities of most microalgae that were available either commercially or could be grown in-house and to employ accurate and reproducible masses in the various experiments, freeze-dried material was employed in all experiments. To ensure that similar results would be obtained with fully hydrated material that had not been freeze-dried, an experiment was conducted with *Chlamydomonas reinhardtii* that was grown in-house and concentrated into a paste using centrifugation. Half the centrifugate was charged volumetrically (based on a 10% solids gravimetric determination) into the reactor as a slurry. The other half was freeze-dried, the dried algal material charged into the reactor and rehydrated. Identical chars were obtained from the two experiments in terms of % mass yield and %C values under the same HTC conditions. This experiment indicated that no irreversible effects were caused by freeze-drying the microalga, and the accuracy of weighing the dried substrate compared to volumetric transfer of slurries was greatly preferred and experimentally appropriate.

Temp. (°C)	Time (h)	% Solids	% Mass	%C	% Carbon recovered[a] in the char product
190	0.5	5	28.4	64.1	40
190	0.5	25	45.7	61.5	62
190	2.0	5	29.3	67.1	40
190	2.0	25	42.9	64.7	61
200	1.25	15	39.3	64.4	55
200	1.25	15	39.0	65.9	56
200	1.25	15	37.4	65.7	55
200	1.25	15	38.1	66.0	55
210	0.5	5	27.9	65.7	40
210	0.5	25	42.1	64.9	60
210	2.0	5	25.3	69.4	38
210	2.0	25	38.8	66.8	57

[a] = This value was obtained by dividing the amount of carbon in the char by the amount of carbon in the starting alga.

Figure 3. Designed experiment examining the effects of reaction temperature, time and *Dunaliella salina* concentration.

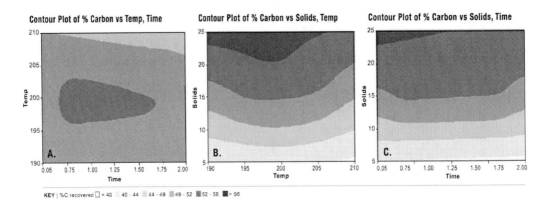

Figure 4. Contour plots examining designed experiment input variables of reaction temperature, time and algal concentration.

Reaction Temperature, Reaction Time and Algal Concentration

Rather than conduct many experiments in which one variable at a time was examined, it was decided to conduct a designed experiment in which multiple input variables of reaction time, temperature and algae concentration could be examined in a systematic fashion. A three-variable, two-level factorial experiment with replicated centerpoints was conducted.

Temperatures examined were: 190, 200, and 210 °C; reaction times: 0.50, 1.25, and 2.00 h; and % solids levels: 5, 15, and 25% *Dunaliella salina* by weight. Principal output results were char mass yield and % C recovered in the char (%C in the char x its mass ÷ %C in the starting microalga x its mass). The data from this designed experiment are contained in Figure 3.

One very important conclusion determined from the centerpoint results was that the quality of the experimental data obtained was quite high, providing standard deviations of only 0.8% and 0.6% for % Mass Yield and %C values, respectively. A linear regression equation was developed from the orthogonal factorial design:

% Carbon Recovered = $51.54 - 1.375X_1 - 0.375X_2 + 9.875X_3$

where X_1 = dimensionless temperature; X_2 = dimensionless time; and X_3 = dimensionless % solids.

The magnitude of the coefficient for the % solids input variable certainly attests to its importance in the system. Another more pictorial method of capturing the relative importance of these three variables is to plot temperature vs. time, % solids vs. temperature and % solids vs. time as in Figures 4a, 4b and 4c. The darker colored regions indicate the percentage of carbon recovered in the char, and quantitative information is contained within each Figure.

In contour mapping plots of the type in Figure 4, the % solids input variable was again clearly shown to be the most important because its variation caused the most change in the plots, and the conclusions from the linear regression analysis were confirmed visually. Figures 4B and 4C also indicated that excessive temperature and reaction time appeared to "overcook" the product such that carbon was being depleted from the char (in the upper right regions of both plots) under prolonged reaction times and high temperatures. This suggested that effective carbonization might be achieved at 210 °C at 25% solids for a shorter time than 30 minutes, e.g.,15 minutes. Under these conditions, a char was obtained in 45.2 % yield having a %C = 64.1 and a recovered carbon level of 60.2 % in 15 minutes at 210 °C. This result even more strongly indicated that HTC processing of microalgae could be conducted by continuous methods.

Catalytic Agents

Yields obtained have been moderate, e.g., only 39% for the centerpoints of the designed experiment employing crude *Dunaliella salina* and 51% if corrected for masses of moisture and water soluble material present. As mentioned earlier, catalysts and acidic reaction conditions had been utilized repeatedly with lignocellulosic materials, and it was desired to examine whether additives would function to provide increased char yields and carbonization levels with microalgal substrates.

An early Japanese report [18] examined the effects of added inorganic salts on the development of turbidity in sucrose solutions at 100 – 120 °C. While several metal salts examined such as Hg, Pb, Sn and Ni salts were environmentally unsuitable, others such as Mg, Ca, and Fe were reported to be effective and could possibly be implemented into the HTC process.

Additive	% Char Yield	%C of Char
None	37.1	64.8
$CaCl_2$	39.7	64.5
$MgCl_2$	37.9	66.2
$Fe(NH_4)_2(SO_4)_2$	40.0	62.8
Oxalic Acid	38.3	65.2
Citric Acid	39.9	65.4

Figure 5. Examination of char yields and %C values with possible catalysts for the HTC of *Dunaliella salina*.

These and other additives were examined with *Dunaliella salina* at 200 °C, 15 % solids and for 2 hours. Molar concentrations of the additives (0.54 mole percent) were employed that was based on 2.0 weight percent $CaCl_2$ examined in the first additive experiment. The results are given in Figure 5.

None of the additives, salts or carboxylic acids, provided a decisive benefit compared to no additive at all.

Re-Employment of Filtrates as Suspending Medium

It was noted that the filtrates on standing at 4 °C often deposited a black film of char on the walls of containers. This suggested that colloidal carbonized material had passed through the filter during workup and later deposited on the container wall.

One method that was examined in an attempt to recover this material and increase yields was to re-employ the filtrate as the suspending medium for microalgae in a subsequent HTC reaction. *Dunaliella salina* was examined at 15% solids, 200 °C and for 2 h in distilled water. The "0" run # in Figure 6 below was the initial reaction conducted in distilled water. Subsequent runs were conducted by employing the filtrate from the previous run as suspending medium.

The data indicated that char yields increased while levels of carbonization (%C) decreased somewhat with filtrate reuse. The overall conclusion was that approximately a 10% gain in yield could be achieved while still maintaining a respectable level of carbonization (64%) in the char.

Run #	%Char Yield	%C	%H	%N
0	40.0	68.4	7.4	6.2
1	42.8	68.0	7.6	6.5
2	47.7	64.1	7.3	7.1
3	48.3	63.9	7.7	6.3
4	49.9	64.2	7.9	6.9

Figure 6. Comparative yield and elemental analysis data for filtrate reuse.

Effect of High Ionic Strength in the Suspending Medium

Another factor that was examined to improve char yields was to increase the ionic strength of the suspending medium and literally "salt out" more of the colloidal carbon suspended in the medium onto the char. When *Dunaliella salina* that was grown in a high salt environment was analyzed for salt content at 15% solids in distilled water, a value of 1.4 M was determined by conductivity measurements. When an additional 1.4 M of sodium sulfate was added to the slurry and the HTC conducted at 200 °C for 2 h, a char was obtained in 43.6% yield and having a %C value of 64.5, compared to 40% yield and 68.4%C in Run #0 in Figure 6. Therefore, only a modest benefit in yield was observed at higher ionic strength.

With the possible re-employment of the filtrate as an HTC suspending medium, the net result of these investigations was that only modest improvements in char yields were achieved with significant additional manipulation and additive costs. Therefore, the basic, simple system of just heating the microalgae in water seemed to provide the most economical process and provide acceptable char yields and levels of carbonization.

ANALYSIS OF THE CHAR PRODUCTS

Visual Changes on Conversion to Char

Figure 7 shows SEM images of freeze-dried starting *Chlamydomonas reinhardtii* (7A) and the corresponding char (7B) obtained using the reaction conditions specified in Figure 2. The image of starting material depicts a microorganism having a general diameter of about 7-10 microns. From the cell in the center of the image, it is apparent that the saucer-shaped cells would be very difficult to filter using a direct filtration arrangement because a relatively impermeable film of collected cells would form almost immediately. The char (7B), on the other hand, was easily filtered using direct filtration. The surface is much more tortuous, resembling the structure of coral in appearance. This suggests that constituents of the microalga become dissolved in the hot aqueous medium, undergo carbonization and re-form onto solid components still present in the milieu. This image is very different from images of natural coal (not shown) that basically has the appearance of a rock with only a somewhat flaky surface formed from by a highly compressive process.

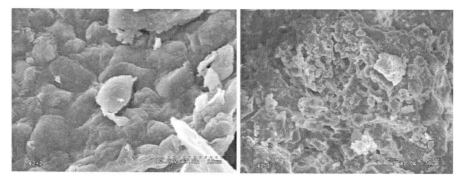

Figure 7. SEM of freeze-dried *Chlamydomonas reinhardtii* (7A) and corresponding Char (7B).

Elemental Analyses, Heats of Combustion, and Ash Contents of Chars Compared to Lignocellulosic Chars and Natural Coal

These data are contained in Figure 8 for algal chars and compared with coal and a char obtained from a lignocellulosic material.

Material	%C	%H	%N	%S	BTU/lb	Ash, %
Natural Coal[a]	69.6	5.7	0.9	0.6	12,293	5.7
Lignocellulosic Char[b]	62.3	5.6	<0.5	--	10,482	--
Algal Char[c]	72.7	9.7	5.2	nd[d]	13,577	--
Agal Char[e]	66.3	7.9	7.3	0.5	13,118	0.3
Algal Char[f]	71.8	9.8	1.5	--	15,064	--

[a] = The natural coal sample is a Powder River Basin coal obtained from Xcel Energy, Inc.
[b] = Char obtained from a monoculture prairie grass (Little Blue Stem) that was finely milled and subjected to conventional hydrothermal carbonization conditions (190 °C for 16h) [12].
[c] = Algal char from *Chlamydomonas reinhardtii* (203 °C, 7.5% solids, 2h).
[d] = nd means none detected.
[e] = Algal char from *Dunaliella salina* (200 °C, 15% solids, 2h).
[f] = Algal char obtained with an unidentified microalga received from Inspired Fuels, Inc. (Austin, TX) (200 °C, 5% solids, 2h).

Figure 8. Elemental analysis, heats of combustion and ash contents of natural coal, lignocellulosic char and algal chars.

Levels of carbonization (%C) values for the algal chars compared very favorably with the dried natural coal sample. Hydrogen contents were substantially higher and certainly contributed as well to the high heat of combustion values of more than 13,000 BTU/lb. Due to its relatively high cellulosic component, the lignocellulosic material required significantly longer reaction time (16 hours) to achieve %C values that were at all comparable to the coal and algal chars, and the heat of combustion value was still decidedly lower than either the natural coal or the algal chars. Nitrogen contents were almost nil with both natural coal and the lignocellulosic char, indicating that natural coal very likely originated from lignocellulosic sources. The difference in formation times for algal char and that proposed for natural coal is noteworthy, in that the former takes place literally in a matter of minutes while the latter requires millions of years. The algal char from Inspired Fuels, Inc. deserves special comment due to the surprisingly high heat of combustion value of > 15,000 BTU/lb. This unidentified microalga was obtained and of interest as a more representative example of microalgae that will be cultivated by the algal oil industry, since the lipid content was purported to be about 30%.

That the heat of combustion was substantially higher than the chars from *Chlamydomonas reinhardtii* or *Dunaliella salina* that possessed relatively low lipid levels (< 5%) suggested that lipids and/or lipid-derived products were contributing significantly to the energy content of the chars. Another noteworthy observation was that the %N of this sample was substantially reduced compared to the other algal chars, and concerns about NOX generation on combustion of algal chars derived from species more commonplace in the developing algal industry may be reduced substantially. The issue of lipid involvement in char formation will be more fully discussed in a later section of this chapter. Additional advantages of the algal chars were low sulfur and ash contents. The low ash content could be especially important and advantageous if employed as a carbon source for making synthesis gas.

One additional practical matter was whether or not the char materials isolated as free-flowing powders could be pelletized and densified for volume reduction and transportation for uses other than for on-site combustion. Contained in Figure 9 are images of a char from *Dunaliella salina* prepared at 15% solids, 1.25 h and 200 °C. The pellets were prepared at room temperature with no added binding agent. The powdery material was subjected to a pressure of 150 MPa in a die having a diameter of 18.8 mm. The bulk density of the pellets was 298.3 +/- 5.7 kg/m^3.

In addition to the encouraging pellet-forming behaviors, the formed pellets have been contained in a glass vial that has been examined and jostled around considerably. The relative toughness of the pellet was indicated by the general lack of sloughing observed with this sample.

Figure 9. Pellets prepared using chars from *Dunaliella salina*.

Process Energy Input Relative to Combustion Output of the Char

An obvious use of the char materials is as a carbon neutral fuel for replacement of coal in electrical power generation. Natural coals are of various types and possess varying carbon contents and heating values [19]. Anthracite coal has a %C range from 86-95% and a fuel value generally of about 15,000 BTU/lb. Bituminous coal has a %C range of 45-86% and a heat of combustion range of 10,500-15,500 BTU/lb. Sub-bituminous coal has a %C range of 25-45% and a heat of combustion range of 8,300-13,000 BTU/lb, while lignite coal trails with 25-35%C and 4,000-8,300 BTU/lb. Obviously, the overlapping heat of combustion ranges with the various types of coal suggests that consideration of the %C value alone is a crude approximation at best for predicting energy contents. Hydrogen and perhaps other elements are certainly involved as well. Nevertheless and as a first approximation, heats of combustion of various coals and other common fuel materials can provide a useful comparative basis for deciding what levels of carbonization are desirable in chars obtained from microalgae. Heats of combustion of various dry, representative materials are reported in Figure 10.

The importance of hydrogen as a contributor to heat of combustion is clearly indicated by the natural gas (methane) entry in the table. The weight percent of carbon in methane is only 75% and, yet, a very high heat of combustion is observed because of the oxidation of the hydrogen present. The importance of hydrogen is further indicated by the slightly reduced energy content of gasoline that is comprised of higher molecular weight alkanes, alkenes and the like. Compared to methane, the carbon content is increased and the hydrogen content decreased in gasoline which contributes to a reduced heat of combustion for gasoline. Charcoal has significant value as a smokeless fuel but possesses only a moderate heat of combustion and requires hours at high temperature for initial drying and more hours at much higher temperatures for carbonization to occur [24]. Sugarcane bagasse also provides a useful comparison, as the combustion of that material provides much of the thermal energy for plant operations and has been of key importance for the sugar industry [25].

Fuel	Heat of Combustion (BTU/lb; dry mass)	Reference
Natural Gas	23,000	20
Gasoline	20,400	20
Coal Samples		23
40%C	6,923	
55%C	9,445	
60%C	10,726	
67%C	12,092	
72%C	12,856	
78%C	14,107	
85%C	14,938	

Figure 10. Listing of various materials and their corresponding measured heats of combustion.

In order to justify the thermal input for the HTC process and moderate char yields that provide reduction in mass from starting microalga to char, an increase of at least 50% in the energy content of the char should be achieved. Starting green microalgae substrates that were more responsive to the HTC process possessed %C values of ca. 50% and heats of combustion of about 7,500 BTU/lb. Corresponding heats of combustion values in chars should, therefore, be in excess of 11,000 BTU/lb to be of practical value, and this level of heat energy should be present in chars that possess %C values of greater than 60%.

Scenario 1: (Combustion of the dry microalga)
- Heat of combustion of the freeze-dried microalga = 7,758 BTU/lb
- Alga collected by centrifugation at 10% solids
- In 10 lbs of collected alga centrifugate, there is 1 lb of alga and 9 lbs of water
- To obtain the dry 1 lb of microalga, 9,900 BTUs are required to evaporate the water
- The net energy loss is 7,758 – 9,900 = -2,142 BTUs

Scenario 2: (Combustion of the dry char)
- Heat of combustion of the char obtained in 40% yield = 13,577 BTU/lb
- Alga collected by centrifugation at 10% solids and used directly for HTC
- To heat suspension from 22 °C to 203 °C, 3,173 BTUs are required. Temperature can be maintained at 203 °C for 2 h without significant additional heat input.
- The filtered algal char product weighs 0.63 lb and is 63% solids, and 223 BTUs are required to dry the char
- To obtain 0.40 lb of dry char, 3,173 + 223 = 3,396 BTUs are required
- The net energy gain is 13,577 (0.4) − 3,396 = +2,035 BTUs

Figure 11. Energy input/output comparisons for combustion of dry *Chlamydomonas reinhardtii* and for combustion of the HTC char obtained from it.

Energy input/output relationships were more rigorously examined in one instance using *Chlamydomonas reinhardtii* and a char derived from it. Using the data obtained from the char in Figure 2, assuming a concentration of 10% solids to make calculations simpler, and utilizing the centrifuged material obtained at that solids level, the comparison energy input/output computations in Figure 11 were determined for combustion of the dry starting microalga and the char obtained in 40% yield.

Despite a relatively good heat of combustion value for the dry microalga, the fundamental problem with Scenario 1 was that significant energy input (9,900 BTU) was required to evaporatively remove 9 pounds of water and obtain 1 pound of dry alga from 10 pounds of 10% solids centrifugate. Rapid evaporation of water was applied in this scenario because drying by the sun over the course of several days does not constitute an industrial process for conversion of a biomass material into fuel products that can economically compete with fuel products of the petroleum and coal industries. Therefore, the overall energy balance for this scenario was negative, i.e., more energy was required to dry the product than was obtained from its combustion.

With the second scenario in which a char was obtained in 40% yield that possessed a very high heat of combustion, i.e., 75% greater than the dry starting microalga, the 10% solids centrifugate was the starting reaction mixture and no removal of water was required. In order to heat the system to 203 °C in a batch system and maintain that temperature for 2 h, 3,173 BTU were required for that operation and could be maintained with proper reactor insulation. When the reaction period was completed, a heat exchanger could have been utilized to recover and effectively recycle a significant portion of this sensible heat. This was omitted from the simple comparative analysis in order to err on the conservative side. The cooled reaction mixture was "de-watered" by filtration to obtain a moist char. An additional 223 BTU were required to dry that material in order to compare its combustion output with that of the dry microalga. Despite being obtained in only 40% yield and the processing required, the energy balance was positive, i.e., an overall output was attained for this process that was 4,000 BTU better than combustion of the microalga. Algal concentration was a very important factor in this comparison, and the two energy scenarios will reach a point of being about equal at higher % solids. A rough calculation revealed that as the algal concentration level of the starting slurry increased and less water was required to be removed evaporatively, the two become about equal at ca. 22% solids. A significant advantage of the HTC method, however, is that microalgae can be processed at relatively low solids, i.e., > 5% solids, that can be obtained by employing conventional techniques such as microfiltration and flocculation/gravity settling methods. Concentration and harvesting techniques are obstacles facing the algal oil industry, and obtaining higher concentrations of algae from native concentrations of generally less than one percent remains a significant challenge.

Potential Utility of Chars in Non-Combustion Applications

Aside from utility as a carbon neutral fuel and replacement of fossil fuels for electrical power generation, other applications envisioned for the char materials include:

1. Due to the low ash contents of algal chars, conversion into synthesis gas and, subsequently, into industrial chemicals and gasoline (Fischer-Tropsch) should be facilitated, as well as possibly more esoteric applications such as a replacement for coal coke in steel manufacturing.
2. The surface of agal chars and other analogous chars [12] is populated by functional groups that enable the relatively hydrophobic particles to be wetted by water. These groups include carboxyl, hydroxyl, and aldehyde groups. These groups can interact through hydrogen bonding with similar groups on organic polymers and concrete, providing reinforcement and improved properties.
3. An algal char has been preliminarily examined as a soil amendment material and was discovered to provide a degradable organic amendment in that both carbon dioxide and nitric oxide were produced with no detrimental effects on microbial communities (unpublished result).
4. The material could also be housed in a suitable storage environment for carbon sequestration purposes. Microalgae are very efficient in metabolizing carbon dioxide and producing microorganisms having 45-50% carbon. HTC can rapidly convert these microoganisms into chars having ca. 70%C.
5. One further, demonstrated utility of the chars discussed in the following section is the ability to adsorb fatty acids. In addition to separating fatty acids from microalgae, this suggests utility for the char as a hydrophobic adsorbent for water and, possibly, air purification.

Fatty Acid Products from HTC of Microalgae Substrates

While investigations were being conducted with microalgae, distiller's grains were also being investigated as HTC substrates. An important discovery from that work which is detailed in a subsequent section of this chapter dealing with the actual nature of the chemical reactants in HTC and also applicable to microalgae was that *lipid materials and lipid-derived products such as fatty acids produced under the hydrolytic conditions of the HTC reaction were not chemically involved in char formation*. Rather, as relatively hydrophobic materials, they became adsorbed onto the most hydrophobic component in the system which was the char. Once separated by filtration with the char, the fatty acids could readily be isolated by solvent extraction and incorporated into green diesel liquid transportation fuels. This observation has also been reported by Savage, et al.,[26] at the University of Michigan. That group examined *Chlorella vulgaris* produced heterotrophically in-house with a lipid content of 50 – 53%. HTC was investigated using a batch process at 250 °C, 25% solids and reaction times of from 15 – 60 minutes. High levels of adsorbed fatty acids (77-90%) were obtained on the hydrolysis solids. These wet solids were converted directly into biodiesel materials using a supercritical transesterification process.

Our efforts in this area have been conducted on a larger experimental scale [27] with the objective of isolating fatty acid products which can serve as a feedstock for green diesel fuel, one of the principal objectives of the developing algal oil industry. The initial process that was employed to isolate fatty acid products involved the following process steps:

1. Conduct HTC of microalgae at 5-15% solids in order to minimize concentration requirements and maximize energy output scenarios;
2. Filter the product mixture to separate filtrate and char;
3. Acidify the filtrate to pH < 4 and convert fatty acid carboxylate anions into fatty acids that can be extracted; and
4. Extract fatty acid products from both char and acidified filtrate.

These steps provided for the determination of total yields of fatty acids, as well as their distribution on the char and in the filtrate. Figure 12 shows the results of extraction studies with microalgae having varied fatty acid contents, as measured by conversion into fatty acid methyl esters (FAMEs) and analyzed by gas chromatography (GC) or NMR [27]. Methyl t-butyl ether (MTBE) was employed as extraction solvent.

Microalga	FAME [a,b] Yields by GC	Weight Average[c] Molecular Weight	Yields, Extracted [a,d] Fats	FAMEs, [a,e] 1H-NMR
Dunaliella Salina	4%	278	9%	4%
Chlamydomonas reinhardtii	9%	284	18%	12%
Inspired Fuels	30%	290	33%	28%

[a] = All % values are based on starting microalgae mass.
[b] = Results were obtained at Medallion Laboratories, Inc., (Minneapolis, MN) employing an acid-catalyzed hydrolysis, FAME formation using BF_3/MeOH and comparison of GC peak areas with a standard solution of FAMEs derived from fatty acids in foods [28].
[c] = Values were computed from FAME profile analyses compiled by Medallion Labs.
[d] = These values were gravimetric masses obtained using MTBE.
[e] = Internal results using dimethyl terephthalate as internal standard and methyl ester formation via BF_3/MeOH.

Figure 12. Results obtained with various microalgae.

Note that the values for % fatty acids, quantified as FAMEs, present in the starting microalgae and in the extracts are virtually the same. Furthermore, in data not shown with the Inspired Fuels microalga having ca. 30% lipid content, 92% of the fatty acids were isolated from extraction of the char, with only 8% being obtained from the acidified aqueous filtrate. This result indicated that acidification and extraction of the filtrate was probably not cost effective and could cause deleterious effects in applications of filtrates, *vide infra*. Importantly, fatty acids that are the most highly prized product of the algal oil industry can be obtained in high yield by the HTC process. The presence of the fatty acids adsorbed onto the chars certainly contributes to heat of combustion, but not overly so, as the Inspired Fuels char when extracted provided a heat of combustion approximately 80% of that obtained with the fatty acids still present on the char.

EXAMINATION OF THE MECHANISM OF CARBONIZATION

Efforts were made also to determine the distribution of carbon among the HTC reaction products, with specific attention being given to measuring the carbon dioxide present in the headspace of the reactor and dissolved in the aqueous suspending medium. With *Chlamydomonas reinhardtii*, 55% of the carbon originally present in the microalga was contained in the char, and 45% was contained in solutes (no carbonate) dissolved in the aqueous filtrate. Based on reaction headspace pressure, the solubility of carbon dioxide in water, and assuming that all the gases contained in the headspace were carbon dioxide, less than 7% carbon was present in carbon dioxide. This experiment clearly indicated that the major carbonization pathway was not by loss of carbon dioxide. While dehydration and loss of water as a carbonization mechanism was supported by this experiment, other possibilities such as an oxidation-reduction process whereby carbonization could occur by some other process such as by formation of carboxylic acid products among the solutes in the filtrate could not be eliminated without further study, although that possibility seems somewhat remote.

Analysis and Utility of the Aqueous Filtrate

To establish the commercial potential of the HTC process of microalgae, it will be important to define a use for the aqueous filtrate. Some of the characteristics of the filtrate, again from *Chlamydomonas reinhardtii*, are provided in Figure 13.

% Solids = 3.55   pH = 6.13   % Ash = 4.65

$[NH_4^+]$ = 0.067 M $[NO_3^-]$ = 0.001 M $[H_2PO_4^-]$ = 0.023 M $[CO_3^{-2}]$ = <0.03%

Elemental composition: %C = 46.0; %H = 7.1; %N = 13.1; %S = 2.0; %P = 4.4

Metal analysis by ICP (1:50 dilution in ppm):

Mg^{160} > K^{80} > Si^{38} > Na^{12} > Ca^8 > Mn^2 = Mo^2

GC and HPLC analysis: Detection of > 200 solute compounds

Many nitrogen heterocyclic compounds and fatty acids by GC-MS

Piperazinediones derived from combinations of glycine, alanine,

Leucine, proline, phenylalanine, and valine

Figure 13. Characteristics and composition of the aqueous filtrate from the HTC of *Chlamydomonas reinhardtii*.

As was apparent from the complexity of the chemical makeup of the filtrate having literally hundreds of compounds present, consideration of the isolation of useful compounds from the filtrate was quickly abandoned. The presence of piperazinediones (six-membered ring dimers of amino acids) was an interesting discovery. There did not appear to be a general breakdown of protein structure to amino acids, as significant quantities of the latter could not be detected by ninhydrin treatment of the filtrate. Another explanation for formation of the piperazinediones was that only certain hydrolytically sensitive peptide linkages in the proteins were hydrolyzed, creating polypeptides having N-terminal amine groups. Formation of the piperazinediones could result from nucleophilic attack by these newly formed terminal amine groups on an adjacent peptide group, displacement of the residual polypeptide, and formation of the piperazinedione. Some support for this conjecture was obtained from an experiment in which phenylalanine was subjected to HTC conditions and was isolated unchanged, with no formation of a piperazinedione.

Significantly, the amounts of nitrogen and phosphorous present in solutes in the filtrate were about 80 and 100%, respectively, of those elements originally present in the microalga. After thermal treatment at ca. 200 °C, the filtrate was sterile, and the pH was nearly neutral. These characteristics suggested that the filtrate could be utilized as a nutrient for growing additional algae. One of the problems with microalgae and obstacles for growth of the algal oil industry is that microalgae are relatively expensive to grow, requiring ca. five times the amount of nitrogen compared to terrestrial plants [29]. Also, conservation of water in the algal oil industry, as with other industries, is of paramount importance. Therefore, evaluation of the filtrate as a nutrient amendment for growing additional *Chlamydomonas* reinhardtii was examined as an application for the aqueous filtrate.

Employing ca. a 20-fold dilution of the filtrate, it was determined that this growth medium functioned about half as well as the optimum medium recommended for growth of this microalga [30]. Higher levels of replacement resulted in inhibition and reduced growth. This result is preliminary and it is not known whether additional optimal medium or one or a few of its components would provide additional growth for this microalga. Also, whether the effects of increasingly higher concentrations of Maillard heterocyclic compounds [31] (*vide infra*) in the growth media upon continued recycle of filtrates will be a problem requires further examination and may prove to be specific to each species of microalgae employed in the industry. This preliminary result, however, was positive in the sense that the filtrate did function as an effective growth medium for microalgae.

DISTILLER'S GRAINS

A significant issue with microalgae as a biomass substrate for HTC is that microalgae are not presently industrial commodities. The whole algal oil industry needs to develop and achieve an economy of scale such that vastly increased quantities of microalgae become available at signifiantly reduced cost before any processing operation like HTC can become of commercial importance. This is not the case with distiller's grains. The fuel ethanol industry in the US produced 10.7 million gallons of ethanol in 2009 [32]. Corn dry milling operations constitute by far the largest process for manufacturing ethanol, and for every bushel of corn that is subjected to fermentation, 2.7 gallons of ethanol and 16 pounds of fermentation residue known as distiller's dry grains with solubles (DDGS) are produced [33]. This process is predicted to generate 16 million tons of DDGS by the end of 2010 [34], and DDGS is exclusively used as an animal feed, primarily for ruminants such as cattle. Utility as an animal food is an important use and economical if the material can be transported to the farm as obtained directly after distillation of ethanol from the residue in the beer well. At that point, the material is referred to as wet distiller's grains (WDG) and typically contains about 65% moisture. However, if the local feed market is saturated, the WDG needs to be dried to obtain a shelf-stable product. The energy costs to accomplish this drying operation are substantial, with estimates approaching up to 30% of a plant's operating budget [35]. While utilization of a basic food for other purposes is controversial, having options that can improve the economics of ethanol plant operation can facilitate this important industry [36].

Evaluation of Reaction Conditions

From a chemical reactant perspective, DDGS is a lignocellulosic material with a cellulose content of < 20 weight percent [37] and could be responsive to HTC employing similar conditions as with microalgae. The fairly large quantity of DDGS available to us eliminated lot-to-lot variability concerns, and the same designed experimental conditions were applied to DDGS as with *Dunaliella salina* in Figure 3. The data are contained in Figure 14.

These data provided the following linear regression equation from the orthogonal design:

For % Carbon $y_1 = 64.98 + 1.78 X_1 + 1.32 X_2 + 0.08 X_3$

For % Mass Yield $y_2 = 37.44 - 1.34 X_1 - 0.34 X_2 + 6.01 X_3$

where X_1 = dimensionless temperature; X_2 = dimensionless time; and X_3 = dimensionless % solids

Temp (°C)	Time h	% Solids	% Mass Yield	%C	% Carbon recovered[a] in the char product
190	0.5	5	61.6	31.1	39.8
190	0.5	25	61.2	45.6	58.0
190	2.0	5	65.0	31.7	42.8
190	2.0	25	64.3	44.4	59.3
200	1.25	15	65.7	39.1	53.4
200	1.25	15	65.2	39.0	52.8
200	1.25	15	65.5	38.9	52.9
210	0.5	5	65.2	30.2	40.9
210	0.5	25	65.9	41.9	57.4
210	2.0	5	67.1	30.4	42.3
210	2.0	25	68.1	39.6	56.0

[a] = This value was computed by dividing the amount of carbon in the char by the amount of carbon in the starting DDGS.

Figure 14. Designed experiment examining the effects of reaction time, temperature and DDGS concentration.

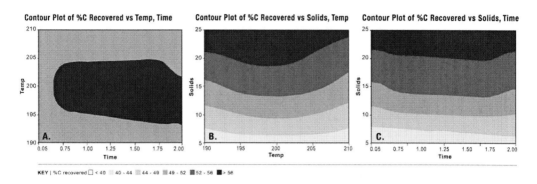

Figure 15. Contour plots examining designed experiment input variables of reaction temperature, time and DDGS concentration.

	DDGS				Char			
Sample	Protein[a]	Fat[b]	Carb.[c]	Ash[d]	%C	%H	%N	Yield
S191	31.7	13.3	50.9	4.1	66.8	7.8	4.1	37.5
S201	31.1	13.4	50.5	nd[e]	67.0	7.9	3.9	38.2
S001	26.4	10.5	57.7	5.4	66.9	8.0	3.8	37.6
S200	26.5	13.1	55.4	nd	65.4	7.9	3.8	39.1
S022	31.3	8.8	52.3	3.7	66.7	7.7	3.8	42.4
S037	31.3	11.3	52.3	5.0	67.4	8.1	4.1	37.2
S026	32.6	12.3	48.8	6.3	66.6	7.8	4.7	37.1
CVE[f]	35.8	10.1	50.4	3.7	67.0	7.8	4.2	40.0
MLN[g]	nd	nd	nd	nd	65.5	7.6	3.9	38.1

[a] = Protein levels for the DDGS substrates were computed by multiplying the %N value of the DDGS by 6.25.
[b] = Fat content was measured using ether extraction.
[c] = Carbohydrate content was computed by subtracting the values for protein, fat, and ash from 100%.
[d] = Ash content was measured gravimetrically.
[e] = nd means "not determined" and a value of 5% was assumed to compute carbohydrate values.
[f] = Chippewa Valley Ethanol (Benton, MN).
[g] = Munson Lake Nutrition (Howard Lake, MN).

Figure 16. Compositions of various DDGS materials and elemental analyses of chars derived from them.

As with microalgae, the magnitudes of the coefficients for the % solids variable were informative. Percent solids was not very important for carbonization but quite important for yield. Contour plots of the data are given in Figure 15.

As was evident from the contour plots, % solids, again, was the most important input variable for recovered carbon in the char. When % solids was plotted with both temperature (15B) and time (15C), much change was occurring and higher carbon levels were obtained with increasing % solids levels. The overall process window for DDGS was "wider" and less sensitive to changes in the input variables compared to microalgae. Also, there was no indication of "overcooking" the material at longer reaction times and higher temperatures as with microalgae. The odor of the DDGS reaction products was similar to microalgae reactions but more intense with DDGS. The odor was not objectionable and actually was reminiscent of a kitchen, having roasted coffee and carmel aromas. This odor was likely derived from similar organic compounds formed by Maillard reactions during the carmelization process in cooking [38].

Examination of a Range of DDGS Materials of Variable Composition

A broad range of DDGS samples in terms of chemical composition was available from the University of Minnesota [39], and these materials are identified along with their corresponding chars obtained by HTC at 15% solids, 200 °C and for 2 h in Figure 16.

Very little actual differences were observed among the chars despite relatively significant compositional differences in starting DDGS, with protein levels ranging from 26.4-35.8%, fats from 8.8-13.4%, and carbohydrates from 48.8-57.7%. This observation was supported also by computation of standard deviation values for the compositions, with values of 0.7 for %C, 0.2 for %H, and 0.3 for %N. In fact, the chars were compositionally quite similar to those obtained from microalgae, except that the %N values were reduced compared to most of the microalgae examined which was likely attributed to lower levels of protein present in the starting DDGS.

Based on the elemental compositions of DDGS chars, approximately 55% of the carbon, 33% of the nitrogen and essentially none of the phosphorous was contained in the char. That the nitrogen and phosphorous solutes were predominantly or exclusively contained in the filtrate also suggests utility of the filtrate as a plant fertilizing solution. Reaction pressures and carbon dioxide levels created were slightly higher than those observed with microalgae.

17A Starting DDGS

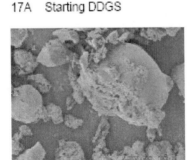

17B Char at 200 °C, 2h, 15% solids

Figure 17. SEMs of DDGS and corresponding char.

As indicated by SEMs in Figure 17 compared to starting DDGS "chunks", chars were substantially altered into coral-like structures. The transformation appeared to be even more dramatic and controlled in the sense that a much more uniform product structure was obtained than with *Chlamydomonas reinhardtii* of Figure 7.

Comparison of the Energy Input/Output Relationships of Chars Relative to Starting WDG

The energy balance was very dependent on char yield and % solids levels of starting WDG. With the presence of centrifuges that are common in the ethanol industry and capable of providing WDG @ 35% solids, the combustion energy output for drying the WDG and burning the residue was actually greater by about 1950 BTU/lb compared to burning a char obtained in 40% yield, despite the higher heat of combustion for the char (12,751 BTU/lb) compared to DDGS (8,650 BTU/lb). These scenarios would be more comparable if char yields could be improved. In any event, the alternative higher value applications mentioned in connection with algal chars should also apply to chars obtained from the HTC processing of WDG.

Experiments Conducted to Elucidate the Actual Reactants That Are Involved in HTC

Reported reactions that occur under HTC conditions chiefly involve carbohydrates and include dehydrations, rearrangements, β-eliminations, and fragmentations [40] [41] [42]. Proteins and amino acids can also become involved by the classical Maillard reaction [31] when carbohydrates possessing a reactive aldehyde groups are also present. As mentioned previously, the kitchen-like odors characteristic of reaction products also support involvement of proteins in reactions.

To systematically investigate the roles of various chemical components that are chemically responsive to HTC conditions, a "synthetic" analog of distiller's grain was constructed in which each component was investigated singly and in combination with others to empirically determine the actual chemical reactants in these reactions. DDGS was a better substrate to model than microalgae because the overall compositions of DDGS materials are relatively constant from sample to sample. Of the dry matter (89%, 2% coefficient of variation {CV}), the basic components are protein (30%, 6% CV), fat (11%, 8% CV), ash (6%, 15% CV), and neutral detergent fiber (NDF) (42%, 6% CV) which is the lignocellulosic portion [43].

The major material present in DDG is NDF, and a process [44] has been reported for obtaining NDF by removing extraneous materials from DDG using an extraction process. However, when NDF was isolated, dried, and rehydrated, a concentration of only about 7.5% solids could be achieved that was sufficiently homogeneous and transferrable as a component in the various combinations, so the concept of precisely modeling the chemical composition of DDGS was abandoned.

We had conducted a fairly large amount of work examining the HTC behavior of cheese whey. A useful property of cheese whey is that it is comprised of an essentially constant composition of lactose (6%) and protein (1%) [45] and could be utilized in the synthetic constructions whenever these two components are needed. An additional advantage is that whey proteins can be purchased from nutrition stores and employed when proteins are examined singly or in combinations not involving a carbohydrate (lactose). Therefore, the synthetic DDGS composition utilized the following components (concentrations are indicated in Figure 18 and were adjusted to maximize the affect of a particular component):

Carbohydrates – Lactose

Proteins – Whey Proteins

Fats – Corn Oil (polyunsaturated) and Canola Oil (monounsaturated)

Neutral Detergent Fiber (NDF) – Extract product

An early reference [1] had described the "hydrocarbonification" of lignocelluslosic woody products as providing a combination of α-coal and β-coal, with the former being soluble in organic solvents. To ascertain whether significant quantities of extractible materials were present in the chars, selected chars were freeze-dried and then extracted with methyl t-butyl ether (MTBE) to determine the amount of mass removed by the organic solvent. Both corn and canola oils were employed in several experiments because a common product formed during HTC of carbohydrates is 4-hydroxymethylfurfural that is an active diene in Diels-Alder reactions [46], and the potential for greater involvement of corn oil due to its polyunsaturated makeup compared to canola oil in char formation could be investigated in this manner. Whenever the oils were employed as components, the biphasic mixtures were microemulsified before reaction to provide better mixing and reactant contact. All HTC reactions were conducted at 200 °C for 2 h. Mass yields and elemental analyses of chars are given in Figure 18.

Concerning the impact of carbohydrates on char formation, lactose by itself provided a highly carbonized char under the reaction conditions and also contributed to char mass in essentially all the combinations in which it was included. NDF when employed singly also provided an extracted char in 35 % yield (entry 4). This likely involved the hemicellulose and lignin components of NDF.

That a hemicellulose could provide a char under the reaction conditions was independently verified by subjecting xylan to HTC and isolating a char in 24% yield having a % C = 67.0.

Proteins by themselves evidently contributed to char mass but not to increased levels of carbonization. Contribution to char mass was evident from measurable %N values in chars containing proteins as components and also with protein alone (entry 1) in which a char was isolated. That proteins do not undergo carbonization was indicated by entry 1 in which the isolated char possessed a %C value of only 38%. SEM images of that material (not shown) also suggested that a precipitation or denaturation of the protein had occurred without increasing the carbon content, as images of the material were devoid of any spherical coral-like microstructure common to other carbonized materials.

Triacylglycerides by themselves and in any combination did not contribute to char mass yields, were adsorbed on the chars, and could be isolated by solvent extraction. This was particularly significant and evident with both corn and canola oils (entry 3) in which no chars were formed at all and in all combinations involving the oils by comparing extracted yields with yields obtained in the absence of the oils, e.g., entries 1 and 6, 2 and 8, 4 and 10.

Entry	Material	Char % Yield	Extracted Char % Yield	Elemental Anal. %C	%H	%N
	SINGLE COMPONENTS					
1	Protein (6% solids)	9	--	38.2	5.3	7.0
	Extracted	--	8	--	--	--
2	Lactose	15	--	66.0	4.5	<0.5
	Extracted	--	14	--	--	--
3	Canola Oil and Corn Oil (40% microemul.)	0 & 0	--	--	--	--
4	NDF	48	--	64.8	7.9	2.3
	Extracted	--	35	57.6	6.2	0.7
	BINARY COMBINATIONS					
5	Lactose + Protein	31	--	64.4	4.7	3.9
	Extracted	--	30	--	--	--
6	Protein (6%) + Canola Oil (33% of Protein)	8	--	54.3	8.1	5.5
7	NDF + Protein (25% of NDF)	53	--	61.2	7.6	3.9
	Extracted	--	41	57.7	6.5	4.9
8	Lactose + Corn Oil (25% of Lactose)	23	--	71.5	7.6	<0.5
	Extracted	--	17	69.5	7.2	<0.5
9	Lactose + NDF	78	--	65.0	6.4	2.2
	Extracted	--	68	61.0	7.5	7.6
10	NDF + Canola Oil (20% of NDF)	58	--	66.9	8.4	2.2
	Extracted	--	36	58.0	6.3	3.2
	TERNARY COMBINATIONS					
11	Lactose + Protein + Canola Oil (50% of Lactose/Protein)	55	--	70.6	8.6	1.7
	Extracted	--	22	60.7	5.3	3.7
12	Lactose + Protein + Corn Oil (20% of lactose)	45	--	70.4	8.4	1.9
	Extracted	--	18	60.5	4.7	4.0
13	Lactose + Protein + NDF	81	--	62.9	6.5	2.7
	Extracted	--	71	58.5	7.2	7.6
14	Lactose + NDF + Corn Oil (20% of lactose/NDF)	98	--	65.7	7.0	1.6
	Extracted	--	81	63.7	6.3	<0.5
	QUATERNARY COMBINATION					
15	Lactose + Protein + NDF + Corn Oil (20% of Lactose/Protein/NDF)	85	--	63.9	7.2	2.5
	Extracted	--	66	60.4	5.9	3.0

Figure 18. HTC of constituent materials of synthetic DDG.

There were also no differences noted between corn and canola oils that might have been attributed to Diels-Alder activity. It was apparent that the extraction procedure to prepare NDF did not remove triacylglycerides, as significant mass was extracted (entry 4) that was verified to have significant fatty acid content by IR. Triacylglycerides did appear to have a positive carbonization effect with proteins, however. In entry 6, the %C of the char was increased to 54% in the presence of an oil, whereas the char from protein alone only had a %C value of 38%. This may be attributed to the bipolar nature of fatty acids created under the reaction conditions that may provide a micellular, vesicular or other structure in which the proteins can undergo a carbonization event.

BREWER'S GRAINS AND OTHER FERMENTATION RESIDUES

Brewer's grains are slightly different being derived from other cereal grains than corn. Examination of yeast and other fermentation residues were also of interest, as the disposition of fermentation residues especially involving genetically modified microorganisms is not routine and is expected to increase in the future. Since application as a food product is generally not available with these materials, conversion into a char product could be a very important alternative. HTC conditions were 15% solids, 200 °C, for 2 h and the data for obtained with these residues is contained in Figure 19. All materials examined were converted effectively into useful char materials.

OTHER BIOMASS SUBSTRATES THAT DID NOT RESPOND TO HTC CONDITIONS

Several biomass materials were examined without success in terms of achieving significantly increased carbon contents during HTC at reaction temperatures < 225 °C and reaction times of 2h. These included: prairie grasses, sugar beet tops and pulp, kelp, water hyacinth, Eurasian milfoil, and duckweed.

Substrate	Char, % Yield	%C	%H	%N
Yeast[a]	31	67.6	7.2	7.4
Brewer's[b] Grains	49	62.5	7.2	3.7
Pichia[c] Residue	33	64.3	7.1	7.7
Actinomyces[c] Residue	27	62.9	7.5	7.1

[a] = Yeast sample was Ethanol Red.
[b] = Sample was obtained from Anheiser Bush and was derived from barley.
[c] = Samples were received from Dr. Frederick Schendel, Head of the Bioprocess Resource Center, BioTechnology Institute, University of Minnesota.

Figure 19. HTC of other fermentation residues.

CONCLUSION

HTC has great appeal as a potential industrial process due to its simplicity. It basically is a process in which a biomass substrate is heated in water under pressure at temperatures < 250 °C and is particularly well-suited for conversion of wet biomass materials. Under appropriate conditions, energy contents of these substrates can increase on the order of 50% and higher, while corresponding thermochemical operations such as torrefaction require significant heat energy to first evaporate water and more heat to accomplish carbonization. Even so, the energy contents of torrefied products are generally increased on the order of only 30%.

The principal objective of our research was to broadly explore the range of biomass materials that could be processed effectively using HTC conditions and accomplish high degrees of carbonization in relatively short reaction times amenable to continuous processing operations. Considerable research has been conducted and reported in the literature with lignocellulosic materials, but when cellulose is a principal component of the biomass, reaction temperatures > 230 °C are required with overnight reaction times. Our approach was to examine biomass materials that contained relatively minor amounts of cellulose, employ reaction temperatures of < 225 °C, and reaction periods of < 2 h to have any chance of the process being conducted continuously.

Microalgae were especially suitable as HTC substrates. With *Dunaliella salina*, for example, conversion was accomplished in 15 minutes into char in 43% yield having a carbon content of 64%, compared to 48% carbon in the starting microalga. This char possessed a heat of combustion of 13,177 BTU/lb (which is equivalent to a high quality bituminous coal) and could be effectively pelletized without addition of binding agents and heat to reduce material volume. Algal chars also have applications beyond combustion as carbon neutral fuels that include: carbon sources for syngas production for industrial chemical, gasoline, coke production; reinforcing agents in polymers and concrete; adsorbents in air and water purification; and soil amendment/carbon sequestration.

Despite having significant cellulose contents, distiller's grains and other fermentation residues responded well to HTC conditions. Chars could be obtained in moderate yields (ca.40-50%) that were remarkably similar in composition and appearance to chars obtained from microalgae.

During the conduct of the char research, it was determined that fatty acids derived from lipids present in the biomass materials could also be isolated in almost quantitative yield. These are the products that are now being sought by the developing algal oil industry and can be converted into conventional liquid transportation fuels. Therefore, the process with microalgae (and fermentation residues) can provide the principal object of that industry's attention in high yield, as well as char and algal nutrient co-products that have value and can reduce the cost of microalgal production.

ACKNOWLEDGEMENTS

We would like to express our appreciation to very talented co-workers for their valuable contributions to aspects of this work. These included Professor H. Ted Davis, Professor

Michael J. Sadowsky, Professor Paul A. Lefebvre, Dr. Fredrick J. Schendel, Dr. Laurie A. Harned, and Ms. Lindsey R. Jader. The artistic talents of Mr. Timothy Montgomery in the preparation of the figures are also highly appreciated. We appreciate the gift of a high lipid-containing microalga from Dr. Marc Ferguson of Inspired Fuels, Inc., of Austin, TX. Funding was provided from the Institute for Energy and the Environment of the University of Minnesota and the Biocatalysis Initiative of the BioTechnology Institute.

REFERENCES

[1] Bergius, F.; Erasmus, P. *Naturewissenschaften* 1928,16,1-10.
[2] Funke, A.; Ziegler, F. *Biofuels, Bioproducts and Biorefining* 2010, 4,160-177.
[3] Kubikova, J.; Zemann, A.; Krkoska, P.; Bobleter, O. *Tappi J.* 1996, 79,163-169.
[4] Sevilla, M. ; Fuertes, A.B. *Carbon* 2009, 47, 2281-2289.
[5] White. R.J. ; Budarin, V. ; Luque, R. ; Clark, J.H. ; Macquarrie, D.J. *Chem. Soc. Rev.* 2009, 38, 3401-3418.
[6] Hu, B.; Wang, K.; Wu, L.; Yu, S.H.; Antonietti, M.; Titiricii, M.M. *Adv. Mat.* 2010, 22, 813-828.
[7] Heilmann, S.M.; Davis, H.T.; Jader, L.R.; Lefebvre, P.A.; Sadowsky, M.J.; Schendel, F.J.; von Keitz, M.G.; Valentas, K.J. *Biomass and Bioeng.* 2010, 34, 875-882.
[8] Chisti, Y. *Biotechnol. Adv.* 2007, 25, 294-306.
[9] Sheehan, J.; Dunnahay,T.; Benemann, J.; Roessler, P. NREL (TP-580-24190) 1998, Technical Review Section, 36-40.
[10] Sheehan, J.; Dunnahay, T.; Benemann, J.; Roessler, P. NREL (TP-580-24190) 1998, Technical Review Section, 96-103.
[11] Pulz, O.; Gross, W. *Appl. Microbiol. Biotechnol.* 2004, 65, 635-48.
[12] Titiricci, M.M.; Thomas, A.; Yu, S.H.; Muller, J.O.; Antonietti, M. *Chem. Mat.* 2007, 19, 4205-4212.
[13] Schumacher, J.P.; Huntjens, F.J.; van Krevelen, D.W. *Fuel* 1960, 39, 223-234.
[14] Blokker, P.; Schouten, S.; Van Den Ende, H.; De Leeuw, J.W.; Hatcher, P.G.; Snninghe Damste, J.S. *Org. Geochem.* 1998, 29, 1453-68.
[15] Voigt, J.; Frank, R.; Wostemeyer. J. *FEMS Microbiol. Lett.* 2009, 291, 209-215.
[16] Becker, E.W. *Microalgae: biotechnology and microbiology.* Baddiley, J.; Ed.; Cambridge, New York, NY,1994, p.177.
[17] Zelazny, A.M.; Shaish, A.; Pick, U. *Plant Physiol.* 1995, 109,1395-1403.
[18] Tachibana, T.; Ichikawa, H. *Nippon Kagaku Zasshi* 1948, 69, 81-84.
[19] www.ket.org/trips/coal/agsmm/agsmmtypes.html.
[20] www.wikipedia.org/wiki/Heat_of_combustion.
[21] Fuwape, J.A. *Agroforestry Syst.* 1993, 22, 175-179.
[22] www.scribd.com/doc/7027056/Bagasse_as_Alternate_Fuel.
[23] Somermeier, E.E.; *Coal: its composition, analysis, utilization and valuation*, McGraw-Hill, New York, NY,1912, p.53.
[24] Brocksiepe, H.G. *Charcoal*, In *Ullmann's Encycl. Industrial Chem.*, 6th ed., 2003, 7, pp 307-312.

[25] Clarke, S.J.; Polack, J.A. *Sugar production*, In *Encycl. Chem. Processing and Design*, McKetta, J.J.; Ed.;1996, 55, pp 238-263.
[26] Levine, R.B.; Pinnarat, T.; Savage, P.E. *Energy and Fuels* 2010, 24, 5235-5243.
[27] Heilmann, S.M.; Jader, L.R.; Harned, L.A.; Sadowsky, M.J.; Schendel, F.J.; Lefebvre, P.A.; von Keitz, M.G.; Valentas, K.J. *Applied Energy*, 2011, 88, 3286-3290.
[28] House, S.D.; Larson, P.A.; Johnson, R.R.; DeVries, J.W.; Martin, D.L. *J. AOAC Int.* 1994, 72, 960-965.
[29] Brezinski, M.A. *J. Phycol.*, 2004,21, 345-357.
[30] Gorman, D.S.; Levine, R.P. *Proc. Natl. Acad. Sci. USA*, 1965, 54, 1665-1669.
[31] http://www.slv.se/upload/heatox/documents/Maillard%2013%20Sep%2007.xls.
[32] http://www.ethanolrfa.org/industry/statistics/.
[33] Rausch, K.D.; Belyea, R.l. *Appl. Biochem. Biotechnol.* 2006, 128, 47-86.
[34] Weiss, B.; Eastridge, M.; Shoemaker, D. St-Pierre, N. *Distillers grains-FACT SHEET*, The Ohio State University Extension, June (2007) pp.1-4 (available at http://ohioline.osu.edu.).
[35] *Christiansen, R.C. DDGS: Ethanol Producer Magazine* 2009, March issue, available at http://www.ethanolproducer.com/article.jsp?aricle.id=5346andq=andpage=all.
[36] Heilmann, S.M., Jader, L.R., Sadowsky, M.J., Schendel, F.J., von Keitz, M.G., Valentas, K.J. *Biomass & Bioeng.* 2011, 35, 2526-2533._
[37] Bals, B.; Dale, B.; Balan, V. *Energy and Fuels* 2006, 20, 2732-2736.
[38] McGee, H. *On food and cooking: the science and lore of the kitchen,* Revised ed. Scribner, New York, NY, 2004, p 778.
[39] Samples were obtained from a collection owned by Dr. Jerry Shurson, Department of Animal Science, University of Minnesota.
[40] Sun, X.; Li, Y. *Angew. Chem. Int. Ed.,* 2004, 43, 597-601.
[41] Baugh, K.D.; McCarty, P.L. *Biotechnol. and Bioeng.,* 1988, 31, 50-61.
[42] Luijkx, G.C.A.; van Rantwijk, F.; van Bekkum, H.; Antal, M.J. *Carbohydrate Res.*, 1995, 272, 191-202.
[43] Speihs, M.J.; Whitney, M.H.; Shurson, G.C. *J. Anim. Sci.* 2002, 80, 2639-2645.
[44] ANKOM Technologies (Macedon, NY). *Neutral detergent fiber in feeds—filter bag technique*, method 6 (08-16-06).
[45] www.dairyforall.com/whey.php.
[46] Demir-Cakan, R.; Baccile, N.; Antonietti, M.; Titirici, M. *Chem. Mat.* 2009, 21, 484-490.

In: Microalgae: Biotechnology, Microbiology and Energy
Editor: Melanie N. Johnsen

ISBN 978-1-61324-625-2
© 2012 Nova Science Publishers, Inc.

Chapter 6

INVESTIGATIONS ON THE USE OF MICROALGAE FOR AQUACULTURE

José Antonio López Elías[1], Luis Rafael Martínez Córdova[1] and Marcel Martínez Porchas[2]

[1]Departamento de Investigaciones Científicas y Tecnológicas de la Universidad de Sonora, Blvd. Luis Donaldo Colosio s/n entre Reforma y Sahuaripa, Edificio 7G, Hermosillo, Sonora, 83000 México
[2]Centro de Investigación en Alimentación y Desarrollo.
Km 0.6 Carretera a La Victoria, Hermosillo, Sonora, México

INTRODUCTION

Despite of its high cost, the use of live feed is essential and frequently irreplaceable in the aquaculture of mollusks, fishes, crustaceans, and some other aquatic organisms, especially in the larviculture and nursery phases (Lin et al. 2009). The use of microalgae during these phases seems to be a universal practice, because some microalgae species have adequate physical and nutritional characteristics for the early development of aquatic organisms, and the operative costs for their production is commonly lower compared to the production of other organisms or formulated feeds (Martínez Córdova et al. 1999; Lovatelli et al. 2004).

A great number of microalgae species have been used worldwide for the larviculture of aquatic organisms, and the selection of a suitable species to feed a particular organism (mollusk, fish or crustacean), depends on many factors such as: size, habitat (plankton or benthos), nutritional value, replication and growth rates, nutrimental requirements, tolerance to environmental factors, etcetera.

Many experimental researches have been done worldwide, including our institution (University of Sonora), related to the use of microalgae for the farming of diverse aquatic organisms. Those researches have covered many aspects of the culture process, including: production systems, evaluation of particular species and their nutritional value for selected organisms (mollusks and crustaceans), effect of physical and chemical environmental factors on the production and nutritional value, evaluation of commercial laboratories of microalgae, evaluation of alternative mediums for their culture, etcetera. All these aspects are widely discussed in the five sections of this chapter.

The authors hope this information can be useful for researchers, aquaculturists, students, and other people related to the microalgae culture.

1. A Brief History

Soeder (1986) hypothesized that our ancestors used to collect the cyanobacteria *Nostoc* as a gelatinous mass from the soil, as well as *Spirulina* from lakes in order to complement their nutrition. He also mentioned that *Nostoc* was consumed in Mongolia and China, and that *Spirulina* was an essential part of Africans and Aztecs (Mexico ancestors) diets; moreover, Africans still at present, consuming the microalgae.

The investigations on microalgae began around 1890, being Beijerinck the pioneer microbiologist who purified the freshwater microalgae *Chlorella vulgaris*. After that, Otto Warburg (1919) produced the same species at higher concentrations, and conducted some studies regarding to its photosynthesis process (Abalde et al. 1995). However, the first massive production of microalgae under controlled conditions occurred in Germany in the early 1920´s and had the objective to produce great microalgae biomasses rich in lipids. Such research was subsequently recuperated in a document entitled "Algal Culture: From Laboratory to Pilot Plant" (Burlew 1953), and was the first evidence of the possibility to produce microalgae at commercial scale (Pipes and Gotaas 1960).

In the 1940´s the massive culture of microalgae was initiated; the first cultures were performed outdoor, because of the high costs of artificial light. In this type of culture, sunlight, carbon dioxide (CO_2), and environmental temperature were used, so the production depended on the natural variation of these parameters (Richmond 1986).

The technical bases for the industrial culture were established in the early 1950´s in Japan, Germany, USA, Israel, and other countries. By that time, Oswald and colleagues from the University of California, produced massively microalgae with the objectives of treating wastewaters and producing protein (Abalde et al. 1995).

The biotechnology for the massive culture of microalgae has been developed successfully in many countries. At first, the production for human consumption was an interesting expectative, however, the high production costs decreased the interest in this type of production, which is now focused on animal nutrition, soil treatment, bioremediation, (Becker, 1994), production of renewable biofuels (Chisti 2008), and extraction of bioactive compounds for human use, such as: antioxidants, lipids, proteins, vitamins, and others (Cleber Bertoldi et al. 2006; Seon Jin and Melis 2003; Matsunaga et al. 2006).

2. Production Systems and Routine Modifications

The phytoplankton culture has been traditionally done in three main types of systems: static, semi-continuous and continuous (Vonshak 1988). The most commonly used in commercial hatcheries is the static system arranged in steps, which consists in starting the culture at low volumes (assay tubes or Erlenmeyer flasks), passing to bottles (19 L or more), then to columns, and finally to pools, big plastic bags or tanks. The commercial medium f/2 is the most frequently used for microalgae culture (Guillard and Ryther, 1962), which is prepared with highly purified chemicals. However, there are several other mediums based on highly purified chemical or commercial products such as plant fertilizers, which diminish productive costs.

The selection of containers for production of microalgae is a key aspect. The form (rounded, oval, squared, rectangular or cylindrical), depth (from 60 cm to 2 m) and material are the most important characteristics. These factors could have an important influence on the water column dynamics and thus on the microalgae contact with light, which is strongly related to the cell development and reproduction (Richmond 1986; Oswald 1988).

In aquaculture, the stock strains of the different species are maintained under controlled conditions (continuous illumination, and constant temperature, using air conditioners), using assay tubes, Erlenmeyer flasks or bigger bottles for great volumes. However, the massive production is commonly done outdoor and rarely indoor (López-Elías et al. 2003). Despite photobioreactors can produce much higher amounts of microalgae (sometimes by one order of magnitude), they also require greater economical investments; therefore, many farms cover their microalgae requirements using the traditional outdoor production systems (Zhang and Richmond 2003; López-Elías et al. 2005a). Thus, microalgae culture for the aquaculture industry does not always require highly technified systems; however, in some particular cases, the use of technology could be necessary.

Regarding to the production routines, a relatively constancy has been observed among commercial laboratories (López-Elías et al. 2005a); such laboratories usually report their highest productions during warm months due to the better environmental conditions, while during cold months the production is almost null. Despite the simplicity of the traditional systems for microalgae production, several adjustments and modifications can be done in order to increase the quantity and/or quality of microalgae. For instance, slight modifications in environmental factors or nutrients can result in high changes in the productive response of microalgae (Medina-Reyna and Cordero-Esquivel 1998).

Some experimental studies have been made in our institution to evaluate the different production systems and the effectiveness of the routines used in those systems, including the evaluation of some modifications on the traditional systems.

2.1. An experimental study focused on the evaluation of an alternative system for the outdoor culture of *Chaetoceros muelleri* and *Dunaliella* sp. during winter and spring, was done in Northwest Mexico (Becerra-Dorame et al. 2010). The alternative system had the same structure than the traditional outdoor systems, but it also had a recirculation cascade in which the water was pumped up from the bottom of the plastic containers by means a pump of 1/8HP at a flux rate of 3.68 $L \cdot min^{-1}$, and thereafter flowed down over a transparent plastic sheet arranged in steps, to improve the exposure of microalgae to the light (Figure 1). The hydraulic retention time (h) was 87 min, and the hydraulic loading rate (flow/horizontal area of cascade structure) was 11.5 $L \cdot mm^{-2}$.

The modification on the structure increased the microalgae exposure to light and the movement of microalgae within the system. The diatom *C. muelleri* showed a greater final cell density with such modifications compared to those cultured in the traditional system (Figure 2).

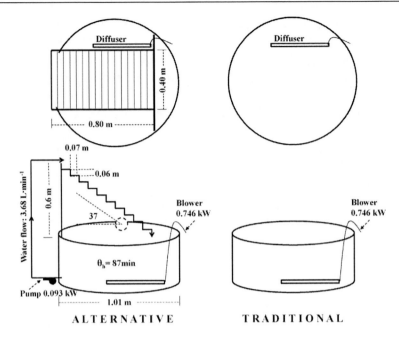

Figure 1. Schemes of the traditional and alternative systems used to culture *C. muelleri* and *Dunaliella* sp. The traditional system is commonly used in shrimp farms or commercial laboratories to fed shrimp larvae. Source: Becerra-Dórame et al. (2010), *Aquacultural Engineering*.

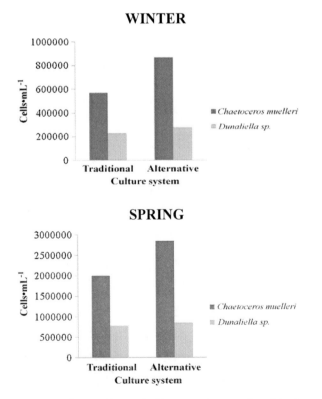

Figure 2. Cell densities of *C. muellerii* and *Dunaliella* sp. outdoor produced in the traditional and alternative systems, during winter and spring (Becerra-Dorame et al. 2010).

No differences on final biomass of *C. muellerii* were found among systems in the winter trial, but *Dunaliella* sp. had higher biomass in the alternative system. Contrarily, in the spring trial, *C. muellerii* had a greater biomass in the alternative system, but *Dunaliella sp.* did not show differences.

It was concluded that the new system is a good alternative for the culture of these two microalgae, especially during spring.

2.2. An experimental study was performed to evaluate the effect of management routines such as: inoculation hour, inoculum concentration, and culture medium on the growth and final biomass of *C. muelleri, Dunaliella* sp. and *Tetrasemis chuii* cultured outdoor during spring, in 250-L plastic tanks (López-Elías, et al., 2008; López-Elías, 2010a, Becerra-Dórame, et al., 2010).

The inoculation for the three species was done at 0600 and 1200 h; the inoculum concentrations were 0.2 and 0.4 x 10^6 cells·mL^{-1} for *C. muelleri*; 0.04 and 0.08 x 10^6 cells·mL^{-1} for *Dunaliella* sp. and *T. chuii*; the f and the f/2 mediums (Guillard, 1975) were used for *C. muelleri*, while f/2 and 2f were used for the two other species.

Results showed that the inoculation time had an effect on the productive response of microalgae, while some were also affected by the culture medium and/or the inoculums density.

Final cellular density of *C. muelleri* at 72 h varied from 1.21 to 2.83 x 10^6 cells·mL^{-1} with the best results obtained when inoculation was made at 0600 h, at a concentration of 0.4 x 10^6 cells·mL^{-1}, and using the f medium. The maximum growth rate was similar among treatments (0.82 and 1.27 divisions·day^{-1}), but the cumulated growth was higher with an inoculum concentration of 0.2 x 10^6 cells·mL^{-1}, in which 2.59 to 3.09 total divisions were obtained (Table 1).

For *Dunaliella* sp., the final cellular density was greater when the inoculum concentration was 0.08 x 10^6 cells·mL^{-1} and the 2f medium was used (0.75 and 0.83 x 10^6 cells·mL^{-1} at the end of the culture) (Figure 3). The inoculation hour did not have an effect on the final cellular density.

Tetraselmis chuii, showed a higher final cellular density when inoculated at 0600 h with a concentration of 0.65 x 10^6 cells·mL^{-1}, as compared when inoculated at 1200 h with 0.56 x 10^6 cells·mL^{-1} (Table 2).

The greater duplication rates were 3.92 and 3.99 divisions·day^{-1}, when the inoculum concentration was 0.04 x 10^6 cells·mL^{-1} and was done at 0600 h. The culture medium did not have a significant effect on the cellular density.

As conclusion of the study it was established that the higher cellular densities of microalgae were obtained in the cultures initiated at 0600 h with an inoculum concentration of 0.08 x 10^6 cells·mL^{-1}.

For *C. muellerii*, the greater biomasses were achieved in general with the higher inoculum concentration, initiated at 0600 h and the f medium (Table 1). For *Dunaliella* sp., the biomass and organic matter were greater in the cultures initiated at 1200 h, using the inoculum of highest concentration, and independently of the culture medium (Table 3). *Tetraselmis chuii*, had a greater final cellular density when inoculated at 0600 h, with a concentration of 0.08x10^6 cells·mL^{-1} and using the 2f medium. (Table 2).

Table 1. Mean ± SD of final cellular density, maximum growth rate (MGR), and cumulated growth rate (CGR) of *C. muellerii* at 48 y 72 hours, inoculated at 0600 and 1200 h with inoculums of 0.2 y 0.4 x 10^6 cells.mL^{-1} in the f and f/2 medium
α = 0.05; a < b < c

Treatment	Cells·mL^{-1} x 10^6 48 h	72 h	MGR (Divisions·day^{-1})	CGR (Total Divisions)	Dry matter (g·m^{-3})	Organic matter (g·m^{-3})
Inoculum 0600 h						
0.2 x 10^6 f/2	0.75 ± 0.01^a	1.52 ± 0.11^a	1.24 ± 0.17^b	2.93 ± 0.10^{bc}	0.087 ± 0.025^a	0.055 ± 0.003^a
0.2 x 10^6 f	0.82 ± 0.04^a	1.71 ± 0.17^a	0.90 ± 0.07^a	3.09 ± 0.14^c	0.131 ± 0.017^b	0.060 ± 0.029^a
0.4 x 10^6 f/2	1.31 ± 0.13^b	1.75 ± 0.02^a	1.15 ± 0.16^{ab}	2.13 ± 0.01^a	0.085 ± 0.006^a	0.064 ± 0.004^a
0.4 x 10^6 f	1.55 ± 0.04^c	2.83 ± 0.07^b	0.96 ± 0.05^{ab}	2.80 ± 0.04^b	0.186 ± 0.010^c	0.099 ± 0.007^b
Inoculum 1200 h						
0.2 x 10^6 f/2	0.73 ± 0.18^a	1.21 ± 0.11^a	0.82 ± 0.36^a	2.59 ± 0.04^b	0.121 ± 0.012^{ab}	0.069 ± 0.016^a
0.2 x 10^6 f	0.92 ± 0.05^{ab}	1.41 ± 0.17^b	1.10 ± 0.12^a	2.82 ± 0.08^c	0.155 ± 0.033^{bc}	0.065 ± 0.011^a
0.4 x 10^6 f/2	1.13 ± 0.12^{bc}	1.39 ± 0.02^{ab}	1.21 ± 0.21^a	1.79 ± 0.01^a	0.110 ± 0.003^a	0.066 ± 0.002^a
0.4 x 10^6 f	1.28 ± 0.03^c	2.78 ± 0.07^c	1.27 ± 0.07^a	2.80 ± 0.06^c	0.173 ± 0.023^c	0.099 ± 0.015^b

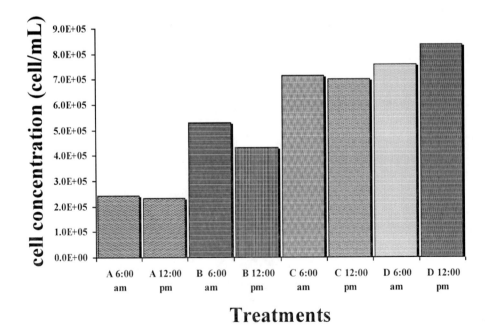

Figure 3. Cellular density of *Dunaliella* sp. at 72 h cultured outdoor with inoculums at 0600 and 1200 (A: inoculum 0.04 x 10^6 cell·mL^{-1} in medium f/2; B: inoculum 0.04 x 10^6 cell·mL^{-1} in medium 2f; C: inoculum 0.08 x 10^6 cell·mL^{-1} in medium f/2; B: inoculum 0.08 x 10^6 cell·mL^{-1} in medium 2f).

Table 2. Means ± SD of cellular density and biomass of *T. chuii* at 72 hours in mass culture, with different inoculum hours, concentration, and culture mediums

	Inoculation hour	Inoculum concentration (x10⁶ cells·mL⁻¹)	Culture medium	Cells·mL⁻¹ x 10⁵	Biomass (g·L⁻¹)
A	0600	0.04	f/2	$6.128^{abc} \pm 0.360$	$0.35^d \pm 0.12$
B	1200	0.04	f/2	$4.202^d \pm 0.350$	$0.50^{bcd} \pm 0.03$
C	0600	0.04	2f	$6.401^{abc} \pm 0.652$	$0.63^b \pm 0.17$
D	1200	0.04	2f	$6.345^{abc} \pm 1.087$	$0.39^{cd} \pm 0.11$
E	0600	0.08	f/2	$7.489^a \pm 0.311$	$0.40^{cd} \pm 0.09$
F	1200	0.08	f/2	$6.821^a \pm 0.360$	$0.57^{bcd} \pm 0.05$
G	0600	0.08	2f	$5.920^{abcd} \pm 1.390$	$0.94^a \pm 0.03$
H	1200	0.08	2f	$4.938^{cd} \pm 1.123$	$0.41^{cd} \pm 0.15$

Different letters means significant differences ($p<0.05$).

Table 3. Mean dry weight, organic matter, and ash of *Dunaliella* sp. in the four treatments

Treatment	Day	Dry weight (g·L⁻¹)	Organic matter (g·L⁻¹)	Ash (g·L⁻¹)
0600 f/2	2	0.079^a	0.054^{ab}	0.025^a
0600 f/2	3	0.094^{ab}	0.065^{bc}	0.029^{ab}
0600 2f	2	0.098^{bc}	0.070^c	0.028^{ab}
0600 2f	3	0.102^{bc}	0.071^c	0.031^{ab}
1200 f/2	2	0.078^a	0.050^a	0.028^{ab}
1200 f/2	3	0.104^{bc}	0.069^c	0.035^b
1200 2f	2	0.093^{ab}	0.064^{bc}	0.029^{ab}
1200 2f	3	0.114^c	0.067^c	0.047^c

Different letters mean significant differences at $P < 0.05$.

3. EVALUATION OF NUTRITIONAL VALUE OF MICROALGAE FOR THE CULTURE OF PARTICULAR SPECIES

Despite the wide use of microalgae in aquaculture and the high nutritional value of such natural feed, the nutritional profiles of microalgae are different not only between different species, but also among the same species (Volkman et al. 1989). As mentioned above, sight changes in the environmental conditions could modify the nutritional composition of microalgae (Sánchez-Saavedra and Voltolina 1996).

Biochemical composition of microalgae varies widely because of diverse factors inherent to the culture systems, such as culture age, pH, salinity, light, temperature and management (Voltolina and López Elías 2002). In addition, it is almost impossible to identify all the factors influencing the microalgae composition in an outdoor system (Voltolina et al. 1999). However, some modifications in the system or management procedures may improve the quantity and quality of microalgae. The nutritional quality of microalgae is a mains aspect for the success of aquaculture.

Some experiments are mentioned below, regarding to the use of microalgae cultured in different mediums and their effectiveness for zooplankton production.

3.1. The effectiveness of *Nanochloropsis oculata* and *C. muellerii*, as feed for rotifers and copepods, was evaluated in a 7-week experiment, at mesocosms level. Both microalgae were cultured in the conventional f/2 medium, and two alternative mediums, one agricultural (mono ammonium phosphate) and one aquacultural (Nutrilake). *Nanochloropsis oculata* was used as feed for the rotifer *Brachionus rotundiformis* and *C. muellerii* for the copepod *Calanus pacificus*.

Results showed that the culture mediums had an effect on the nutritional quality of microalgae; such microalgae quality had a subsequent effect on the productive response and chemical composition of rotifers and copepods when was used as feed source. Rotifer recorded differences on the production parameters as well as on the chemical proximate composition when fed on *N. oculata* cultured in the alternative mediums, as compared to the control. The best final density of rotifers and eggs, as well as the higher content of protein, were found in the organisms fed on microalgae produced in the aquacultural fertilizer, followed by those fed on microalgae produced in f/2 and MAP (Table 4). Carbohydrate and lipid contents were higher in the MAP treatment.

Table 4. Production parameters and chemical proximate composition of *B. rotundiformis* fed *N. oculata* cultured on the alternative mediums and the control

Medium	Final density Rotifers mL^{-1}	Carbohydrate (%)	Lipid (%)	Protein (%)
MAP	96.75±10.56a	25.92±1.72c	26.34±2.39b	47.74±3.47a
NLK	132.5±6.35c	12.95±1.10a	22.79±3.33a	64.27±4.42b
f/2	111.5±9.29b	18.79±2.15b	23.08±2.62a	58.13±2.82b

Table 5. Production parameters and chemical proximate composition of *C. pacificus* fed on *C. muellerii* cultured on alternative and conventional mediums

Mediums	Final density copepods mL^{-1}	Carbohydrate (%)	Lipid (%)	Protein (%)
MAP	1.41±1.79b	18.68±1.44a	25.80±2.56a	55.52±1.21a
NLK	1.62±1.22c	15.51±0.80ab	25.81±1.16a	58.68±1.60a
GF/2	0.97±2.6a	15.91±2.31a	24.83±0.48a	62.26±1.92b

The chemical proximate composition of the rotifers showed significant differences on the carbohydrate content among treatments, with higher levels on the organisms fed on microalgae cultured in the agricultural fertilizer compared to those fed cultured in the aquacultural one. The lipid content was higher in the rotifers fed microalgae from the agricultural medium as compared to the two other. The protein content was greater in the organisms fed microalgae farmed in the aquacultural and conventional f/2 medium (Table 4).

The copepod *C. pacificus* fed on *C. muellerii* which was previously cultured in the agricultural fertilizer, showed a higher final density (Table 5). In addition, significant differences were found on the protein content of the copepods fed microalgae produced in the different mediums. The higher levels were found in those fed on microalgae from the conventional medium (f/2), compared to the agricultural and aquacultural medium. No differences in carbohydrate and lipid content were recorded.

Table 6. Nutrient fraction (%) of shrimp larvae (*L. vannamei*) fed monospecific diets (*C. muelleri* and *Isochrysis* sp.) and the mixture of both. Two similar experiments were done (Exp I and Exp II)

Larvae	Protein (%)		Carbohydrate (%)		Lipid (%)	
	Exp. I	Exp. II	Exp. I	Exp. II	Exp. I	Exp. II
Nauplii	40.58 (0.00)	37.06 (9.74)	31.64 (0.00)	32.13 (1.07)	12.64 (0.00)	10.16 (0.08)
Chaetoceros Zoea III	39.94^a (6.53)	45.94^a (6.73)	30.77^a (0.09)	30.84^a (0.49)	15.87^a (4.75)	15.12^a (9.80)
Isochrysis Zoea III	43.68^a (7.69)	42.34^a (4.73)	31.37^a (0.39)	30.82^a (0.94)	18.54^a (3.18)	12.05^a (5.78)
Mixture Zoea III	38.94^a (2.36)	48.63^a (6.95)	31.20^a (0.11)	30.57^a (0.11)	26.28^b (3.11)	13.75^a (6.72)

Table 7. Fatty acid profile (%) of *L. vannamei* larvae fed with Diet I (*Chaetoceros muelleri*), Diet II (*Isochrysis* sp.) and Diet III (mixture) in two experimental runs

Fatty Acids	Diet I	Diet II	Diet III
14:0	31.62	11.70	66.41
16:0	41.30	----	----
16:1	----	----	6.46
18:0	----	25.75	2.24
18:1w9	----	13.45	----
18:1w7	----	----	1.52
18:2w6	----	7.16	1.52
20:1w9	----	----	6.4
20:5w3	11.01	7.53	4.01
22:1w11	----	----	2.48
22:1w9	----	----	4.83
22:5w3	10.26	11.59	8.89
22:6w3	----	13.84	----
24:1w9	5.79	15.69	----

3.2. The nutritional value of microalgae *Isochrysis* sp. (Diet 1), *Chaetoceros muelleri* (Diet 2), and the combination of both (Diet 3), was evaluated as feed source for larviculture and chemical proximate composition of *Litopenaeus vannamei*. The protein levels of shrimp larvae were similar when for all treatments (~43 – 44%), (Table 6). Polyunsaturated fatty acids (PUFA) 20:5w3 and 22:5w3, were more abundant in larvae fed on Diet 1; 22:6w3 was more abundant in organisms fed on Diet 2. However, the profile of fatty acids was more complete in those larvae fed on Diet 3. No significant differences on larvae survival were found among diets (> 60%) (Rodríguez, et al., 2010). The best growth was observed in larvae feed Diet 2 and 3 (Table 7).

These results differed from those reported by D´Souza and Loneragan (1999) who found that a monoalgal diet based on *Isochrysis* was unsatisfactory for shrimp larvae nutrition. Also, Piña, et al. (2006), found that monoalgal diet with *Isochrysis* sp. did not improve survival rate and rate of development in *L. vannamei* protozoea larvae. In this study however, *Isochrysis* was cultivated in a greenhouse, being able to synthesize a high amounts of DHA and other important cell constituents, which had a positive effect on the larval culture. Thus, microalgae species that are not adequate for shrimp or fish rearing, can be modified in their nutritional quality in order to cover the nutritional requirements of any aquatic organism.

4. Effect of Physical, Chemical and Management Factors, on Production and Composition of Microalgae

As mentioned, the production response and biochemical composition of microalgae, both in natural or aquaculture environments, are strongly influenced by abiotic factors such as: light intensity, temperature, salinity, pH, and nutrients availability.

The salinity is one of the most important environmental factors in the culture of microalgae, because it affects their growth and replication rate. In the natural environment, this parameter determines the distribution and abundance of phytoplankton species in the aquatic ecosystems. Additionally, it has a determinant influence on the density, viscosity and solubility of diverse gases in the water column.

Variations on salinity cause physiological responses in microalgae related to the cellular volume and osmotic adjustment. These responses imply the release of ions, synthesis of organic compounds, diminution of CO_2 fixation, and modification on the nitrogen metabolism (Kirst, 1989).

Most of the microalgae have a high tolerance to salinity variations, and can growth well in a wide range (Brown et al. 1996). The hypo osmotic conditions tend to be more adverse than the hyper osmotic ones. Growth rate and metabolic activity of many marine microalgae decline significantly when salinity drop under 20 ppt (Kirst 1989).

The optimum growth and tolerance to variations on salinity, depends mainly on the species and its natural habitat. For instance, Thessen et al. (2005) argued that the optimal salinity for microalgae growth depends on the species, and documented that *Pseudo-nitzschia delicatissima* had a maximum growth rate at low salinities (10-30 ppt), while *Pseudo-nitzschia pseudodelicatissima* had the best growth performance at higher salinities (25-40 ppt) and *Pseudo-nitzschia multiseries* showed an intermediate behavior (25-30 ppt). In addition, the estuarine species are more tolerant to low salinities than the oceanic species (Kirst, 1989). Renaud and Parry (1994), reported that salinities from 10 to 35 ppt do not have any effect on the growth rates of *Isochrysis* sp. and *Nitzschia frustulum*, however, the growth rate of *N. oculata*, decreased at 35 ppt. In some aquaculture hatcheries, the salinity may be difficult to control, because the water sources are estuaries or the ocean, thus, the salinity levels are 35 ppt or higher if evaporation is considered.

Temperature is an abiotic factor that affects the cellular metabolism and, in consequence, the growth and chemical proximate composition of microalgae (Richmond, 2004). Most of the species used for aquaculture purposes growth well on the range of 10 °C to 35 °C, with an optimum within 16 °C and 24 °C (Voltolina et al. 1989). In commercial hatcheries, temperature is controlled by air conditioned apparatus for indoor cultures but, outdoor systems are exposed to the environmental variation of the region.

The pH is one of the most important factors to consider for microalgae production, because their membranes are completely permeable to H^+ and OH^- ions, and the concentration of such ions may affect some cellular functions, causing death at extreme concentrations of H^+ (very low pH). High levels of OH^- (high pH), can affect microalgae development, but are not as lethal as a low pH condition. The optimum pH for most of the microalgae ranges from 7 to 8 (Abalde et al. 1995).

Another aspect to consider for microalgae culture is the nitrogen: phosphorus ratio. Changes in biochemical functions can be attributed to N:P ratio; however, the ratio may also depend upon the N or P sources. Some of the mediums used for microalgae culture can be modified in their N:P or C:N ratios in order to improve production and quality of particular microalgae species.

4.1. A study conducted under laboratory conditions, evaluated the effect of four salinities (25, 30, 35, 40, 45 and 50 ppt) on the growth and chemical proximate composition of the microalgae *T. weissflogii* at three culture phases. The best growth rates and final cellular densities were observed at 25 ppt. Diminutions in size, as well as morphological changes, were observed at high salinities ($\geq$ 40 ppt).

The most drastic changes occurred at 50 ppt (Table 8). Protein and carbohydrate contents were greater at 25 and 30 ppt, and the higher levels were recorded at the stationary phase. Lipid contents were higher at low salinities, but did not change during any of the three growth phases (López-Elías, et al., 2009).

Table 8. Means and standard deviations of maximum cellular density, specific growth rate and cell volume of *Thalassiosira weissflogii* at different salinities

Salinity (ppt)	Maximum cellular density ($\times 10^4$ mL^{-1})	Specific growth rate (μ) day^{-1}	Cell volume (mμ^3)
25	43 ± 0.5^{ab}	1.24 ± 0.025^a	1594.3 ± 145.1^a
30	42 ± 1.2^b	1.12 ± 0.023^b	1489.0 ± 151.6^a
35	44 ± 1.2^a	1.16 ± 0.039^b	1401.4 ± 57.2^a
40	40 ± 0.9^c	0.99 ± 0.030^c	1013.9 ± 145.0^b
45	37 ± 0.6^d	0.82 ± 0.038^d	700.0 ± 14.8^c
50	35 ± 1.0^e	0.81 ± 0.007^d	563.7 ± 29.9^c

Different letters show significant differences (α=0.05).

Lipids and carbohydrates are considered as stored energy products, and their decrease can negatively affect the growth and metabolic activities of microalgae (Brown et al. 1997). In addition, the osmotic challenge at higher or lower salinities increase the energy demand of cells, leading to an enhanced use of lipids and carbohydrates to cope with the energy demand.

Several species of microalgae are tolerant to great variations in salinity, although, their chemical proximate composition may be affected by such factor (Renaud and Parry 1994; Brown et al. 1996). Protein, lipids and carbohydrates in most of the microalgae species seem to be affected by high salinities (Richmond 1986). In some species, a considerable increase in ash and lipid contents have been documented when salinity increase (Kirst 1989).

4.2. An other laboratory study was conducted to evaluate the growth rate of *Thalassiosira pseudonana* in a static system, at salinities of 15, 25 and 35 ppt, under two conditions: continuous lighting or photoperiod of 12L:12D; the f/2 medium was used. It was found that salinity and light affected the final cellular density and the growth rate of the microalgae (Table 9). The highest growth rate (1.71 divisions·day^{-1}) and the greatest cellular density (5.6 x 10^5 cells·mL^{-1}) were recorded at 35 ppt, in the cultures using continuous illumination. Those values decreased when photoperiod was used (1.0 divisions·day^{-1} and 4.8 x 10^5 cells·mL^{-1}), but decreased even more at 15 ppt, in either continuous lighting or photoperiod. The longer light period increased the microalgae production, such results may explain the lower efficiency of outdoor systems compared to those indoor ones. Tzovenis et al. (2003), documented that light has an important roll in the development of microalgae. Similar results were reported by Brown et al. (1996), who recorded, a growth rate of 1.9 divisions·day^{-1} at continuous illumination and 1.0 divisions·day^{-1}, using a 12:12 photoperiod for the same species (López-Elías, et al., 2009). Moreover, it was observed that the salinity can potentiate or diminish the positive effect of light on the growth performance of the microalgae.

Table 9. Means ± SD of final cellular density, duplication rate (μ maximum, μ mean, μ cummulated) of *T. pseudonana* at three salinities, using photoperiod (P) or continuous lighting (CL)

Salinity (ppt)	Light condition	Final cellular density (Cells·mL^{-1} x10^5)	Specific growth rate (Divisions·day^{-1})		
			μ maximum	μ mean	μ cummulated
15	P	3.0 ± 1.5 a	0.68 ± 0.26a	0.33 ± 0.05a	2.30 ± 0.36a
25	P	4.2 ± 3.0 b	0.95 ± 0.18a	0.38 ± 0.01ab	2.74 ± 0.16ab
35	P	4.5 ± 3.5 bc	1.00 ± 0.20ab	0.46 ± 0.04bc	3.23 ± 0.29b
15	CL	3.9 ± 1.8 ab	0.71 ± 0.10a	0.35 ± 0.06a	2.50 ± 0.34a
25	CL	4.5 ± 1.8 bc	1.39 ± 0.09bc	0.43 ± 0.03abc	3.21 ± 0.28b
35	CL	5.5 ± 1.4c	1.71 ± 0.15c	0.49 ± 0.05c	3.43 ± 0.37b

Different letters mean significant differences at $P < 0.05$.

Table 10. Means ± SD of final cellular density dry weight, organic matter and ash of *Isochrysis* sp. at different salinities

Salinity (‰)	Density (cells·mL^{-1} x 10^6)	Dry weight (pg·cell^{-1})	Organic matter (pg·cell^{-1})	Ash (pg·cell^{-1})
20	9.02± 0.85 [d]	27.2±17.56 [a]	15.76±7.53 [a]	11.45±2.94 [a]
30	8.77±2.06 [d]	40.97±19.39 [b]	23.28±16.73 [a]	17.69±8.42 [b]
40	3.45±0.36 [c]	47.96±14.45 [b]	26.31±3.62 [b]	23.90±14.01 [b]
55	1.84±0.17 [b]	97.07±14.12 [c]	38.03±8.54 [c]	59.04±12.95 [c]
60	0.87±0.06 [a]	226.92±30.92 [d]	89.66±2.42 [d]	137.56±29.03 [d]

Different letters in a column jeans significant differences at α = 0.05.

4.3. A similar investigation was done to evaluate the growth, cellular density, and biomass production of *Isochrysis* sp. under controlled laboratory conditions, at salinities of 20, 30, 40, 55 and 60 ppt, in a static system. The number of cells was counted every 12 h and the biomass production was evaluated every 24 h. The higher cellular densities and growth rates were recorded at the lower salinities (20 and 30 ppt), with final values of 9.0 and 8.8 x10^6 cells·mL^{-1}, respectively. The pH increased over the culture period at the salinities of 20, 30 and 40 ppt, and remained without changes at 55 and 60 ppt. The lowest values of dry weight, organic matter and ash were observed at salinities of 20 and 30 ppt, and increased as salinity did (Table 10). It was concluded that low salinities favored duplication rate of the species, and high salinities increase cellular volume and diminish the growth rate (López-Elías et al., 2004a).

Richmond (1986) and Abalde et al. (1995), documented that microalgae equilibrate the osmotic pressure by enhancing the synthesis of organic compounds osmoprotectants or incorporating inorganic salts. That suggests that *Isochrysis* sp. in our study used one or both of these strategies.

4.4. In a related investigation, the growth of *T. chuii* at five salinities (20, 30, 40, 50 and 60 ppt), and two nitrogen:phosphorous rates (15:1 and 30:1), was evaluated. An inoculum of 1.0 x 10^5 cells·mL^{-1} was used for the treatments. The pH increased over time from 7.2 at the beginning to 8.8 at the end of the trial. The greatest final cellular density was obtained at 30 and 40 ppt, with values of ~5.0 x 10^6 cells·mL^{-1} at the N:P ratio of 30:1. The final cellular concentrations were also higher at 30 and 40 ppt, at N:P ratio of 15:1, however, they were lower (~4.2 x 10^6 cells·mL^{-1}) than those observed for microalgae cultured in 30N:1P. A greater decrease in growth was observed in microalgae cultured in 15N:1P at higher salinities (50 and 60 pps) (< 3.7 x 10^6 cells·mL^{-1}), (Figure 4) (López-Elías et al., 2010b). Independently of salinity, the higher means of daily duplication rate were observed at the higher N:P ratio (30:1; 0.6 divisions·day^{-1}) compared to the 15:1 (0.5 divisions·day^{-1}). Domínguez and Guevara (1994) and López-Elías et al. (2006) also found even higher growth performances for the same species at a higher N:P ratios.

4.5. In shrimp hatcheries, is important to feed the first larvae phases of aquatic organisms with phytoplankton, however, cultures of green flagellated are commonly not efficient, because of the inadequate formulation and use of mediums, which subsequently leads to lower productions. The growth of *T. chuii* was evaluated in outdoor mass cultures using different culture mediums. These cultures were maintained for two days in tanks of 250 L. The mediums used were f/2, f, 2f, 4f and two alternative mediums with modifications in their N:P ratio [$N_{30}:P_1(1)$ and $N_{30}:P_1(2)$]. $N_{30}:P_1(1)$ medium was formulated with sodium nitrate (150 g·L^{-1}) and monobasic sodium phosphate (8.09 g·L^{-1}), while $N_{30}:P_1(2)$ was prepared with sodium nitrate (183.6 g·L^{-1}) and monobasic sodium phosphate (10 g·L^{-1}). Final cell density varied from 2.7 to 4.3 x 10^5 cells·mL^{-1} and dry weight ranged between 32 and 51 g.m^3 (Table 11). The highest growth was observed both modified mediums [$N_{30}:P_1(1)$ and $N_{30}:P_1(2)$] compared to those considered as conventional. It was concluded that and adequate management of N:P ratio canenhance cell density of outdoor mass culture of *T.chuii* at less cost (López-Elías et al. 2006). In commercial hatcheries form Sinaloa, Mexico, mean cellular densities around 0.2 x 10^6 cél mL^{-1} were obtained in 3-days culture (López-Elías et al., 2003), which means that the modification of the culture medium has a positive effect on the growth of the species as reported by Lourenco et al. (1997).

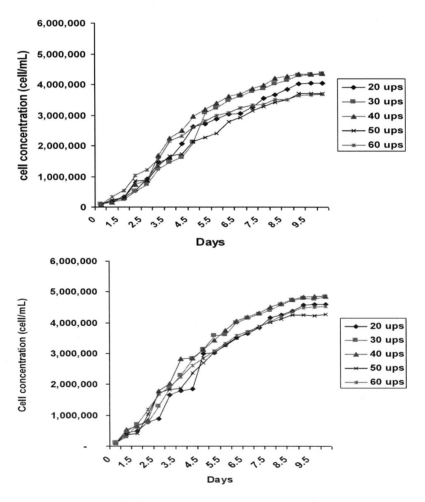

Figure 4. Growth of *T. chuii* at salinities of 20, 30, 40, 50 y 60 pps a and N:P rates of 15:1 (A) and 30:1 (B).

Table 11. Means ± SD of cellular density, growth rate and dry weight production in the outdoor culture of *T. chuii* in the mediums f/2, f, 2f, 4f, $N_{30}:P_1(1)$ and $N_{30}:P_1(2)$

Day	Cells·mL^{-1} x 10^5	Variation (%)	µ (Divisions·day^{-1})	Final biomass (g·m^3)
f/2 medium				
0	0.60	0.00	0.92	34.42 ± 5.69 [a]
1	1.13 ± 0.33	29.20		
1.5	0.93 ± 0.29	31.18	1.44	
2	3.08 ± 0.65 [a]	21.10		
f medium				
0	0.60	0.00	0.56	32.44 ± 3.24 [a]
1	0.89 ± 0.21	23.60		
1.5	0.93 ± 0.15	16.13	1.60	
2	2.99 ± 0.21 [a]	7.02		
2f medium				
0	0.60	0.00	1.14	51.24 ± 5.67 [b]
1	1.32 ± 0.30	2.27		
1.5	1.18 ± 0.50	4.24	1.17	
2	2.97 ± 0.48 [a]	16.16		
4f medium				
0	0.60	0.00	0.93	49.80 ± 2.96 [b]
1	1.14 ± 0.16	14.04		
1.5	11.4 ± 0.11	9.65	1.31	
2	2.84 ± 0.26 [a]	9.15		
30N:1P (1)				
0	0.60	0.00	0.81	47.97 ± 6.04 [b]
1	1.05 ± 0.11	10.48		
1.5	1.29 ± 0.08	6.20	1.62	
2	3.24 ± 0.05 [a]	1.54		
30N:1P (2)				
0	0.60	0.00	1.18	48.37 ± 5.58 [b]
1	1.36 ± 0.10	7.35		
1.5	1.39 ± 0.03	2.16	1.66	
2	4.30 ± 0.42 [b]	9.77		

5. EVALUATION OF PRODUCTION PROTOCOLS IN COMMERCIAL LABORATORIES

Although the traditional protocols for microalgae production are similar among laboratories, some routines can vary in diverse aspects such as: type of containers (size, form, material and etcetera), size and hour of inoculum, illumination intensity, culture medium, time for harvest, and others. Those differences may have a significant influence in the production results in terms of cellular concentration, duration of each one of the development phases, final biomass and chemical proximate composition.

5.1. The protocols for microalgae outdoor production were evaluated in different commercial hatcheries from Sinaloa and Sonora states, Northwest Mexico, during three years. It was found that in Sinaloa, the containers for microalgae production were pools from 2 to 4 m^3 and fiberglass transparent cylinders of 1 m^3; the harvests were made in a period from 1 to 4 days. In Sonora, some hatcheries used pools of 4 m^3 and the microalgae were harvested in 2 to 3 days; other laboratories used opaque cylinders of 0.8 m³, and harvested at 3 days, while others used pools of 2.5 m³ and harvested microalgae at days 2 and 3 (López-Elías et al. 2003). The total of evaluated laboratories performed outdoor cultures to reduce costs, and because the biomass obtained was greater than that needed.

5.2. The growth rate, final biomass, and proximate composition of massive outdoor culture of *C. muelleri*, were evaluated in two commercial hatcheries of Sonora, Mexico: El Camarón Dorado (CD), and Aqualarvas (AL) (Table 12). The first one maintained its intermediate cultures in a laboratory multi step system at temperatures between 20 and 22 °C, and a continuous photon flux of 7.9 mol·m^{-2}·d^{-1}; and for higher volume cultures, they used outdoor ponds of 2.5 m³ and 1 m depth. The second hatchery maintained similar conditions but used tanks of 3.3 m³ and 0.5 m depth for the last phase.

In CD hatchery, the indoor culture (300 L) achieved a cellular density of 0.77 x 10^6 cells·mL^{-1} and a duplication rate of 0.74 divisions·day^{-1} in three days; however, the growth performance significantly increased when microalgae were transferred to outdoor tanks (2.04 x 10^6 cells·mL^{-1}; 1.4 divisions·day^{-1}).

In AL hatchery, microalgae reached in two days a mean cellular density of 1.38 x 10^6 cells·mL^{-1}, but during the warm months the harvest was done after 1 day with a density of 1.84 x 10^6 x 10^6 cells·mL^{-1}, while in the cold months microalgae were harvested after 3 days with a concentration of only 0.72 x 10^6 cells·mL^{-1}.

Table 12. Mean cell concentrations (10^6 cells·mL^{-1}) and growth rates (divisions·day^{-1}) of *C. muelleri* cultured in 300-L cylinders (A) and 2.5-3.3 m³ tanks (B) in two commercial hatcheries

	Acualarvas (n=26)		Camarón Dorado (n=25)	
	A			
Day	cells·mL^{-1} x10^6	Growth rate μ	cells·mL^{-1} x10^6	Growth rate μ
0	0.36 ± 0.19	-	0.21 ± 0.07	-
1	1.34 ± 0.40	2.01 ± 0.45	0.38 ± 0.10	0.98 ± 0.39
2	2.04 ± 0.46	0.65 ± 0.54	0.56 ± 0.17	0.54 ± 0.19
3	-	-	0.77 ± 0.21	0.46 ± 0.25
X̄μ	-	1.39 ± 0.36	-	0.74 ± 0.24

	B							
	(n=6)		(n=12)		(n=16)		(n=6)	
Day	N	μ	N	μ	N	μ	N	μ
0	0.51	-	0.44	-	0.22 ± 0.13	-	0.17	-
1	1.84	1.89 ± 0.32	1.08	1.36 ± 0.52	0.58 ± 0.26	1.32 ± 0.45	0.27	0.70 ± 0.48
2	-	-	1.38	0.41 ± 0.28	0.88 ± 0.33	0.54 ± 0.65	0.48	0.81 ± 0.24
3	-	-	-	-	-	-	0.72	0.59 ± 0.27
μ̄	-	1.89 ± 0.32	-	0.88 ± 0.30	-	0.94 ± 0.35	-	0.70 ± 0.19

Table 13. Mean of ash-free biomass (AFW), proteins, carbohydrates, and lipids on the outdoor culture of *C. muelleri* after 1, 2, and 3 days in two commercial hatcheries

Hatchery	Days	n	AFW g·m^{-3}	Protein g·m^{-3}	Carbohydrate g·m^{-3}	Lipid g·m^{-3}
AL	1	19	39.98 ± 10.7b	29.54 ± 7.3b	5.40 ± 2.45a	5.06 ± 1.76a
	2	65	48.65 ± 16.3c	29.47 ±12.8b	10.15 ± 6.38b	10.13 ± 5.25b
CD	2	65	32.84 ± 6.9a	17.60 ± 3.7a	5.57 ± 1.96a	9.66 ± 3.57b
	3	14	33.62 ± 6.5a	26.36 ±6.5ab	4.85 ± 1.31a	2.40 ± 0.61a

Different letters indicate significant differences between values in the same column (two way nonparametric analysis of variance and Dunn's multiple comparison tests. α = 0.05). a≤ab≤b and a<b.

The final biomass was significantly greater in the pools from CD hatchery (0.5 m depth), with values of 40 g·m^{-3} after 1 day and 49 g·m^{-3} with the normal routine of 2 days. In the tanks of AL hatchery (1 m depth), the production ranged from 32.8 to 33.6 g·m^{-3} independently of the time of harvest.

Table 14. Production parameters of *C. muelleri* at inoculation and harvest (x 10^6 cells·mL^{-1}) cultured outdoor in hatcheries of Sinaloa and Sonora. Cellular variation coefficient (CVC), hour of inoculum, replication rate (RR, divisions/day), days of culture (DC), and growth rate (GR. 24 h) outdoor cultured in pools in hatcheries of Sinaloa and Sonora

Hatchery	Inoculum concentration	CVC (%)	RR	DC	GR	Harvest
Tanks, year 1999						
Maricultura Lab. 1	0.157 (0.065)	41.4	1.21	3	2.43	0.798 (0.195)
Maricultura Lab. 2	0.080 (0.039)	48.9	1.59	3	3.02	0.597 (0.096)
Generación 50	0.419 (0.039)	9.3	1.95	1	1.95	1.620 (0.145)
Camarón Dorado	0.237 (0.071)	30.2	1.50	3	2.49	1.187 (0.077)
Tanks, year 2000						
Maricultura Lab. 1	0.288 (0.119)	41.3	1.08	3	1.84	1.065 (0.148)
Maricultura Lab. 2	0.230 (0.025)	10.9	0.93	3	1.84	0.835 (0.167)
Maricultura Lab. 2 (Tinas de 1.0 m^3)	0.486 (0.223)	45.9	1.19	2	2.06	1.836 (0.173)
Generación 50	0.452 (0.056)	12.4	1.03	2-3	1.76	1.534 (0.367)
Camarón Dorado	0.187 (0.066)	35.4	1.08	2	1.45	0.822 (0.406)
Acualarvas	0.372 (0.268)	72.1	1.01	2	1.61	1.068 (0.821)
Aremar (Cylinders 800 L)	0.305 (0.071)	23.4	1.27	3	2.79	2.093 (0.130)
DICTUS (Tanks 200 L)	0.434 (0.038)	8.8	1.95	2	2.42	2.350 (0.467)

Table 15. Mean cell density and organic yield of indoor and outdoor (CREMES), indoor and greenhouse (UEK) cultures of *C. muelleri* during spring, summer and winter

	Cell density (cells·ml^{-1} x10^6)		Organic yield (g·m^{-3})	
CREMES(300 L)	Indoor	Outdoor	Indoor	Outdoor
Spring	2.2 ± 0.1	4.9 ± 0.2	0.101 ± 0.001	0.159 ± 0.012
Summer	1.4 ± 0.2	1.8 ± 0.2	0.076 ± 0.003	0.124 ± 0.007
Winter	1.8 ± 0.1	1.1 ± 0.7	0.066 ± 0.012	0.036 ± 0.100
CREMES (3000 L)				
Spring	1.0 ± 0.1	1.3 ± 0.0	0.036 ± 0.011	0.046 ± 0.007
Summer	0.7 ± 0.1	1.5 ± 0.3	0.049 ± 0.003	0.076 ± 0.002
Winter	0.6 ± 0.1	1.1 ± 0.1	0.045 ± 0.002	0.050 ± 0.002
UEK (300 L)				
Spring	1.3 ± 0.5	1.4 ± 0.4	0.057 ± 0.012	0.068 ± 0.026
Summer	1.1 ± 0.6	0.9 ± 0.2	0.060 ± 0.007	0.084 ± 0.018
Winter	1.2 ± 0.5	1.4 ± 0.4	0.084 ± 0.031	0.072 ± 0.027

Different letters indicate significant differences (two way ANOVA and Tukey multiple comparison test; a=0.05). Standard deviations in parenthesis.

The harvest time, had a significant effect on the chemical proximate composition of the microalgae (Table 13). In AL hatchery, the protein content in dry basis was 75.0% at day 1 and 60.5% at day 2. The content of lipids plus carbohydrates ranged from 5.1 to 5.4 g·m^{-3} after 1 day and 10.1 g·m^{-3} after 2 days. In CD hatchery, the protein obtained was 17.6 (53.6 %) at day 2, but increased to 26.4 (78.8 %) in the following 24 hours (day 3) (López-Elías, et al., 2005a).

5.3. The microalgae production routines of 7 shrimp hatcheries of Sonora (3) and Sinaloa (4) were described, using the field data of each one of the hatcheries. The cellular density of inoculum and during each one of the growth phases was evaluated, considering the environmental conditions. The inoculum density varied in both, within the same hatchery and among different hatcheries (0.08–0.60 x 10^6 cells·mL^{-1}). In general, it was found that production routines were carried out in long periods, leading to late harvests when microalgae were at the stationary phase; i.e., they lost time trying to obtain higher productions (without results) instead of make a harvest and perform a semi-continuous culture. In the last step of production, the cellular density varied from 0.82 to 2.35 x 10^6 cells·mL^{-1} (Table 14). The conclusion of the study was that it is necessary to reduce the routine times and standardize the concentration of inoculums at all levels of production (López-Elías, et al., 2005b).

5.4. In a similar investigation, the cellular concentration and the final biomass of *C. muelleri* cultured indoor and outdoor at levels of 300 and 3000 L, during spring, summer, and winter, were evaluated (Table 15). Temperature varied from 8 °C during the winter to more than 45 °C at summer. Outdoor systems showed a higher variability in terms of production, such results were associated to environmental conditions and water quality. In general, the cellular density and biomass production were significantly better in the outdoor systems (except for indoor culture at 300 L) (López-Elías, et al., 2005a). Regarding to indoor systems, the production was constant in all seasons and hatcheries because of the higher level in the control of variables; such productions were similar to those obtained at laboratory conditions (López-Elías et al. 1999). It was concluded that producers can choice to maintain their microalgae cultures outdoor with a higher but variable production, or indoor with a constant but lower production. Otherwise, they can practice outdoor cultures during warm months and indoor during cold months.

6. EVALUATION OF ALTERNATIVE CULTURE MEDIUMS

The mediums commonly used to culture microalgae are those considered chemically complete, such as f, f/2, Walne and others (Tzovenis et al. 1997). However some alternative mediums have been proposed and evaluated in order to improve production or decrease costs (Piña et al., 2007; Lopez-Elias et al. 2008). The use of agricultural fertilizers seems to be an adequate alternative since they have nutrients that can be used by microalgae; in addition, their costs are commonly lower compared to the conventional mediums. The feasibility of using such alternative has been demonstrated; moreover it was observed that microalgae cultures in these mediums had similar production and quality parameters compared to those cultured in f/2 medium (López-Elías and Voltolina 1993; Piña et al. 2007). Thus, the conventional mediums can be substituted by alternative and cheaper mediums.

6.1. An experimental study was conducted to evaluate effectiveness of two alternative mediums for the culture of microalgae *N. oculata* and *C. muellerii*, in terms of production response and nutritional value. The mediums evaluated were: mono ammonium phosphate (MAP, used in agriculture) and Nutrilake (NLK, used in aquaculture). The control consisted in the conventional medium Guillard f/2 (Guillard 1975). The cultures were initiated in 80-L fiberglass columns at a density of 4×10^6 cells mL^{-1}; a light intensity of 128-192 $\mu mol \cdot m^{-2} s^{-1}$ was maintained during the experiment, by using white light lamps, and an environmental temperature of 18-20 °C as suggested by Lopez-Elías et al. (2004).

Some differences on the production parameters of *N. oculata* (Table 16) and *C. muellerii* were recorded (Table 17). For both species, the best final cell density was obtained with the agricultural medium. *N. oculata* had similar protein content when farmed in the three mediums (28-32%), while *C. muelleri* ranged from 24-27% in protein content. Carbohydrate content increased in the alternative medium MAP, but the lipid content decreased. Results suggested that the medium used for agricultural purposes (MAP) was an adequate alternative to substitute the conventional medium, not only because it increased the production of both microalgae, but also because it had a quite lower price than the conventional medium.

Table 16. Production parameters and chemical proximate composition of *N. oculata* cultured in the two alternative mediums and the control

Medium	K	TD	Final density (cells mL^{-1})	Carbohidrates (%)	Lípids (%)	Proteins (%)
MAP	0.69±0.81a	1.23±0.57a	2.05±0.67b	36.83±1.49b	34.54±1.49a	28.63±2.00a
NLK	0.46±0.29a	1.38±0.67a	1.26±0.49a	31.63±0.69b	37.58±0.69a	32.12±0.74a
GF/2	0.71±0.65a	1.35±0.75a	1.76±0.80b	28.96±1.83a	42.01±1.83b	29.02±2.25a

K: growth rate (divisions day^{-1}), DT: duplication time (days). The proximate composition is expressed in organic fraction. Different letters in a row means significant differences at $P < 0.05$.

Table 17. Production parameters and chemical proximate composition of *C. muelleri* in the two alternative mediums and the control

Medium	K	DT	Final density (cells mL^{-1})	Carbohidrates (%)	Lípids (%)	Proteins (%)
MAP	1.00±0.27b	0.89±0.46a	3.69±0.76b	35.52±2.64b	40.45±2.10a	24.03±2.70a
NLK	0.61±0.22a	1.33±0.45b	2.28±0.67a	34.31±1.32ab	41.67±0.69a	24.02±1.00a
GF/2	0.86±0.29ab	0.83±0.31a	2.94±0.09ab	30.76±1.43a	41.68±2.78a	27.56±1.48b

K: growth rate (divisions day^{-1}), DT: duplication time (days). The proximate composition is expressed in organic fraction. Different letters in a row means significant differences at $P < 0.05$.

6.2. An investigation at laboratory and massive level was performed to evaluate the effectiveness of two agricultural fertilizers (FERTIMEX, México, and Peters, U.S.A.) on the growth and chemical proximate composition of *I. galbana* (Table 18).

At laboratory level, culture was done in 2.0-L Erlenmeyer flasks, and it was observed that final cellular density was significantly higher in the stationary phase using FERTIMEX (7.47 x 10^6 cells·mL^{-1}), compared to that using Peters (5.21 x 10^6 cells·mL^{-1}), however no differences were found for dry biomass among treatments (~112 g.L^{-1}). Chemical proximate composition varied depending on the culture age and the medium used. The lipid content tended to increase over time in both treatments; contrarily, the protein content decreased over time. Apparently, the microalgae at their low growth phase, had a more balanced nutrient content; however, it also depends on the requirements of the species that is pretended to feed with the microalgae (López-Elías et al., 2008). The conclusion was that the agricultural fertilizers were adequate for the culture of *I. galbana* microalgae, and that higher contents of protein are found in the exponential phase of growth.

6.3. A similar study was conducted to evaluate the aquacultural fertilizer Nutrilake, compared to urea and the traditional f medium, as nitrogen source for the culture of the microalgae *C. muelleri, T. weissflogii, Isochrysis* sp. and *T. suecica*. Final cellular concentration, duplication rate, dry weight, biomass, and chemical proximate composition were monitored. *C. muelleri* showed the highest cellular concentrations and biomass using urea as culture medium, and no significant differences among mediums were found with respect to the chemical proximate composition. *T. weissflogii* observed greater cellular concentrations with Nutrilake.

For *Isochrysis* sp. y *T. suecica* no significant differences were found in cellular concentrations among mediums. Their greatest protein contents were obtained with urea. Carbohydrate content of *T. weissflogii* was significantly higher with Nutrilake and f medium, while higher limpid contents were found using urea. No changes in dry weight were recorded for *T. weissflogii* and *T. suecica,* but *Isochrysis* sp. showed higher values with Nutrilake. The replication rate was higher using urea and Nutrilake for *T. weissflogii, T. suecica* and *Isochrysis* sp., but not for *C. muelleri*. (Piña et al., 2007).

Table 18. Cellular density, dry biomass and other constituents of *I. galbana* cultured with two agricultural ferilizers in different growth phases (EXP= exponential, LG= low growth and STAT= stationary) in a static system

	Fertimex			Peters (FP)		
	EXP	LG	STAT	EXP	LG	STAT
Cellular density	1.56 [b]	3.90 [a]	7.47 [b]	0.98 [a]	3.47 [a]	5.21 [a]
(x 10^6 cél·mL^{-1})	(0.50)	(0.68)	(1.01)	(0.32)	(0.54)	(1.12)
Dry biomass	0.049 [a]	0.093 [a]	0.122 [b]	0.055 [a]	0.099 [a]	0.112 [a]
(g.L^{-1})	(0.008)	(0.015)	(0.028)	(0.005)	(0.015)	(0.013)
Proteins	4.80 [a]	4.45 [a]	3.12 [a]	9.47 [b]	4.22 [a]	3.48 [a]
(pg cell^{-1})	(1.79)	(0.73)	(0.18)	(2.60)	(0.54)	(0.64)
Carbohydrate	2.22 [a]	2.44 [ab]	2.42 [ab]	2.90 [b]	2.55 [ab]	2.44 [ab]
(pg cell^{-1})	(0.32)	(0.51)	(0.20)	(0.46)	(0.29)	(0.46)
Lipids	2.23 [a]	2.26 [a]	3.83 [c]	3.01 [b]	2.68 [ab]	3.97 [c]
(pg cell^{-1})	(0.40)	(0.16)	(0.23)	(0.49)	(0.23)	(0.07)

REFERENCES

Abalde, J., Cid, A., Hidalgo, J. P., Torres, E. and Herrero, C. 1995. Microalgas: Cultivo y Aplicaciones. Universidad de Coruña, España, 209 pp.

Arroyo-Pacheco, L.E. y Martínez-Baldenebro F. 1994. Producción de biomasa y composición química de dos especies de microalgas marinas a diferentes salinidades. Tesis de licenciatura. Depto. de Ciencias Químico Biológicas, Universidad de Sonora. 78 pp.

Becerra-Dórame, M., López-Elías, J.A., Enríquez-Ocaña, F., Huerta-Aldaz, N., Voltolina, D., Osuna-López, I. and Izaguirre-Fierro, G. 2010. The effect of initial cell and nutrient concentrations on the growth and biomass production of outdoor cultures of *Dunaliella* sp. Ann. Bot. Fennici., 47:109-112.

Becker, E.W. 1994. Microalgae. Biotechnology and microbiology. Cambridge University Press. Great Britain. 293 pp.

Becker, W. 2004. Microalgae for Aquaculture: The Nutritional value of Microalgae for Aquaculture. En Richmond A. Handbook of Microalgal culture: Biotechnology and Applied Phycology. Blackwell Publishing. U.S.A. 380-391 pp.

Brown R.M., Dunstan, A.G., Norwood, J.A. and Miller, A.K.. 1996. Effects of harvest stage and light on the biochemical composition of the diatom *Thalassiosira pseudonana*. *Journal of Phycology*, 32, 64-73.

Brown M.R., Jeffrey, S.W., Volkman, J.K. and Dunstan, G.A. 1997. Nutritional properties of micro algae for mariculture. *Aquaculture*, 151, 315-331.

Burlew, J.S. 1953. Current status of the large-scale culture of algae. In: Burlew, J.S. (ed.). Algal Culture form Laboratory to Pilot Plant. Carnegie Institution or Washington, USA. No. 600, Pp 3-33

Chisti Y. 2008. Biodiesel from microalgae beats bioethanol. *Trends Biotechnol* 26:126–131.

Cleber Bertoldi, F., Sant'Anna, E., Villela da Costa Braga, M. and Barcelos Oliveira, J.L.. 2006. Lipids, fatty acids composition and carotenoids of *Chlorella vulgaris* cultivated in hydroponic wastewater. *Grasas y Aceites*. 57 (3): 270-274.

Domínguez, R.L. M. and Guevara E.B.L. 1994. Optimización de la razón nitrógeno:fósforo sobre el crecimiento y biomasa de dos especies de microalgas. Tesis de licenciatura. Universidad de Sonora, México, 91 pp.

D´Souza, F.M. and Loneragan, N.R.. 1999. Effects of monospecific and mixed-algae diets on survival, development and fatty acid composition of penaeid prawn (*Penaeus* sp.) larvae. *Marine Biology*. 133: 621-633.

Gallegos-Simental, G., Voltolina, D., López-Elías, J.A., Enríquez-Ocaña, F. y P., Piña. 2002. Producción a la intemperie y en el laboratorio de la diatomea *Chaetoceros muelleri* en Bahía Kino, Sonora, México. *Oceánide*. 17 (2): 85-91.

García-Lagunas, N., López-Elías, J.A., Miranda-Baeza, A., Martínez-Porchas, M., Huerta-Aldaz, N. and García-Triana A. 2010. Effect of salinity on growth and chemical composition of the diatom *Thalassiosira weissflogii* in three phases of culture. (Enviada).

Guillard, R.L. and Ryther, J. H. 1962. Studies on marine planktonic diatoms I. *Cyclotella nana* Husted and *Detonula confervacea* (Cleve) Gran. *Canadian Journal of Microbiology*, 8: 229-239.

Guillard, R.R.L. 1975. Culture of the phytoplankton for feeding marine invertebrate larvae. In: Smith, W.L., Chanley, M.E. (Eds.), Culture of Marine Invertebrate Animals. Plenum Publishing Co., New Cork, pp. 296-360.

Kirst, G.O. 1989. Salinity tolerance of eukaryotic marine algae. *Annu. Rev. Plant Physiol. Plant Mol. Biol.* 40: 21-53.

López Elías, J.A. y Voltolina, D. 1993. Cultivos semicontinuos de cuatro especies de microalgas con un medio no convencional. *Ciencias Marinas* 19 (2): 169-180.

López-Elías, J.A., Encinas-Arreola, A. R., García-Valenzuela, A.C., Váldez, J. and Hoyos-Chaires, F. 1999. Producción anual de dos especies de microalgas en un centro acuícola en Bahia Kino, Sonora. *OCEÁNIDES*, 14 (1): 59-65.

López-Elías, J.A., Voltolina, D., Chavira-Ortega, C.O., Rodríguez-Rodríguez, B.B., Sáenz-Gaxiola, L.M., Cordero-Esquivel, B. y Nieves-Soto, M. 2003. Mass production of microalgae in six commercial shrimp hatcheries of the Mexican northwest. *Aquacultural engineering* 29: 155-164.

López Elías J.A., Huerta Aldaz N, Estrada Durán G.J., Celis Salgado M. P., De la Re Vega E., Quintero Arredondo N., Estrada Quintero J. A., Niebla Larreta J. L., Miramontes Higuera N., García Quiroz K. M., Niebla Rodríguez S. J., Carvajal Sánchez I. S. y Velazco Rameños J. 2004a. Efecto de la salinidad en el crecimiento de *Isochrysis sp.* Bajo condiciones de cultivo estático. Biotécnia Vol. 6 No. 3: 10-15.

López-Elías, J. A., Voltolina, D., Nieves-Soto, M. y Figueroa-Ortiz, L. 2004b. Producción y Composición de Microalgas en Laboratorios Comerciales del Noroeste de México. In: Cruz Suárez, L.E., Ricque Marie, D., Nieto López, M.G., Villarreal, D., Scholz, U. y González, M. Avances en Nutrición Acuícola VII. Memorias del VII Simposium Internacional de Nutrición Acuícola.16-19 Noviembre, 2004. Hermosillo, Sonora, México.

López Elías, J.A., Voltolina, D., Enríquez Ocaña, F. and Gallegos Semental, G. 2005a. Indoor and outdoor mass production of the diatom *Chaetoceros muelleri* in Mexican commercial hatchery. *Aquacultural Engineering* 33 (3): 181-191.

López Elías, J.A., Cortes González, I., Nieves Soto, M., Enríquez Ocaña, F., Piña Valdéz, P., Voltolina, D. y Pablos Mitre, N.M. 2005b. Seguimiento de cultivos masivos de microalgas en siete laboratorios productores de larvas de camarón. *Biotécnia* Vol. 7 No. 3: 36.43.

López-Elías, J.A., Badilla-Flores, K.D., Estrada-Raygoza, L.A., Fimbres Olivarría, D., Ochoa-Castillo, L. de J., Ramos-Brito, L. y Huerta-Aldaz, N.. 2006. Manejo de medios de cultivo en el crecimiento de *Tetraselmis chuii* en cultivos al exterior. *Biotecnia*. Vol 8 No. 3: 14-21.

López-Elías, J.A., Carvallo-Ruiz, G., Enríquez-Ocaña, F., Huerta-Aldaz, N. y Aguirre-Rosas, J.C.. 2008. Crecimiento y composición bioquímica de *Isochrysis galbana* cultivada en fertilizantes agrícolas. *Biotecnia*, vol. X No. 2: 22-32.

López-Elías, J.A., Enríquez-Ocaña, F., Pablos-Mitre, M.N., Huerta-Aldaz, N., Leal, S., Miranda-Baeza, A., Nieves-Soto, M. and Vásquez- Salgado, I. 2008. Growth and biomass production of *Chaetoceros muelleri* in mass outdoor cultures: Effect of the hour of the inoculation, size of the inoculum and culture medium. *Rev. Invest. Mar*. 29 (2): 171-177.

López-Elías, J.A., García-Lagunas, N., Jiménez-Gutiérrez, L.R. y Huerta-Aldaz, N. 2009. Crecimiento de la diatomea *Thalassiosira pseudonana* en cultivos estáticos con iluminación continua y fotoperiodo a diferentes salinidades. *Biotecnia*, Vol. XI No.1: 11-18.

López-Elías, J.A, Esquer-Miranda, E., Martínez-Porchas, M., Garza-Aguirre, M. del C., Rivas-Vega, M., Huerta-Aldaz, N., Aguirre-Hinojosa, E. y Nieves-Soto, M. 2010a. Evaluación del efecto de la hora y la concentración de inóculo en el desarrollo de cultivos estáticos masivos de *Tetraselmis chuii* (Butcher,1958*)* en los medios f/2 y 2 f. (en preparación).

López-Elías, J.A., Gonzalez-Bello, G. E.., Huerta-Aldaz, N., Murguia-López, A., Mercado-Castillo, L. y Miranda-Baeza, A. 2010b. Efecto de cinco salinidades y dos razones nitrógeno-fósforo sobre el crecimiento de la microalga *Tetraselmis chuii*. XII Congreso de la Asociación de Investigadores del Mar de Cortés y Simposium Internacional sobre el Mar de Cortés. Marzo 2 a 5. 2010. Guayamas, Son., Mex.

Lourenco, S.O., Lanfer Marquez, U.M., Mancini-Filho, Barbarino, E. and Aidar, E. 1997. Changes in biochemical profile of *Tetraselmis gracilis* I. Comparison of two culture media. Aquaculture. 148: 153-168.

Lovatelli, A. 2004. Hatchery Culture of bivalves. A practical manual. *FAO Technical paper* 471. 177 pp.

Martínez-Córdova. 1999. Cultivo de Camarones Peneidos. Principios y Prácticas. AGT Editor, México, D.F. 102 p.

Matsunaga, T., Takeyama, H., Miyashita, H. and Yokouchi, H. 2006. Marine microalgae. *Adv. Biochem. Engin. Biotechnol*. 96:165 – 188.

Medina-Reyna C.E. and Cordero-Esquivel, B. 1998. Crecimiento y composición bioquímica de la diatomea *Chaetoceros muelleri* (Lemerman), mantenida en cultivo estático con un medio comercial. *Ciencia y Mar* 6, 19-26.

Oswald, W.J. 1988. Large-scale algal culture systems (engineering aspects). En Borowitzka, M.A. y Borowitzka, L. J. (eds.). Microalgal Biotechnology. Cambridge University Press. pp. 357-394.

Piña, P., Voltolina, D., Nieves, M. and Robles, M. 2006. Survival, development and growth of the Pacific White Shrimp *Litopenaeus vannamei* protozea larvae, fed with monoalgal and mixed diets. *Aquaculture* 253:523-530.

Piña, P. Medina A.M., Nieves, M., Leal, S., López-Elías, J.A.y Guerrero, M.A.. 2007. Cultivo de cuatro especies de microalgas con diferentes fertilizantes utilizados en acuicultura. *Rev. Invest. Mar.* 28 (3): 225-236.

Pipes, W.O. and Gotaas, H.B. 1960. Utilization of organic matter by *Chlorella* grown in seawage. *Appl. Microbiol.* 8:163-169.

Richmond, A. 1986. Handbook of microalgal mass culture. CRC Press, Boca Raton Florida. 528 pp.

Richmond, A, 2004. Handbook of Microalgal Culture. *Biotechnology and Applied Phycology.* Blackwell Publishing, USA, 566 pp.

Renaud S.M. and Parry D.L. 1994. Microalgae for use in tropical aquaculture II: Effect of salinity on growth, gross chemical composition and fatty acid composition of three species of marine microalgae. *Australian Journal of Botany.* 51: 703-713.

Rodríguez, E. O., López-Elías, J.A., Aguirre-Hinojosa, E., Garza-Aguirre, M. del C., Constantino-Franco, F., Miranda-Baeza, A. and Nieves-Soto, M. 2010. Evaluation of the nutritional quality of *Chaetoceros muelleri* Schütt (Chaetocerotales: Chaetocerotaceae) and *Isochrysis* sp. (Isochrysidales: Isochrysidaceae) grown outdoors for the larval development of *Litopeneaus vannamei* (Boone, 1931) (Decapoda: Penaeidae). (En preparación).

Rosales, N., Ortega, J., Mora, R. and Morales, E. 2005. Influencia de la salinidad sobre crecimiento y composición bioquímica de la cianobacteria *Synechococcus* sp. Ciencias Marinas. 31(2): 349-355.

Seon Jin. E. and A., Melis. 2003. Microalgal biotechnology: Carotenoid production by the green algae *Dunaliella salina*. *Biotechnology and Bioprocess Engineering*. 8: 331- 337.

Soeder, C.J. 1986. An historical outline of applied algology. Pp. 25-41. In Richmond, A. (ed.). Handbook of Microalgal Mass Culture. CRC.

Thessen, A., Dortch, Q., Parsons, M. and Morrison, W. 2005. Effect of salinity of *Pseudonitzschia* species (bacillariophyceae) growth and distribution. *Journal of Phycology*. 41: 21-29.

Tzovenis, I., De Pauw N. and Sorgeloos P. 2003. Optimization of T-ISO biomass production rich in essential fatty acids I. Effect of different light regimes on growth and biomass production. *Aquaculture,* 216;203-222.

Tzovenis, I., DePauw, N. y Sorgeloos, P. 1997. Effect of different light regimes on the docohexaenoic acid (DHA) content of *Isochrysis* aff. *galbana* (clone T-ISO). *Aquaculture International*, 5: 489-507.

Sánchez-Saavedra, M.P. and Voltolina, D. 1996. Effects of blue-green light on growth rate and chemical composition of three diatoms. *Journal of Applied Phycology*. 8:131-137.

Voltolina, D., Bückle Ramírez, L.F. y Morales Guerrero, E. L. 1989. Manual de metodologías y alternativas para el cultivo de microalgas (2ª Ed.) Centro de Investigación Científica y de Educación Superior de Ensenada, B. C., México. Informe Especial OC-89-01, 67 pp.

Voltolina, D. and López-Elías, J.A. 2002. Cultivos de apoyo para la acuacultura: Tendencias e innovaciones. Pp. 23-41. En Martínez-Córdova L.R. (ed.). Camaronicultura. Avances y tendencias. AGT Editor, S.A., Mex., 167 pp.

Voltolina, D., Piña, P. and Nieves, M. 1999. Fertilizers as cheap growth media: a mexican point of view. *Rivista Italiana di Acquacoltura*, 34: 43-45.

Volkman, J.K., Jeffrey, S.W., Nichols, P.D., Rogers, G.I. and Garland, C.D. 1989. Fatty acid composition of 10 species of microalgae used in mariculture. *Journal of Experimental Marina Biology and Ecology*. 128:219-240.

Vonshak, A. 1988. Microalgas: técnicas de cultivo en laboratorio y para producción de biomasa a la intemperie. p. 155- 165. En Coombs, J., May, D.O., Long, S.P. y Scurlock, J.M. (eds.) Técnicas en Fotosíntesis y Bioproductividad. Colegio de Graduados, Chapingo, Estado de México, México, 258 pp.

Zhang C.W., Richmond A. 2003. Sustainable, high-yielding outdoor mass cultures of *Chaetoceros muelleri* var. *subalsum* and *Isochrysis galbana* vertical plate reactors. *Marine Biotechnology*. 5:302-310.

In: Microalgae: Biotechnology, Microbiology and Energy
Editor: Melanie N. Johnsen

ISBN 978-1-61324-625-2
© 2012 Nova Science Publishers, Inc.

Chapter 7

MICROALGAE: THE FUTURE OF GREEN ENERGY

K. K. I. U. Arunakumara
Department of Crop Science,
Faculty of Agriculture,
University of Ruhuna, Sri Lanka

ABSTRACT

Carbon neutral renewable source of energy is needed to displace petroleum-derived fuels, which contribute to global warming and are of limited availability. Biodiesel and bioethanol, in this context, are the two potential renewable fuels that have gained substantial attraction. However, sustainability of biodiesel and bioethanol production from conventional agricultural crops is still questionable. Microalgae, a source of biodiesel are at the center of new research conducted with the aim of completely displacing fossil-based diesel. With special reference to biodiesel, the present article reviews prospects and constraints of microalgae as a source of biofuel.

There are at least 30,000 known species of microalgae, of which only a handful are currently of commercial significance due to their non-energy products such as nutraceuticals, pigments, proteins and functional foods. Though may vary with the species, microalgal biomass can be rich in proteins or rich in lipids or have a balanced composition of lipids, sugars and proteins. Under laboratory conditions, some microalgae strains were reported to generate 70 % lipid in their biomass. The fundamental chemical reaction required to produce biodiesel is the esterification of lipids, either triglycerides or oil, with alcohol, which results in a fatty acid alkylester called biodiesel (Fatty acid methyl-ester). As the fastest growing photosynthesizing organisms, biomass harvest of microalgae (158 tons/ha) is significantly higher than that of crop species such as sugarcane (75 tons/ha) used for bioethanol production. Under optimum growing conditions, a hectare of microalgae may potentially yield about 8,000 liters of biodiesel, which is 10 to 1000 times as much liquid fuel per year per hectare as conventional crops.

However, achieving the capacity to inexpensively produce biodiesel from microalgae is still challengeable. It could therefore be concluded that though microalgae are considered to be a potential source of green energy, the sustainability will largely depend on development of cost effective culture and processing techniques. Screening and collecting strains of algal species to access their potential for high oil production with high biomass productivity, investigating the physiology and biochemistry of the algae, use of molecular-biology and genetic engineering techniques to enhance the oil yield and development of advance processing techniques of cost competitive are considered to be the priority areas of research concern.

Keywords: Renewable fuels, microalgae, esterification, biodiesel.

INTRODUCTION

Energy consumption across the world is predicted to be increased substantially over the next couple of decades (Crookes, 2006), thus the world is entering to a period of declining non-renewable energy resources. The issue is particularly acute in the transportation sector, where reliable alternatives to fossil-based fuels are yet to be declared. Reduction of crude oil reserves and difficulties in their extraction and processing has lead to increase of its cost also (Laherrere, 2005). The reliance on fossil-based fuels has caused carbon dioxide (CO_2) enrichment of the atmosphere, which is considered to be the primary contributor to the generally-accepted phenomenon called "global warming". In fact, many countries and regions around the world have established targets for CO_2 reduction in order to meet the sustainability goals agreed under the Kyoto Protocol. Therefore, taking measures for minimizing transportation emissions through gradual replacement of fossil-based fuels by renewable energy sources is of paramount important. Discovering viable renewable energy sources, however, ranks as one of the most challenging tasks facing mankind in the medium to long term.

Among the possible alternatives, solar and wind energy, thermal or photovoltaic, hydroelectric, geothermal, biofuels, and carbon sequestration are being studied and implemented in practice, with different degrees of success (Dewulf and Van Langenhove, 2006; Gilbert and Perl, 2008). In addition, other non-renewable sources of energy such as coal and uranium are also available (Campbell, 2008). Coal is likely to be the immediate candidate for replacing oil as an energy supply, because it can be converted to liquid fuel and is still very abundant. It is energy rich and is particularly plentiful in countries such as the United States, China, and India, where energy demand is ever-increasing due to heavily industrialized nature (Clayton, 2004). However, due to the fact that coal produces even greater CO_2 emissions than oil, regardless the depletion of fossil-based fuels, the CO_2 enrichment would continue to be happened. Furthermore, the stocks are limited and will also inevitably decline in availability (Campbell, 2008). It is in this context that sources of renewable energy emerge as viable contributors, which could meet the world demand while mitigating climate change. Ironically, most renewable energy initiatives are focused on electricity generation, while the majority of world energy consumption, about two thirds, is derived from liquid fuels (Hankamer *et al*., 2007). The need for renewable sources of portable liquid fuel is starting to receive greater attention, and much of this attention has been focused on biomass-derived liquid fuels, or biofuels (Haag, 2006; Schneider, 2006). Large-scale introduction of biomass energy could contribute to sustainable development on several fronts, environmentally, socially and economically (Turkenburg, 2000). In particular, biomass has gained much attention as it fixes CO_2 in the atmosphere through photosynthesis. Thus if biomass is grown in a sustained way, its combustion has no impact on the CO_2 balance in the atmosphere, because, CO_2 emitted by the burning of biomass is offset by the CO_2 fixed by photosynthesis (Hossain et al., 2008).

In this regard, a variety of sources of biomass with potential feedstock such as corn, small grains, soybeans, cane sugar, switchgrass and organic waste have been tested. However, ethanol, hydrogen and biodiesel produced from conventional feedstocks fail to be cost competitive with petroleum (Scott and Bryner, 2006). In addition, due to the other limitations associated with the current biofuel production from conventional agricultural crops, its sustainability is still arguable. It is under this background that both the government and private sector are examining alternative sources of biofuel. In this context, algal biofuels are found to have enormous potential and offer a breakthrough solution to both energy security and global warming concerns. Algae (macro and microalgae) usually have a higher photosynthetic efficiency than other biomass producing plants (Hossain et al., 2008). Approximately half of the dry weight of the microalgal biomass is carbon (Sa´nchez Miro´n et al., 2003), which is typically derived from carbon dioxide. Thus, producing 100 tons of algal biomass means the reduction of roughly about 183 tons of carbon dioxide in the atmosphere.

However, commercialization of algal biofuel production is extremely challenging, due to the fact that factors attributed to its economy are highly variable. Therefore, understanding and addressing algal biofuel economic drivers is vital in developing culturing, harvesting and processing technologies. Microalgae, in particular, are at the center of many research because, with the acceptable technical maturity and commercial viability, microalgae, seem to be a candidate which can completely displace fossil-based fuels. Microalgae can provide several different types of renewable biofuels (Hossain et al., 2008). These include methane produced through anaerobic digestion of the algal biomass (Spolaore et al., 2006), biodiesel derived from microalgal oil (Banerjee et al., 2002; Thomas, 2006) and photobiologically produced biohydrogen (Fedorov et al., 2005; Gavrilescu and Chisti, 2005; Kapdan and Kargi, 2006). The present article reviews prospects and constraints of microalgae as a source of biofuel.

MICROALGAE

Microalgae are prokaryotic or eukaryotic photosynthetic microorganisms that can grow rapidly and live in harsh conditions due to their unicellular or simple multicellular structure (Mata et al., 2010). Cyanobacteria (Cyanophyceae) are the examples of prokaryotic microorganisms, whereas examples for eukaryotic microalgae are green algae (Chlorophyta) and diatoms (Bacillariophyta) (Li et al., 2008b). Microalgae can be found virtually in all most all the existing earth ecosystems, representing a variety of species living in a wide range of environmental conditions. It is estimated that more than 50,000 species exist, but only a limited number, of around 30,000, have been studied and analyzed (Richmond, 2004). The worldwide annual production of algal biomass is estimated to be 5 million kilograms per year with a market value of about 330 USD per kilogram (Pulz, 2004). High-value microalgal products include nutritional supplements, aquaculture feeds, biofertilizers, pharmaceuticals, β-carotene and cosmetics, and they also have the potential to be used as edible vaccines through genetic recombination (Christi, 2006; Rosenberg et al., 2008). Microalgae can effectively be used in bioremediation and wastewater treatments, because they can eliminate heavy metals, uranium, nitrogen, phosphorous and other pollutants from wastewater and they can degrade carcinogenic polyaromatic hydrocarbons and other organics. Furthermore, algae are accountable for at least 50 % of the photosynthetic biomass production on our planet and they are great sources of biofuels because they can accrue 70 % or more of their dry biomass as hydrocarbons (Christi, 2006).

Microalgae complete entire growth cycle every few days and can convert solar energy into chemical energy through photosynthesis (Sheehan et al., 1998). They have the ability to grow almost anywhere, provided the location receives enough sunlight and some simple nutrients. The growth rate of microalgae can be accelerated through altering the nutrient content and aeration of the growing media (Renaud et al., 1999; Pratoomyot et al., 2005; Aslan and Kapdan, 2006). Different microalgae species can adapt to live in a variety of environmental conditions. Thus, it is possible to find species best suited to local environments or specific growth characteristics, which is not possible to do with other current biodiesel feedstocks (e.g. soybean, rapeseed, sunflower and palm oil) (Mata et al., 2010).

BIOFUELS

Biofuels are a wide range of fuels, which are in some way derived from biomass. The term covers solid biomass, liquid fuels and various biogases (Demirbas, 2009a). Photosynthesis is responsible for converting sunlight into chemical energy and hence generates the feedstock needed for bioenergy synthesis: protons and electrons for biohydrogen, starch and sugar for bioethanol, biomass for biomethane, and oil for biodiesel (Hankamer, 2007). Biofuels, as a renewable source of energy, gain increased public and scientific attention, driven by factors such as oil price spikes, the need for increased energy security, and concern over greenhouse gas emissions from fossil-based fuels. Though fuels constitute approximately 67 % of the present global energy demand, only 33 % of the demand is met with all the renewable sources of energy including solar, wind, hydroelectric (Hankamer, 2007; Schenk, 2008). At today's consumption rate of about 85 million barrels per day of oil and 260 billion cubic feet per day of natural gas, the reserves represent 40 years of oil and 64 years of natural gas (Vasudevan, 2008). Biofuels are thus being inspected and developed rapidly, representing renewable energy derived from biological materials through photosynthesis. The main biofuels currently being produced include biohydrogen, bioethanol, biomethane and biodiesel.

BIODIESEL

Biodiesel is a biofuel consisting of monoalkyl esters that are derived from organic oils, plant or animal, through the process called "tranesterification" (Demirbas, 2007). It is a type of clean-burning diesel replacement fuel that can be used in compression-ignition (CI) engines (Bowman et al., 2006). The increasing competitive advantages of biodiesel are raising interest among investors and consumers alike. This interest has been expressed through a booming market. Today the global biodiesel industry is among the fastest-growing markets (Scott and Bryner, 2006), where it is available as a blend with conventional diesel fuel, typically in concentrations from 2 – 20 %. However blends up to 100 % can also be used and are available commercially.

The biodiesel transesterification reaction is very simple:

$$\begin{array}{c} CH_2-OCOCR_1 \\ | \\ CH-OCOCR_2 \\ | \\ CH_2-OCOCR_3 \end{array} + 3\ HOCH_3 \xrightleftharpoons{\text{Catalyst}} \begin{array}{c} CH_2-OH \\ | \\ CH-OH \\ | \\ CH_2-OH \end{array} + \begin{array}{c} R_1-COOCH_3 \\ R_2-COOCH_3 \\ R_3-COOCH_3 \end{array}$$

Triglyceride (Oil)   Methanol (Alcohol)   Glycerol   Methyl Esters (Biodiesel)

As indicated by the equilibrium reaction, biodiesel can be produced with an organic oil, or triglyceride, in the presence of a catalyst, usually potassium or sodium hydroxide (Christi, 2007; Demirbas, 2007). At 60 °C, the reaction can complete in 90 minutes (Campbell, 2008). The entire biodiesel production process can basically be subdivided into following steps (Xu et al., 2006):

A. The triglycerides, methanol and catalyst are placed in a controlled reaction chamber to undergo transesterification,
B. The initial product is placed in a separator to remove the glycerine by-product,
C. The excess methanol is recovered from the methyl esters through evaporation,
D. The final biodiesel is rinsed with water, pH neutralized, and dried.

The quality of biodiesel can vary with fatty acid profiles of the different sources. In turn, the properties of the various fatty esters are determined by the structural features of the fatty acid and the alcohol moieties that comprise a fatty ester (Yamane et al., 2001; Knothe and Steidley, 2005). Structural features that influence the physical and fuel properties of a fatty ester molecule are chain length, degree of unsaturation and branching of the chain. Important fuel properties of biodiesel that are influenced by the fatty acid profile are cetane number with relation to combustion and exhaust emissions, heat of combustion, cold flow, oxidative stability, viscosity and lubricity.

The oxygen content of biodiesel (10 - 12 wt %) is the most important compositional difference between diesel and biodiesel (Graboski and McCormick, 1998). According to Kousoulidou et al. (2008), except of oxygen content, biodiesel differs from petroleum-based diesel in the following properties:

- No sulphur or ultralow sulphur content
- No aromatic contents and no polycyclic aromatic hydrocarbons
- Higher cetane value
- Lower heating value
- Better lubricity
- Higher viscosity
- Higher flash point
- Biodegradability
- No toxicity or low toxicity

Biodiesel can be made from virtually any source of organic oil (Campbell, 2008). Large commercial producers often use seed oils, such as soybean, rapeseed, palm and corn, while in small scale, restaurant waste, animal fats etc., are used. However, biodiesel derived from seed oil is a matter of great concern as it may compete with food supply resulting biodiesel, to become increasingly expensive (Campbell, 2008). Therefore, if biodiesel is to become a true replacement for petroleum, more economical and sustainable source of oil is needed (Scott and Bryner, 2006; Christi, 2007). In this context, microalgae offer many advantages over conventional land plants. Depending on the species and growth conditions, the lipid content of microalgae can be 2 – 75 % of total cell dry matter (Wijffels, 2006), as membrane components, storage products, metabolites and storages of energy. Tri-glycerides and free fatty acids, a fraction of the total lipid content of microalgae can be converted into biodiesel. Table 1 shows the lipid content and lipid productivities of various marine and freshwater microalgae species.

Table 1. Lipid content and lipid productivity of different microalgae species (Derived from Mata *et al.*, 2010)

Microalgae species	Lipid content (% dry wt biomass)	Lipid productivity (mg/L/day)
Ankistrodesmus sp.	24.0 - 31.0	-
Botryococcus braunii	25.0 - 75.0	-
Chaetoceros muelleri	33.6	21.8
Chaetoceros calcitrans	14.6 - 16.4	17.6
Chlorella emersonii	25.0 - 63.0	10.3 - 50.0
Chlorella protothecoides	14.6 - 57.8	12-14
Chlorella sorokiniana	19.0 - 22.0	44.7
Chlorella vulgaris	5.0 - 58.0	11.2 - 40.0
Chlorella sp.	10.0 - 48.0	42.1
Chlorella pyrenoidosa	2.0 - 2.90	-
Chlorella	18.0 - 57.0	18.7 - 3.50
Chlorococcum sp.	19.3	53.7
Crypthecodinium cohnii	20.0 - 51.1	-
Dunaliella salina	6.0 - 25.0	116.0
Dunaliella primolecta	23.1 - 0.09	-
Dunaliella tertiolecta	16.7- 71.0	-
Dunaliella sp.	17.5 - 67.0	33.5
Ellipsoidion sp.	27.4	47.3
Euglena gracilis	14.0- 20.0	-
Haematococcus pluvialis	25.0 - 0.05	-
Isochrysis galbana	7.0 - 40.0	-
Isochrysis sp.	7.1 - 33	37.8
Monodus subterraneus	16.0	30.4
Monallanthus salina	20.0 - 22.0	-
Nannochloris sp.	20.0 - 56.0	60.9 -76.5
Nannochloropsis oculata	22.7 - 29.7	84.0
Nannochloropsis sp.	12.0 - 53.0	37.6 - 90.0

Microalgae species	Lipid content (% dry wt biomass)	Lipid productivity (mg/L/day)
Neochloris oleoabundans	29.0 - 65.0	90.0 - 134.0
Nitzschia sp.	16.0 - 47.0	-
Oocystis pusilla	10.5	-
Pavlova salina	30.9	49.4
Pavlova lutheri	35.5	40.2
Phaeodactylum tricornutum	18.0 - 57.0	44.8
Porphyridium cruentum	9.0 - 18.8	34.8
Scenedesmus obliquus	11.0 - 55.0	-
Scenedesmus quadricauda	1.9 - 18.4	35.1
Scenedesmus sp.	19.6 - 21.1	40.8 - 53.9
Skeletonema sp.	13.3 - 31.8	27.3
Skeletonema costatum	13.5 - 51.3	17.4
Spirulina platensis	4.0 - 16.6	-
Spirulina maxima	4.0 - 9.0	-
Thalassiosira pseudonana	20.6	17.4
Tetraselmis suecica	8.5 - 23.0	27.0 - 36.4
Tetraselmis sp.	12.6 - 14.7	43.4

(Michiki, 1995; Minowa et al., 1995; Zhu and Lee, 1997; Renaud et al., 1999; Sancho et al., 1999; Sawayama et al., 1999; Grima et al., 2000; Illman et al., 2000; Lee, 2001; Peng et al., 2001; Scragg et al., 2002; Richmond, 2004; Moheimani, 2005; Miao and Wu, 2006; Moheimani et al., 2006; Chisti, 2007; De Morais and Costa, 2007; Huntley and Redalje, 2007; Leathers et al., 2007; Natrah et al., 2007; Teixeira and Morales, 2007; Eriksen, 2008; Li et al., 2008a; Raja et al., 2008; Ugwu et al., 2008; Chiu et al., 2009; Demirbas, 2009b; Gouveia and Oliveira, 2009; Poisson et al., 2009; Rodolfi et al., 2009).

As can be seen in Table 1, the lipid content of different microalgae species shows significant differences from 2 % (*Chlorella pyrenoidosa*) to 75 % (*Botryococcus braunii*). Interestingly, most common algae (*Chlorella, Crypthecodinium, Cylindrotheca, Dunaliella, Isochrysis, Nannochloris, Nannochloropsis, Neochloris, Nitzschia, Phaeodactylum, Porphyridium, Schizochytrium, Tetraselmis*) possess the oil levels between 20 and 50 %. However, low productivity of some species stresses the need of biological and technological innovations in order to make the cultivation economically viable. Although the microalgae oil yield is strain-dependent, it is generally much greater than other vegetable oil crops. The Table 2 compares the microalgae and land-based oil crops in terms of oil content in dry weight basis and the oil yield per hectare, per year. Biodiesel production efficiency and land use efficiency of microalgae and land-based oil crops are depicted in Table 3.

Though microalgae offer a sound source of biomass for biodiesel production, a lot depends on the cultivation, harvesting and processing techniques. The procedure and technologies used for the other biofuel feedstocks are employed for microalgae also. The major steps include a production unit where cells are grown, followed by the separation of the cells from the growing media and subsequent lipids extraction. However, instead of transesterification reaction, other possibilities such as thermal cracking are also being investigated recently for biofuel production (Babu, 2008; Boateng et al., 2008).

Table 2. A comparison of microalgae and land-based oil crops in terms of oil content in dry weight basis and the oil yield per hectare, per year (Derived from Mata et al., 2010)

Plant source	Oil content (% oil by wt in biomass)	Oil yield (L oil/ha year)
Corn/Maize (*Zea mays* L.)	44	172
Hemp (*Cannabis sativa* L.)	33	363
Soybean (*Glycine max* L.)	18	636
Jatropha (*Jatropha curcas* L.)	28	741
Camelina (*Camelina sativa* L.)	42	915
Canola/Rapeseed (*Brassica napus* L.)	41	974
Sunflower (*Helianthus annuus* L.)	40	1070
Castor (*Ricinus communis*)	48	1307
Palm oil (*Elaeis guineensis*)	36	5366
Microalgae (low oil content)	30	58,700
Microalgae (medium oil content)	50	97,800
Microalgae (high oil content)	70	136,900

(Peterson and Hustrulid, 1998; Rathbauer et al., 2002; Zappi et al., 2003; Callaway, 2004; Kulay and Silva, 2005; Chisti, 2007; Teixeira and Morales, 2007; Vollmann et al., 2007; Kheira and Atta, 2008; Nielsen, 2008; Reijnders and Huijbregts, 2008).

Table 3. Biodiesel production efficiency and land use of microalgae and land-based oil crops (Derived from Mata et al., 2010)

Plant source	Land use (m^2 year/kg biodiesel)	Biodiesel productivity (kg biodiesel/ha year)
Corn/Maize (*Zea mays* L.)	66	152
Hemp (*Cannabis sativa* L.)	31	321
Soybean (*Glycine max* L.)	18	562
Jatropha (*Jatropha curcas* L.)	15	656
Camelina (*Camelina sativa* L.)	12	809
Canola/Rapeseed (*Brassica napus* L.)	12	862
Sunflower (*Helianthus annuus* L.)	11	946
Castor (*Ricinus communis*)	09	1156
Palm oil (*Elaeis guineensis*)	02	4747
Microalgae (low oil content)	0.2	51,927
Microalgae (medium oil content)	0.1	86,515
Microalgae (high oil content)	0.1	121,104

(Peterson and Hustrulid, 1998; Rathbauer et al., 2002; Zappi et al., 2003; Callaway, 2004; Kulay and Silva, 2005; Chisti, 2007; Teixeira and Morales, 2007; Vollmann et al., 2007; Kheira and Atta, 2008; Nielsen, 2008; Reijnders and Huijbregts, 2008).

BIOETHANOL

Bioethanol can be used as a biofuel, which can replace part of the fossil-derived petrol. It can be used in a number of different ways (Kousoulidou *et al.*, 2008):

- As a blend with gasoline (from 5 to 85 %). If used as a 5 % blend, all most all petrol engines can be operated with no or little modification.
- As a direct substitute for petrol in cars with appropriately modified engines.
- As a blend with diesel in diesel engines, also known as "E-diesel" fuel blends
- As a blend with biodiesel in diesel engines, also known as "BE-diesel" fuel blends

The production of bioethanol first uses enzyme amylase to convert a feedstock into fermentable sugars. Yeast is then added to the mash to ferment the sugars to alcohol and carbon dioxide, the liquid fraction being distilled to produce ethanol. Currently bioethanol is produced in large scale by fermenting sugars from starch crops such as corn. According to Balat *et al.* (2008), the major problem associated with bioethanol production is the availability of raw materials, due to which the price of the raw materials is also highly variable affecting the production costs of bioethanol.

In algae, a starch content of over 50 percent has been reported. With new technologies, cellulose and hemicellulose can also be hydrolysed to sugars (Hamelinck *et al.*, 2005), creating the possibility of converting an even larger part of algal dry matter to ethanol. Algae have some beneficial characteristics compared to woody biomass, the traditional target for this technology. Most notable is the absence of lignin in algae, making its removal needed for woody material redundant. Furthermore, algae composition is generally much more uniform and consistent than biomass from terrestrial plants, because algae lack specific functional parts such as roots and leaves. Algal cell walls are largely made up of polysaccharides, which can be hydrolyzed to sugar.

MICROALGAE CULTIVATION SYSTEMS

1) Open Systems

The conventional large-scale cultivation takes place in outdoor open systems such as ponds and lagoons, which may be circular or raceway-designed (Chisti, 2006). Open culture systems comprise a relatively cost-effective method for growing microalgae. The simplest open algae cultivation systems are shallow, unstirred ponds of different extents ranging from a few square meters to several hundreds hectares. Dissolution of CO_2 from air into water limits the growth rate of algae resulting in low productivity of the systems. Slow diffusion of nutrients, flotation and sedimentation of dead and living algae and limiting usage of available sunlight also contribute to the low biomass yield per hectare. In raceway ponds, paddle wheels are employed to drive water flows, which can create gas bubbles through the medium ensuring better supply of CO_2. The major bottlenecks of these open systems are loss of water by evaporation, difficulties in controlling cultivar parameters, and susceptibility to competition and contamination by bacteria, viruses and invasive algae. There is almost no possibility to control the temperature also. Consequently, only a limited number of species is dominant enough to maintain itself in an open system (Pulz, 2001; Carlsson et al., 2007; Chisti, 2007; Rodolfi et al., 2009). In this regard, extremophiles such as *Spirulina* and *Dunaliella* that favor highly alkaline and saline environments are better suited for open systems (Chisti, 2006; Schenk, 2008).

2) Photobioreactors (PBRs)

Microalgae can be grown on a large scale in photobioreactors (Pulz, 2001; Janssen et al., 2003; Carvalho et al., 2006). This closed system receives sunlight or artificial illumination and the algae are contaminant-free. Many different designs of photobioreactors (such as tubular reactors, vertical alveolar panels, flat panels, bubble column reactors etc.,) have been developed. Among them the tubular photobioreactors have proven to be the most successful for producing algal biomass on the scale needed for biofuel production (Chisti, 2008). They are made of one or more small-diameter straight transparent tubes that can be designed in several ways and the tubes represent the reactor's solar collectors (Chisti, 2006). These transparent tubes are usually made of plastic or glass of less than 0.1 m in diameter, thus enable easy penetration and distribution of light over a large surface area to prevent photoinhibition. A photobioreactor is typically operated as a continuous culture (Chisti, 2007), as the fresh culture medium is fed at a constant rate while the same quantity of microalgal broth is withdrawn continuously (Molina Grima et al., 1999). In addition, a mechanism for constant mixing is operated in order to prevent biomass sedimentation and to distribute photosynthetic gases (Schenk, 2008). However it has been noticed that substantial amount of the biomass (25 %) produced during daytime could be consumed during the night to sustain the cells until sunrise (Sanchez Miron et al., 2002; Chisti, 2007). The extent of this loss depends on the light level under which the biomass is grown, the growth temperature and the temperature at night.

The carbon assimilation process releases oxygen and a high concentration of dissolved oxygen in combination with intense sunlight produces photooxidative damage to algal cells. The maximum dissolved oxygen level should not exceed 400 % of air saturation value to prevent photosynthesis inhibition and cell damage (Molina et al., 2001). As the dissolved oxygen is hard to remove within a photobioreactor tube, the maximum length of a continuous tube run should be limited. In addition, the algal culture can be periodically returned to a degassing zone in which it is aerated to strip out the accumulated oxygen (Chisti, 2006). On the other hand, closed systems can get too hot and require cooling, which can be done with heat exchangers (Chisti, 2007) or spraying water on the outside (Chini Zittelli et al., 1999).

3) Hybrid Systems

Hybrid systems bring a combination of open ponds and photobioreactors, thus can avoid most of the limitations of both open ponds and photobioreactors. In a hybrid system, contaminant-free algal cells are first grown in photobioreactors, where continuous cell division takes place ensuring sufficient algal growth enters to the open pond. The cells are then exposed to grow under environmental stress conditions in open ponds to maximize the lipid content (Huntley and Redalje, 2007). The hybrid systems result in high-yield cultures at a comparative cost. However, it is worth to highlight that the yield is low compared to the potential yield based on the quantum limits of photosynthetic efficiency, as well as compared to other means for harnessing solar energy (Vasudevan, 2008).

Figure 1 illustrates a model for integrated algal biofuel production. The selected microalgae species is grown in the production unit, which provides the basic growth requirements; water, nutrients, carbon dioxide, light and space adequately. In order to maximize algal growth, CO_2 needs to be provided as the source of Carbon. The concentration should be much higher than that can be attained under natural conditions thus has a direct impact on cost of production.

In order to ensure adequate sunlight, which is necessary for photosynthesis, the production unit should be placed in a geographic location with abundant and uninterrupted sunshine, in particular, when open ponds are used. However, with bioreactors, sunlight quantity and quality can be further enhanced through the use of solar collectors, solar concentrators, and fibre optics in a system called photo-bioreactors (Scott and Bryner, 2006; Christi, 2007). However, depending on local specific conditions and the scale of cultivation, the design and the structure of the production unit may vary.

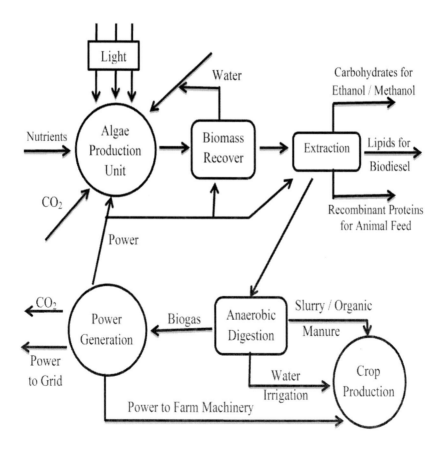

Figure 1. A model for integrated microalgal biofuel production.

The most common harvesting processes used in the present day aquaculture are microfiltration, flocculation, sedimentation and centrifugation. Each method has advantages as well as disadvantages. Thus selecting the appropriate harvesting method is dependent upon many factors such as nature of the species, scale and method of cultivation etc. Harvested algal biomass is then directed to the extraction unit, where lipids, carbohydrates and proteins are extracted. Lipids and carbohydrates are used for the production of biodiesel and bioethanol/methanol respectively. After lipid extraction, the remaining biomass fraction can be used as a high protein feed for livestock (Schneider, 2006; Haag, 2007). This gives further value to the process and reduces waste. The rest can be used in generating biogas through anaerobic digestion. Power generated through biogas is primarily used in meeting the energy requirement of algal production and processing and the excess may have other uses depending on the operational system.

Though, fresh water is generally used with salts and minerals needed, microalgae cultivation can be coupled to an environmental remediation technique that would enhance productivity while mitigating pollution. For this purpose, nutritionally rich wastewater from domestic or industrial sources can be added to the algal growth media directly (Schneider, 2006). This allows algae production unit to be operated bit cheaply, while simultaneously treating wastewater. Van Harmelen and Oonk (2006) estimated a global potential of about 30 million tons of algal biomass production, and a similar level of CO_2 abatement credits, using municipal wastewaters.

The prospects of use of municipal wastewater for algal biomass production can be listed as follows,

- Municipal wastewater contains substantial amount of Carbon, thus could enhance the algal growth considerably.
- The relatively high Nitrogen (30-40 mg/L) and Phosphorus (5-10 mg/L) contents reduce the supply of nutrients to the growth media.
- Apart from the N and P, municipal wastewater contains some essential micronutrients too.
- It is produced in substantial quantities (100 gallons/person/day)
- Duo to the ability of taking up nutrients, algal plant can act as a treatment unit also.

By diverting the CO_2 fraction of the flue gases from industrial processes, and in particular from power plants, through the algal cultivation process, the CO_2 can be diverted back into the energy stream while increasing the rate of algal production (Pulz, 2007). The practical implication of this is being investigated using both bioreactor designs and open ponds (Schneider, 2006).

WHAT ADVANTAGES DO MICROALGAE BRING?

- Microalgae are easy to culture and can grow with little or even no attention (Mata *et al.*, 2010).
- Water unsuitable for human consumption also can be used in culturing microalgae (Mata *et al.*, 2010).

- Microalgae are the fastest growing photosynthesizing organisms. They grow extremely rapidly and commonly double their biomass within 24 h (Chisti, 2008). In fact, the biomass doubling time for microalgae during exponential growth can be as short as 3.5 h (Chisti, 2007), which is significantly quicker than the doubling time for land-based oil crops. Algae growth is dependent on climatic conditions, not essentially on the seasons. They do not exhibit an annual growth cycle of sprouting in the spring and harvest in autumn, thus year round production is ensured (however, in temperate and subtropical regions, algae may have a growth season).
- By contrast, many microalgae species contain much more energy per unit of weight (Banerjee et al., 2002).
- They can be grown in a closed system almost anywhere, including deserts or even rooftops, and there is no competition for food or fertile soil.
- Because algae are grown in water, the cultivation systems have much lower land quality requirements than Agriculture (FAO, 2009).
- Biofuel production from land plants possess sustainability issues, in particular, food security and agro-biodiversity are concerned (Rossi and Lambrou, 2008). Microalgae can avoid these issues by using non-arable lands and even improving food security through co-production of food.
- Soil fertility is not an issue at all. Thus microalgae allow a huge area of land to be used, such as deserts, infertile saline soils, polluted land and other land with low economic (and ecologic) value. Glenn et al. (1998) indicate that 43 percent of the earth's total land surface is arid or semi-arid and estimate that 15 percent of undeveloped land has sufficient access seawater (max 100 km), which amounts to 130 million ha.
- Algae are also easier to harvest because it has no roots or fruit and grows dispersed in water. Algal body is much more uniform than higher plants, microalgae are in many cases unicellular thus completely uniform. This way the entire biomass can be processed, rather than just parts like seeds or roots in higher plants.
- Algae can utilize CO_2 much more effectively than land plants. Though the amount of CO_2 captured by algae varies with the species, generally about 1.8 tonnes CO_2 is integrated in 1 tonne algal biomass (Chisti, 2007).
- Their solar energy conversion rate is very high; about 3-8 % of solar energy can be converted to biomass whereas observed yields for terrestrial plants are about 0.5 % (Huntley and Redalje 2007; Li et al., 2008a). Therefore, their potential productivity (in terms of oil production per ha and per year) is far higher than those of land plants (Chisti, 2008).
- Microalgal biomass production can potentially make use of some of the carbon dioxide that is released in power plants by burning fossil fuels (Sawayama et al., 1995; Yun et al., 1997; Rosenberg et al., 2008). This carbon dioxide is often available at little or no cost. Placing algae plants near carbon producing facilities like regular power plants, or manufacturing plants could sequester the CO_2 they create and use those emissions to help algae growth. Algae can grow directly on combustion gas (typically containing 4 -15 percent CO_2), whereas plants take up CO_2 from the atmosphere (open air concentration 0.036 percent). Therefore, a bioreactor

built in the right way can have the added benefit of preventing carbon dioxide emissions, nitrogen oxide and sulfur oxide, from entering the atmosphere.
- Microalgae can provide feedstock for several different types of renewable fuels such as biodiesel, methane, hydrogen and ethanol (Mata *et al.*, 2010).
- Algae biofuel is non-toxic and contains no sulfur (Aresta *et al.*, 2005; Rakopoulos *et al.*, 2006; Demirbas, 2007).
- Microalgae biofuel is highly biodegradable (Aresta *et al.*, 2005).
- Existing engines can use microalgal biofuel without or with little modification. Microalgal biofuel can be mixed with conventional petroleum at any ratio. As a result, this biofuel can use existing distribution infrastructure (Crookes, 2006).
- Microalgal biodiesel has twice the viscosity of petroleum diesel, its lubrication properties can thus actually improve engine life (Bowman *et al.*, 2006).
- Algae can contain many desired components in large concentrations. Therefore, after oil extraction, the resulting algal biomass can be processed into ethanol, methane, livestock feed, used as organic fertilizer with high N:P ratio, or simply burned for energy generation (Wang *et al.*, 2008). Depending on the microalgae species, other compounds such as fats, polyunsaturated fatty acids, natural dyes, sugars, pigments, antioxidants, high-value bioactive compounds and other fine chemicals and biomass may also be extracted (Li *et al.*, 2008a; Raja *et al.*, 2008).
- Because of the variety of high-value biological derivatives, with many possible commercial applications, microalgae can potentially revolutionize a large number of biotechnology areas including biofuels, cosmetics, pharmaceuticals, nutrition and food additives, aquaculture and pollution prevention (Raja *et al.*, 2008; Rosenberg *et al.*, 2008).

WHAT ARE THE LIMITATIONS OF MICROALGAE AS A SOURCE OF BIOFUEL?

- The operating costs (labor, electricity, replacement parts, fertilizers, CO_2, flocculent, water, hexane, sterilizers, etc.) for microalgae production are estimated to be very high. Thus the price of biodiesel is typically higher than petroleum diesel (Campbell, 2008).
- Microalgae too need nutrients (especially phosphates), which are expected to experience strong inflation over the coming decades.
- In addition to the nutrients that must be fed to any crop, the algae must be fed concentrated CO_2 also to ensure better carbon assimilation.
- If open ponds are used for algae production, the water loss by means of evaporation is substantially high. This would be a major drawback in arid regions, where water loss is apparently enormous.
- Open ponds are susceptible to local strains as they grow better than those inoculated.
- Not like other conventional crops, technological barriers could play a decisive role in assuring high productivity from a given area of land.

- To be economically viable, large scale algae cultivation is always proposed (Wijffels, 2008), which prevents small scale farmers entering to the industry. Furthermore, plots of this size with low current economic and ecologic value can be scarce.
- Excess rain can cause water disposal problems in closed systems built on hard surfaces (BCIC, 2009) while in open systems this may lead to high nutrient and excess water.
- Regardless the biomass source used to produce biodiesel, it does contribute increased NOx emissions, relative to petroleum diesel, owing to the higher compression ratios typically used in biodiesel engines (Crookes, 2006; Pradeep and Sharma, 2007).

FUTURE CHALLENGERS

The future research and development strategies must focus first on the biological aspect of microalgae. Screening and collecting strains of algal species to access their potential for high oil production with high biomass productivity, investigating the physiology and biochemistry of the algae, and use of molecular-biology and genetic engineering techniques to enhance the oil yield can be considered as the key areas to be worked on.

If the cultivation is done in large open ponds, selection of stable algae strains, which can be maintained, is identified as one of the main technical issues. Algal species dominance can be challenged from invasion by 'weed' algal strain, grazing by zooplankton or others, often unknown, factors resulting into a pond crash. Current commercial technologies are based on 'extremophiles'; species that thrive in extreme environments for example *Spirulina* grows in highly alkaline waters. This extreme environment avoids algal pond contamination but leads also to low productivities.

In deciding the method of harvesting, two aspects need to be considered. The size of algal body is only a few micrometers. Microalgae concentrations always remain very low while growing, typically 0.02 to 0.05 percent dry matter in raceways and between 0.1 to 0.5 percent dry matter in tubular reactors (Tredici, 2009) (this means 1 tonne dry biomass has to be recovered from 200 - 5000 m^3 water). These two aspects make the harvesting and further concentration of algae difficult and therefore expensive. Harvesting has been claimed to contribute 20 - 30 percent to the total cost of producing the biomass (Grima *et al.*, 2003). In present-day algal aquaculture, the most common harvesting processes are microfiltration, flocculation and centrifugation. Though pure sedimentation is also practiced for microalgae, it needs more space and time thus not advisable to be used in biodiesel production. Some acceptable results for microfiltration have been obtained for colonial microalgae, but not for unicellular species (Grima *et al.*, 2003). Furthermore, filtration is a slow process (Sazdanoff, 2006), thus a very large total capacity system would be required to keep up with the production unit. Adding flocculants is also considered not sustainable despite less expensive. In this context, cell self-flocculation has been recently studied by regulating carbon and the pH. Centrifugation is often used for the concentration of high-value algae. Though generally considered to be expensive and electricity consuming, centrifugation is so far the best known method of concentrating small unicellular algae (Grima *et al.*, 2003). However, it is obvious that in addition to selecting easy-to-harvest strains, harvesting method should be energy-efficient and cost effective.

According to Lardon *et al.* (2009), about 90 % of the energy consumption of the entire biodiesel production process is dedicated to lipid extraction (70 % when considering the wet extraction), thus any improvement of oil extraction technique would have a direct impact on the sustainability of this production. Future research should focus on new means of lipid recovering with limited drying of the biomass. The dry extraction is economically viable only with a sound method for drying the algae. Practical feasibility of solar drying in large scale is yet to be demonstrated, while lipid stability during solar drying is also questionable. Lipid extraction is facilitated by combining methyl esterification with the use of immobilized lipases. Mechanical crushing of biomass followed by squeezing can also be practiced. A modern technique used to disrupt the cells is electroporation, where a strong electric field is applied to the biomass in order to perforate the cell wall resulting better extraction of lipids. However, not only the high lipid content, but easy oil recovery should also be taken into consideration in screening the species for the cultivation.

The lipid production of microalgae is found to be high in the cultures with low nitrogen, though such culture conditions strongly affect the growth rate, and ultimately to the net productivity (Rodolfi *et al.*, 2009). Therefore, species which can maintain a high productivity under nitrogen-limiting conditions should be given priority if other key requirements are also met by such species. The properties of the biodiesel can vary with the various fatty acid profiles of the different feedstocks. The quality of the various fatty esters is determined by the structural features of the fatty acid and the alcohol moieties that comprise a fatty ester (Knothe and Steidley 2005; Yamane *et al.*, 2001). It is therefore important to enrich certain fatty esters with desirable properties in order to improve the quality of the fuel. Thus genetic engineering can play a significant role in improving the quality of biodiesel through enrichment of certain fatty acids, possibly oleic acid, to the parent feedstock.

Since microalgae represent a much simpler genetic make up than land-based oil crops, genetic manipulations aiming at increasing its content of high-value compounds is considered to be very tempting. Nevertheless, progress in the genetic engineering of algae was extremely slow until recently. Also, these promising advances should be viewed with caution because transgenic algae potentially pose a considerable threat to the ecosystem and thus will most likely be banned from outdoor cultivation systems.

CONCLUSION

Algal biofuels are found to have enormous potential and offer a breakthrough solution to both energy security and global warming concerns. Algal biofuels can minimize the dependency of the current biofuel production from conventional agricultural crops also. Microalgal biodiesel is appeared to be the only renewable biodiesel that has the potential to completely displace liquid transport fuels derived from petroleum. Both open and closed microalgae culturing systems are being assessed in this regard. However, achieving the capacity to inexpensively produce biodiesel from microalgae is still challengeable. The mass-culture methods of microalgae so far have been developed, targeting commercial production of high value-added substances such as bioactive compounds. One of them is the vertical type photobioreactor. The reactor allows a rapid and high-density cultivation of microalgae by illuminating light from both sides of the panel. However, its operation consumes a lot of energy, thus production of biofuel using this system is not economically viable. Therefore, large scale microalgal biomass generation in photobioreactors needs a rigorous assessment of the economics of production to be competitive with petroleum-derived fuels.

If the price of crude oil is about $100 per barrel, microalgal biomass with an oil content of 55 % should be produced at less than $340 ton^{-1} to be competitive with petroleum diesel. Literature suggests that, currently, microalgal biomass can be produced for around $3000 ton^{-1}. Therefore, the price of producing the biomass needs to decline by a factor of 9, through advances in production technology and algal biology. With time and experience, significant improvements in algal density, growth rates and oil content of algae should be able to achieve. This will require improved growing methods, species selection, cultivation techniques, and bio-engineering. However, such integrated processes and systems for mass production of biofuel from microalgae are still premature.

REFERENCES

Aresta, M., Dibenedetto, A., Carone, M., Colonna, T. and Fagale, C. (2005). Production of biodiesel from macroalgae by supercritical CO_2 extraction and thermochemical liquefaction. *Environmental Chemistry Letters*, 3, 136-139.

Aslan, S. and Kapdan, I. K. (2006). Batch kinetics of nitrogen and phosphorus removal from synthetic wastewater by algae. *Ecological Engineering*, 28(1), 64-70.

Babu, B. V. (2008). Biomass pyrolysis: a state-of-the-art review. *Biofuels Bioproducts Biorefinin*, 2(5), 393-414.

Balat, M., Balat, H. and Oz, C. (2008). Progress in bioethanol processing. *Progress in Energy and Combustion Science*, 34, 551-573.

Banerjee, A., Sharma, R., Yusuf, C. and Banerjee, U. C. (2002). *Botryococcus braunii*: a renewable source of hydrocarbons and other chemicals. *Crit. Rev. Biotechnol.*, 22, 245-279.

BCIC (2009). *Microalgae Technologies and Processes for Biofuels – Bioenergy Production in British Columbia*.

Boateng, A. A., Mullen, C. A., Goldberg, N., Hicks, K. B., Jung, H. J. G. and Lamb, J. F. S. (2008). Production of bio-oil from alfalfa stems by fluidized-bed fast pyrolysis. *Industrial and Engineering Chemistry Research*, 47, 4115-4122.

Bowman, M., Hilligoss, D., Rasmussen, S. and Thomas, R. (2006). Biodiesel: A renewable and biodegradable fuel. *Hydrocarbon Processing*, 85, 103-106.

Callaway, J. C. (2004). Hempseed as a nutritional resource: an overview. *Euphytica*, 140, 65-72.

Campbell, M. N. (2008). Biodiesel: Algae as a renewable source for liquid fuel. *Guelph Engineering Journal*, 1(2-7), 1916-1107.

Carlsson, A., Beilen van, J., Möller, R., Clayton, D. and Bowles, D. E. (2007). *Micro and macroalgae - utility for industrial applications*. Bioproducts, CNAP, University of York.

Carvalho, A. P., Meireles, L. A. and Malcata, F. X. (2006). Microalgal reactors: a review of enclosed system designs and performances. *Biotechnol. Prog.*, 22, 1490-1506.

Chini Zittelli, G., Lavista, F., Bastianini, A., Rodolfi, L., Vincenzini, M. and Tredici, M. R. (1999). Production of eicosapentaenoic acid by *Nannochloropsis* sp cultures in outdoor tubular photobioreactors. *J. Biotechnol.*, *70(1-3)*, 299-312.
Chisti, Y. (2006). Microalgae as sustainable cell factories. *Environmental Engineering and Management Journal*, *5(3)*, 261-274.
Chisti, Y. (2007). Biodiesel from microalgae. *Biotechnol. Adv.*, *25*, 294-306.
Chisti, Y. (2008). Biodiesel from microalgae beats bioethanol. *Trends in Biotechnology*, *26(3)*, 126-131.
Chiu, S. Y., Kao, C. Y., Tsai, M. T., Ong, S. C., Chen, C. H. and Lin, C. S. (2009). Lipid accumulation and CO_2 utilization of *Nannochloropsis oculata* in response to CO_2 aeration. *Bioresource Technology*, *100*, 833-838.
Clayton, M. (2004). *New coal plants bury 'Kyoto*, Christian Science Monitor, December 23, 2004.
Crookes, R. J. (2006). Comparative biofuel performance in internal combustion engines. *Biomass and Bioenergy*, *30*, 461-468.
Demirbas, A. (2007). Importance of biodiesel as transportation fuel. *Energy Policy*, *35*, 4661-4670.
Demirbas, A. (2009a). Political, economic and environmental impacts of biofuels: A review. *Applied Energy*, *86*, 108-117.
Demirbas, A. (2009b). Progress and recent trends in biodiesel fuels. *Energy Conversion and Management*, *50*, 14-34.
De Morais, M. G. and Costa, J. A. V. (2007). Carbon dioxide fixation by *Chlorella kessleri, C. vulgaris, Scenedesmus obliquus* and *Spirulina* sp. cultivated in flasks and vertical tubular photobioreactors. *Biotechnology Letters*, *29(9)*, 1349-1352.
Dewulf, J. and Van Langenhove, H. (2006). *Renewables-based technology: sustainability assessment*. John Wiley and Sons, Ltd.
Eriksen, N. T. (2008). The technology of microalgal culturing. *Biotechnology Letters*, *30*, 1525-1536.
FAO (2009). *Algae-based biofuels; a review of challenges and opportunities for developing countries.* Rome, Italy, FAO.
Fedorov, A. S., Kosourov, S., Ghirardi, M. L. and Seibert, M. (2005). Continuous H_2 photoproduction by *Chlamydomonas reinhardtii* using a novel two stage, sulfate-limited chemostat system. *Appl. Biochem. Biotechnol.*, *124*, 403-412.
Gavrilescu, M. and Chisti, Y. (2005). Biotechnology - a sustainable alternative for chemical industry. *Biotechnol. Adv.*, *23*, 471-499.
Gilbert, R. and Perl, A. (2008). *Transport revolutions: moving people and freight without oil.* Earthscan.
Glenn, E. P., Brown, J. J. and O'Leary, J. W. (1998). Irrigating crops with seawater. *Scientific American*, *279(2)*, 76-81.
Gouveia, L. and Oliveira, A. C. (2009). Microalgae as a raw material for biofuels production. *Journal of Industrial Microbiology and Biotechnology*, *36*, 269-274.
Graboski, M. and McCormick, R. (1998). Combustion of fat and vegetable oil derived fuels in diesel engines. *Progress in Energy and Combustion Science*, *24*, 125-164.
Grima, E. M., Ferna´ndez, F. G. A., Camacho, F. G., Rubio, F. C. and Chisti, Y. (2000). Scale-up of tubular photobioreactors. *Journal of Applied Phycology*, *12*, 355-368.

Grima, E. M., Belarbi, E. H., Fernandez, F. G. A., Medina, A. R. and Chisti, Y. (2003). Recovery of microalgal biomass and metabolites: process options and economics. *Biotechnol. Adv.*, *20(7-8)*, 491-515.

Haag, A. L. (2007). Algae bloom again. *Nature*, *447*, 520-521.

Hankamer, B. (2007). Photosynthetic biomass and H_2 production by green algae: from bioengineering to bioreactor scale-up. *Physiologia Plantarum*, *131*, 10-21.

Hamelinck, C. N., van Hooijdonk, G. and Faaij, A. P. C. (2005). Ethanol from lignocellulosic biomass: techno-economic performance in short-, middle- and long-term. *Biomass Bioenergy*, *28(4)*, 384-410.

Hankamer, B., Lehr, F., Rupprecht, J., Mussgnug, J. H., Posten, C. and Kruse, O. (2007). Photosynthetic biomass and H_2 production by green algae: From bioengineering to bioreactor scale-up. *Physiologia Plantarum*, *131*, 10-21.

Hossain, A. B. M. S., Salleh, A., Boyce, A. N., Chowdhury, P. and Naqiuddin, M. (2008). Biodiesel fuel production from algae as renewable energy. *Am. J. Biochem. and Biotech.*, *4(3)*, 250-254.

Huntley, M. E., Redalje, D. G. (2007). CO_2 mitigation and renewable oil from photosynthetic microbes: a new appraisal. *Mitigation and Adaptation Strategies for Global Change*, *12(4)*, 573-608.

Illman, A. M., Scragg, A. H. and Shales, S. W. (2000). Increase in *Chlorella* strains calorific values when grown in low nitrogen medium. *Enzyme and Microbial Technology*, *27*, 631-635.

Janssen, M. (2001). Photosynthetic efficiency of *Dunaliella tertiolecta* under short light/dark cycles. *Enzyme and Microbial Technology*, *29*, 298-305.

Janssen, M., Tramper, J, Mur, L. R. and Wijffels, R. H. (2003). Enclosed outdoor photobioreactors: light regime, photosynthetic efficiency, scale-up, and future prospects. *Biotechnol. Bioeng.*, *81*, 193-210.

Kapdan, I. K. and Kargi, F. (2006). Bio-hydrogen production from waste materials. *Enzyme Microbiol. Technol.*, *38*, 569-582.

Kheira, A. A. A. and Atta, N. M. M. (2008). Response of *Jatropha curcas* L. to water deficit: yield, water use efficiency and oilseed characteristics. *Biomass and Bioenergy*, *33*, 1343-1350.

Knothe, G. and Steidley, K. (2005). Kinematic viscosity of biodiesel fuel components and related compounds, influence of compound structure and comparison to petrodiesel fuel components. *Fuel Processing Technology*, *84*, 1059-1065.

Kousoulidou, M., Fontaras, G., Mellios, G. and Ntziachristos, L. (2008). Effect of biodiesel and bioethanol on exhaust emissions. ETC/ACC Technical Paper 2008/5, The European Topic Centre on Air and Climate Change (ETC/ACC).

Kulay, L. A. and Silva, G. A. (2005). Comparative screening LCA of agricultural stages of soy and castor beans. In: 2nd international conference on life cycle management - LCM, 2005. pp. 5-7.

Laherrere, J. (2005). Forecasting production from discovery. In: ASPO; 2005.

Lardon, L., He´lias, A., Sialve, B., Steyer, J. P. and Bernard, O. (2009). Life-cycle assessment of biodiesel production from microalgae. *Environ. Sci. Technol.*, *3*, 1-6.

Leathers, J., Celina, M., Chianelli, R., Thoma, S. and Gupta, V. (2007). Systems analysis and futuristic designs of advanced biofuel factory, concepts. SANDIA Report, SAND 2007-6872.

Lee, Y. K. (2001). Microalgal mass culture systems and methods: their limitation and potential. *Journal of Applied Phycology, 13*, 307-315.

Li, Y., Horsman, M., Wu, N., Lan, C. Q. and Dubois-Calero, N. (2008a). Biofuels from microalgae. *Biotechnology Progress, 24(4)*, 815-820.

Li, Y., Wang, B., Wu, N. and Lan, C. Q. (2008b). Effects of nitrogen sources on cell growth and lipid production of *Neochloris oleoabundans*. *Applied Microbiology and Biotechnology, 81(4)*, 629-636.

Mata, T. M., Martins, A. A. and Caetano, N. S. (2010). Microalgae for biodiesel production and other applications: A review. *Renewable and Sustainable Energy Reviews, 14*, 217-232.

Miao, X. and Wu, Q. (2006). Biodiesel production from heterotrophic microalgal oil. *Bioresource Technology, 97(6)*, 841-846.

Michiki, H. (1995). Biological CO_2 fixation and utilization project. *Energy Conversion and Management, 36(6-9)*, 701-705.

Minowa, R., Yokoyama, S., Kishimoto, M. and Okakurat, T. (1995). Oil production from algal cells of *Dunaliella tertiolecta* by direct thermochemical liquefaction. *Fuel, 74(12)*, 1735-1738.

Moheimani, N. R. (2005). The culture of Coccolithophorid algae for carbon dioxide bioremediation. PhD thesis. Murdoch University.

Moheimani, N. R. and Borowitzka, M. A. (2006). The long-term culture of the coccolithophore *Pleurochrysis carterae* (Haptophyta) in outdoor raceway ponds. *Journal of Applied Phycology, 18*, 703-712.

Molina Grima, E., Acién Fernández, F. G., García Camacho, F. and Chisti, Y. E. (1999). Photobioreactors: light regime, mass transfer, and scale up. *J. Biotechnol., 70*, 231-247.

Molina, E., Fernandez, J., Acien, F. G. and Chisti, Y. (2001). Tubular photobioreactor design for algal cultures. *J. Biotechnol., 92*, 113-131.

Natrah, F., Yoso, V. F. M., Shari, V. M., Abas, F. and Mariana, N. S. (2007). Screening of Malaysian indigenous microalgae for antioxidant properties and nutritional value. *Journal of Applied Phycology, 19*, 711-718.

Nielsen, D. C. (2008). Oilseed productivity under varying water availability. In: Proceedings of 20th annual central plains irrigation conference and exposition; 2008. pp. 30-33.

Peng, W., Wu, Q., Tu, P. and Zhao, N. (2001). Pyrolitic characteristics of microalgae as renewable energy source determined by thermogravimetric analysis. *Bioresource Technology, 80*, 1-7.

Peterson, C. L. and Hustrulid, T. (1998). Carbon cycle for rapeseed oil biodiesel fuels. *Biomass and Bioenergy, 14(2)*, 91-101.

Poisson, L., Devos, M., Pencreac'h, G. and Ergan, F. (2009). Benefits and current developments of polyunsaturated fatty acids from microalgae lipids. *OCL – Oleagineux Corps Gras Lipides, 9(2-3)*, 92-95.

Pradeep, V. and Sharma, R. K. (2007). Use of HOT EGR for Nox control in a compression ignition engine fuelled with biodiesel from Jatropha oil. *Renewable Energy, 32*, 1136-1154.

Pratoomyot, J., Srivilas, P. and Noiraksar, T. (2005). Fatty acids composition of 10 microalgal species. *Songklanakarin Journal of Science and Technology, 27(6)*, 1179-1187.

Pulz, O. (2001). Photobioreactors: production systems for phototrophic microorganisms. *Appl. Microbiol. Biotechnol.*, *57*, 287-293.

Pulz, O. (2004). Valuable products from biotechnology of microalgae. *Applied Microbiology and Biotechnology*, *65*, 635-648.

Raja, R., Hemaiswarya, S., Kumar, N. A., Sridhar, S. and Rengasamy, R. (2008). A perspective on the biotechnological potential of microalgae. *Critical Reviews in Microbiology*, *34(2)*, 77-88.

Rakopoulos, C. D., Antonopoulos, K. A., Rakopoulos, D. C., Hountalas, D. T. and Giakoumis, E. G. (2006). Comparative performance and emissions study of a direct injection diesel engine using blends of diesel fuel with vegetable oils or biodiesels of various origins. *Energy and Conservation Management*, *47*, 3272-3287.

Rathbauer, J., Prankl, H. and Krammer, K. (2002). Energetic use of natural vegetable oil in Austria. Austria: BLT - Federal Institute of Agricultural Engineering.

Reijnders, L. and Huijbregts, M. A. J. (2008). Biogenic greenhouse gas emissions linked to the life cycles of biodiesel derived from European rapeseed and Brazilian soybeans. *Journal of Cleaner Production*, *16*, 1943-1948.

Renaud, S. M., Thinh, L. V. and Parry, D. L. (1999). The gross chemical composition and fatty acid composition of 18 species of tropical Australian microalgae for possible use in mariculture. *Aquaculture*, *170*, 147-159.

Richmond, A. (2004). *Handbook of microalgal culture: biotechnology and applied phycology*. Blackwell Science Ltd.

Rodolfi, L., Zittelli, G. C., Bassi, N., Padovani, G., Biondi, N. and Bonini, G. (2009). Microalgae for oil: strain selection, induction of lipid synthesis and outdoor mass cultivation in a low-cost photobioreactor. *Biotechnology and Bioengineering*, *102(1)*, 100-112.

Rosenberg, J. N., Oyler, G. A., Wilkinson, L. and Betenbaugh, M. J. (2008). A green light for engineered algae: redirecting metabolism to fuel a biotechnology revolution. *Current Opinion in Biotechnology*, *19(5)*, 430-436.

Rossi, A. and Lambrou, Y. (2009). Making sustainable biofuels work for smallholder farmers and rural households - issues and perspectives, Food and Agriculture Organization of the UN (FAO).

Sanchez Miron, A., Ceron Garca, M. C., Garca Camacho, F., Molina Grima, E. and Chisti, Y. (2002). Growth and biochemical characterization of microalgal biomass produced in bubble column and airlift photobioreactors: Studies in fed-batch culture. *Enzyme Microb. Technol.*, *31*, 1015-1023.

Sa´nchez Miro´n, A., Cero´n Garcıa, M. C., Contreras Go´mez, A., Garcıa Camacho, F., Molina Grima, E. and Chisti, Y. (2003). Shear stress tolerance and biochemical characterization of Phaeodactylum tricornutum in quasi steady-state continuous culture in outdoor photobioreactors. *Biochem. Eng. J.*, *16*, 287-297.

Sancho, M. E. M., Castillo, J. M. J. and El Yousfi, F. (1999). Photoautotrophic consumption of phosphorus by *Scenedesmus obliquus* in a continuous culture. Influence of light intensity. *Process Biochemistry*, *34(8)*, 811-818.

Sawayama, S., Inoue, S., Dote, Y. and Yokoyama, S. (1995). CO_2 fixation and oil production through microalga. *Energy Convers. Manage.*, *36*, 729-731.

Sawayama, S., Minowa, T. and Yokoyama, S. Y. (1999). Possibility of renewable energy production and CO_2 mitigation by thermochemical liquefaction of microalgae. *Biomass and Bioenergy, 17(1)*, 33-39.

Sazdanoff, N. (2006). Modeling and simulation of the algae to biodiesel fuel cycle, College of Engineering, Department of Mechanical Engineering, The Ohio State University.

Schenk, P. M. (2008). Second generation biofuels: high-efficiency microalgae for biodiesel production. *Bioenergy Research Journal, 1*, 20-43.

Schneider, D. (2006). Grow your own: Would the wide spread adoption of biomass-derived transportation fuels really help the environment. *American Scientist, 94*, 408- 409.

Scott, A. and Bryner, M. (2006). Alternative Fuels: Rolling out Next-Generation Technologies. Chemical Week, December 20-27, 2006, pp. 17-21.

Scragg, A. H., Illman, A. M., Carden, A. and Shales, S. W. (2002). Growth of microalgae with increased calorific values in a tubular bioreactor. *Biomass and Bioenergy, 23(1)*, 67-73.

Sheehan, J., Dunahay, T., Benemann, J. and Roessler, P. (1998). A look back at the U.S. Department of Energy's aquatic species program: biodiesel from algae. NREL/TP-580-24190, National Renewable Energy Laboratory, USA.

Spolaore, P., Joannis-Cassan, C., Duran, E. and Isambert, A. (2006). Commercial applications of microalgae. *J. Biosci. Bioeng., 101*, 87-96.

Teixeira, C. M. and Morales, M. E. (2007). Microalga como mate´ ria-prima para a produc‚a˜o de biodiesel. Revista: Biodiesel o Novo combustı´vel do Brasil; pp. 91-96.

Thomas, F. R. (2006). Algae for liquid fuel production Oakhaven Permaculture center. Retrieved on 2006-12-18. *Permaculture Activist, 59*, 1-2.

Turkenburg, W. C. (2000). Renewable energy technologies. In: Goldemberg, J. (Ed). World Energy Assessment, Preface. United Nations Development Programme, New York, USA, pp: 219-272.

Ugwu, C. U., Aoyagi, H. and Uchiyama, H. (2008). Photobioreactors for mass cultivation of algae. *Bioresource Technology, 99(10)*, 4021-4028.

Van Harlem, T. and Oonk, H. (2006). Micro-algae Bio-fixation Processes: Applications and Potential Contributions to Greenhouse Gas Mitigation Options, Report for the International Network on Bio-fixation of CO_2 and Greenhouse Gas Abatement with Micro-algae operated under the International Agency Greenhouse Gas RandD Programme.

Vasudevan, P. T. (2008). Biodiesel production - current state of the art and challenges. *Journal of Industrial Microbiology and Biotechnology, 35*, 421-430.

Vollmann, J., Moritz, T., Karg, C., Baumgartner, S. and Wagentrist, H. (2007). Agronomic evaluation of camelina genotypes selected for seed quality characteristics. *Industrial Crops and Products, 26*, 270-277.

Wang, B., Li, Y., Wu, N. and Lan, C. Q. (2008). CO_2 bio-mitigation using microalgae. *Applied Microbiology and Biotechnology, 79(5)*, 707-718.

Wijffels, R. (2006). Energie via microbiologie: Status en toekomstperspectief voor Nederland. Utrecht, SenterNovem.

Wijffels, R. H. (2008). Potential of sponges and microalgae for marine biotechnology. *Trends Biotechnol., 26(1)*, 26-31.

Xu, H., Miao, X. and Wu, Q. (2006). High quality biodiesel production from a microalga *Chlorella protothecoides* by heterotrophic growth in fermenters. *Journal of Biotechnology*, *126*, 499-507.

Yamane, K., Ueta, A. and Shimamoto, Y. (2001). Influence of physical and chemical properties of biodiesel fuels on injection, combustion and exhaust emission characteristics in a direct injection compression ignition engine. *International Journal of Engine Research*, *2*, 249-261.

Yun, Y. S., Lee, S. B., Park, J. M., Lee, C.I. and Yang, J. W. (1997). Carbon dioxide fixation by algal cultivation using wastewater nutrients. *J. Chem. Technol. Biotechnol.*, *69*, 451-455.

Zappi, M., Hernandez, R., Sparks, D., Horne, J., Brough, M. and Swalm, D. C. (2003). A review of the engineering aspects of the biodiesel industry. MSU E-TECH Laboratory Report ET-03-003.

Zhu, C. J. and Lee, Y. K. (1997). Determination of biomass dry weight of marine microalgae. *Journal of Applied Phycology*, *9*, 189-194.

In: Microalgae: Biotechnology, Microbiology and Energy ISBN 978-1-61324-625-2
Editor: Melanie N. Johnsen © 2012 Nova Science Publishers, Inc.

Chapter 8

REAL-TIME SPECTRAL TECHNIQUES FOR THE DETECTION OF BUILDUP OF VALUABLE COMPOUNDS AND STRESS IN MICROALGAL CULTURES: IMPLICATIONS FOR BIOTECHNOLOGY

Alexei Solovchenko[1], Inna Khozin-Goldberg[2] and Olga Chivkunova[3]
[1]Biophysics, Faculty of Biology,
Moscow State University, Russia
[2]The Microalgal Biotechnology Laboratory,
French Associates Institute for Agriculture and Biotechnology of Drylands,
The Jacob Blaustein Institutes for Desert Research,
Ben-GurionUniversity of the Negev, Midreshet Ben-Gurion84990, Israel
[3]Bioengineering, Faculty of Biology, Moscow State University, Russia

ABSTRACT

Single-cell algae (microalgae) are among the most promising resources for the production of biofuels and bioactive compounds, as well as for CO_2 biomitigation and bioremediation. Improvement of microalgal photobiotechnologies for the production of value-added products such as long-chain polyunsaturated fatty acids, storage triacylglycerols and carotenoids, requires fast and reliable, and preferably non-destructive techniques for on-line monitoring of the target product's content and the physiological condition of the algal culture. These techniques can provide essential information for timely and informed decisions on adjusting illumination conditions and medium composition, and on the optimal time for biomass harvesting. Often, such decisions must be taken within hours, and mistakes can lead to a significant reduction in productivity or a total loss of the culture. A promising approach for real-time non-destructive monitoring of laboratory and upscaled microalgal cultures is based on measuring the optical properties of algal suspensions, such as absorption, scattering and reflection of light by microalgal cells in certain spectral regions. To this aim, the following criteria should be met: i) reliable spectral measurements, ii) efficient algorithms for the processing of spectral data, and iii) a thorough understanding of the relationships between changes in physiological condition and/or biochemical composition of the algal culture and accompanying changes in its optical properties. This chapter presents a review of recent experimental work in this area, with an emphasis on investigations conducted by the authors and their colleagues in the fields of physiology, biochemistry and spectroscopy which have implications for the cultivation of biotechnologically important microalgal species.

ABBREVIATIONS

AA—arachidonic acid;
Car—carotenoid(s);
Chl—chlorophyll(s);
DGLA—dihomo-γ-linolenic acid;
DMSO—dimethyl sulfoxide;
(T)FA—(total) fatty acid(s);
FTIR—Fourier Transform Infrared;
FC—flow cytometry;
GC—gas chromatography;
IS—integrating sphere(s);
NIR—near-infrared;
NMR—nuclear magnetic resonance;
NR—Nile Red;
PLS—partial least squares (regression);
TAG—triacylglycerol(s).

INTRODUCTION

In recent years, the world community of researchers and commercial ventureshave shown a rapidly growing interest in microalgal biotechnology due, to a large extent, to an urgent need for the development of novel renewable oil resources for next-generation biofuels such as biodiesel [1-3]. Biodiesel is currently produced by transesterification of neutral storage lipids (mainly triacylglycerols—TAG) obtained mainly from non-food (non-edible) oil-seed plants [4]. However, oleaginous microorganisms such as bacteria, yeasts [5], and microalgae [1-3], have been recently recognized as an important and promising oil feedstock. Microalgal oil production is considered a valuable alternative to higher plant oil for several reasons: i) microlagae can be grown on diverse water sources, including sea and brackish water; ii) their biomass production does not compete with higher plants for arable lands and, in the case of marine species, for water resources; iii) microalgal cultures are superior to higher plants in biomass-accumulation rates (for details, refer to [2] and refs. therein).Hence, oleaginous microalgal species are being intensively studied in respect to their tolerance to adverse environmental conditions and to further develop sustainable cultivation processes which will enable rapid growth rates along with high oil productivity and. It should be noted, however, that several technological difficulties need to be overcome to reduce the energy overhead of phototrophic microalgal cultivation and biomass harvesting, toward a cost-efficient process [6]. Microalgae also constitute a valuable source for a wide array of value-added "health" products such as carotenoids (Car) [7] and essential long-chain polyunsaturated fatty acids [8], for both aquaculture and human nutrition.The demand for lipophilic microalgal products is expected to increase in the decades to come, emphasizing the need to develop more advanced processes for their production.

Upscaling of microalgal photobiotechnology for the efficient production of microalgal oil and value-added products requires, in particular, a fast and accurate method to quantify the target compound(s) in the biomass. Ideally, such a method should support on-line monitoring of the target product's content and the physiological state of the algal culture:information that is essential for timely and informed decisions on adjusting illumination conditions and medium composition, and on the optimal time for biomass harvesting. Such decisions must often be taken within hours and mistakes may lead to a significant reduction in productivity or in total loss of the culture [9]. However, the mainstream traditional methods, based on solvent extraction with chloroform:methanol mixtures [10] or the non-polar solvent *n*-hexane, followed by gravimetrical yield determinations, are laborious and time-consuming, and of insufficient accuracy for an estimation of lipid content. Moreover, lipid extracts carry additional extractable compounds, aside from TAG. Determination of the total fatty acid (TFA) content by gas chromatography (GC) in combination with flame-ionization detector (GC-FID) or mass spectrometry (GC-MS) is precise but also time-consuming and expensive, and therefore less suitable for express analyses of biomass composition.

In view of these complications, more attention has recently been paid to advances in rapid *in-situ* techniques for neutral lipid assay in microalgal biomass, including those based on (vital) fluorescent staining [11-13], and optical [14-16], infrared (IR) [17, 18] or other spectrometric approaches. These techniques provide a number of distinct advantages over traditional extraction-based methods allowing a rapid non-destructive assay of valued compounds in the microalgal biomass. More importantly, these techniques might be applicable for on-line monitoring of biomass quality and culture conditions in photobioreactors. The demand for modern biotechnological solutions based on cultivation of microalgae makes these novel approaches increasingly important as substitutes for the traditional techniques of neutral lipid quantification. The recent advances in non-destructive methods for the analysis of microalgae are reviewed below, with a particular emphasis on optical spectroscopy-based approaches investigated and developed by the authors.

FLUORESCENT-STAINING-BASED NEUTRAL LIPID ASSAY

This section provides a brief review of the non-destructive methods for neutral lipid quantification in microalgal cells based on their vital staining with fluorescent dyes. One common approach utilizes a lipid-soluble fluorescent dye, Nile Red (NR, 9-diethylamino-5H-benzo[α]phenoxazine-5-one). In hydrophobic environments, such as non-polar organic solvents and oil-storing lipid droplets, NR is generally applied to samples of algal suspensions as a stock solution in acetone, methanol or dimethyl sulfoxide (DMSO); it stains the neutral lipid-containingintracellular inclusions and emits bright-yellow fluorescence upon excitation at 490 nm [11, 19-21]. NR staining allows for visualization of oil droplets in the cells for microscopic observation of neutral lipid formation [11], as well as for the development of rapid screening, detection and quantification methods for TAG production in microalgal cells [12, 13]. Several algorithms and experimental procedures have been developed for a variety of microalgal species, including members of the Chlorophyceae—*Botryococcusbraunii*, *Chlorellavulgaris*, *Chlorella zofingensis*, *Pseudochlorococcum* sp. [12, 22], the Eustigmatophyceae—*Nannochloropsis* sp. [11], the Haptophyceae—*Isocrysisgalbana* and *Emilianahuxleyi*, and the Prasinophyceae—*Tetraselmissuecica*[23]. The intensity of the NR fluorescence generally correlates with the gravimetrical determination of biomass or cell lipid content [11-13]. A recent modification of the method was introduced by Chen et al. [13], toward developing a high-throughput assay to quantify lipid content in algal samples.

A major limitation of the NR-staining method, in our opinion, stems from the species-dependent and uneven permeation of NR solution via the algal cell wall. Furthermore, algal species with a strong cell wall, containing the solvent- and acid-resistant polymer sporopollenin (a common feature of green microalgae), require an additional heating step or even microwave-assisted treatment to achieve homogeneous and efficient staining [12]. Another drawback is the quenching of NR fluorescence at higher than 5% final concentrations of DMSO, while DMSO is often used for lipid extraction from green algae with a strong cell wall, many of which are regarded as candidate strains for biofuel production. Protocols have been recently reported that further attempt to improve NR staining of live cells, permitting efficient vital staining of algal cells for flow cytometry (FC) and single-cell sorting [24]. For example, it was recently claimed that a combination of NR staining and FC was recently used to select and isolate cells with high lipid content from wild populations of the green microalga*Tetraselmissuecica*[23].

IR- AND NUCLEAR MAGNETIC RESONANCE (NMR)-BASED METHODS

As already noted, rapidly growing interest in neutral lipid production by microalgae has paved the way for the use of sophisticated spectrometric methods to quantify TAG production [17, 18, 25, 26]. For instance, Raman spectroscopy was adopted to detect TAG formation in microalgaeand specific Raman signals were identified in the nitrogen-starved cellsof *Chlorella sorokiniana* and *Neochlorisoleoabundans* due to the presence of storage TAG [27].NMR is considered a useful diagnostic methodthat is complementary to common analytical tools for selective TAG identification in algal cells [26, 28]; a ^{13}C liquid-state NMR approach was shown to have potential in the detection of TAG formation in cells of *N. oleoabundans* induced by nitrogen starvation.

Near-infrared (NIR) and Fourier transform infrared (FTIR) spectroscopy show promise as rapid and relatively inexpensive methods of monitoring neutral lipid content in microalgal cells. Dean et al. [17] demonstrated the use of FTIR to estimate changes in lipid and carbohydrate contents of two freshwater microalgae (Chlamydomonasreinhardtii and Scenedesmussubspicatus), grown under nitrogen starvation. The authors reported significant correlations between the FTIR- and NR-based lipid measurements. A recent work by Laurens and Wolfrum[18] demonstrated the potential use of NIR and FTIR spectroscopic fingerprinting of algal biomass to predict lipid content and composition. The authors attempted to build prediction models able to quantify the lipid content in algal biomass and distinguish between TAG and phospholipid contents. They analyzed the NIR spectra of homogenized, dried biomass of four representative classes of microalgae: Chlorophyceae, Eustigmatophyceae, Bacillariophyceae, and the blue-green algae (Cyanobacteria). By using a chemometric approach—partial least squares (PLS) regression method—on biomass samples spiked with exogenous lipids, they found that TAG and phospholipids have sufficiently different NIR spectral fingerprints to contribute independently to a PLS2 calibration model. The single-species calibration models accurately predicted spiked levels of both types of lipids, and the combined models were robust in their prediction across the different biomass samples. However, these experiments and conclusions were based on lipids that were exogenously added to algal biomass. Hence, before these models and methods can be used on a routine basis for strain screening or monitoring lipid production in cultures, they must be further developed, and a robust calibration performed, before they can be applied to unknown samples.

OPTICAL SPECTROSCOPY-BASED METHODS

Optical methods provide a quantitative description of the interaction between optical radiation and cells of cultivated algae. This information is essential for understanding utilization of light energy in photosynthesis [29], estimating their physiological condition and productivity, detecting stress conditions [30, 31], and monitoring their growth and the accumulation of value-added compounds during mass cultivation, particularly in real time [14, 16, 32].

Development of optical spectroscopy-based methods for the non-destructive monitoring of algal cultures requires a deep understanding of the relationships between changes in light absorption by algal cells and the underlying dynamics of their pigment and lipid contents. The spectral analysis of photosynthesizing microorganisms presumes obtaining information on both radiation absorption by different pigments and their light-scattering characteristics *in vivo*[33, 34]. An elaborated methodology of optical measurements and their interpretation, including remote-sensing applications [35-37], has been developed for plankton microalgae [38, 39] however, approaches to the optical monitoring of cultivated microalgae appear to be somewhat less studied, although there are a number of notable exceptions (see e.g. [9, 40]). In these latter studies, the focus was on light absorption and, less frequently, scattering, to follow both qualitative and quantitative changes in pigment content [14, 16, 41].

Absorption spectrophotometry in the visible range is widely employed in the analysis of microalgal cultures. The basic relationship between the concentration of the absorber and the attenuation of light by its solution is defined by the Lambert-Beer law:

$$A = -\log \frac{I}{I_0} = -\log(T) = \varepsilon \cdot c \cdot l \quad , \tag{1}$$

where I_0 denotes the intensity of the incident measuring beam, I is the intensity of the beam passed through the sample and attenuated by it, T is the transmittance of the sample, ε is an extinction coefficient that depends on wavelength and the substance under investigation, c is the concentration of the light-absorbing compound, l is optical path length, and A is optical density or absorbance. A plot of A or T as a function of wavelength, λ, is called absorbance or transmittance spectrum, $A(\lambda)$ or $T(\lambda)$, respectively. Valuable information can be deduced from the sample's spectral characteristics, such as position, half-width and amplitude of its absorption maxima, etc. However, the Lambert-Beer law works only if i) the chromophore exists in the form of a molecular or true solution and ii) the basic assumption of independent light absorption by each chromophore is not violated, and this is not the case with microalgal cultures (which are essentially suspensions of pigment-containing cells). The true chromophore absorption is normally of primary importance for qualitative and quantitative analyses of microalgal cultures via optical methods. On the other hand, valuable information can, in many cases, be extracted from the scattering spectra of algal culture samples [33, 34]. Therefore, one needs a method to decrease or at least estimate the influence of light scattering on algal spectra. One of the methods for deconvolution of the spectra of a microalgal cell suspension into contributions by 'true' absorbance and scattering is discussed in the following section.

CORRECTION OF MICROALGAL CELL SUSPENSION SPECTRA FOR LIGHT SCATTERING

In samples representing true or molecular solutions, such as diluted pigment extracts (such as chlorophylls (Chl) and Car) in a non-polar solvent, all chromophore particles (represented by molecules) absorb light independently of each other. However, in comparison to pigment solutions, a suspension of microalgal cells is a dramatically different and optically more complex system. In particular, apart from the absorption by pigments, its optical properties are considerably influenced by the cells' scattering of light. As a result, in a spectrophotometer, many of the photons of the measuring beam are diverted from their initial path and can therefore miss the detector (Figure 1; see also [34]). This phenomenon appears as an increase in the optical density of the sample. Then, in dense suspensions containing a large number of cells, multiple scattering is possible, which also leads to an increase in the effective optical path length and hence to an additional increase in measured optical density (for additional details on the phenomenon of optical path lengthening due to scattering see [42] and references therein).

Generally, the cells of microorganisms, including microalgae, display featureless non-selective scattering which depends on the size and shape of the particles and results from the changes in the refractive index at the partition boundaries between phases (medium, cell wall, cytosol, intracellular structures, etc.[38]). In cyanobacteria and microalgae, the spectra are also characterized by the presence of selective scattering with pronounced spectral features, related to sharp changes in the refractive index during light quanta absorption by aggregated pigments incorporated in the thylakoid membranes [33, 42-44]. Scattering-related phenomena can significantly distort spectral curves.

First, due to the loss of measuring light (see Figure 1), the spectral curve appears to rise above the baseline. This rise is more pronounced in the blue part of the spectrum since scattering increases with decreasing wavelength.

Second, the regions of weak absorption are more affected by the apparent increase in optical density. This leads to flattening and broadening of the peaks in the spectra of algal cell suspensions [33, 34].

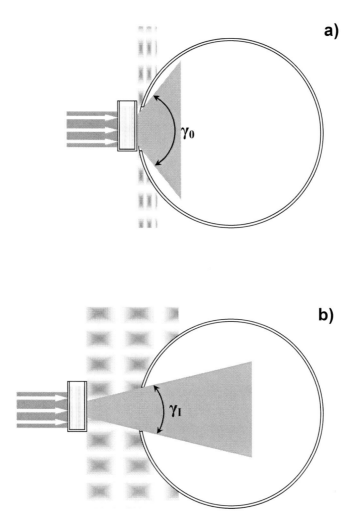

Figure 1. A scheme of light collection by an integrating sphere with a scattering sample placed close to (*a*) and at a distance from (*b*) the integrating sphere. For simplicity, only the light leaving the rear cuvette surface is shown. With kind permission from Springer Science+Business Media: Russian Journal of Plant Physiology. Light absorption and scattering by cell suspensions of some cyanobacteria and microalgae. V. 55 (3). 2008. Merzlyak et al. 421. Figure 1.

In biological systems such as eukaryotic microalgal cells, light-absorbing compounds are characterized by non-uniform distribution. For example, photosynthetic pigments do not occupy the whole volume of the cell but are localized predominantly to corresponding cell structures (chloroplast thylakoid membranes). The beams of light that hit these pigment-containing structures as they travel through the sample are strongly attenuated; at the same time, the beams that miss these structures pass through the sample virtually unaffected. This phenomenon is generally known as 'sieve' or 'packaging' effect [45]. The 'package' effect is especially pronounced in photosynthesizing microorganisms, with their high pigment content and cell dimensions that are comparable to the wavelength of the measuring light [46-48]. It should be noted that the 'package' effect causes a decrease in the optical density of the bands with strong absorption relative to a similar sample unaffected by packaging [49].

Due to these phenomena and some other circumstances, pigment absorption in cellsuspensions differs considerably from that in true solutions. Moreover, variable wavelength-dependent contributions of scattering, both non-selective and selective, introduce considerable uncertainty into the measurements of optical properties of microalgal suspensions. To estimate and/or compensate for the effect of scattering, a number of approaches have been developed based on different principles.

A common method aimed to reduce the contribution of scattering is based on the use of integrating spheres (IS) with a cuvette placed at the IS-input window (Figure 1). A simple method for measuring turbid biological samples is the technique developed by Shibata [50] based on the use of a diffuser such as opal glass which strongly scatters light. A variation of this technique, using a wet glass-fiber tissue, was applied by Solovchenko et al. [16] to obtain reliable spectra of microalgal suspensions in a simple spectrophotometer lacking IS. Other approaches involve microspectrophotometry[51, 52] and integrating cavity absorption measurements [53].

Despite these efforts, with strongly scattering samples, such configurations do not completely eliminate scattering-dependent light loss [34, 54]. As shown further on, the influence of scattering can be decreased (but not completely eliminated) using spectrophotometers fitted with IS, but these are expensive and not readily available for routine analysis. Moreover, data obtained by Merzlyak et al. [33] suggest that even with the use of IS, the contribution of scattering to overall light attenuation can be considerable and the loss in light should be taken into account in quantifying pigment absorption. At the same time, measurements in the 750 to 800 nm range are often ignored and the attenuation spectra, obtained in different laboratories are dependent on the geometry and efficiency of light collection by the IS employed in the particular experiment, preventing direct comparisons [34].

With this in mind, a method that allows approaching the "true" absorption spectrum of cell suspensions was suggested [34, 55]: the spectra are measured in the cuvette placed as close as possible to, and at a certain distance from the IS, i.e. at larger and smaller solid angles of light collection, γ_0 and γ_1, respectively (Figure 1). Briefly, when $\gamma_0 \gg \gamma_1$, the use of attenuation as a correcting factor in the spectral region in which pigment light absorption is negligible makes it possible to obtain the spectrum $\tilde{A}(\lambda) \equiv \tilde{A}(\lambda;\pi)$ as an approximation of the light leaving the rear wall of the cuvette. This simple approach made it possible to obtain essentially scatter-free $\tilde{A}(\lambda)$ spectra of a considerable number of taxonomically different biotechnologically important microalgae: the cyanobacterium *Anabaena variabilis*[56], the green algae *Parietochlorisincisa*[41] and *Haematococcuspluvialis*[57], the diatom *Thalassiosiraweisflogii*[58] and some other species [59]. Absorption spectra compensated for scattering were calculated as

$$\tilde{A}(\lambda) = D(\lambda;\gamma_I) - \frac{D_{NIR}^{\gamma_I}}{D_{NIR}^{\gamma_I} - D_{NIR}^{\gamma_0}} \times \left[D(\lambda;\gamma_I) - D(\lambda;\gamma_0)\right]$$

(2)

where D_{NIR} is light attenuation in the spectral range 760 to 800 nm, in which pigments do not possess measurable absorption (see [34] for more details). This approach allows obtaining spectra of microalgal suspensions that are essentially free of the influence of scattering. As can be seen e.g. in the spectra presented in [34], even in the cuvette position close to IS (γ_0) in the NIR, scattering-related losses of light occurred. The resulting spectra were essentially flat and close to the baseline in the NIR region where pigment absorption is nearly absent, and the peaks attributable to Car and Chl were resolved. The use of scatter-free absorption spectra opens up many possibilities for reliable monitoring of the physiological condition of microalgal cultures (Figure 2). For example, marked changes in the shapes of the spectra of nitrogen-deficient *P. incisa* cell suspensions were recorded in experiments carried out by Merzlyak et al. [41] (Figure 3). It is important to stress that such an analysis of stress-induced changes in the fine structure of microalgal cell suspension spectra became possible only after scattering compensation, which removed the large uncertainty related with changes in size and refractive properties characteristic of cells cultivated under stressful conditions. It should be noted, however, that the scattering signal also carries important information on cell size, shape and number. After calculation of the scattering-compensated $\tilde{A}(\lambda)$ spectrum, spectral contribution of scattering to a given measured spectrum $D(\lambda;\gamma)$ could be easily deduced from the latter:

$$S(\lambda;\gamma) = D(\lambda;\gamma) - \tilde{A}(\lambda;\gamma) \qquad (3)$$

Spectra could be corrected for the 'package' effect using an approach based on Duysens' theory of spectrum flattening [48]. It should be emphasized that, strictly speaking, Duysens' treatment can be applied only if the clusters of pigmented particles are all identical and multiple scattering is absent. At the same time, the clusters of microalgal cells within a suspension can hardly be of the same shape and size. Nevertheless, the results of the study by Merzlyak et al. [42] indicated that Duysens' relations for sieving, though derived for systems in which multiple scattering is absent, could still be used to compare the spectrum of a leaf with that of a suspension of isolated chloroplasts. Taking into account this finding, as well as the original work by Duysens[48] with *Chlorella* suspensions, one can speculate that this approach might be applicable for correction of the sieving effect in other microalgal species.

DEPOSITION OF MICROALGAL CELLS ON FILTERS

There are a number of cases in which recording reliable spectra of microalgal suspensions poses certain difficulties. For instance, large cells of certain microalgal species are characterized by rapid sedimentation, as with stressed *P. incisa* and *H. pluvialis* cells. As a result, the number of cells in the light path can change in the course of the spectrum scanning, making it impossible to obtain a reliable spectrum via traditional measurements in cuvettes. Another problem can occur at the initial stages of cultivation, when cell density might be too low for reliable spectrum recording.

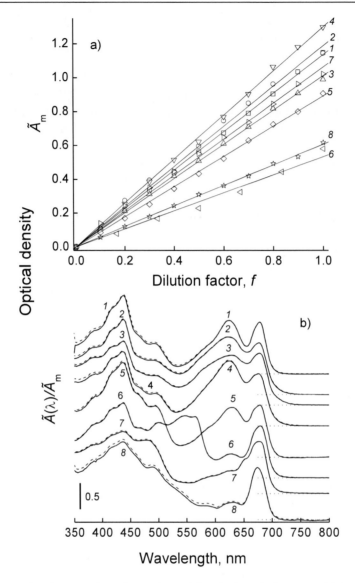

Figure 2. Compensation for scattering of cyanobacterial and microalgal spectra. (*a*) Dependence of optical density in the red chlorophyll *a* absorption maximum ($\tilde{A}_m$, m = 674-678 nm) on dilution factor. (*b*) Absorption spectra normalized to the chlorophyll *a* absorption maximum. Solid lines – average of the spectra for dilutions presented in (*a*); broken lines – average + standard deviation. *1 – Anacystisnidulans, 2 – Anabaena variabilis, 3–Chlorogloeopsisfritschii, 4 – Cyanidiumcaldarum, 5 – Porphyridiumcruentum* 520, *6 – Porphyridiumcruentum* 273, *7– Dunaliellamaritima, 8 – Thalassiosiraweisflogii*. Note that after compensating for scattering, the $\tilde{A}(\lambda)$ curves in the near-infrared region are flat and absorption is close to zero and in a wide dynamic range, and $\tilde{A}$ in the red Chl*a* maximum is strictly proportional ($r^2 \approx 0.99$) to the dilution factor (cell density) for all species examined. With kind permission from Springer Science+Business Media: Russian Journal of Plant Physiology. Light absorption and scattering by cell suspensions of some cyanobacteria and microalgae. V. 55 (3). 2008. Merzlyak et al. 424. Figure 5.

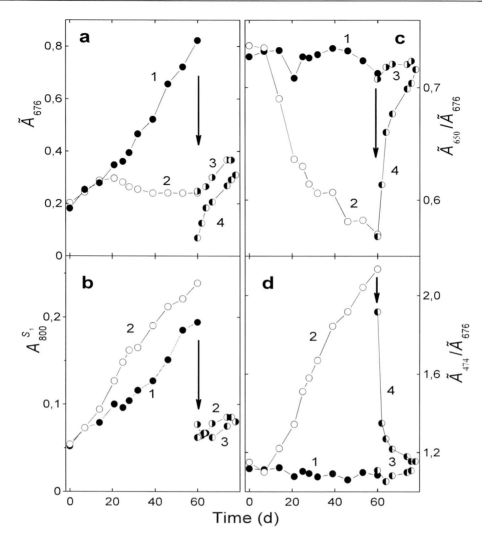

Figure 3. Time course of spectral absorption (*a*, *c* and *d*) and scattering (*b*) changes for *P. incisa* (+N) and (–N) cell suspensions (1 and 2, closed and open symbols, respectively) at an irradiance of 1.24 W m^{-2}. After 60 d (shown by arrows), the cultures were nitrogen-replenished (curves 3 and 4, shown using semi-closed symbols). Reprinted from (41) with kind permission from John Wiley and Sons.

In the above situations, a potentially effective way of circumventing the problem is deposition of algal cells on glass-fiber filters, such as Whatman GF/F, prior to their measurement. In this case, the wet filters are mounted as overheads (deposited cells facing the detector) on the output window of the spectrophotometer's cuvette compartment. The deposition technique was initially suggested by Mitchell [39] for plankton microalgae. Solovchenko et al. successfully used this approach to record the spectra of cultivated microalgae such as *P. incisa*[14, 16] and *Nannochloropsis* sp. [15]. Nevertheless, there is some question as to the relationship between the optical characteristics of algal cells suspended in media and those deposited on a filter. According to Mitchell's work [39] which is extensively used in practice, the relationship between the optical densities of certain green and diatom algae, as well as cyanobacteria, in suspension and on a filter is non-linear and can be empirically described using the quadratic equation

$$A_s(\lambda) = \alpha \cdot [-\lg T(\lambda)] + \beta \cdot [-\lg T(\lambda)]^2, \qquad (4)$$

where is the optical density of the cell suspension and T is the transmittance of the filter carrying the deposited cells. As already mentioned, this approach allowed reliable recording of *Nannochloropsis* sp. suspension spectra with a simple spectrophotometer lacking IS [15]. It should be noted that compensation for scattering is often not required for spectra obtained with the use of the GF/F filter-based approach (though it is essential for obtaining scattering-free spectra via measurements in cuvettes). The shape of the spectra and the ratio of the maxima in the red and blue regions of the spectra recorded using this technique appear to be close to those of spectra measured with the use of an IS [41].

To date, the transmittance-based approach has been predominantly employed to analyze plankton microalgae deposited on filters [39]. These measurements do not take into account light losses due to reflection (backscattering) by the filter. Tassan and Ferrari [60] described a modification of the light-transmission method that corrects for backscattering. This technique combines light-transmission (T) and light-reflection (R) measurements, carried out using an IS attached to a dual-beam spectrophotometer (for additional details, see [60, 61]. On the other hand, according to our recent findings, the analysis of optical properties of microalgae deposited on filters can be carried out solely via reflectance measurements (Solovchenko and Merzlyak, unpublished). These results are in agreement with the theory of diffuse reflectance developed by Atherton and co-workers [62, 63], according to which the following relationship holds for strongly scattering media such as glass-fiber filters:

$$\frac{1}{R(\lambda)} - \frac{1}{R_0(\lambda)} = \sum c_i \cdot \varepsilon_i(\lambda) \qquad (5)$$

where R_0 is the reflectance of the background (a blank glass-fiber filter in this case), and c is the concentration and ε is the absorption coefficient of an absorbent (pigment) contained in the cells.

Indeed, the relationship A_s vs. reciprocal reflectance ($1/R$) of filters carrying algal cells was close to linearity with the contribution of the quadratic term (coefficient β in Eq. 4) being 20 times lower than α. At the same time, the dynamic range of $1/R$ values was ca. one order of magnitude higher than that of transmittance-based optical parameters. It should be noted that the reciprocal reflectance-based approach has been successfully used in the non-destructive quantification of pigments in higher plants [64-66].

OPTICAL SENSING OF MICROALGAL CULTURE CONDITION

Biomass Accumulation

The afore-described method of scatter-correction of absorption spectra (Eq. 2; see also [33, 34]) has been successfully applied to a number of prokaryotic and eukaryotic microalgae, including those of biotechnological importance, differing in cell size and shape as well as in pigment composition and localization (see e.g. Figure 2). In a wide dynamic range, $\tilde{A}$ in the red Chl$a$ maximum increased linearly ($r^2 \approx 0.99$) with an increase in the dilution factor (which is proportional to cell density) for all species examined (Figure 2). These relationships also held during cultivation of *Nannochloropsis*sp. under stressful conditions [15]. All of these observations indicate that the $\tilde{A}(\lambda)$ characteristics of cyanobacterial and microalgal cell suspensions are independent of the quantity of absorbing and scattering particles. Therefore, the scattering-corrected absorption spectra in the NIR can be employed to monitor the growth of microalgae, as apparent from an increase in volumetric pigment (mainly Chl) content in the cell suspension. It should be noted that the attenuation of light due to scattering, after its separation from the contribution of absorption by the algal pigments (Eq. 3), appears to be an efficient indicator of cell number in culture. Simultaneous monitoring of the changes in absorption by Chl and attenuation of light due to scattering makes it possible to follow the growth of the culture, based on cell density (biomass accumulation) and increasing pigment content, on-line using a single spectral channel.

Coordinated Buildup of Secondary Carotenoids and Storage Lipids Under Stress

An increase in the Car/Chl ratio under various stresses, including nitrogen starvation and high light, is characteristic of many microalgal species (see [14, 15, 67, 68] and references therein). Within the thylakoid membranes, the protection afforded by Car is related to deactivation of the Chl's excited state, quenching of singlet oxygen, interception of free radicals and dissipation of the excess absorbed light [69, 70]. In addition, extra-thylakoid Car in oil bodies provide photoprotection by trapping harmful radiation in the blue range [14, 71, 72]. It should be noted that the induction of carotenogenesis is more strongly expressed in nitrogen-deficient algal cells under high irradiance: in this case, the molar Car content exceeds that of Chl and the Car contribute strongly to absorption between 400 and 500 nm [14, 15].

An increase in TFA as percentage of dry weight (an approximate measure of total lipid, mainly TAG, content) in stressed microalgae often occurs on the background of a marked decline in Chl and a small decline (*Nannochloropsis*sp.[15]) or increase (*P. incisa*[16]) in Car, resulting in a gross increase in the Car/Chl ratio (Figure 4 and Figure 5). This observation is in line with numerous reports of declining Chl levels under stressful conditions in many algal species, including those from the genus *Nannochloropsis*[73, 74], obviously indicating photoacclimation[75]. The stress-induced increase in Car characteristic of many microalgal species, such as chlorophytes from the genera *Dunaliella*[67, 76-80], *Haematococcus*[7, 81-85] and *Parietochloris*[16, 41], was not detected in *Nannochloropsis*sp. under the experimental conditions used [15].

An important consequence of the ultimate linkage between carbon and nitrogen metabolism [86, 87] is the existence of a tight correlation between the stress-induced increase in the Car/Chl ratio and a buildup of TFA percentage (Figure 6; $r^2 > 0.80$). One might assume that predominant channeling of photosynthates to storage lipids, together with Car accumulation and reduction of Chl content, play a protective role under nitrogen starvation and the excess irradiation stress that renders microalgae prone to photooxidative damage.

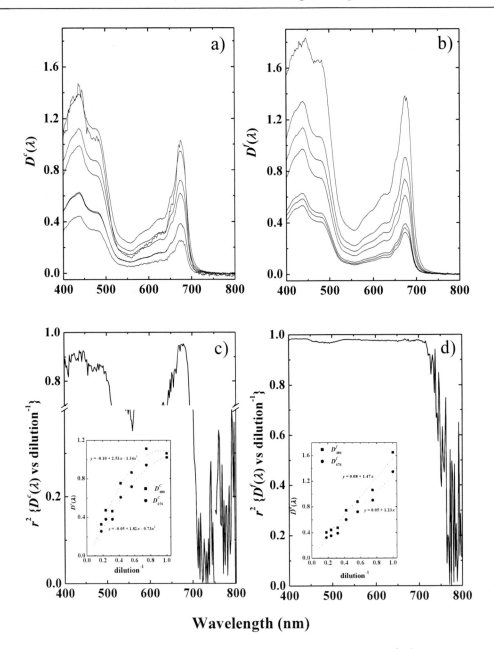

Figure 4. Absorbance spectra of *Nannochloropsis* sp. culture (75 μmol photons m^{-2} s^{-1} PAR, complete f/2 medium) measured (*a*) in suspension or (*b*) on GF/F filters, as well as their relationship with volumetric chlorophyll content (*c* and *d*, respectively). Changes in optical properties, pigment and fatty acid content of *Nannochloropsis* sp.: implications for non-destructive assay of total fatty acids. 2011. Solovchenko et al. DOI 10.1007/s10126-010-9323-x Figure 1.

Accordingly, the Car/Chl ratio per se gives a good indication of the physiological status of many biotechnologically important algal species, such as *Nannochloropsis*[88], *Dunaliella bardawil* (*D. salina*) [68, 79, 89, 90], and *H. pluvialis*[7, 84, 91]. It was also found that the increase in Car/Chl ratio in the unique microalga *P. incisa,* which accumulates long-chain polyunsaturated FA in TAG reserves, is closely related with biosynthesis and accumulation of certain specific FA such as arachidonic acid (AA) in *P. incisa*[16, 92] or dihomo-γ-linolenic acid (DGLA) in its mutant P127 [14].

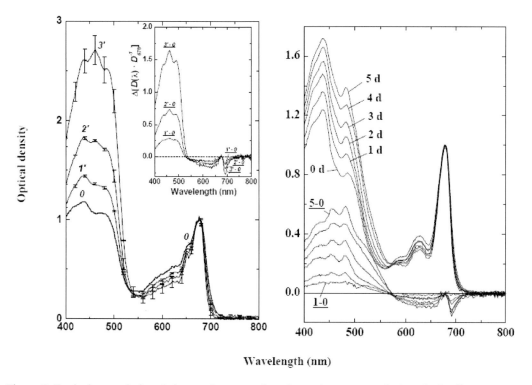

Figure 5. Typical stress-induced changes in scatter-free absorption spectra of microalgal cell suspensions. Note the three-band signature of carotenoids on the differential spectra. Left panel: Spectra of optical density of P127 cell inoculum deposited on GF/F glass-fiber filters (*0*) and from cultures grown for 14 d on nitrogen-free media at low (*1*, *1'*), medium (*2*, *2'*) or high (*3*, *3'*) irradiance. Inset: corresponding difference spectra. Reprinted from (14) with kind permission from John Wiley and Sons. Right panel: Spectral changes, normalized to the red chlorophyll maximum absorption spectra, of *Nannochloropsis*sp. cultures grown under stressful conditions and corresponding difference spectra (the culture age is indicated on the figure). The spectra were recorded after deposition of the cells on GF/F filters. Changes in optical properties, pigment and fatty acid content of *Nannochloropsis* sp.: implications for non-destructive assay of total fatty acids. 2011. Solovchenko et al. DOI 10.1007/s10126-010-9323-x Figure 5.

Development of Algorithms for Non-Destructive Assay of TFA

The changes in algal metabolism, both developmental and stress-induced, are often accompanied by specific changes in pigment content and composition, such as the abovementioned dramatic shift of Car/Chl ratio linked with storage lipid accumulation under nitrogen-deficiency stress. Such changes subsequently affect the optical properties of algal cells and cell suspensions [16, 41, 47, 93].

In many green algae, the nitrogen-deficiency-induced decrease in Chl content likely affects light absorption and utilization by these organisms [29]. As in some other species of microalgae, during nitrogen deficiency, no significant changes in Chl*a/b* ratio were found in *P. incisa*. This is in keeping with observations that, during nitrogen starvation, the composition of the light-harvesting pigment-protein complex is conserved [29]. At the same time, one of the spectral regions sensitive to and closely correlated with the changes in Chl content and hence FA content is usually situated in the orange-red region, dominated solely by Chl[14-16]. A plausible explanation for the close relationship between changes in FA percentage and decreasing cellular Chl content involves the afore-mentioned 'package' effect: the decrease in Chl content brings about a profound weakening of this effect, seen as a progressive narrowing of the red Chl absorption peak in the course of a decrease in Chl, similar to that found in *P. incisa*[41], see also Figure 5). Thus, the normalized

spectrum exhibited a strong negative correlation with Car/Chl ratio (not shown), TFA and DGLA contents in the region governed solely by Chl absorption (Figure 6). A similar effect was found in *Dunaliellatertiolecta*[93], and in *P. incisa* grown under low [41] and high [16] light.

A close positive correlation between absorbance of cell suspensions, Car/Chl ratio and FA content was observed in a broad band between 400 and 500 nm in which the contribution of Car to light absorption progressively increased with duration of nitrogen starvation in a number of microalgal species. In our recent work, this signature of Car absorption was detected in the different spectra of stressed cultures of *P. incisa*[16] and its Δ5 desaturase-deficient mutant [14], *Nannochloropsis*sp. [15] and *H. pluvialis*[57].

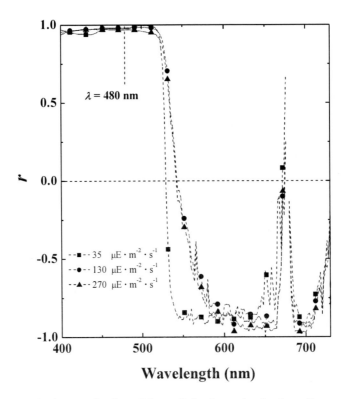

Figure 5. The increase in the contribution of Car to light absorption by the cell suspension is evident on spectra normalized to the Chl red maximum as a three-band signature characteristic of xanthophylls and carotenes.

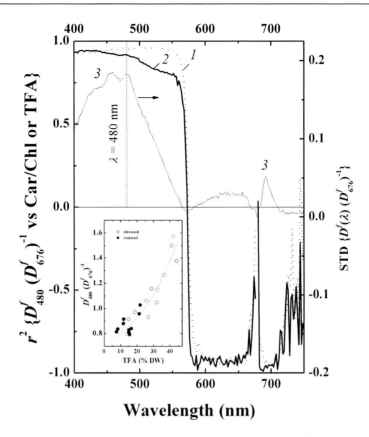

Figure 6. The relationships between scatter-free absorbance of microalgal cell suspensions grown under stressful conditions and their fatty acid content. Left panel: Spectra of correlation coefficient between total fatty acid content and spectral absorption, normalized to the red chlorophyll maximum, of *Parietochlorisincisa* mutant P127 cells grown on nitrogen-free medium under the PAR irradiances specified on the figureand deposited on glass-fiber (GF/F) filters. Adapted from (14)] with kind permission from John Wiley and Sons. Right panel: Spectra of correlation coefficient between carotenoid-to-chlorophyll ratio (curve 1) or total fatty acid (%DW, curve 2) and OD spectra, normalized to the red chlorophyll absorption maximum, of the *Nannochloropsis*sp. cells deposited on GF/F filters. Curve 3 – Standard deviation spectrum, revealing the spectral range influenced by carotenoid absorption. Spectra 1–3 were calculated for all samples studied. Inset: Relationships between the index $D^f_{480} \cdot D_{676}^{-1}$ and total fatty acid dry weight percentage. With kind permission from Springer Science+Business Media: Marine biotechnology. changes in optical properties, pigment and fatty acid content of *nannochloropsis* sp.: implications for non-destructive assay of total fatty acids. 2011. Solovchenko et al. DOI 10.1007/s10126-010-9323-x Figure 6.

Interestingly, the absorption changes, mimicking the consequences of accumulation of high amounts of Car, could even be detected in non-carotenogenic microalgae such as *Nannochloropsis*sp. during a considerable decline in Chl[15]. In this case, the high correlation between FA content and the normalized absorption in the blue-green region was retained (Figure 6).

It should be emphasized that the correlation between extensive accumulation of Car (and hence TFA content) and the changes in spectral absorption in the blue region of the spectrum was observed only in algae grown under high irradiances (see e.g.Figure 6). Interestingly, growth under low-light conditions in the absence of nitrogen also induced an increase in the relative contribution of Car to *P. incisa* absorption in the blue region, but this was due to a pronounced decline in Chl content. In this case, only a weak correlation was found between TFA or AA content and absorption in the blue-green region; TFA content was much more closely related to the spectral changes in the red region, which were ascribed to a change in the so-called 'package' effect (see above). This could, at least in part, explain the existence of the correlation peak near 725 nm.

The above findings were employed to develop algorithms for a non-destructive assay of the contents of TFA and certain specific FA such as AA or DGLA, constituents of TAG in *P. incisa*. The ratio of absorption in the blue region (460-520 nm) to that in the red region (at the long-wave Chl absorption maximum) correlated closely with (T)FA dry weight percentages and volumetric contents, and was suggested as a proxy for the FA assay. Several algorithms for non-destructive estimation of FA under a broad range of illumination intensities and conditions of nitrogen availability were developed (Table 1).

Table 1. Algorithms for the assay of fatty acids in microalgal cell cultures via absorbance of the cells deposited on glass-fiber filter

Species	Growth conditions	Estimated parameter	Range	Estimation algorithm	RMSE	r^2	Reference
Parietochlorisinc isa (wild type)	N-free BG-11; 350 µE PAR	TFA, % DW	0.09–3.04	$[TFA]=3.74\,\tilde{A}(510)\cdot[\tilde{A}(678)]^{-1}-1.96$	0.07	0.84	(16)
	N-free BG-11; 350 µE PAR	AA, % DW	0.04–1.7	$[AA]=1.94\,\tilde{A}(510)\cdot[\tilde{A}(678)]^{-1}-0.95$	0.02	0.90	(16)
P. incisa (Δ5-desaturase mutant)	N-free BG-11; 350 µE PAR	TFA, % DW	8.0–38	$[TFA] = 44.01 + 15.31 \cdot \ln(\tilde{A}_{480}/\tilde{A}_{678} - 0.93)$	1.58	0.95	(14)
		TFA, g·L^{-1}	0.1–1.7	$[TFA] = 1.04 \cdot (\tilde{A}_{480}/\tilde{A}_{678} - 0.95)$	0.12	0.96	(14)
	N-free and complete BG-11; 350 µE PAR	DGLA, % DW (via TFA%DW)	1.2–13	$[DGLA] = 0.35 \cdot [TFA] - 1.84$	1.44	0.90	(14)
Nannochloropsis sp.	N-free f/2; 350 µE PAR	TFA, % DW	5.0–45	$[TFA] = 61.71 \cdot (\tilde{A}_{480}/\tilde{A}_{676}) - 39.97$	2.75	0.93	(15)

CONCLUSION AND OUTLOOK

Numerous advances in cultivation technology and management are needed for sustainable upscaling of microalgal biotechnology and cost-competitive production of valuable products, such as oils for biodiesel, long-chain polyunsaturated FA and carotenoid pigments. The development of innovative methods for the accurate quantification of target products and detection of stressful events in real time will play an essential role in quality improvement and in lowering the cost of biomass production by providing timely information on composition, nutritional and physiological state of microalgal cells. Furthermore, implementation of these techniques will accelerate the fine-tuning of cultivation conditions for better growth and prevention of damage, as well as higher target product accumulation.

Collectively, the findings reviewed in this chapter strongly suggest the possibility of employing spectral methods, especially those based on optical spectroscopy, to obtain essential information on the physiological condition of algal cells in culture. Despite the complicated optics of turbid, light-absorbing systems such as microalgal cell suspensions, valuable information on these organisms' composition and physiological condition can be obtained quickly and non-destructively via absorption spectrophotometry. The use of this approach tocorrelatechanges in key pigments and TFA content allowed us to develop simple algorithms for a non-destructive assay of volumetric and biomass contents of TFA and certain LC-PUFA with reasonable accuracy. Obviously, similar approaches could be used to assay valuable Car in the biomass. Such algorithms could potentially be used in routine laboratory protocols, but are of particular importance for upscaled microalgal biotechnology applications. Optical spectroscopy-based techniques appear to be among the best candidates for express analyses of microalgal mass-production systems.

We believe that algorithms similar to those presented in this chapter, following calibration and spectral fine-tuning, will find extensive use with optical sensors developed for on-line monitoring of algae grown in photobioreactors. Such sensors will become indispensable for on-line monitoringof growth, early detection of damage and selection of optimal timing for biomass harvesting. Moreover, the non-destrucivespectral techniquesare expected, in most cases, to be able to replace tedious and time-consuming chromatographic analyses and minimize the use of expensive solvents and sophisticated equipment.

ACKNOWLEDGMENTS

The financial support of the Russian Foundation of Basic Research (project # 09-04-00419-a) and the Ministry of Education and Science of Russian Federation (contract #16.512.11.2182; AES and OBC) is greatly appreciated Authors would like to thank Jacob Blaustein Center for Scientific Cooperation and Microalgal Biotechnology Laboratory, Jacob Blaustein Institutes for Desert Research, Ben-Gurion University of the Negev, for the support of the collaborative research, as well as Ms. Camille Vainstein for professional language editing.

REFERENCES

[1] Chisti Y. Fuels from microalgae. *Biofuels*. 2010;1(2):233-5.
[2] Tredici M. Photobiology of microalgae mass cultures: understanding the tools for the next green revolution. *Biofuels*. 2010;1(1):143-62.
[3] Mutanda T, Ramesh D, Karthikeyan S, Kumari S, Anandraj A, Bux F. Bioprospecting for hyper-lipid producing microalgal strains for sustainable biofuel production. *Bioresource Technology*. 2011;102(1):57-70.

[4] Kondili EM, Kaldellis JK. Biofuel implementation in East Europe: Current status and future prospects. *Renewable and Sustainable Energy Reviews.* 2007;11(9):2137-51.
[5] Kosa M, Ragauskas A. Lipids from heterotrophic microbes: advances in metabolism research. *Trends in Biotechnology.* 2010.
[6] Ratledge C, Cohen Z. Microbial and algal oils: Do they have a future for biodiesel or as commodity oils? *Lipid Technology.* 2008;20(7):155-60.
[7] Boussiba S. Carotenogenesis in the green alga *Haematococcus pluvialis*: cellular physiology and stress response. *Physiologia Plantarum.* 2000;108(2):111-7.
[8] Cohen Z, Khozin-Goldberg I. Searching for PUFA-rich microalgae. In: Cohen Z, Ratledge C, editors. Single Cell Oils. 2 ed. Champaign IL: American Oil Chemists' Society; 2010. p. 201-24.
[9] Gitelson A, Grits Y, Etzion D, Ning Z, Richmond A. Optical properties of Nannochloropsis sp and their application to remote estimation of cell mass. *Biotechnology and Bioengineering.* 2000;69(5):516-25.
[10] Bligh E, Dyer W. A rapid method of total lipid extraction and purification. *Canadian Journal of Physiology and Pharmacology.* 1959;37(8):911-7.
[11] Elsey D, Jameson D, Raleigh B, Cooney M. Fluorescent measurement of microalgal neutral lipids. *Journal of microbiological methods.* 2007;68(3):639-42.
[12] Chen W, Sommerfeld M, Hu Q. Microwave-assisted Nile red method for in vivo quantification of neutral lipids in microalgae. *Bioresource Technology.* 2011;102(1):135-41.
[13] Chen W, Zhang C, Song L, Sommerfeld M, Hu Q. A high throughput Nile red method for quantitative measurement of neutral lipids in microalgae. *Journal of microbiological methods.* 2009;77(1):41-7.
[14] Solovchenko A, Merzlyak M, Khozin-Goldberg I, Cohen Z, Boussiba S. Coordinated carotenoid and lipid syntheses induced in *Parietochloris incisa* (Chlorophyta, Trebouxiophyceae) mutant deficient in Δ5 desaturase by nitrogen starvation and high light. *Journal of Phycology.* 2010;46(4):763-72.
[15] Solovchenko A, Khozin-Goldberg I, Recht L, Boussiba S. Stress-Induced Changes in Optical Properties, Pigment and Fatty Acid Content of Nannochloropsis sp.: Implications for Non-destructive Assay of Total Fatty Acids. *Marine Biotechnology.* 2010:1-9.
[16] Solovchenko A, Khozin-Goldberg I, Cohen Z, Merzlyak M. Carotenoid-to-chlorophyll ratio as a proxy for assay of total fatty acids and arachidonic acid content in the green microalga *Parietochloris incisa*. *Journal of Applied Phycology.* 2009;21(3):361-6.
[17] Dean A, Sigee D, Estrada B, Pittman J. Using FTIR spectroscopy for rapid determination of lipid accumulation in response to nitrogen limitation in freshwater microalgae. *Bioresource Technology.* 2010;101(12):4499-507.
[18] Laurens L, Wolfrum E. Feasibility of Spectroscopic Characterization of Algal Lipids: Chemometric Correlation of NIR and FTIR Spectra with Exogenous Lipids in Algal Biomass. *BioEnergy Research.* 2011;4:1-14.
[19] Greenspan P, Mayer E, Fowler S. Nile red: a selective fluorescent stain for intracellular lipid droplets. *Journal of Cell Biology.* 1985;100(3):965-73.
[20] Greenspan P, Fowler S. Spectrofluorometric studies of the lipid probe, nile red. *The Journal of Lipid Research.* 1985;26(7):781-9.

[21] Cooksey K, Guckert J, Williams S, Callis P. Fluorometric determination of the neutral lipid content of microalgal cells using Nile Red. *Journal of microbiological methods*. 1987;6(6):333-45.

[22] Lee S, Yoon B, Oh H. Rapid method for the determination of lipid from the green alga Botryococcus braunii. *Biotechnology Techniques*. 1998;12(7):553-6.

[23] Montero M, Aristizábal M, García Reina G. Isolation of high-lipid content strains of the marine microalga Tetraselmis suecica for biodiesel production by flow cytometry and single-cell sorting. *Journal of Applied Phycology*. 2011:1-5.

[24] Doan T, Obbard J. Improved Nile Red staining of Nannochloropsis sp. Jo*urnal of Applied Phycology*. 2010:1-7.

[25] Huang Y, Beal C, Cai W, Ruoff R, Terentjev E. Micro Raman spectroscopy of algae: Composition analysis and fluorescence background behavior. *Biotechnology and bioengineering.* 2010;105(5):889-98.

[26] Beal C, Webber M, Ruoff R, Hebner R. Lipid analysis of Neochloris oleoabundans by liquid state NMR. *Biotechnology and bioengineering*. 2010;106(4):573-83.

[27] Huang G, Chen G, Chen F. Rapid screening method for lipid production in alga based on Nile red fluorescence. *Biomass and Bioenergy*. 2009;33(10):1386-92.

[28] Gao C, Xiong W, Zhang Y, Yuan W, Wu Q. Rapid quantitation of lipid in microalgae by time-domain nuclear magnetic resonance. *Journal of microbiological methods*. 2008;75(3):437-40.

[29] Osborne B, Raven J. Light absorption by plants and its implications for photosynthesis. *Biological Reviews*. 1986;61(1):1-60.

[30] Vonshak A. Microalgae: Laboratory growth techniques and outdoor biomass production. In: Coombs J, Hall D, Long S, Scurlock J, editors. Techniques in Bioproductivity and Photosynthesis. Oxford: Pergamon Press; 1985. p. 188–203.

[31] Gitelson A, Qiuang H, Richmond A. Photic volume in photobioreactors supporting ultrahigh population densities of the photoautotroph Spirulina platensis. *Applied and Environmental Microbiology*. 1996;62(5):1570-3.

[32] Solovchenko AE, Merzlyak MN, Pogosyan SI. Light-induced decrease of reflectance provides an insight in the photoprotective mechanisms of ripening apple fruit. *Plant Science*. 2010;178(3):281-8.

[33] Merzlyak M, Chivkunova O, Maslova I, Naqvi K, Solovchenko A, Klyachko-Gurvich G. Light absorption and scattering by cell suspensions of some cyanobacteria and microalgae. *Russian Journal of Plant Physiology*. 2008;55(3):420-5.

[34] Merzlyak MN, Naqvi KR. On recording the true absorption spectrum and the scattering spectrum of a turbid sample: application to cell suspensions of the cyanobacterium *Anabaena variabilis*. *J. Photochem. Photobiol. B.* 2000;58(2-3):123-9.

[35] Dall'Olmo G, Gitelson A. Effect of bio-optical parameter variability on the remote estimation of chlorophyll-a concentration in turbid productive waters: experimental results. *Applied optics*. 2005;44(3):412-22.

[36] Schalles J, Gitelson A, Yacobi Y, Kroenke A. Estimation of chlorophyll a from time series measurements of high spectral resolution reflectance in an eutrophic lake. *Journal of Phycology*. 1998;34(2):383-90.

[37] Gilerson A, Gitelson A, Zhou J, Gurlin D, Moses W, Ioannou I, et al. Algorithms for remote estimation of chlorophyll-a in coastal and inland waters using red and near infrared bands. *Optics Express*. 2010;18(23):24109-25.

[38] Bricaud A, Morel A. Light attenuation and scattering by phytoplanktonic cells: a theoretical modeling. *Applied optics*. 1986;25(4):571-80.

[39] Mitchell B. Algorithms for determining the absorption coefficient for aquatic particulates using the quantitative filter technique. *Proc SPIE, Ocean Optics, X*. 1990;1302:137-48.

[40] Gitelson A, Laorawat S, Keydan G, Vonshak A. Optical properties of dense algal cultures outdoors and their application to remote estimation of biomass and pigment concentration in *Spirulina platensis* (Cyanobacteria). *Journal of Phycology*. 1995;31(5):828-34.

[41] Merzlyak M, Chivkunova O, Gorelova O, Reshetnikova I, Solovchenko A, Khozin-Goldberg I, et al. Effect of nitrogen starvation on optical properties, pigments, and arachidonic acid content of the unicellular green alga *Parietochloris incisa* (Trebouxiophyceae, Chlorophyta). *Journal of Phycology*. 2007;43(4):833-43.

[42] Merzlyak M, Chivkunova O, Zhigalova T, Naqvi K. Light absorption by isolated chloroplasts and leaves: effects of scattering and 'packing'. *Photosynthesis Research*. 2009;102(1):31-41.

[43] Parkash J, Robblee J, Agnew J, Gibbs E, Collings P, Pasternack R, et al. Depolarized resonance light scattering by porphyrin and chlorophyll a aggregates. *Biophysical Journal*. 1998;74(4):2089-99.

[44] Pasternack R, Collings P. Resonance light scattering: a new technique for studying chromophore aggregation. *Science*. 1995;269(5226):935.

[45] Naqvi K, Mely T, Raji B. Assaying of chromophore composition of photosynthetic systems by spectral resolution: application to the light-harvesting complex (LHC II) and total pigment content of higher plants. *Spectrochimica Acta*. 1997;53:2229–34.

[46] Das M, Rabinowitch E, Szalay L, Papageorgiou G. " Sieve-effect" in Chlorella suspensions. *The Journal of Physical Chemistry*. 1967;71(11):3543-9.

[47] Berner T, Dubinsky Z, Wyman K, Falkowski P. Photoadaptation and the "packace" effect in *Dunaliella tertiolecta* (Chlorophyceae). *Journal of Phycology*. 1989;25(1):70-8.

[48] Duysens L. The flattering of the absorption spectrum of suspensions, as compared to that of solutions. *Biochimica et Biophysica Acta*. 1956;19:1-12.

[49] Osborne B, Geider R. Problems in the assessment of the package effect in five small phytoplankters. *Marine Biology*. 1989;100(2):151-9.

[50] Shibata K. Dual wavelength scanning of leaves and tissues with opal glass. *Biochimicaet Biophysica Acta*. 1973;304(2):249.

[51] Evangelista V, Frassanito A, Passarelli V, Barsanti L, Gualtieri P. Microspectroscopy of the photosynthetic compartment of algae. *Photochemistry and Photobiology*. 2006;82(4):1039-46.

[52] Barsanti L, Evangelista V, Frassanito A, Vesentini N, Passarelli V, Gualtieri P. Absorption Microspectroscopy, theory and applications in the case of the photosynthetic compartment. *Micron*. 2007;38(3):197-213.

[53] Röttgers R, Häse C, Doerffer R. Determination of the particulate absorption of microalgae using a point-source integrating-cavity absorption meter: verification with a photometric technique, improvements for pigment bleaching and correction for chlorophyll fluorescence. *Limnol. Oceanogr: Methods*. 2007;5:1-12.

[54] Merzlyak MN, Chivkunova OB, Melo TB, Naqvi KR. Does a leaf absorb radiation in the near infrared (780-900 nm) region? A new approach to quantifying optical reflection, absorption and transmission of leaves. *Photosynthesis Research.* 2002;72(3):263-70.

[55] Latimer P, Eubanks C. Absorption Spectrophotometry of Turbid Suspensions: A Method of Correcting for Large Systematic Deviations. *Arch. Biochem. Biophys.* 1962;98:274–85.

[56] Baulina OI, Chivkunova OB, Merzlyak MN. Destruction of pigments and ultrastructural changes in cyanobacteria during photodamage. *Russian Journal of Plant Physiology.* 2004;51(6):761-9.

[57] Solovchenko A. Pigment composition, optical properties, and resistance to photodamage of the microalga *Haematococcus pluvialis* cultivated under high light. *Russ. J. Plant Physiol.* 2011;58(1):9-17.

[58] Voronova E, Volkova E, Kazimirko Y, Chivkunova O, Merzlyak M, Pogosyan S, et al. Response of the Photosynthetic Apparatus of the Diatom *Thallassiosira weisflogii* to High Irradiance Light. *Russian Journal of Plant Physiology.* 2002;49(3):311-9.

[59] Naqvi KR, Merzlyak MN, Melo TB. Absorption and scattering of light by suspensions of cells and subcellular particles: an analysis in terms of Kramers-Kronig relations. *Photochemical and Photobiological Sciences.* 2004;3(1):132-7.

[60] Tassan S, Ferrari G. An alternative approach to absorption measurements of aquatic particles retained on filters. *Limnology and Oceanography.* 1995;40(8):1358-68.

[61] Mitchell B, Bricaud A, Carder K, Cleveland J, Ferrari G, Gould R, et al. Determination of spectral absorption coefficients of particles, dissolved material and phytoplankton for discrete water samples. *NASA Tech Memo.* 2003;209966:125–53.

[62] Atherton E. The relation of the reflectance of dyed fabrics to dye concentration and the instrumental approach to colour matching. *Journal of the Society of Dyers and Colourists.* 1955;71(7):389-98.

[63] Aldeeson J, Atherton E, Derbyshire A. Modern physical techniques in colour formulation. *Journal of the Society of Dyers and Colourists.* 1961;77(12):657-69.

[64] Gitelson A, Gritz Y, Merzlyak M. Non destructive chlorophyll assessment in higher plant leaves: algorithms and accuracy *Journal of Plant Physiology.* 2003;160:271-82.

[65] Merzlyak M, Gitelson A, Chivkunova O, Solovchenko A, Pogosyan S. Application of reflectance spectroscopy for analysis of higher plant pigments. *Russian Journal of Plant Physiology.* 2003;50(5):704-10.

[66] Gitelson A, Keydan G, Merzlyak M. Three-band model for noninvasive estimation of chlorophyll, carotenoids, and anthocyanin contents in higher plant leaves. Geophysical Research Letters. 2006;33:L11402.

[67] Young E, Beardall J. Photosynthetic function in *Dunaliella tertiolecta* (Chlorophyta) during a nitrogen starvation and recovery cycle. *J. Phycol.* 2003;39(5):897-905.

[68] Pick U. *Dunaliella*: a model extremophilic alga. Israel *Journal of Plant Sciences.* 1998;46(2):131-9.

[69] Wilhelm C, Selmar D. Energy dissipation is an essential mechanism to sustain the viability of plants: The physiological limits of improved photosynthesis. *Journal of Plant Physiology.* 2011;168:79-87.

[70] Muller P, Li X, Niyogi K. Non-photochemical quenching. A response to excess light energy. *Plant Physiology*; 2001. p. 1558-66.

[71] Solovchenko A, Merzlyak M. Screening of visible and UV radiation as a photoprotective mechanism in plants. *Russian Journal of Plant Physiology.* 2008;55(6):719-37.

[72] Solovchenko A. Localization of Screening Pigments Within Plant Cells and Tissues. Photoprotection in Plants: Springer; 2010. p. 67-88.

[73] Renaud S, Parry D, Thinh L, Kuo C, Padovan A, Sammy N. Effect of light intensity on the proximate biochemical and fatty acid composition of Isochrysis sp. and Nannochloropsis oculata for use in tropical aquaculture. *Journal of Applied Phycology.* 1991;3(1):43-53.

[74] Sukenik A, Carmeli Y, Berner T. Regulation of fatty acid composition by irradiance level in the eustigmatophyte *Nannochloropsis* sp. 1. *Journal of Phycology.* 1989;25(4):686-92.

[75] Falkowski PG, LaRoche J. Acclimation to spectral irradiance in algae. *Journal of Phycology.* 1991;27(1):8-14.

[76] Oren A. A century of *Dunaliella* research: 1905-2005. Adaptation to Life at High Salt Concentrations in Archaea, Bacteria, And Eukarya. 2005:493.

[77] Borowitzka M, Siva C. The taxonomy of the genus *Dunaliella* (Chlorophyta, Dunaliellales) with emphasis on the marine and halophilic species. *Journal of Applied Phycology.* 2007;19(5):567-90.

[78] Jahnke L. Massive carotenoid accumulation in *Dunaliella bardawil* induced by ultraviolet-A radiation. *Journal of Photochemistry and Photobiology, B: Biology.* 1999;48(1):68-74.

[79] Rabbani S, Beyer P, Lintig J, Hugueney P, Kleinig H. Induced □-carotene synthesis driven by triacylglycerol deposition in the unicellular alga *Dunaliella bardawil*. *Plant Physiology.* 1998;116(4):1239-48.

[80] Salguero A, de la Morena B, Vigara J, Vega JM, Vilchez C, Leyn R. Carotenoids as protective response against oxidative damage in *Dunaliella bardawil*. *BiomolecularEngineering.* 2003;20(4-6):249-53.

[81] Fabregas J, Dominguez A, Maseda A, Otero A. Interactions between irradiance and nutrient availability during astaxanthin accumulation and degradation in *Haematococcus pluvialis*. *Applied microbiology and biotechnology.* 2003;61(5):545-51.

[82] Kobayashi M, Katsuragi T, Tani Y. Enlarged and astaxanthin-accumulating cyst cells of the green alga *Haematococcus pluvialis*. *Journal of Bioscience and Bioengineering.* 2001;92(6):565-8.

[83] Vidhyavathi R, Venkatachalam L, Sarada R, Ravishankar G. Regulation of carotenoid biosynthetic genes expression and carotenoid accumulation in the green alga *Haematococcus pluvialis* under nutrient stress conditions. *Journal of Experimental Botany.* 2008;59(6):1409-18.

[84] Zhekisheva M, Boussiba S, Khozin-Goldberg I, Zarka A, Cohen Z. Accumulation of oleic acid in *Haematococcus pluvialis* (Chlorophyceae) under nitrogen starvation or high light is correlated with that of astaxanthin esters. *Journal of Phycology.* 2002;38(2):325-31.

[85] Steinbrenner J, Linden H. Light induction of carotenoid biosynthesis genes in the green alga *Haematococcus pluvialis*: regulation by photosynthetic redox control. *Plant Molecular Biology.* 2003;52(2):343-56.

[86] Raven J, Cockell C, De La Rocha C. The evolution of inorganic carbon concentrating mechanisms in photosynthesis. *Philosophical Transactions B*. 2008;363(1504):2641.

[87] Thompson G. Lipids and membrane function in green algae. *Biochimica et Biophysica Acta (BBA)/Lipids and Lipid Metabolism*. 1996;1302(1):17-45.

[88] Flynn K, Davidson K, Cunningham A. Relations between carbon and nitrogen during growth of Nannochloropsis oculata (Droop) Hibberd under continuous illumination. *New Phytologist*. 1993;125:717-22.

[89] Mendoza H, Martel A, Jimenez del Rio M, Garcia Reina G. Oleic acid is the main fatty acid related with carotenogenesis in *Dunaliella salina*. *Journal of Applied Phycology*. 1999;11(1):15-9.

[90] Ben-Amotz A, Katz A, Avron M. Accumulation of β-carotene in halotolerant alge: purification and characterization of ☐-carotene-rich globules from *Dunaliella bardawil* (Chlorophyceae). *Journal of Phycology*. 1982;18(4):529-37.

[91] Wang B, Zarka A, Trebst A, Boussiba S. Astaxanthin accumulation in *Haematococcus pluvialis* (Chlorophyceae) as an active photoprotective process under high irradiance. *Journal of Phycology*. 2003;39(6):1116-24.

[92] Solovchenko A, Khozin-Goldberg I, Didi-Cohen S, Cohen Z, Merzlyak M. Effects of light intensity and nitrogen starvation on growth, total fatty acids and arachidonic acid in the green microalga *Parietochloris incisa*. *Journal of Applied Phycology*. 2008;20(3):245-51.

[93] Sosik H, Mitchell B. Absorption, fluorescence, and quantum yield for growth in nitrogen-limited *Dunaliella tertiolecta*. *Limnology and Oceanography*. 1991;36:910-21.

In: Microalgae: Biotechnology, Microbiology and Energy
Editor: Melanie N. Johnsen

ISBN 978-1-61324-625-2
© 2012 Nova Science Publishers, Inc.

Chapter 9

MICROALGAE AS AN ALTERNATIVE FEED STOCK FOR GREEN BIOFUEL TECHNOLOGY

G. S. Anisha[1]** *and Rojan P. John*[2]

[1]Department of Zoology, Government College Chittur,
Palakkad 678 104, Kerala, India
[2]Institut National de la Recherche Scientifique-Eau Terre Environnement,
490, rue de la Couronne, Québec (QC), G1K 9A9, Canada

ABSTRACT

The worldwide fossil fuel reserves are on the decline but the fuel demand is increasing remarkably. The combustion of fossil fuels needs to be reduced due to several environmental concerns. Biofuels are receiving attention as alternative renewable and sustainable fuels to ease our reliance on fossil fuels. Biodiesel and bioethanol, the two most successful biofuels in the transport sector, are currently produced in increasing amounts from oil or food crops, but their production on a large scale competes with world food supply and security. Microalgae offer a favourable alternative source of biomass for biofuel production without compromising land and water resources since they can be easily cultured on waste land which cannot support agriculture. This chapter focuses on the potential of microalgal biomass for production of the transport fuels, biodiesel and bioethanol and the bottlenecks and prospects in algal fuel technology.

Keywords: Microalgae, Biomass, Biofuel, Biodiesel, Bioethanol, Biorefinery.

INTRODUCTION

The fossil fuel reserves are on the decline and the available fossil reserves cannot continue to support the global fuel need for long. Moreover, the environmental concerns raised by fossil fuel consumption also necessitates that we must reduce our dependence on them. Researchers all over the world are in the search for alternative feedstocks to meet the fossil fuel demand of the global population. Biomass, being renewable, potentially sustainable and relatively environmentally source of energy, can serve as an excellent alternative feedstock to meet the present and future fuel demands (Harun et al., 2010; Saxena et al., 2009). The entire living matter on earth constitutes the biomass and the fuels produced from biomass are generally termed as biofuels (Demirbas, 2008). The biofuels include ethanol, methanol, biodiesel, biohydrogen, biomethane etc. (Nigam and Singh, 2011). Of these, the two most common and successful biofuels in the transport sector are bioethanol and biodiesel which can replace the conventional liquid fuels like petrol and diesel (John et al., 2011).

An array of biomass, including plant and microbial sources, has been reported as suitable feedstocks for biofuel production. This biomass used for biofuel generation ranges from various kinds of bio-wastes, energy crops, and various aquatic plants identified as bio-oil sources. Depending upon the type of raw material and technology used for the production of biofuels, they are generally classified into first, second and third generation biofuels. First generation biofuels are those obtained from food and oil crops (viz. rapeseed oil, palm oil, sugarcane, sugar beet, wheat, barley, maize, etc.) as well as animal fats using conventional technology. Second generation biofuels comprises those generated from agricultural lignocellulosic biomass, which are either non-edible residues of food crop production or non-edible whole plant biomass which includes grasses or trees grown specifically for energy purpose. The biofuels derived specifically from microbes and algal biomass constitutes the 'third generation biofuels' (Nigam and Singh, 2011; John et al., 2011; Singh et al., 2011).

BIOFUELS: PROS AND CONS

The ultimate displacement of petroleum-derived transport fuels needs carbon neutral renewable liquid fuels (Chisti, 2008). From an environmental perspective, the prime benefit of replacing fossil fuels with biomass-based fuels is that the energy obtained from biomass does not add to global warming. The combustion of all fuels, whether from biomass or fossils, releases carbon dioxide into the atmosphere. Since plants use carbon dioxide from the atmosphere for photosynthesis, the carbon dioxide released during combustion of plant biomass feedstock is balanced by that absorbed during their annual growth (Sivakumar et al., 2010; Nigam and Singh, 2011). On the contrary, burning fossil fuels releases carbon dioxide captured billions of years ago. Furthermore, petroleum, diesel and gasoline consist of melange of hundreds of different hydrocarbon chains many of which are toxic, volatile compounds such as benzene, toluene, and xylenes, responsible for environmental pollution and health hazards. Carbon monoxide, nitrogen oxides, sulfur oxides and particulates, are other specific emissions of concern. Using biofuels as an additivie to petroleum-based transportation fuels benefits the reduction of these harmful emissions. Both bioethanol and biodiesel are used as fuel oxygenates to improve combustion characteristics. Adding oxygen results in more complete combustion, which reduces carbon monoxide emissions. This is another environmental benefit of replacing petroleum fuels with biofuels. Using land for agro-based energy offers greater monetary returns to farmers. Biofuels can be utilized as transportation fuels with no or very little engine modification (Harun et al., 2010; Singh et al., 2011; Carere et al., 2008). Since industrial residues generally discarded as wastes can be utilized for biofuel production, it provides value addition to these residues. Moreover, the residues obtained after biofuel production from biomass can provide valuable by-products such as biofertilizers (Nigam and Singh, 2011). Energy security, foreign exchange savings and socioeconomic issues related to the rural sector are other reasons for considering biofuels as relevant technologies by both developing and industrialized countries (Demirbas, 2008).

The worldwide production of biofuels has witnessed an increase by the order of magnitude from 4.4 to 50.1 billion litres between 1980 and 2005 and the growth is expected to rise dramatically in future (Nigam and Singh, 2011). Despite their outstanding benefits, the growing biofuel industry has raised serious doubts about 'how much green are the biofuels'. These concerns need to be addressed satisfactorily in order to realize the goal of developing a completely green sustainable technology to meet the fuel demands of the increasing the population. The most serious accusation against first generation biofuels is that they can have potential impact on food supply and security (John et al., 2011). The production of biofuels from food crops will starve people by taking scarce land and water supplies that should be used to grow food. Using any food source as a fuel is controversial and there is simply not sufficient land to grow enough crops for both food and fuel. The boom in the use of food-derived biofuels will result in a spurt in world food prices and this will levy the burden on the poor (John et al., 2011). Whether to grow crops for food or for fuel is a controversy. There are still other concerns such as extensive cultivation of energy crops can lead to pollution of agricultural land and water resources with fertilizers and pesticides, eutrophication, soil erosion, reduced crop biodiversity and biocontrol ecosystem service losses (Singh et al., 2011; Subhadra and Edwards, 2010; Donner and Kucharik, 2008; Fargione et al., 2008; Hill et al., 2006; Landis et al., 2008; Searchinger et al., 2008; Tillman et al., 2006). Environmentalists warmly welcome the production of biofuels since they will help fight climate change, while there is a growing consensus worldwide that biofuels can cause hunger and trash the environment.

The second generation biofuels are also not free from limitations. The production of biofuels from lignocellulosic biomass needs that the lignin be removed and the carbohydrates are broken down to simpler sugars by enzymatic treatment for easy metabolism during fermentation (John et al., 2011). In this process the cost of enzymes involved makes the technology cost-ineffective and they do not produce sufficiently high biofuel yield for large scale supply. In helping the motoring public to embrace new greener technologies, the scientific world has now diverted their attention to alternative sustainable sources of biomass for biofuel production.

MICROALGAL BIOFUELS AS AN ALTERNATIVE SUSTAINABLE GREENER TECHNOLOGY

Third generation biofuels derived from algal biomass are currently judged as the only viable alternative energy source to overcome the problems faced by first and second generation biofuels. Microalgae are thought to be the earliest life forms on earth (Falkowski et al., 2008). They are microscopic prokaryotic or eukaryotic photosynthetic organisms, mostly single-celled, that convert sunlight, water and carbon dioxide to algal biomass constituting carbohydrates, proteins and oils (Chisti, 2008). They are the fastest growing plants in the world and have high energy content (John et al., 2011). Many of the microalgae are oleaginous in nature and are exploited for the production of biodiesel. The most important advantage that makes microalgae an exceptionally attractive candidate for biofuel technology is that they can generate remarkably large quantities of biomass without compromising land used for growing food crops, thus ensuring food security. In this respect microalgal biofuels are more potent candidates to completely displace petroleum-derived transport fuels than the currently used best oil crops. Chisti (2008) and Gray et al. (2006) advocate that microalgal biodiesel is a better alternative than bioethanol from sugarcane, which is currently the most widely used transport biofuel.

Microalgae, both marine and freshwater, have higher photosynthetic efficiencies than terrestrial plants and are more efficient in capturing carbon Subhadra and Edwards, 2010; Packer, 2009). Microalgae may have an autotrophic or heterotrophic mode of nutrition. Most of the autotrophic microalgae assimilate carbon mainly in the form of carbohydrate and heterotrophic microalgae assimilate carbon mainly as fats or oils and proteins (John et al., 2011). The microalgae have shorter harvesting cycle, which may be approximately 1 – 10 days; where as biomass from terrestrial plants can be harvested only once or twice a year (Harun et al., 2010; Schenk et al., 2008). They can be easily cultivated in marginal lands that are unsuitable for crop growth (Sivakumar et al., 2010; Chisti, 2007; Hu et al., 2008). Unlike terrestrial plants, year round supply of microalgal biomass can be ensured since they can be easily grown in photobioreactors (John et al., 2011; Matsumoto et al., 2003). Since microalgae are minute organisms, culturing them in liquid culture systems provides better control of growth conditions and continuous productivity (Walker et al., 2005). The biofuel generation process from the microalgal biomass is simplified by the structural simplicity of microalgae which excludes the presence of complex biopolymers such as hemicellulose and lignin. This rather eliminates the chemical and enzymatic pre-treatment steps which are other wise needed for terrestrial plant biomass.

Microalgae can grow in diverse climates and environments and can survive in extreme environmental conditions. The cultivation of microalgae can be exploited for bio-mitigation of CO_2 (Brennan and Owende, 2010). Chisti (2008) pointed out that approximately 50% of the dry weight of microalgal biomass is carbon derived from CO_2 and henceforth, producing 100 tons of algal biomass fixes roughly 183 tons of CO_2. The ability to tolerate high concentrations of CO_2 helps microalgae to utilize CO_2 in the flue gas emitted from petroleum-based power stations and industrial sources and thus facilitates reduction of green house gas emissions (Nigam and Singh, 2011). An increase in CO_2 concentration was demonstrated to have a favourable influence on high biomass production and lipid accumulation by the marine unicellular alga *Nannochloropsis oculata* (Chiu et al., 2009). Seambiotic, Tel Aviv, Israel is the first company in the world to utilize flue gas from petroleum-based power stations as source of CO_2 for microalgae cultivation (Goh and Lee, 2010).

Without competing with oil and food crops for fresh water irrigation, many microalgae can flourish in highly saline or municipal/industrial wastewater (Sheehan, 2009; Chinnasamy et al., 2010). This can have very ambitious outcomes since it offers a sustainable method for the bioremediation of wastewater, over and above yielding biomass for biofuel. This phyco-remediation process of coupling algal biomass production with wastewater treatment can also yield treated water for other uses (Singh et al,, 2010).

Microalgae have very simple growth requirements (Dismukes et al., 2008) and there is no need to supplement nutrients for the growth of microalgae since they can very well utilize nutrients available in wastewater from various sources (Subhadra and Edwards, 2010; Shilton et al., 2008). Microalgae cultivation also exempts the need for pesticides (Hu et al., 2008;

Mata et al., 2010). Valuable co-products such as biopolymers and proteins for animal feed supplement can also be obtained during the processing of algal biomass for biofuels (John et al., 2011).

Biodiesel from Microalgae

Biodiesel refers to renewable fuel comprising of monoalkyl esters of long chain fatty acids which can be derived from renewable lipid feedstock such as vegetable oil, animal fat or algal oil. Currently biodiesel is produced by transesterification of vegetable oils and animal fats. Vegetable oils have several merits of being used as diesel such as renewable and inexhaustible nature with energy content close to petroleum diesel (Demirbas, 2011). Most of the biodiesel made from vegetable oils currently use soybean, rapeseed and palm oils which arouses competition with food supply. Moreover, high viscosity, lower volatility and the reactivity of unsaturated hydrocarbon chains (Usta et al., 2005; Demirbas, 2009) and recurring engine problems resulting from the use of vegetable oils as biodiesel make its economic viability questionable (Bajpai and Tyagi, 2006).

Algal cells can naturally synthesize and store lipids and they can be used as minifactories in the process of biodiesel production. Oleaginous microalgae can accumulate rich amounts of lipid (Spiertz and Ewert, 2009) which can be as high as 80% of their dry biomass (Nigam and Singh, 2011). Microalgae can yield several folds higher lipid content than that can be achieved with any plant system (Nigam and Singh, 2011; Rittmann, 2008; Chisti, 2008). The US DOE (2002) statistics also backs up this statement according to which microalgae have the potential to synthesize 100 times more oil per acre of land than any other plant. High lipid concentrations can be achieved by optimizing the growth parameters for microalgae such as nitrogen level (Weldy and Huesemann, 2007; Wu and Hsieh, 2008; Widjaja et al., 2009; Gouveia and Oliveira, 2009), light intensity (Weldy and Huesemann, 2007; Qin 2005), temperature (Qin 2005), salinity (Wu and Hsieh, 2008; Qin 2005), CO_2 concentration (Chiu et al., 2009; de Morais and Costa, 2007) and harvesting procedure (Chiu et al., 2009; Widjaja et al., 2009).

The microalgae lipid accumulation can be most effectively improved by nitrogen limitation (Gouveia and Oliveira, 2009; Miao and Wu, 2006). In addition to improving accumulation of lipids, it can also bring about a change in the lipid profile with more amounts of triglycerides (Meng et al., 2009) which are more useful for conversion to biodiesel than free fatty acids (John et al., 2011; Tsukahara and Sawayama, 2005). Gouveia and Oliveira (2009) reported an increase in the oil yield to about 50% of the dry mass using *Ettlia oleoabundans* and *Nannochloropsis* sp. when deprived of nitrogen. Lipid productivity is a more useful term for assessing the potential of microalgae as source for biodiesel, since lipid accumulation and biomass productivity are not necessarily correlated. Lipid accumulation refers to the amount of lipid concentrated within the cells without considering the biomass, where as lipid productivity considers both these criteria (Brennan and Owende, 2010). Huntley and Redalje (2007) claims that by using only 7.3% of the surplus arable land projected to be available by 2050, biodiesel can be feasibly produced from *Chlorella* at a rate of 3200 GJ/ha/yr which is sufficient to replace the current fossil fuel usage equivalent to about 300 EJ/yr. It can also eliminate about 6.5 gigatons of carbon (Wu and Merchuk, 2004) reported the production of 11,000 L of biodiesel using *Chlorella protothecoides*.

Various methods are in practice for extraction of lipids from microalgae, such as expeller/oil press, solvent extraction, supercritical fluid extraction and ultrasound techniques (Singh and Gu, 2010). To recover algal oil, the algal biomass harvested from the broth is extracted with any of these techniques and the algal oil is then converted to biodiesel by any of the existing technologies. The process of making biodiesel from algal oil is simple which involves a 'transesterification' process. In this process the algal oil containing triglycerides is reacted with an alcohol in presence of a catalyst resulting in the production of a mixture of methyl esters and glycerol. The catalyst used can be acid, alkali or enzyme. The methyl esters form the biodiesel and glycerol forms a valuable co-product (Um and Kim, 2009). Being immiscible, the glycerol can be easily separated from biodiesel.

Biodiesel can be used in existing diesel engines without modification and it is suitable for blending in at any ratio with petroleum diesel (Singh and Gu, 2010). Biodiesel is usually blended with petroleum diesel to form a B20 blend (20% biodiesel and 80% petroleum diesel), although other blend levels can be used up to B100 (pure biodiesel). The blend of biodiesel and petroleum is termed BXX, where XX denotes the percentage of biodiesel in the blend (Amin, 2009). Unlike petroleum diesel, biodiesel is biodegradable and can be used neat without toxic emissions on combustion and it is safer than petroleum or gasoline (Demirbas, 2011).

The quality of biodiesel produced is an important parameter considering the commercialization potential since it determines the stability and performance of the fuel and is dependent on the quality of oil used as raw material (Nigam and Singh, 2011). Each microalgal species possess its own lipid profile and the most suitable strain needs to be selected (Schenk et al., 2008). The up-regulation or down-regulation of fatty acid metabolism by genetic engineering approach is a feasible option to improve the quantity and quality of microalgal lipids (Mussgnug et al., 2007; Polle et al., 2002). Microalgae too have excellent potential for genetic modification of their lipid metabolism. Genetic engineering can help to achieve commercially viable levels of fuel generation.

A drawback of using biopolymers such as triglycerides from algae for biodiesel generation is that it involves costly oil extraction and separation steps since a considerable energy is expended to create and destroy biopolymers normally needed for cell structure or energy storage. A viable solution for this could be direct or *in situ* transesterification which involves simultaneous extraction and transesterification of fatty acid containing lipids. Johnson and Wen (2009) conducted a comparative study on *in situ* transesterification of dry and wet algae with conventional transesterification of extracted oil. They accomplished higher crude biodiesel yield with *in situ* transesterification of dry algae than with the conventional transesterification process. However, *in situ* transesterification of wet algae was not productive in comparison to the conventional process. Ehimen et al. (2010) also reported that *in situ* transesterification of microalgae lipids was inhibited when the biomass water content was greater than 115% w/w. Further studies are required in this regard to improve this advantageous method.

Bioethanol from Microalgae

Currently bioethanol is the chief biofuel that is produced and used worldwide. Demirbas (2011) pointed out that from 2004 to 2008 the global ethanol production boosted from 10.7

billion gallons to 20.4 billion gallons and right now US ethanol production has surpassed that of Brazil. It also expected that the next decade will witness incredible growth in the global ethanol market. The most important bioethanol production countries in the world are Brazil, US and Canada (Chiramonti, 2007). About 94% of the global biofuel production is taken hold of by ethanol (Demirbas, 2011). Sugar cane is the main feedstock for bioethanol production in Brazil, while corn and sugar beet are the major resources in United States and Europen Union respectively (Chiramonti, 2007). Ethanol is well established as fuel for use in transport and industries in several countries, in particular Brazil. Most of the currently produced automobiles in Brazil and United States have flex-fuel facility which allows the users to switch between petrol/gasoline and ethanol (Demirbas, 2011).

Bioethanol is produced by alcoholic fermentation of carbohydrates available in the biomass (Harun et al., 2010). It can be used as a petrol additive or substitute (Demirbas, 2008). Much effort has been devoted to commercializing cellulosic biofuels made from energy crops, such as fast-growing hybrid poplar and switchgrass (Sheehan, 2009). Currently (a) simple sugar-containing feedstocks such as sugar cane, sugar beet, sorghum, whey and molasses (b) starch-containing food grains such as maize, corn, wheat, root crops like cassava, and (c) lignocellulosic feedstock such as wood, straw, agroindustrial residues are used for ethanol production, but their production in large quantities is not sustainable. As mentioned earlier the major hurdles in the bioethanol technology from terrestrial plants is the competition with food crops for land and water for irrigation and the thought-provoking question 'whether to use crops for food or fuel?'. The high cost of pre-treatment steps also limits the utilization of lignocellulosic wastes. If all of the corn grown in the United States was used to make fuel, only 15% of current US fuel needs would be satisfied which shows that we will need more than corn ethanol to fuel our cars.

Microalgae are currently catching wide attention as the sustainable and renewable source of fermentation feedstock for bioethanol production. The absence of lignin makes the process of microalga-based bioethanol production cheaper than from their terrestrial relatives where lignin removal is a pre-requisite. The high starch/cellulose accumulating microalgae can be excellent renewable sources of biomass intended for bioethanol. The prokaryotic microalgae like cyanobacteria and eukaryotic microalge like the green alga *Chlorella* and diatoms are examples from the group of microalgae used for bioethanol production (Li et al., 2008). Recent estimates show that approximately 5000–15,000 gal of ethanol/acre/year (46,760–140,290 L/ha) can be produced from microalgae (Cheryl, 2008). Even switchgrass can yield only a fraction of this estimated microalgal yield (Mussatto et al., 2010). The microalga *Chlorella vulgaris* is considered as an excellent feedstock for bioethanol production. Doucha and Lívanský (2009) claimed that in their study under conditions of protein suppression *Chlorella* sp. was able to accumulate starch up to 70% of the dry algal biomass.

Several methods are being experimented for the production of bioethanol from microalgal biomass. The polymeric carbohydrates, starch and cellulose in the algal biomass can be extracted to produce simple sugars which can then be subjected to microbial fermentation to yield ethanol. There are even microalgae capable of acting as mini factories for the production of ethanol during dark fermentation (John et al., 2011). Genetic engineering of microalgae for direct production of ethanol is also attempted (Deng and Coleman, 1999; Wahlund et al., 1996). Figure 1. shows the potentials of microalgae for biofuel production in various ways.

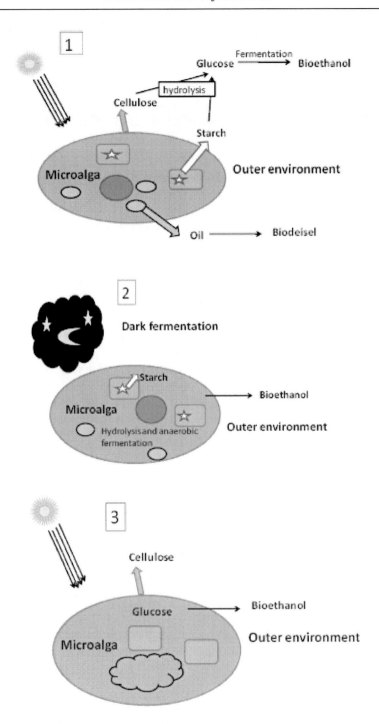

Figure 1. Microalgae as cell factory: 1. Photosynthesis yields reserve food materials and cell components such as oil, fat and carbohydrate which can be processed for biofuel production 2. Dark fermentation: the stored food material, mainly starch, is converted to ethanol in dark; 3. Genetically engineered microalgae (blue green algae) produce ethanol during photosynthesis

:Day time (sun), :Night time, :Circular DNA, :Nucleus, :Plastid.

The utilization of microalgae as fermentation feedstock for bioethanol production requires the pre-treatment of algal biomass for the extraction of fermentable sugars. Each step in this bioprocess is influenced by various parameters optimization of these steps is essential to ensure the highest bioethanol yield (Harun et al., 2010). The starch can be extracted from the cells by mechanical means or using cell disrupting enzymes. There are several algae, particularly green algae capable of accumulating cellulose as the cell wall carbohydrate. The cellulose can also be extracted and used for alcoholic fermentation. The residual algal biomass obtained after oil extraction may also be used for ethanol fermentation (Harun et al., 2010). The ethanol production from algal starch/cellulose involves saccharification and fermentation. Saccharification is process of conversion of complex polysaccharides like starch/cellulose to simple sugars with the aid of acid or enzymatic hydrolysis or using microbes. Microbes can break down starch/cellulose to produce sugar efficiently, but during the process they also consume the sugar. This loss can be avoided by using enzymes instead of microbes. Enzymatic hydrolysis using amylolytic or cellulolytic enzymes allows the saccharification process to be carried out under milder condition than acid hydrolysis and produces higher yields of sugars with little degradation (Mussatto et al., 2010). The saccharification step is followed by fermentation of the resultant simple sugars to yield ethanol. This process is carried out by a suitable ethanol producing microorganism. Conventionally the yeast *Saccharomyces cerevisiae* is used for industrial ethanol fermentation due to its capacity to grow rapidly under anaerobic conditions in large-scale fermentation vessels and high tolerance to ethanol and other inhibitors. Nonetheless, *S. cerevisiae* can utilize only the hexose sugars and the other simple sugars like xylose and arabinose obtained by the breakdown of microalgal cell walls are lost. Microorganisms capable of converting these sugars to ethanol, such as yeasts of the genera *Candida* and *Pichia* or genetically modified *S. cerevisiae* can be exploited for better utilization of the microalgal biomass (Agbogbo and Coward-Kelly, 2008; Watanabe et al., 2007). The bacterial microbes used for ethanol fermentation include *Zymomonas mobilis* (Rogers et al., 2007) and *Escherichia coli* (Jarboe et al., 2007). The process can be made cost effective by using ethanol fermenting microbes having the machinery for starch/cellulose hydrolysis so that the saccharification and fermentation process can be coupled in a single step process. The fermented broth containing ethanol is then fed to distillation unit for concentration and purification of ethanol by removal of water and other impurities. The concentrated ethanol is condensed to liquid form which can then be used directly as fuel or in combination with gasoline or petrol (John et al., 2011; Mussatto et al., 2010; Brennan and Owende, 2010). Moen (1997) reported that brown seaweeds yield higher bioethanol than other algal species. Recently Harun et al. (2010) reported fermentative production of ethanol at the level of 38 % w/w using the microalgae *Chlorococcum* sp. as fermentation feedstock.

The cost incurred for the extraction, saccharification and fermentation of algal starch can be evaded if the algal species can themselves produce ethanol. Several heterotrophic algal species can conduct ethanol fermentation in dark which is a comparatively simpler process with shorter fermentation time than the conventional fermentation (Singh and Gu, 2010; Chen et al., 2009). In the absence of sunlight certain algae can utilize starch or glycogen stored in the cells by oxidative decomposition to CO_2. Under dark and anaerobic conditions, the oxidative decomposition of starch is incomplete ethanol is produced. Ueno et al. (1998) reported that the marine green alga *Chlorococcum littorale* produced 450 µmol/g dry weight ethanol by dark fermentation. Genetic engineering approach can be adopted to enable microalgae to directly convert CO_2 to ethanol so that the cost and energy expended for the synthesis and destruction of complex polymers can be excluded (Deng and Coleman, 1999; Wahlund et al., 1996).

The ethanol generated from microalgal biomass through any of the above means, can be used in pure form or blended with gasoline or petrol. The bioethanol comes in two blends E5 and E85 denoting the percentage blend of ethanol with petrol. E85 is a mixture of petrol and bioethanol in the ratio 85% bioethanol and 15% petrol. It is used to power "flex-fuel" vehicles, which are also capable of running on standard unleaded petrol. E5, a low blend of bioethanol and petrol in the ratio 5% bioethanol and 95% petrol is used in conventional petrol engines and is usually marketed as normal unleaded petrol. A 100% pure ethanol is not suitable for use in cold countries since ethanol does not vaporize well and hence petrol is needed to aid cold starting. Ethanol can also be blended with gasoline. Typically it is blended with gasoline to form an E10 blend (5%-10% ethanol and 90%-95% gasoline), but it can be used in higher concentrations such as E85 or in its pure form. Akin to biodiesel, bioethanol is also clean and renewable and can effectively replace oil (Mussatto et al., 2010; Bai et al., 2008). Combustion of bioethanol in place of petrol or gasoline reduces carbon dioxide emissions by 80% and eliminates the release of sulphur dioxide which is the cause for acid rain (Mussatto et al., 2010).

ECONOMIC FEASIBILITY OF MICROALGAL BIOFUELS

In order to make the microalgal biofuel market profitable, the bioprocess technology for biofuel generation should be made cost-effective. To ensure reliable high productivity of algal biomass it is indispensable that the microalgal strains are mass cultured and the influencing factors are optimized (Borowitzka, 2008). The most common ways to grow microalgae for fuel are open pond system, closed pond systems, and engineered photobioreactors. Some of the cultivation systems are represented in Figure 2.

The most widely used system for large-scale outdoor cultivation of microalgae is open ponds. The closed pond systems differ from open ponds in that they are kept closed during cold climates.

The ponds can be constructed in marginal land where growth of food or other crops cannot be supported. Algae culturing plants for biofuel generation may be installed on land adjacent to fossil fuel-based power stations or industrial areas where there is emission of flue gas containing CO_2 so that there will be efficient reduction of green house gas emissions. Open ponds are cheaper, easy to build and operate, have low capital and maintenance costs (Brennan and Owende, 2010) but they have low productivity, high harvesting cost, water loss through evaporation and lower carbon dioxide use efficiency, diurnal fluctuations in temperature and chances of contamination by undesired algae species and protozoa (Chisti, 2007; Lee, 2001, Shen et al., 2009).

Several photobioreactor designs are developed to achieve high cell densities of microalgal cells such as horizontal tubes, vertical tubes and thin films. The capital and maintenance costs of photobioreactors are higher than for open pond cultivation.

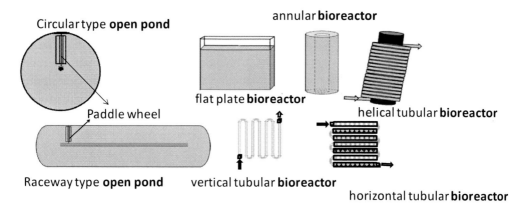

Figure 2. Different cultivation systems used for microalgal biomass production.

Akin to any bioprocess technology, the commercial feasibility of algae-based biofuels is reliant on its cost-effectiveness for which several factors including costs associated with growth and harvesting of algal biomass, oil/carbohydrate extraction and engineering, infrastructure, plant installation costs etc. are to be considered (Singh and Gu, 2010). Selection of the ideal microalgal strain which can either produce biofuels directly or through accumulation of lipid/carbohydrate/protein biomass is very pertinent. The harvesting cost can be lowered by increasing the algal biomass production. The cost-efficiency can be further improved by using cheaper materials for the culturing of algae such as wastewater and flue gas containing CO_2. This will not only lower the production cost and improves algal biomass yield, but also helps in pollution abatement and reduction of green house gas emissions. The metabolic pathways regulating lipid/carbohydrate/protein biosynthesis can be engineered to modify their accumulation as desired.

BIO-REFINERY APPROACH TO REVITALIZE MICROALGAL FUEL ECONOMY

Chisti (2007) defined bio-refinery as the production of a wide range of chemicals and biofuels from biomasses by the integration of bioprocessing and appropriate low environmental impact chemical technologies in a cost-effective and environmentally sustainable manner. The concept of bio-refining is derived from petroleum refining, where in multiple fuels and petrochemicals finding diverse applications are produced from petroleum. The IEA Bioenergy Task 42 document (2009) defines bio-refining as "the sustainable processing of biomass into a spectrum of marketable products and energy". Typically, algae feedstock is composed of 20-40% proteins, 20% carbohydrates, 30-50% lipids and 10% other valuable compounds like pigments, anti-oxidants, fatty acids, vitamins etc (Singh and Gu, 2010; Chen et al., 2009). Apart from biofuels, there is a vast portfolio of products that can be obtained from the microalgal biomass, ranging from nutraceuticals for human consumption, animal feed, pharmaceuticals, pigments, biofertilizers etc. Microalgae are also sources of polyunsaturated fatty acids (PUFA) essential for human development and physiology (Hu et al., 2008). The carotenoids from microalgae can act as antioxidants and can have immunostimulating role (Waldenstedt et al., 2003).

The biomass residue that remains after extraction of oil/carbohydrates and other valuable products can be sold as high protein animal feed. The biomass obtained after extraction of oil can be subjected to fermentation for ethanol production. The algal biomass residue from oil extraction can be used for anaerobic digestion to produce biomethane (biogas). This energy generated may be used within the system for further production and processing of algal biomass there by making the technology partially self-reliant in terms of energy. The CO_2 generated in the process of biofuel production and from the combustion of biogas can be recycled directly to microalgae cultivation ponds or bioreactors for the production of the microalgae biomass. The technology for biogas production from waste biomass (Lantz et al., 2007) and converting biogas to electrical/mechanical power (Gokalp and Lebas, 2004) are well established.

By employing bio-refinery approach a diverse range of products can be generated from the microalgal biomass there by effecting maximum and efficient utilization of biomass with promising economic and environmental benefits. Henceforth, income can be generated from a single biomass feedstock in diverse routes so that the expenditure incurred for biofuel production could be counter-balanced. An integrated bio-refinery approach would be a much more profitable venture than a single product-based industry. It would certainly enhance the overall cost-efficiency of microalgal biofuel technology. Chen et al. (2009) suggested that development of new processes, design of the system, and life cycle analysis are necessary for the development and implementation of algae based biorefineries.

BOTTLENECKS IN ALGAL FUEL TECHNOLOGY AND PROSPECTS

Albeit microalgae have the potential to meet the energy needs of the world, its successful commercialization is challenged by several techno-economic impediments. The cost of algal biofuels is increased several folds due to several energy intensive steps involved in the process such as harvesting and drying of biomass, extraction etc. Theoretically burning biofuels is carbon-neutral because they emit the same amount of carbon they absorbed on assimilation. But in practice it takes a lot of energy to grow, transport and process biofuels and this energy probably comes from burning coal or oil. Manufacturing fertiliser, for instance, requires a lot of energy and if the biofuel is to be made into a liquid to put in petrol tanks, that process is energy-intensive, too. To establish the green credentials of biofuels it is pertinent to consider how much energy is needed to grow and process biofuels.

While it is generally proven that microalgae are suitable candidates for world biofuel production, process development is still at its infancy. Breakthrough technological innovations are expected to make the algal fuel technology economical. Open pond cultivation has the chances of growth of undesired algal species and predators. Algal strains which can fend off other organisms are to be screened for cultivation in open ponds. Though transgenic algae have commercially important traits they may be not be suitable for open cultivation. It is hypothesized that in extreme conditions competitors would be limited to minimum and hence transgenic extremophiles can be more efficiently cultivated in open ponds (Waltz, 2009).

Efficient methods are to be developed to recover algal biomass from the dilute broths from photobioreactors. Besides, drying the algal biomass for oil extraction and pre-treatment of algal biomass for alcoholic fermentation are energy consuming processes and make the cost of algal biofuels several folds higher. More intense extraction methods are needed which can enable biofuel extraction from the wet biomass exempting the need for drying. Direct or *in situ* transesterification offer the option to process wet algae. Enzymatic pre-treatment of algal biomass requires the use of enzymes which are at present supplied from microbial bioreactors. Genetic engineering methods should be developed so that the necessary enzymes can be synthesized by the algal cultures on their own so that the energy and cost expended on producing these enzymes in bioreactors can be saved. Only little progress seems to have been made in genetically engineering algae and it lags well behind that of bacteria, fungi and other eukaryotes (Chisti, 2008). Metabolic engineering should be adopted to up or down regulate the various metabolic pathways within the algal cellular machinery so that the simultaneous production of biofuels can be facilitated without the need for synthesis of complex biopolymers and their subsequent breakdown. Microalgae do experience photoinhibition at high intensity day lights and metabolic engineering methods should find a possible solution to develop algal strains which can tolerate high threshold intensity of day light (Chisti, 2008). Molecular engineering methods should also be used to improve the photosynthetic efficiency of algae.

Growing algae for biofuel production requires enormous amounts of water. Development of high salt and temperature tolerant microalgae is necessary for a better utilization of marine water and trapping sunlight in elevated temperature area for getting higher growth and productivity (John et al., 2011). The social and political implications of microalgae-based biofuel economy need to be critically evaluated. Logan (2008) calls the attention of microbiologists, electrochemists, engineers and politicians in solving the energy crisis using renewable biofuels.

Bio-refinery concept should be advocated more intensely since it is expected to generate maximum revenue than from single product-based industry while at the same time is socially and environmentally sustainable. An integrated technology needs to be developed with minimum production costs and maximum product recovery so as to ensure algal biofuels at low cost.

CONCLUSION

Microalgae have received worldwide attention as an alternative sustainable and renewable biomass resource for production of biofuels with greener credentials. Microalge assure biomass production without compromising land and water resource that will not affect the food price adversely.

Moreover, biofuels production from microalgae has the advantage of coupling CO_2 mitigation process and bioremediation of industrial and municipal wastewater. Regardless of the several merits that make microalgae biofuels attractive as efficient transport fuels to replace fossil fuels, it is still awaiting successful commercialization.

REFERENCES

Agbogbo, FK; Coward-Kelly, G. Cellulosic ethanol production using the naturally occurring xylose-fermenting yeast, *Pichia stipitis*. *Biotechnology Letters*, 2008, 30, 1515–1524.

Amin, S. Review on biofuel oil and gas production processes from microalgae. *Energy Conservation and Management,* 2009, 50, 1834–1840.

Bai, FW; Anderson, WA; Moo-Young, M. Ethanol fermentation technologies from sugar and starch feedstocks. *Biotechnology Advances,* 2008, 26, 89-105.

Bajpai, D; Tyagi, VK. Biodiesel: source, production, composition, properties and its benefits. *Journal of Oleo Science* 2006, 55, 487-502.

Biorefineries: adding value to the sustainable utilization of biomass. *IEA Bioenergy*: T42, 2009 :01.

Borowitzka, MA. Marine and halophilic algae for the production of biofuels. *Jorunal of Biotechnology,* 2008, 136, S7.

Brennan, L; Owende, P. Biofuels from microalgae—A review of technologies for production, processing, and extractions of biofuels and co-products. *Renewable and Sustainable Energy Reviews*, 2010, 14, 557–577.

Carere, CR; Sparling, R; Cicek, N; Levin, DB. Third generation biofuels *via* direct cellulose fermentation. *International Journal of Molecular Science,* 2008, 9, 1342–1360.

Chen, P; Min, M; Chen, Y; Wang, L; Li, Y; Chen, Q; Wang, C; Wan, Y; Wang, X; Cheng, Y; Deng, S; Hennessy, K; Lin, X; Liu, Y; Wang, Y; Martinez, B; Ruan, R. Review of the biological and engineering aspects of algae to fuels approach. *International Journal of Agricultural and Biological Engineering,* 2009, 2, 1-30.

Cheryl. Algae becoming the new biofuel of choice; 2008 (available online bhttp://duelingfuels.com/biofuels/non-food-biofuels/algae-biofuel.php#more-115N.

Chinnasamy, S; Bhatnagar, A; Hunt, RW; Das, KC. Microalgae cultivation in a wastewater dominated by carpet mill effluents for biofuel applications. *Bioresource Technology,* 2010, 101, 3097–3105

Chiramonti, D. Bioethanol: role and production technologies. In: Ranalli P, editor. *Improvement of Crop Plants for Industrial End Uses*. Springer; 2007; 209–251.

Chisti, Y. Biodiesel from microalgae beats bioethanol. *Trends in Biotechnology,* 2008, 26, 126-131.

Chisti, Y. Biodiesel from microalgae. *Biotechnol Advances,* 2007, 25, 294–306.

Chiu SY, Kao CY, Tsai MT, Ong SC, Chen CH, Lin CS. Lipid accumulation and CO2 utilization of *Nanochloropsis oculata* in response to CO_2 aeration. *Bioresource Technology,* 2009, 100, 833–838.

Dale, B. Biofuels: thinking clearly about the issues. *Journal of Agricultural and Food Chemistry*, 2008, 56, 3885–3891.

de Morais, MG; Costa, JAV. Isolation and selection of microalgae from coal fired thermoelectric power plant for biofixation of carbon dioxide. *Energy Conservation and Management,* 2007, 48, 2169–2173.

Demirbas, A. Competitive liquid biofuels from biomass. *Applied Energy,* 2011, 88, 17-28.

Demirbas, A. Progress and recent trends in biodiesel fuels. *Energy Conservation and Management,* 2009, 50, 14-34.

Demirbas, A. The importance of bioethanol and biodiesel from biomass. *Energy Sources Part B: Economics, Planning, and Policy*, 2008, 3, 177 –1.

Deng, M; Coleman, J. Ethanol Synthesis by Genetic Engineering in Cyanobacteria. *Applied and Environmental Microbiology*, 1999, 65, 523-528.

Dismukes, CG; Carrieri, D; Bennette, N; Ananyev, GM; Posewitz, MC. Aquatic phototrophs: efficient alternatives to land-based crops for biofuels. *Current Opinion in Biotechnology*, 2008, 19, 235–240.

Donner, SD; Kucharik, CJ. Corn-based ethanol production compromises goal of reducing nitrogen export by the Mississippi River. *Proceedings of the National Academy of Sciences of the United States of America*, 2008, 105, 4513–4518.

Doucha, J; Lívanský, K. Outdoor open thin-layer microalgal photobioreactor: potential productivity. *Journal of Applied Phycology*, 2009, 21, 111–117.

Ehimen, EA; Sun, ZF; Carrington, CG. Variables affecting the in situ transesterification of microalgae lipids. *Fuel*, 2010, 89, 677−684.

Falkowski, PG; Katz, ME; Knoll, AH; Quigg, A; Raven, JA; Schofield, O; Taylor, FJR. The evolution of modern eukaryotic phytoplankton. *Science*, 2004, 305, 354−360.

Fargione, J; Hill, J; Tilman, D; Polasky, S; Hawthorne, P. Land clearing and the biofuel carbon debt. *Science*, 2008, 319, 1235–1238.

Goh, SC; Lee, KT. A visionary and conceptual macroalgae-based third-generation bioethanol (TGB) biorefinery in Sabah, Malaysia as an underlay for renewable and sustainable development. *Renewable and Sustainable Energy Reviews*, 2010, 14, 842–848.

Gokalp, I; Lebas, E. Alternative fuels for industrial gas turbines (AFTUR). *Applied Thermal Engineering*, 2004, 24, 1655–1663

Gouveia, L; Oliveira, AC. Microalgae as a raw material for biofuels production. *Journal of Industrial Microbiology and Biotechnology*, 2009, 36, 269–274.

Gray, KA; Zhao, L; Emptage, M. Bioethanol. *Current Opinion in Chemical Biology*, 2006, 10, 141–146.

Harun, R; Danquah, MK; Forde, GM. Microalgal biomass as a fermentation feedstock for bioethanol production. *Journal of Chemical Technology and Biotechnology*, 2010, 85, 199–203.

Harun, R; Jason, WSY; Cherrington, T; Danquah, MK. Microalgal biomass as a cellulosic fermentation feedstock for bioethanol production. *Renewable and Sustainable Energy Reviews*, 2010, doi:10.1016/j.rser.2010.07.071

Hill, J; Nelson, E; Tilman, D; Polasky, S; Tiffany, D. Environmental, economic and energetic costs and benefits of biodiesel and ethanol biofuels. *Proceedings of the National Academy of Sciences of the United States of America*, 2006, 103, 11206–11210.

Hu, C; Li, M; Li, J; Zhu, Q; Liu, Z. Variation of lipid and fatty acid compositions of the marine microalga *Pavlova viridis* (Prymnesiophyceae) under laboratory and outdoor culture conditions. *World Journal of Microbiology and Biotechnology*, 2008, 24, 1209–1214.

Hu, Q; Sommerfeld, M; Jarvis, E; Ghirardi, M; Posewitz M, Seibert M, Darzins A., Microalgal triacylglycerols as feedstocks for biofuel production: perspectives and advances. *Plant Journal*, 2008, 54, 621–639.

Huntley, ME; Redalje, DG. CO_2 mitigation and renewable oil from photosynthetic microbes: a new appraisal. *Mitigation and Adaptation Strategies for Global Change,* 2007, 12, 573–608.

Jarboe, LR; Grabar, TB; Yomano, LP; Shanmugan, KT; Ingram, LO. Development of ethanologenic bacteria. *Advances in Biochemical Engineering/ Biotechnology,* 2007, 108, 237–261.

John, RP; Anisha, GS; Nampoothiri, KM; Pandey, A. Micro and macro algal biomass: a renewable source for bioethanol. *Bioresource Technology,* 2011, 102, 186-193.

Johnson, MB; Wen, ZY. Production of biodiesel fuel from the microalga *Schizochytrium limacinum* by direct transesterification of algal biomass. *Energy Fuels,* 2009, 23, 5179−5183.

Landis, DA; Gardiner, MM; van der Werf, W; Swinton, SM. Increasing corn for biofuel production reduces biocontrol services in agricultural landscapes. *Proceedings of the National Academy of Sciences of the United States of America,* 2008, 105, 20552–20557.

Lantz, M; Svensson, M; Björnsson, L; Börjesson, P. The prospects for an expansion of biogas systems in Sweden - incentives, barriers and potentials. *Energy Policy,* 2007, 35, 1830–1843

Lee, YK. Microalgal mass culture systems and methods: Their limitation and potential. *Journal of Applied Phycology,* 2001, 13, 307−315.

Li, Y; Horsman, M; Wu, N; Lan, CQ; Dubois-Calero, N. Biofuels from microalgae. *Biotechnol Progress,* 2008, 24, 815–820.

Logan, B. Microbial Fuel Cells. In: Wall JD, Harwood CS, Demain A, editors. *Bioenergy,* ASM Press; 2008.

Mata, TM; Martins, AA; Caetano, NS. Microalgae for biodiesel production and other applications: a review. *Renewable and Sustainable Energy Reviews,* 2010, 14, 217–232.

Matsumoto, M; Yokouchi, H; Suzuki, N; Ohata, H; Matsunaga, T. Saccharification of marine microalgae using marine bacteria for ethanol production. *Applied Biochemistry and Biotechnology,* 2003, 105–108, 247-254.

Meng, J; Yang, X; Xu, L; Zhang, Q; Nie Xian, M. Biodiesel production from oleaginous microorganisms. *Renewable Energy,* 2009, 34, 1-5.

Miao, XL; Wu, QY. Biodiesel production from heterotrophic microalgal oil. *Bioresource Technology,* 2006, 97, 841-846.

Moen, E. Biological degradation of brown seaweeds. Doctoral thesis submitted to Norwegian University of Science and technology, 1997.

Mussatto, SI; Dragone, G; Guimarães, PMR; Silva, JPA; Carneiro, LM; Roberto, IC; Vicente, A; Domingues, L; Teixeira, JA. Technological trends, global market, and challenges of bio-ethanol production. *Biotechnology Advances,* 2010, 28, 817-830.

Mussgnug, JH; Thomas, HS; Rupprecht, J; Foo, A; Klassen, VMA; Schenk, PM; Kruse, O; Hankamer, B. Engineering photosynthetic light capture: impacts on improved solar energy to biomass conversion. *Plant Biotechnology Journal,* 2007, 5, 802–814.

Nigam, PS; Singh, A. Production of liquid biofuels from renewable resources. *Progress in Energy and Combustion Science,* 2011, 37, 52-68.

Packer, M. Algal capture of carbon dioxide; biomass generation as a tool for greenhouse gas mitigation with reference to New Zealand energy strategy and policy. *Energy Policy,* 2009, 37, 3428-3437.

Polle, JEW; Kanakagiri, S; Jin, E; Masuda, T; Melis, A. Truncated chlorophyll antenna size of the photosystems - A practical method to improve microalgal productivity and hydrogen production in mass culture. *International Journal of Hydrogen Energy,* 2002, 27, 1257–64.

Qin, J. Bio-hydrocarbons from algae—impacts of temperature, light and salinity on algae growth. Barton, Australia: Rural Industries Research and Development Corporation 2005.

Rittmann, BE. Opportunities for renewable bioenergy using microorganisms. *Biotechnology and Bioengineering,* 2008, 100, 203–212.

Rogers, PL; Jeon, YJ; Lee, KJ; Lawford, HG. *Zymomonas mobilis* for fuel ethanol and higher value products. *Advances in Biochemical Engineering/ Biotechnology,* 2007, 108, 263–288.

Saxena, RC; Adhikari, DK; Goyal, HB. Biomass-based energy fuel through biochemical routes: a review. *Renewable and Sustainable Energy Reviews,* 2009, 13, 167–178.

Schenk, P; Thomas-Hall, S; Stephens, E; Marx, U; Mussgnug, J; Posten, C; Kruse O; Hankamer B, Second generation biofuels: high-efficiency microalgae for biodiesel production. *Bioenergy Research,* 2008, 1, 20–43.

Searchinger, T; Heimlich, R; Houghton, RA; Dong, F; Elobeid, A; Fabiosa, J; Tokgoz, S; Hayes, D; Yu, TH. Use of US croplands for biofuels increases greenhouse gases through emissions from land-use change. *Science,* 2008, 319, 1238–1240.

Sheehan, J. Engineering direct conversion of CO_2 to biofuel. *Nature Biotechology,* 2009, 27, 1128-1129.

Shen, Y; Yuan, W; Pei, ZJ; Wu, Q; Mao, E. Microalgae mass production methods. *Transactions of the ASABE (American Society of Agricultural and Biological Engineers)* 2009, 52, 1275-1287.

Shilton, AN; Powell, N; Mara, DD; Craggs, R. Solar-powered aeration and disinfection, anaerobic co-digestion, biological CO_2 scrubbing and biofuel production: the energy and carbon management opportunities of waste stabilization ponds. *Water Science Technology,* 2008, 58, 253–258.

Singh, A; Nigam, PS; Murphy, JD. Renewable fuels from algae: An answer to debatable land based fuels. *Bioresource Technology,* 2011, 102, 10-16.

Singh, J; Gu, S. Commercialization potential of microalgae for biofuels production. *Renewable and Sustainable Energy Reviews,* 2010, 14, 2596-2610.

Sivakumar, G; Vail, DR; Xu, J; Burner, DM; Lay, JO Jr.; Ge, X; Weathers, PJ. Bioethanol and biodiesel: Alternative liquid fuels for future generations. *Engineering Life Science,* 2010, 10, 8-18.

Spiertz, JHJ; Ewert, F. Crop production and resource use to meet the growing demand for food, feed and fuel: opportunities and constraints. *NJAS –Wageningen Journal of Life Science,* 2009, 56, 281–300.

Subhadra, B; Edwards, M. An integrated renewable energy park approach for algal biofuel production in United States. *Energy Policy,* 2010, 38, 4897-4902.

Tilman, D; Hill, J; Lehman, C. Carbon-negative biofuels from low-input high-diversity grassland biomass. *Science,* 2006, 314, 1598–1600.

Tsukahara, K; Sawayama, S. Liquid fuel production using microalgae. *Journal of the Japanese Petroleum Institute,* 2005, 48, 251-259.

Ueno, Y; Kurano, N; Miyachi, S. Ethanol production by dark fermentation in the marine green alga, *Chlorococcum littorale*. *Journal of Fermentation and Bioengineering,* 1998, 86, 38-43.

Um, BH; Kim, YS. Review: A chance for Korea to advance algal-biodiesel technology. *Journal of Industrial and Engineering Chemistry,* 2009, 15, 1–7.

USDOE.U.S. Department of Energy. A national vision of americas transition in a hydrogen economy e in 2030 and beyond. U.S. Department of Energy; February 2002.

Usta, N; Ozturk, E; Can, O; Conkur, ES; Nas, S; Con, AH. Combustion of biodiesel fuel produced from hazelnut soapstuck/waste sunflower oil mixture in a diesel engine. *Energy Conservation and Management,* 2005, 46, 741-755.

Wahlund, TM; Conway, T; Tabita, FR. Bioconversion of CO_2 to ethanol and other products. *American Chemical Society, Division of Fuel Chemistry,* 1996, 41, 1403–1406.

Waldenstedt, L; Inborr, J; Hansson, I; Elwinger, K. Effects of astaxanthin-rich algal meal (*Haematococcus pluvalis*) on growth performance, caecal campylobacter and clostridial counts and tissue astaxanthin concentration of broiler chickens. *Animal Feed Science Technology,* 2003, 108(1–4), 119–132.

Walker, TL; Purton, S; Becker, DK; Collet, C. Microalgae as bioreactors. *Plant Cell Reports,* 2005, 24, 629–641.

Waltz, E. Biotech's green gold? *Nature Biotechnology,* 2009, 27, 15-18.

Watanabe, S; Saleh, AA; Pack, SP; Annaluru, N; Kodaki, T; Makino, K. Ethanol production from xylose by recombinant *Saccharomyces cerevisiae* expressing protein engineered NADP+-dependent xylitol dehydrogenase. *Journal of Biotechnology,* 2007, 130, 316–319.

Weldy, CS; Huesemann, M. Lipid production by *Dunaliella salina* in batch culture: effects of nitrogen limitation and light intensity. *US Department of Energy Journal of Undergraduate Research,* 2007, 7, 115–122.

Widjaja, A; Chien, C-C; Ju, YH. Study of increasing lipid production from fresh water microalgae *Chlorella vulgaris*. *Journal of Taiwan Institute of Chemical Engineers,* 2009, 40, 13–20.

Wu, WT; Hsieh, CH. Cultivation of microalgae for optimal oil production. *Journal of Biotechnology,* 2008, 136(Suppl. 1), S521–1521.

Wu, X; Merchuk, JC. Simulation of algae growth in a bench scale internal loop airlift reactor. *Chemical Engineering Science,* 2004, 59, 2899–2912.

In: Microalgae: Biotechnology, Microbiology and Energy
Editor: Melanie N. Johnsen
ISBN 978-1-61324-625-2
© 2012 Nova Science Publishers, Inc.

Chapter 10

A CRITICAL REVIEW: MICROALGAL CO_2 SEQUESTRATION, WHICH STRAIN IS THE BEST?

Yanna Liang
Department of Civil and Environmental Engineering,
Southern Illinois University Carbondale, Carbondale, Illinois, US

ABSTRACT

While human beings are combating against global warming, fuel shortage, resource depletion, and economic downturn, microalgae, the oldest plants on earth, are gaining intensive and unprecedented attentions. The broad variety, wide distribution, and versatile growth conditions allow microalgae to be used in various fields. To be more specific, microalgae can assist humans in solving many of the challenges we are facing. But taking advantages of their unique capabilities requires better knowledge of them. Different microalgae thrive in different environment. This review focuses on identifying the best species/strains for sequestering CO_2 from flue gas released from stationary sources. Though no complete studies have been conducted for selected strains, this review helps to narrow the range and pave the way for future in-depth investigations of well-suited microalgal species in terms of capturing CO_2 and developing value-added products.

INTRODUCTION

As governments and the general public have realized how deeply we are addicted to fossil fuels and how limited this resource is and will be, a fight for human being's future has started. Microalgae, the oldest plants phylogenetically, have been focused under the spotlight. During recent years, the attentions given to microalgae from academe, industry, and government are seriously beyond description. Various potential applications have been proposed for microalgae. Firstly, microalgae are enthusiastically pursued for producing biofuels which include bioethanol, biodiesel, biomethane, biohydrogen, gasoline, and jet fuels. A series of reviews have already discussed this exciting topic [1-12]; Secondly, microalgae have been deeply and broadly explored for producing high value commodity products in addition to many historical uses. Microalgal biomass can be directly used for human consumption, as feed to animals and aquaculture and as biofertilizer. Products extracted from microalgae include antioxidants, polysaccharides, polyunsaturated fatty acids, coloring and food coloring products (astaxanthin, β-carotene, lutein, zeaxantin, and canthaxantin), proteins, and functional food. Detailed reviews on this aspect have also been published [12, 13].

To produce fuels or consumer products, three microalgal growth modes have been investigated: photoautotrophic, heterotrophic, and mixotrophic. Though the latter two growth conditions do offer several advantages [14-18], the target of this review is on photoautotrophic. The driving force is the CO_2 concern which is addressed below. Regarding CO_2 capture using microalgae, numerous publications have covered different subtopics, for example, strain isolation, selection, and screening; culture growth optimization and maintenance; bioreactor selection and design; harvesting; lipid extraction; biofuel production (biochemical or thermochemical); life cycle assessment, and commercialization evaluation. Despite the enormous efforts poured into this research and development, a large number of questions remain. For an investigator entering this new field, the first question that will be asked is what microalgal species should be chosen for study. For an experienced researcher in this arena, the recurring doubt is whether the species or strains being evaluated so diligently are the best one. In another word, is there a better strain somewhere? To help answer these critical questions and fill the gap in the current literature, this review strives to compare and identify the optimal microalgal strains that can be adopted to fix CO_2 effectively.

THE CO_2 CONCERN- WHY NOT GEOLOGICAL SEQUESTRATION

Anthropogenic activities have contributed significantly to elevated CO_2 concentrations in the atmosphere. Annually, approximately 7.6 Gigatons (Gt) of this chemical is released from burning fossil fuels and 1.4 Gt from land-use change. Among these huge amounts of CO_2, 55% is absorbed by natural processes and the left is deposited in the atmosphere [19]. As a result of this annual deposition increase of 2 ppm, the present CO_2 concentration in the air is 385 ppm which is much higher than the pre-industrial level of 280 ppm. The link between increased atmospheric CO_2 levels and increased global temperatures has been well established and accepted by the world [20]. The human society has responded to the CO_2 issue vigorously. Copious efforts have been devoted to reducing the CO_2 emission including the well-known Kyoto protocol. On September 22, 2009, the US EPA's mandatory greenhouse gas (GHG) reporting rule was signed into effect. Facilities that emit 25,000 metric tons or more per year of GHGs are required to submit annual reports to EPA beginning in the year of 2011. Based on EPA's estimation, approximately 10,000 facilities will be covered by this rule. These facilities account for around 85% of GHG emissions in the US. Though facing great resistance from the industry now, it is reasonable to predict that EPA will make reduction of CO_2 emission obligatory in the near future.

Various strategies, either by chemical, physical, or biological means, have been explored to combat CO_2 release. Existing chemical and physical CO_2 fixation technologies include wet scrubbing, dry regenerable sorbents, membranes, cryogenics, pressure and temperature swing adsorption, etc. [20]. Among all of these methods, chemical reaction-based CO_2 fixation has been broadly adopted [21]. This approach consists mainly of three major steps: separation, transportation, and sequestration. To separate CO_2 from other gases, a cyclic carbonation/decarbonation reaction is required. Two kinds of chemicals, solid metal oxide (e.g. CaO) or aqueous amine solution (e.g. monoethanolamine) have been tested for this purpose.

After CO_2 is separated from a gas mixture, it needs to be compressed to 110 bars for transportation to a specific location for sequestration. In terms of one ton of CO_2, the separation and compression cost is estimated to be $30-50. The costs for transportation and sequestration are evaluated as $1-3 per 100 km and $1-3 per ton of CO_2, respectively [22, 23]. Therefore, to sequester one ton of CO_2 at a site 10 miles away from the CO_2 release location, $41-101 is needed. 100 miles away, $209-536 has to be spent. Therefore, this process is exceptionally expensive and energy-consuming. Besides cost, there are other issues associated with geological sequestration. First, the CO_2 mitigation benefits are marginal. Second, the wasted absorbents have to be disposed properly [24, 25]. Third, plenty issues are difficult to resolve, such as site selection, storage capacity calculation, safety concerns (possibilities of leakage to the surface and induction of seismic activity), monitoring and verification, etc.

BIOFIXATION OF CO_2

Biological-based CO_2 capture through photosynthesis has attracted much attention during recent years [26-28]. Biological CO_2 fixation is accomplished by plants and photosynthetic microorganisms, takes place at mild conditions without the need for further disposal of trapped CO_2. By plants, the potential for increased CO_2 capture has been appraised to contribute only 3-6% of the total fossil fuel emissions due to their slow growth rates [29].

Compared to CO_2 uptake by plants, photosynthetic microorganisms provide many advantages: 1) photosynthetic efficiency is 10-20% higher than that of the fastest-growing switch grass [30, 31]; 2) CO_2 fixation efficiency has been calculated as 10-50 times greater than that of plants [6, 32, 33]; 3) CO_2 mitigation can be made more economically and environmentally sustainable when it is combined with wastewater treatment; 4) water consumption is low [34]; 5) cultivation units can be placed in barren areas and do not compete with agriculture land [35]; and 6) CO_2 can be completely recycled since it is converted into chemical energy which in turn can be transformed to various fuels using existing technologies.

Microalgae are good examples of photosynthetic microorganisms. Over the years, several microalgal species have been examined for CO_2 capture. *Chlorella vulgaris* [34, 36], *Scenedesmus obliquus* [28, 37], *Dunaliella* [38], *Haematococcus pluvialis* [30], *Spirulina* sp [28], and *Botryococcus braunii* [39] have been demonstrated to possess CO_2 capturing abilities in the range of 0.03 to 1.0 g CO_2 /l-day. However, most of the studies have been conducted in a controlled laboratory environment with air enriched with CO_2.

Coupling a microalgal farm with a power plant or other sources of CO_2 provides a way to reduce CO_2 emission, improve the environment, and produce liquid transportation fuels as well as other commodity products at the same time. Microalgae essentially recycle the CO_2 from the stationary source's stack gases into a secondary energy product inside of the algal cells. Although this carbon dioxide is eventually released when the fuel is burned or product is used, the process effectively doubles the amount of energy generated from a given quantity of waste CO_2 [40].

Direct use of flue gas reduces the cost for pretreatment, but imposes extreme conditions on microalgae owing to the high CO_2 concentration (10-20%) and the presence of inhibitory compounds, such as NO_x and SO_x [41]. The NO_x and SO_x themselves may not be toxic to microalgae, but the acidification resulting from the solubilization of these two compounds can be a major factor for inhibiting microalgal growth. Besides affecting the culture pH, flue gas typically has a temperature of 115-120°C which is way above the normal temperature range of most microalgal species. Therefore, the best microalgal strains that can be used for capturing CO_2 in the released flue gas mixture need to have these characteristics: 1) tolerance to high CO_2 concentration and possessing high CO_2 fixation rate; 2) tolerance to a wide pH range; 3) tolerance to a broad temperature range; 4) the resulting algal biomass is easy to be harvested; and 5) the biomass has high commercial values which enables production of fuels and commodity products simultaneously to offset the operation cost. Finding the perfect or well-suited microalgae in nature is not a simple task, but will allow major input reduction and has been the dream of many people involved in this field.

Tolerance to Flue Gas Components

Flue gas typically contains 10-20% of CO_2, 100-400 ppm NO_x and SO_x, soot dust, and trace metals, such as mercury [42]. The effect of soot dust and ash containing heavy metals has not been studied much. Using *Nannochloropsis salina* (NANNP2) and *Phaeodactylum triconutum* (PHAEO2) from National Renewable Energy Laboratory (NREL, USA) which was named Solar Energy Research Institute (SERI) before September 1991, Matsumoto discovered that algal productivity can be influenced if soot dust concentration is greater than 200,000 mg/m^3 [43]. But this concentration is very rare to be observed since soot dust concentration is generally on the order of 50 mg/m^3. The same argument can be applied to the presence of heavy trace metals. Higher concentration can affect algal growth, but only in rare situations will the concentrations exceed those that will result in negative impact [44].

Though greater concentrations of CO_2 are toxic to plants and decrease their photosynthetic levels through narcotic poisoning or acidification of the cell fluid, a great number of microalgal species can grow on CO_2 at a concentration of 10-20%. Examples include: *Botryococcus braunii* [36, 45], *C. vulgaris* 259 [46], *Chlorella* sp. [47], *C. vulgaris* P12 [34], *Chlorella* T-1 [48], *Chlorococcum littorale* [49], *Euglena gracilis* [50], *Phaeodactylum tricornutum* [43, 51], and *Scenedesmus* sp. [36]. Some, for example, *Synechococcus elongatus* (a cyanobacterium though) may even tolerate a CO_2 content as high as 60% [52]. In the case of *Cyanidium caldarium* strains, 100% or pure CO_2 is an excellent substrate for growth [53]. Thus, CO_2 in flue gas is not a limiting factor for microalgal proliferation, but the SO_x and NO_x are. The major components of these two gases are SO_2 and NO, respectively. SO_2 itself does not influence the growth of microalgae. However, compared to CO_2 and NO, SO_2 has high water solubility. At 20°C, the solubility is 11.58 g/100 ml. Thus, when SO_2 concentration is high, pH of the medium will decrease dramatically and the productivity of the microalgae will be lowered. The presence of NO does not affect the growth of microalgae, either. NO absorbed in the medium may have a beneficial effect as it is oxidized to NO_2^- which can be utilized as a nitrogen source by microalgae [43].

A significant number of microalgal species can tolerate low concentrations of SO_x and NO_x. *Chlorella* T-1, a high CO_2 tolerant, can survive the presence of 150 ppm NO, but not SO_2 at a concentration above 10 ppm [48]. Several acidophilic algal strains isolated from hot water spring, *Cyanidium caldarium*, *Galdieria partita*, and *Cyanidioschyzon melorae* exhibit good growth at 50 ppm NO, but only *G. partita* demonstrates growth with 50 ppm SO_2 [49]. Actual industrial sources emit flue gas containing 100-400 ppm SO_x and NO_x as aforementioned above. For some plants burning fuels containing high sulfur content and those that do not have SO_x/NO_x scrubbing, which is 70% of all US power plants at present [54], the SO_2 concentration may be even higher. Thus, better strains need to be identified. The following section describes several strains that have great potential in fixing CO_2 from real flue gas.

One important microalgal strain is *Nannochoris* sp. (NANNO2) obtained from NREL [55]. Out of ten marine and halotolerant microalgae, this strain was tested to be the best for tolerating high CO_2 concentration. Growth of this strain on 15% CO_2 was not affected by pH change when cultured semi-continuously in a cycle of 16 h of light and 8 h of dark. When pH was maintained at 7.0 or 8.0, the growth rate was 290 mg dry weight (DW)/l-day, which was lower than 320 mg DW/l-day when the pH was not controlled. When 50 ppm SO_2 was fed to this algal culture together with 15% CO_2, growth was not impacted. But when the SO_2 concentration was 400 ppm, the pH dropped significantly and growth ceased after 20 h of cultivation. The accumulation of sulfate from SO_2 oxidation and the decreased pH may explain the complete inhibition of algal growth. With 15% CO_2 and 300 ppm NO, this alga grew after a prolonged lag phase. Input of NO resulted in increased concentration of NO_2^- which was then decreased after the start of algal growth. This phenomenon demonstrates that nitrogen oxides can be assimilated by algal cells. The exact form of nitrogen oxide utilized, however, is unknown. Besides the unique growth characteristics, this strain accumulated 56% of cell biomass as lipid. Thus, this microalga can be a candidate for capturing CO_2 from flue gas containing 15% CO_2 with SO_x and NO_x lower than 400 ppm and 300 ppm, respectively.

Chlorella species are freshwater microalgae, ubiquitous in nature, and have been examined by different research groups intensively and broadly over the decades (Table 1). Generally, they tolerate SO_2 and NO at low concentrations. Strain HA-1 is the one that is most tolerant to NO. At 300 ppm NO where *Chlorella* KR-1 failed to grow when aerated with 15% CO_2, HA-1 had a biomass productivity of 1.15 g/l-day. It also survived under a SO_2 concentration of 60 ppm. But at 100 ppm SO_2, the growth completely stopped [41]. In contrast, *Chlorella* KR-1 tolerated SO_2 better. The linear growth rate of the culture bubbled with 100 ppm SO_2 and 15% CO_2 was 0.78 g/l-day. The cell growth was totally suppressed, however, when SO_2 concentration was increased to 150 ppm. Though 300 ppm NO is not beneficial for this algal strain, 100 ppm NO did not affect its growth compared to that without [42].

Two microalgal strains really stand out from the list as shown in Table 1. *Monoraphidium* MONOR02 and *Nannochloropsis* NANNO02 have been tested to capture CO_2 in simulated flue gas which contained SO_2 and NO_x as 400 and 120 ppm, respectively [44].

Table 1. Comparison of mesophilic microalgal species that can grow on high concentration CO_2

Microalgal Species	CO_2 conc. (%)	Fuel for flue gas	SO_x (ppm)	NO_x (ppm)	Biomass prod. (mg DW/(l-d))	CO_2 removel efficiency (%)	Reference
Botryococcus braunii	10	No	NA	NA	26.6	NA	[36]
	5.5	Liquified petroleum gas	NA	NA	77	NA	[36]
Botryococcus braunii 765	20	No	NA	NA	92.4	NA	[45]
Chlorella sp. HA-1	15	No	60	300	1150	NA	[41]
Chlorella sp. KR-1	15	Simulated flue gas	100	100	780	NA	[42]
Chlorella sp. KR-1	15	Llquified natural gas	80	60	710	NA	[42]
Chlorella sp.	15	No	NA	NA	105	17.2 g/l-day	[47]
Chlorella sp. P12	8	natural gas	NA	37	NA	50	[72]
Chlorella T-1	13	Coal-fired Power plant	10	150	400	NA	[48]
Chlorella vulgaris	10	No	NA	NA	104.8	NA	[36]
Chlorella vulgaris 259	15	No	NA	NA	NA	0.62 g/l-day	[46]
Chlorella vulgaris P12	10-13	Municipal waste incinerator	NA	NA	NA	4.4 g/l-day	[34]
Euglena gracilis	11	Kerosene	5	26	NA	0.074 g/l-day	[50]
Monoraphidium MONOR02	13.6	Simulated flue gas	200	150	NA	NA	[56]
Monoraphidium MONOR02	12.1	Simulated flue gas	400	120	1403*	NA	[44]
Nannochloris sp. NANNO2	15	No	NA	NA	320	NA	[55]
Nannochloropsis NANNO02	12.1	Simulated flue gas	400	120	1403*	NA	[44]
Scenedesmus sp.	10	No	NA	NA	217.5	NA	[36]
Scenedesmus sp.	5.5	Liquified petroleum gas	NA	NA	203	24	[36]
Scenedesmus obliquus	12	No	NA	NA	140	4.4-8.6	[70]

*: Calculated value based on information presented by authors.
NA: not available.

These two cultures were initially sparged with simulated flue gas without SO_2 which was added incrementally following a period of acclimation. No signs of inhibition were observed for these two cultures. Analysis of the biomass attained after 3 months revealed that the cells contained 26% of lipids, 41% of proteins, and 33% of carbohydrates. Using the data given by the investigators- estimated algal yield as 2,136 dry ton/day, pond depth of 0.9 m, and pond surface area of 418 acres, the biomass productivity is calculated as 1,403 mg DW/l-day, which is the highest among all microalgae listed in the Table. The researchers have also tested other promising strains, such as *Chlorella* sp. HA-1, *Chlorella* T-1, and *Galdieria* sp. acquired from University of Texas Culture Collection, but no results on them were presented.

Monoraphidium MONOR02 fed by simulated flue gas containing 13.6% CO_2, 200 ppm SO_2, and 150 ppm NO also grew well. No inhibitory effects were observed since the growth rate with flue gas was the same as that from control gas which had no SO_2 and NO. Addition of nitrogen and phosphorous further stimulated cell growth by increasing cell numbers, biomass, and chlorophyll contents. Under high nutrient condition, the ash-free dry biomass was 1.7 g/l, higher than 1.26 g/l with low nutrients [56]. In addition, this strain was found to be a dominant one in an open pond system during winter months from October to March in New Mexico, USA [40].

The above mentioned strains do seem promising for future use. To utilize these microalgae however, it needs to be noted that they are all mesophilic, which means that hot flue gas cannot be directly diverted to the algal pond or photobioreactor. A cooling tank is necessary to lower the flue gas temperature, especially during summer seasons or for plants located at tropical areas. A cooling facility may not be needed though if the selected microalga is thermo-tolerant which is described below.

Tolerance to Flue Gas Temperature

As stated above, temperature of flue gas is around 120°C. If mesophilic microalge are used for CO_2 mitigation, the flue gas must be cooled down first, which can increase the operation cost. To enable direct use of flue gas, thermophilic microalgae are highly sought after. The advantages of using thermophilic microalgal species are: 1) reduced cost for cooling [57] and 2) minimized risk of contamination. The disadvantages are: 1) loss of water may be increased due to evaporation, 2) extremophiles may grow slower than common algae [13], and 3) solubility of CO_2 is lowered at high temperatures.

Several thermophilic microalgae have been studied over the years for CO_2 sequestration. One is the red alga, *Cyanidium caldarium*. As early as 1935 [58], this alga was found to thrive in thermal and acidic environments, such as acidic hot springs throughout the world [59, 60]. It has a morphology and developmental history resembling those of *Chlorella*, but contains C-phycocyanin and no chlorophyll other than chlorophyll *a* [61]. Very impressively, this microalga can grow under pure CO_2 (100%) at pH 1 and has a broad temperature range of 20-60°C [53, 62, 63]. However, when grown at high temperatures, for example 55°C, the total cellular lipid content (20%) is lower than that at 20°C (50%) [64]. The decreased lipid content was observed to be a result of increased thickness of cell wall.

Cyanidium caladarium strain IIID2 was tested on synthetic flue gas containing 11.7% of CO_2, 435 ppm SO_2, 5.57% of O_2, 70.9% N_2, and other minor components. When grown at 35°C, pH 3.5 in Allen's salt medium, the cell number increased from 1×10^6 cells/ml to $1.77 \times$

10^8 cells/ml in 33 days with the onset of stationary phase starting after 20 days. NO was not present in the simulated flue gas [65]. Another strain of *Cyanidium caladarium* isolated from hot spring samples in Japan had different growth characteristics. This strain can flourish with 100% CO_2 containing 50 ppm NO at pH 1 and 50°C, but it did not survive under 50 ppm SO_2 [49].

Another thermo-acidophilic species is *Galdieria sulphuraria* which normally co-exists with *C. caldarium*. Under autotrophic growth condition, these two microalgae are indistinguishable in growth and pigmentation [66]. Recently, *G. sulphuraria* is differentiated from *C. caldarium* considering its capability to grow heterotrophically and mixotrophically on more than 50 different carbon sources [66-69]. Though this microalga shows great potential for CO_2 fixation, in particular, CO_2 sequestration from flue gas given its thermo-tolerance, this microalga has not been studied in details on this aspect. Additionally, the algal biomass produced from CO_2 fixation needs to be studied to enable the evaluation of the commercial value of such a process.

CO_2 Removal Efficiency

For selecting microalgal species dedicated for CO_2 sequestration, another critical determination factor is the maximum amount of CO_2 that can be captured from a given flue gas. Some microalgal strains may be able to propagate on the flue gas, but if the CO_2 utilization rate is low, it will render the whole process useless. However, reports on CO_2 removal efficiency vary widely [54]. For *Scenedesmus obliquus* isolated from the waste treatment ponds of a coal-fired power plant in the Southernmost Brazilian state of Rio Grande Do Sul, the mean CO_2 fixation rate with 12% CO_2 was 4.4-8.6% in 21 days. Maximum daily CO_2 fixation rate for this species was 13.6% on day 9 [28, 70]. With regard to *Scenedesmus* sp. KCTC AG 20831 obtained from the Biological Center of the Korea Research Institute of Bioscience and Biotechnology, the CO_2 removal capability was 24% when a mixture of ambient air and flue gas containing 5.5% CO_2 was fed to the alga. This calculation was based on the CO_2 concentrations at the inlet and outlet of the bioreactor, which was 5.5% and 4.2%, respectively [36]. These reported values are much lower than those of others illustrated below.

In outdoor mass culture ponds, more than 90% of CO_2 removal was achieved by controlling CO_2 delivery on demand as determined by a pH stats [40]. For this system, carbonation sumps transferred CO_2 nearly 100% from the inflowing gas to the pond water. The culture pH was maintained at above 7.8 to eliminate outgassing of CO_2 to the atmosphere. However, flue gas was not tested in this open pond. For flue gas released from the Cogeneration Power Plant at the Massachusetts Institute of Technology which contained 8% CO_2 and 20 ppm NO_x [71], reduction of CO_2 in the gas mixture was reported to be 82.3% on sunny days and 50.1% on cloudy days by *Dunaliella parva* and *Dunaliella tertiolecta* in photobioreactors. In terms of flue gas emitted from a boiler combusting natural gas, a 50% of CO_2 removal by *Chlorella* sp. P12 was demonstrated in an outdoor thin-layer photobioreactor [72].

The documented difference on CO_2 fixation extent could be due to: 1) difference in algal species or strains, 2) difference in operating conditions; or 3) difference in measurement. Certainly, different algal species will have different capability for sequestering CO_2 by nature, but the operation conditions can contribute significantly to the efficiency. In this regard, pH of the culture is a critical factor. As SO_x is introduced into the culture, the pH will drop, which consequently will lead to CO_2 escape from the water to the atmosphere. In summer days, keeping CO_2 in solution with a low pH will be difficult. Thus, though some algal species can tolerate a broad range of pH, from the perspective of efficient CO_2 capture, it may be better to control the pH within a certain narrow and basic range. Specific to CO_2 measurement, most of the publications do not provide the details. Apart from that, the unit representing CO_2 removal is not the same. As shown in Table 1, some studies disclose CO_2 removal as g/l-day. It is absolutely fine for individual study or for comparison if all researchers are using this same unit. But for simplicity, a uniform CO_2 removal efficiency expressed as a percentage may serve better. To share information and allow accurate comparisons, it is imperative that all testings are conducted under similar conditions. Standard US EPA testing procedure prescribed by the Code of Federal Regulations Title 40, Protection of Environment, Part 60, Appendix A may be a good choice. As demonstrated by Vunjak-Novakovie, applying this testing protocol for continuous emission measurement needs a thermoelectric water condenser, a flow controller for gas sample removal, an infrared gas analyzer for CO_2, and a data chart recorder [71].

Harvesting and Extraction

In terms of producing renewable fuels and value-added products from microalgae under the autotrophic mode, harvesting and extraction can account for up to 50% of the total cost due to the fact that microalgal cultures have low biomass density. Centrifugation requires high capital, energy, and operation costs and is only suitable for high value products [73]. Filtration is suitable for filamentous or colony-forming microalgae such as *Spirulina* sp. or *Micractinium* sp. [74]. Chemical flocculation by using multivalent cations does work but may contaminate the resulting algal biomass and is expensive for large-scale use. Bioflocculation in which microalgae clump together and settle out of the medium may offer a low-cost method for harvesting and is recommended to be a promising harvesting technique [75]. In addition to *Micractinium* sp. that can self-flocculate, microalgal flocculation has been observed in activated sludge process where both bacteria and microalgae produce extracellular exopolysaccharides (EPS) under stressed conditions [76]. EPS production has also been reported for *Botryococcus braunii* [45]. While cultivation of *B. braunii* in 16:8 h light/dark cycle yields higher hydrocarbon content, continuous illumination with agitation results in higher amount of EPS [77]. Flocculation is still poorly understood by now. Future research should focus on identifying the strains that flocculate naturally and understanding the molecular factors that trigger flocculation [54].

There are many approaches for extracting products of interest from microalgal biomass. The traditional way is to sacrifice the cells, break the cells open, and use different chemicals to extract different products. Several reviews have already covered this topic [3, 5, 78]. A special case is for *B. braunii*. This algal species accumulates more hydrocarbons than lipids intra-cellularly. Different race of this microalga synthesizes different hydrocarbons: n-alkadienes, mono-, tri-, tetra-, and pentaenes for race A; lycopadiene for race L; and polymethylated botryococcenes for race B [77]. Though the forms of hydrocarbons vary from each other, they can all serve directly as biofuels and dissolve well in hexane which is biocompatible.

Based on this feature, a "milking the algae" process was developed. Basically, this process uses hexane to recover the hydrocarbon molecules from the cells while the algal growth is not subsequently affected even after repeated extractions [79]. A better process which immobilizes or entraps the algal cells in alginate beads and adsorption on polyurethane foams leads to a higher recovery of hydrocarbons.

For other microalgal species that accumulate lipids rather than hydrocarbons, producing biofuels is currently a two-step process: lipid extraction followed by chemical reactions for forming fuels.

No matter what method of lipid extraction is used, whether it is through organic solvents, microwave, or supercritical CO_2, the energy cost is high. Ideally, if biofuel, for example biodiesel or bio-gasoline can be produced inside of the cells and secreted to the extracellular medium, the whole process will be greatly simplified. To accomplish this purpose, genetic and metabolic engineering is unavoidable to redirect the cellular pathways and trick algal species to release the chemicals of importance.

So far, this has not been achieved on microalgae, nor on bacteria. But with *E. coli*, a partial success was reported to produce micro-diesel inside of the cells through metabolic engineering. This was conducted by heterologous expression of three enzymes in *E. coli*: pyruvate decarboxylase and alcohol dehydrogenase from *Zymomonas mobilis* and the unspecific acyltransferase from *Acinetobacter baylyi* strain ADP1. By using this approach and culturing *E. coli* under aerobic conditions in the presence of glucose and oleic acid, ethanol formation is combined with subsequent esterification of the ethanol with the acyl moieties of coenzyme A thioesters of fatty acids.

Ethyl oleate is the major constituent of these fatty acid ethyl esters (FAEEs), with minor amounts of ethyl palmitate and ethyl palmitoleate. FAEE concentrations of 1.28 g/l and a FAEE content as 26 % of the cellular dry mass are achieved by a fed-batch fermentation process using glucose as the carbon source [80]. Though genetically and metabolically engineering microalgae will be extremely difficult given their much more complex genomes than that of *E. coli*, this strategy certainly points to a new direction of developing micro-fuels from microorganisms directly.

In summary, based on available literature information, no microalgal species is perfect and can fully satisfy all the requirements listed above. At the time of writing, the best species for microalgal CO_2 fixation are *Nannochoris* sp. NANNO2, *Monoraphidium* MONOR02, and *Nannochloropsis* NANNO02. All of them are marine microalgae and tolerate high saline concentrations.

Therefore, they can be the ideal candidates for sequestering CO_2 emitted from a stationary source situated along the coastline or other places where saline aquifer is present. However, more research is still needed to determine: 1) their CO_2 fixation rate which is unknown so far; 2) detailed cellular composition in terms of lipids, proteins, and carbohydrates; 3) suitable harvesting techniques for these three algae; and 4) ways for optimal utilization of the biomass. For power plants located in the inland, freshwater microalgae will be suitable. In this case, strains of *C. caladarium* and *Chlorella* would be the ideal choices. Similar to the seawater microalgae, more studies are needed to further characterize the selected species and evaluate their potential.

FUTURE RESEARCH RECOMMENDATION

1. While the search for a superb strain for mitigating CO_2 in flue gas and for producing valuable commodity products is likely to continue, genetic engineering and synthetic biology are still essential to make strains tailored for producing a specific product faster and less expensive.
2. Performances of selected strains that have shown great potential in CO_2 capture need to be further evaluated in laboratory settings and field applications.
3. Mechanisms for bioflocculation need to be further understood to develop an inexpensive and effective harvesting technique.

REFERENCES

[1] Khan SA, Rashmi, Hussain MZ, Prasad S, Banerjee UC: Prospects of biodiesel production from microalgae in India. *Renewable and Sustainable Energy Reviews* 2009, 13:2361-2372.

[2] Li Y, Horsman M, Wu N, Lan CQ, Dubois-Calero N: Biofuels from microalgae. *Biotechnology Progress* 2008, 24:815-820.

[3] Brennan L, Owende P: Biofuels from microalgae-A review of technologies for production, processing, and extractions of biofuels and co-products. *Renewable and Sustainable Energy Reviews* 14:557-577.

[4] Amin S: Review on biofuel oil and gas production processes from microalgae. *Energy Conversion and Management* 2009, 50:1834-1840.

[5] Mata TM, Martins AA, Caetano N: Microalgae for biodiesel production and other applications: A review. *Renewable and Sustainable Energy Reviews* 14:217-232.

[6] Chisti Y: Biodiesel from microalgae. *Biotechnology Advances* 2007, 25:294-306.

[7] Chisti Y: Biodiesel from microalgae beats bioethanol. *Trends in Biotechnology* 2008, 26:126-131.

[8] Lei JI, Zhang L, Yao Z, Min E: Review on the progress of producing biofuel from microalgae. *Acta Petrolei Sinica (Petroleum Processing Section)* 2007, 6.

[9] Gouveia L, Oliveira AC: Microalgae as a raw material for biofuels production. *Journal of Industrial Microbiology and Biotechnology* 2009, 36:269-274.

[10] Hossain A, Salleh A, Boyce AN, Chowdhury P, Naqiuddin M: Biodiesel fuel production from algae as renewable energy. *American Journal of Biochemistry and Biotechnology* 2008, 4:250-254.

[11] Huesemann MH, Benemann JR: Biofuels from Microalgae: Review of products, processes and potential, with special focus on *Dunaliella* sp. Pacific Northwest National Laboratory (PNNL), Richland, WA (US); 2009.

[12] Harun R, Singh M, Forde GM, Danquah MK: Bioprocess engineering of microalgae to produce a variety of consumer products. *Renewable and Sustainable Energy Reviews* 14:1037-1047.

[13] Pulz O, Gross W: Valuable products from biotechnology of microalgae. *Applied Microbiology and Biotechnology* 2004, 65:635-648.

[14] Wen Z, Chen F: Heterotrophic production of eicosapentaenoic acid. *Biotechnology Advances* 2003, 21:273-294.

[15] Chen F: High cell density culture of microalgae in heterotrophic growth. *Trends in Biotechnology* 1996, 14:421-426.

[16] Miao X, Wu Q: High yield bio-oil production from fast pyrolysis by metabolic controlling of *Chlorella protothecoides*. *Journal of Biotechnology* 2004, 110:85-93.

[17] Miao X, Wu Q: Biodiesel production from heterotrophic microalgal oil. *Bioresource Technology* 2006, 97:841-846.

[18] Xu H, Miao X, Wu Q: High quality biodiesel production from a microalga *Chlorella protothecoides* by heterotrophic growth in fermenters. *Journal of Biotechnology* 2006, 126:499-507.

[19] Lal R: Sequestration of atmospheric CO_2 in global carbon pools. *Energy and Environmental Science* 2008, 1:86-100.

[20] White CM, Strazisar BR, Granite EJ, Hoffman JS, HW P: Separation and capture of CO_2 from large stationary sources and sequestration in geological formations--coalbeds and deep saline aquifers. *Journal of Air and Waste Management Association* 2003, 53:643-644.

[21] Wang B, Li Y, Wu N, Lan CQ: CO_2 bio-mitigation using microalgae. *Applied Microbiology and Biotechnology* 2008, 79:707-718.

[22] Gupta DK, Rai UN, Tripathi RD, Inouhe M: Impacts of fly-ash on soil and plant responses. *Journal of Plant Research* 2002, 115:401-409.

[23] Shi M, Shen YM: Recent progresses on the fixation of carbon dioxide. *Current Organic Chemistry* 2003, 7:737–745.

[24] Bonenfant D, Mimeault M, Hausler R: Determination of the structural features of distinct amines important for the absorption of CO_2 and regeneration in aqueous solution. *Industrial and Engineering Chemistry Research* 2003, 42:3179–3184.

[25] Resnik KP, Yeh JT, HW P: Aqua ammonia process for simultaneous removal of CO_2, SO_2 and NO_x. *International Journal of Environmental Technology and Management* 2004, 4:89-104.

[26] Kondili EM, Kaldellisb JK: Biofuel implementation in East Europe: current status and future prospects. *Renewable and Sustainable Energy Reviews* 2007, 11:2137–2151.

[27] Ragauskas AJ, Williams CK, Davison BH, Britovsek G, Cairney J, Eckert CA, Frederick WJ Jr, Hallett JP, Leak DJ, et al: The path forward for biofuels and biomaterials. *Science* 2006, 311:484–489.

[28] de Morais MG, Costa JAV: Biofixation of carbon dioxide by *Spirulina* sp. and *Scenedesmus obliquus* cultivated in a three stage serial tubular photobioreactor. *Journal of Biotechnology* 2007, 129:439-445.

[29] Skjanes K, Lindblad P, Muller J: Bio CO_2 - a multidisciplinary, biological approach using solar energy to capture CO_2 while producing H_2 and high value products. *Biomolecular Engineering* 2007, 24:405-413.

[30] Huntley ME, Redalje DG: CO_2 mitigation and renewable oil from photosynthetic microbes: a new appraisal. *Mitigation and Adaptation Strategies for Global Change* 2007, 12:573–608.

[31] Richmond A: Microalgal biotechnology at the turn of the millennium: a personal view. *Journal of Applied Phycology* 2000, 12:441–451.

[32] Li Y, Horsman M, Wu N, Lan CQ, Dubois-Calero N: Biofuels from microalgae. *Biotechnology Progress* 2008, 24:815-820.

[33] Usui N, Ikenouchi M: The biological CO_2 fixation and utilization project by RITE (1): highly-effective photobioreactor system. *Energy Conversion and Management* 1997, 38:487-492.

[34] Douskova I, Doucha J, Livansky K, Machat J, Novak P, Umysova D, Zachleder V, Vitova M: Simultaneous flue gas bioremediation and reduction of microalgal biomass production costs. *Applied Microbiology and Biotechnology* 2009, 82:179-185.

[35] Brown LM, Zeiler KG: Aquatic biomass and carbon dioxide trapping. *Energy Conversion and Management* 1993, 34:1005–1013.

[36] Yoo C, Jun SY, Lee JY, Ahn CY, Oh HM: Selection of microalgae for lipid production under high levels carbon dioxide. *Bioresource Technology*, 101:S71-S74.

[37] Gomez-Villa H, Voltolina D, Nieves M, Pina P: Biomass production and nutrient budget in outdoor cultures of *Scenedesmus obliquus* (Chlorophyceae) in artificial wastewater, under the winter and summer conditions of Mazatlan, Sinaloa, Mexico. *Vie et Milieu* 2005, 55:121-126.

[38] Kishimoto M, Okakura T, Nagashima H, Minowa T, Yokoyama S, Yamaberi K: CO_2 fixation and oil production using microalgae. *Journal of Fermentation and Bioengineering* 1994, 78:479-482.

[39] Murakami M, Ikenouchi M: The biological CO_2 fixation and utilization project by RITE (2): screening and breeding of microalgae with high capability in fixing CO_2. *Energy Conversion and Management* 1997, 38:493-497.

[40] Weissman JC, Tillett DT: In Aquatic species project report, FY 1989-1990. *NREL, Golden Co, NREL/MP-232-4174* 1992:32–56.

[41] Lee JH, Lee JS, Shin CS, Park SC, Kim SW: Effects of NO and SO_2 on growth of highly- CO_2-tolerant microalgae. *Journal of Microbiology and Biotechnology* 2000, 10:338-343.

[42] Lee JS, Kim DK, Lee JP, Park SC, Koh JH, Cho HS, Kim SW: Effects of SO_2 and NO on growth of *Chlorella* sp. KR-1. *Bioresource Technology* 2002, 82:1-4.

[43] Matsumoto H, Hamasaki A, Sioji N, Ikuta Y: Influence of CO_2, SO_2 and NO in flue gas on microalgae productivity. *Journal of Chemical Engineering of Japan* 1997, 30:620-624.

[44] Stepan DJ, Shockey RE, Moe TA, Dorn R: Carbon dioxide sequestering using microalgal systems. University of North Dakota, 2002.

[45] Ge Y, Liu J, Tian G: Growth characteristics of *Botryococcus braunii* 765 under high CO_2 concentration in photobioreactor. *Bioresource Technology* 2011, 102: 130-134.

[46] Yun YS, Lee SB, Park JM, Lee CI, Yang JW: Carbon dioxide fixation by algal cultivation using wastewater nutrients. *Journal of Chemical Technology and Biotechnology* 1997, 69:451-455.

[47] Chiu S-Y, Kao C-Y, Chen C-H, Kuan T-C, Ong S-C, Lin C-S: Reduction of CO_2 by a high-density culture of *Chlorella* sp. in a semicontinuous photobioreactor. *Bioresource Technology* 2008, 99:3389-3396.

[48] Maeda K, Owada M, Kimura N, Omata K, Karube I: CO_2 fixation from the flue gas on coal-fired thermal power plant by microalgae. *Energy Conversion and Management* 1995, 36:717-720.

[49] Kurano N, Ikemoto H, Miyashita H, Hasegawa T, Hata H, Miyachi S: Fixation and utilization of carbon dioxide by microalgal photosynthesis. *Energy Conversion and Management* 1995, 36:689-692.

[50] Chae SR, Hwang EJ, Shin HS: Single cell protein production of *Euglena gracilis* and carbon dioxide fixation in an innovative photo-bioreactor. *Bioresource Technology* 2006, 97:322-329.

[51] Matsumoto H, Shioji N, Hamasaki A, Ikuta Y, Fukuda Y, Sato M, Endo N, Tsukamoto T: Carbon dioxide fixation by microalgae photosynthesis using actual flue gas discharged from a boiler. *Applied Biochemistry and Biotechnology* 1995, 51-52:681-692.

[52] Miyairi S: CO_2 assimilation in a thermophilic cyanobacterium. *Energy Conversion and Management* 1995, 36:763-766.

[53] Seckbach J, Baker FA, Shugarman PM: Algae thrive under pure CO_2. *Nature* 1970, 227, 744-745.

[54] Milne JL, Cameron JC, Page LE, Benson SM, Pakrasi HB: Report from workshop on biological capture and utilization of CO_2, Charles F. Knight Center, Washington University in St. Louis, September 1-2, 2009.

[55] Negoro M, Shioji N, Miyamoto K, Micira Y: Growth of microalgae in high CO_2 gas and effects of SOX and NOX. *Applied Biochemistry and Biotechnology* 1991, 28-29:877-886.

[56] Zeiler KG, Heacox DA, Toon ST, Kadam KL, Brown LM: The use of microalgae for assimilation and utilization of carbon dioxide from fossil fuel-fired power plant flue gas. *Energy Conversion and Management* 1995, 36:707-712.

[57] Ono E, Cuello JL: Selection of optimal microalgae species for CO_2 sequestration. In: *Proceeding of Second Annual Conference on Carbon Sequestration* 2003.

[58] Geitler L, Ruttner F: The Cyanophyceae of the German Limnological Sunda - Expedition, their morphology, systematics and ecology. *Archive of Hydrobiology* 1935, XVI:371–482.

[59] Bailey RW, Staehelin LA: The chemical composition of isolated cell walls of *Cyanidium caldarium*. *Microbiology* 1968, 54:269.

[60] Brock TD: Microbial growth under extreme conditions. In: *Microbial Growth,* Edited by Meadow PM and Pitt S, London, Cambridge University, 1969, 15-41.

[61] Allen MB: Studies with *Cyanidium caldarium*, an anomalously pigmented chlorophyte. *Archives of Microbiology* 1959, 32:270-277.

[62] Seckbach J, Kaplan IR: Growth pattern and 13C/12C isotope fractionation of *Cyanidium caldarium* and hot spring algal mats. *Chemical Geology* 1973, 12:161-169.

[63] Doemel WN, Brock TD: The upper temperature limit of *Cyanidium caldarium*. *Archives of Microbiology* 1970, 72:326-332.

[64] Kleinschmidt MG, McMahon VA: Effect of growth temperature on the lipid composition of *Cyanidium caldarium*: I. class separation of lipids. *Plant Physiology* 1970, 46:286.

[65] Woodward CA, MacInnis JM, Lewis SN, Greenbaum E: Chemical interaction of flue gas components with the growth of *Cyanidium caldarium*. *Applied Biochemistry and Biotechnology* 1992, 34:819-826.

[66] Gross W, Schnarrenberger C: Heterotrophic growth of two strains of the acido-thermophilic red alga *Galdieria sulphuraria*. *Plant and Cell Physiology* 1995, 36:633.

[67] Rigano C, Aliotta G, Martino Rigano V, Fuggi A, Vona V: Heterotrophic growth patterns in the unicellular alga *Cyanidium caldarium*. *Archives of Microbiology* 1977, 113:191-196.

[68] Rigano C, Fuggi A, Rigano VDM, Aliotta G: Studies on utilization of 2-ketoglutarate, glutamate and other amino acids by the unicellular alga *Cyanidium caldarium*. *Archives of Microbiology* 1976, 107:133-138.

[69] Oesterhelt C, Schnarrenberger C, Gross W: Characterization of a sugar/polyol uptake system in the red alga Galdieria sulphuraria. *European Journal of Phycology* 1999, 34:271-277.

[70] de Morais MG, Costa JAV: Isolation and selection of microalgae from coal fired thermoelectric power plant for biofixation of carbon dioxide. *Energy Conversion and Management* 2007, 48:2169-2173.

[71] Vunjak-Novakovic G, Kim Y, Wu X, Berzin I, Merchuk JC: Air-lift bioreactors for algal growth on flue gas: Mathematical modeling and pilot-plant studies. *Ind Eng Chem Res* 2005, 44:6154-6163.

[72] Doucha J, Straka F, LÃvanskÃ½ K: Utilization of flue gas for cultivation of microalgae (*Chlorella* sp.) in an outdoor open thin-layer photobioreactor. *Journal of Applied Phycology* 2005, 17:403-412.

[73] Becker EW: Microalgae: biotechnology and microbiology. Cambridge University Press, 1994.

[74] Lee A, Lewis D, Ashman P: Microbial flocculation, a potentially low-cost harvesting technique for marine microalgae for the production of biodiesel. *Journal of Applied Phycology* 2009, 21:559-567.

[75] Benemann JR, Oswald WJ: Systems and economic analysis of microalgae ponds for conversion of CO_2 to biomass. *NASA STI/Recon Technical Report N* 1994, 95:19554.

[76] Shipin OV, Meiring PGJ, Phaswana R, Kluever H: Integrating ponds and activated sludge process in the PETRO concept. *Water Research* 1999, 33:1767-1774.

[77] Dayananda C, Sarada R, Usha Rani M, Shamala TR, Ravishankar GA: Autotrophic cultivation of *Botryococcus braunii* for the production of hydrocarbons and exopolysaccharides in various media. *Biomass and Bioenergy* 2007, 31:87-93.

[78] Herrero M, Cifuentes A, Iba ez E: Sub-and supercritical fluid extraction of functional ingredients from different natural sources: Plants, food-by-products, algae and microalgae: A review. *Food Chemistry* 2006, 98:136-148.

[79] Frenz J, Largeau C, Casadevall E: Hydrocarbon recovery by extraction with a biocompatible solvent from free and immobilized cultures of *Botryococcus braunii*. *Enzyme and Microbial Technology* 1989, 11:717-724.

[80] Kalscheuer R, Stolting T, Steinbuchel A: Microdiesel: *Escherichia coli* engineered for fuel production. *Microbiology* 2006, 152:2529.

Chapter 11

USE OF MICROALGAE AS BIOLOGICAL INDICATORS OF POLLUTION: LOOKING FOR NEW RELEVANT CYTOTOXICITY ENDPOINTS

Ángeles Cid, Raquel Prado, Carmen Rioboo,
Paula Suárez-Bregua and Concepción Herrero
Laboratorio de Microbiología, Departamento de Biología Celular y Molecular,
Facultad de Ciencias. Universidad de A Coruña. Spain

ABSTRACT

An important amount of the applied load of pesticides enter into aquatic ecosystems from agricultural runoff or leaching and, as a consequence, have become some of the organic pollutants that appear most frequently on aquatic ecosystems. The assessment of toxic potential in surface water is one of the main tasks of environmental monitoring for the control of pollution. Animal organisms such as fishes or mussels have been examined intensively whereas little information is available on the susceptibility of water plants and plankton organisms.

As primary producers, microalgae constitute the first level of aquatic trophic chains. Due to its microscopic size, it is possible to get sample at population and community levels. Some species can be cultivated in photobioreactors under controlled conditions. Because of their short generation times, microalgae respond rapidly to environmental changes, and any effect on them will affect to higher trophic levels. In addition, microalgae offer the possibility to study the trans-generational effects of pollutant exposure, being a model of choice for the study of the long term effects of pollutant exposure at population level. Furthermore, microalgal tests are generally sensitive, rapid and low-cost effective. For all these reasons, the use of microalgal toxicity tests is increasing, and today these tests are frequently required by authorities for notifications of chemicals and are also increasingly being used to manage chemical discharges. For example, algal toxicity tests of chemicals are mandatory tests for notification of chemicals in the European Union countries. Others fields of use for algae in toxicity assessment are industrial wastewaters and leachates from waste deposits.

Cytotoxic effects of aquatic pollutants on microalgae are very heterogeneous, and they are influenced by the environmental conditions and the test species. Growth, photosynthesis, chlorophyll fluorescence and others parameters reflect the toxic effects of pollutants on microalgae; however, other relevant endpoints are less known because experimental difficulties, especially under in vivo conditions.

During the last two decades, our research group has a high priority scientific objective: study the effect of different aquatic pollutants on freshwater microalgae, with the aim to develop new methods for the detection of contaminants based on the physiological response of microalgae, with the purpose of providing an early warning signal of sublethal levels of pollution.

INTRODUCTION

Water quality issues are a major challenge that humanity is facing in the twenty-first century. Given the importance of water in public health, the growing pollution of aquatic environments as a result of rapid industrialization and intensive use of pesticides in agriculture, is a major threat to life on our planet. In this sense, there is an increasingly evident need for effective methods for assessing the toxicity of various environmental contaminants that can achieve these aquatic systems, thereby attempting to regulate the entry of potentially harmful substances to the ecosystem and, ultimately, for humans.

The problem of the impact of pollutants on ecosystems is complex. Environmental monitoring is necessary to control and reduce this impact, and all aspects (legal, social, economic and biological) should be considered. Given the need for appropriate and effective methods for assessing the toxicity of different pollutants, microorganisms, and in particular microalgae, have begun to be used as biological indicators of pollution in ecotoxicity studies because of its predominant role in the first level of the food chain. The aim of these studies is to control entry into the ecosystem of substances potentially harmful to life in them.

The most common ecotoxicity assays for monitoring aquatic pollutants are short-term lethality tests on fishes, which have been criticized for economic, logistical and ethical reasons (Fentem and Balls, 1993). One alternative proposed is the use of lower level organisms of the aquatic food chain, such as bacteria and microalgae, which would be beneficial from all points of view raised. Microalgal bioassays conducted in the laboratory may help to give valuable information on the effects of these pollutants, so the management or environmental advice in these organisms have an early and appropriate alarm system that would allow decision-making prevent such effects.

The growing awareness of the harmful effects of pollution on the environment has changed the traditional strategy of seeking ways to restore or environmental recovery towards a strategy for preventing the entry of potential environmental contamination. This change of attitude must lead scientific community to investigate new methods to predict the toxicity of various substances that can be released into the environment and can exercise potential damage to the ecosystem. There is also a growing demand for test methods increasingly sensitive.

The history of biological assays began in 1924 when Carpenter (1924) began studying metal toxicity to fish in a mining area. Bioassays with algae began to use more than three decades ago, and the first job held in 1971 (Burrows, 1971). These toxicity tests with algae have been becoming more important, so that algal growth inhibition tests are now included in the toxicity bioassays required for registration and notification of new chemicals in the European Union (Girling et al., 2000; Pascoe et al., 2000).

MICROALGAL (UNIALGAL) BIOASSAYS

Microalgae are a diverse group of phototrophic microorganisms found in most environments, especially in the water bodies, both freshwater and saltwater. In these aquatic environments are the main primary producers and represent, therefore, the main energy input to the ecosystem. Thus, any disturbance in the microalgal population and / or alteration of primary production can impact severely on other organisms in these environments (Campanella et al., 2001; Lürling and Roessink, 2006).

Nowadays, microalgae are considered useful indicators of environmental quality, thanks to some of its characteristics: they are ubiquitous inhabitants of all water bodies; it is undeniable they are representative members of the phytoplankton; in general, they are easily cultivated in the laboratory; and they are sensitive to a broad group of compounds, both organic and inorganic. Therefore, microalgae are commonly used in laboratory bioassays (McCormick and Cairns, 1994; Nie et al., 2009).

However, the algae are still underrepresented as test organisms in standardized methods recommended, and since there is no species of microalgae that are always the most sensitive and ecologically representative, it is necessary to introduce new microalgal species that can be used toxicity bioassays, so that in each case to choose the most appropriate, taking into account the nature of the aquatic environment to be protected and the organisms that live naturally in the middle so that the ecological significance of tests in the laboratory is growing.

Laboratory bioassays using a single microalgal species are the most common, and usually required, to evaluate the toxicity of new substances that seek entry into the market. Thus, laboratory bioassays with microalgae is increasingly used to evaluate the toxicity of chemical compounds and pollutants are part of the strategies recommended by the European Community Commission and the Environmental Protection Agency for U.S. damage assessment of toxic agents (Petersen and Kusk, 2000).

Unialgal bioassays conducted in the laboratory can be performed using batch cultures, in which the toxicant is added to the assay and there is no renewal of the culture medium or regulation of the concentration of pollutant, or continuous cultures, in which there is a renewal of culture medium at a certain rate.

On the other hand, are increasingly being used more batteries of bioassays performed at the same time and independently with different species of microalgae but with the same toxic. These batteries can thus compare the sensitivity of different microalgal species, or even different strains of the same species, to a particular pollutant.

Since different species of microalgae differ markedly in their responses to toxic agents, the bioassays using a single microalgal species are of limited applicability in assessing the effects of these environmental pollutants on algal communities, which consist of several species different sensitivities. However, these toxicity tests have been the source of many biological data for risk assessment of different pollutants and there are authors who believe, like us, that these bioassays can still provide valuable information on this subject (Ma, 2005).

There are relatively few microalgal species have been studied for use in toxicity tests, and those used have been chosen for its ease of cultivation rather than by their sensitivity to pollutants. Therefore, there is a need to seek new and more sensitive species respond to both promoters and growth inhibitors.

Most toxicity tests with microalgae performed in laboratory have been conducted with freshwater microalgae, and there are relatively few bioassays that can be classified as standard in marine environments. It is well known that green algae and cyanobacteria are relatively sensitive to many chemicals. The green microalgae of the genera *Chlamydomonas*, *Chlorella*, *Scenedesmus* and *Selenastrum* are frequently used in bioassays of the toxicity of different contaminants. In particular, the most widely used microalgal bioassay is based on inhibiting the growth of algae *Selenastrum capricornutum* and *Scenedesmus subspicatus* (ISO 8692, 1989). There are also many studies conducted with various species of cyanobacteria, being *Anabaena*, *Nostoc* and *Microcystis* the most common genera in these studies. On the other hand, their ecological position (primary producers) and their essential roles in recycling nutrients are critical for all environments. In addition, knowledge of the processes of absorption, accumulation and metabolism of contaminants by algae is essential, as they play an essential role in the process of biomagnification of these contaminants along the food chain, which may ultimately lead to mortality fish, birds and mammals.

MICROALGAL PARAMETERS APPLIED ON TOXICITY STUDIES

Growth

Growth is the most studied parameter in toxicity tests with microalgae, so that 95% or more of the published works include it. It is a very general parameter, reflecting the physiological state of cells. Microalgal population growth can be monitored directly by counting cells under a microscope in special chambers, or electronic particle counters, or using flow cytometry. Indirect estimates of growth can also be used that can be correlated to the turbidity of the culture, the dry weight or the amount of chlorophyll *a* (by fluorometry or spectrophotometry).

There are different index or rates, usually based on the results of cell density, allowing us to quantify the effect of pesticide on microalgal growth, the most used growth rate (μ) and the median effective concentration (Median Effective Concentration, Effective Concentration 50%, EC_{50}) which is the concentration of toxic compound that reduces population growth by 50%. The easiest way to obtain an EC value is the graphic interpolation.

In contrast, very little is known about growth and proliferation in relation to the cell cycle regulation of algae. The lack of knowledge is even greater when referring to the potential toxic effects of pollutants on microalgal cell division. To assess the effect of terbutryn, a triazine herbicide, on the proliferation of the freshwater microalga *Chlorella vulgaris* we have used a flow cytometric approach; *in vivo* cell division was followed using 5-,6-carboxyfluorescein diacetate succinimidyl ester (CFSE) as staining (Rioboo et al., 2009a). In all *C. vulgaris* cultures, each mother cell had undergone only one round of division through the 96 h of assay and the cell division occurred during the dark period. Cell division of the cultures exposed to the herbicide was asynchronous, and terbutryn altered the normal number of daughter cells (4 autospores) obtained from each mother cell. The number was only two in the cultures treated with 250 nM. The duration of the lag phase after the exposure to terbutryn could be dependent on the existence of a critical cell size to activate cytoplasmic division. The rapid and precise determination of cell proliferation by CFSE staining has allowed develop a model for assessing both the cell cycle of *C. vulgaris* and the *in vivo* effects of pollutants on growth and reproduction at microalgal cell level (Rioboo et al. 2009a).

Cell Viability

An effective method to determine cell viability is to measure the fluorescence of cells stained with propidium iodide by microscopy or flow cytometry. Propidium iodide is a fluorescent dye that penetrates cells when they die and/or the cell membrane integrity is lost, this fluorochrome fluoresces red when excited with blue light. Thus, it can be used to discriminate between viable cells and cells not viable fluorescent fluorescent (Cid et al., 1996; Franqueira et al., 2000).

Some studies have also used staining with fluorescein diacetate as a test to assess cell viability, based on the fluorescein molecule, resulting from the conversion of the compound by nonspecific intracellular esterases, is polar in nature and is retained in cells that have intact membrane; however, cells that have lost membrane integrity showed no green fluorescence characteristic of the fluorophore (Lage et al., 2001).

Taking into account our results, cell viability assayed by flow cytometry was the less sensitive parameter when the marine diatom *Phaeodactylum tricornutum* was exposed to copper concentrations lower than 1 mg l^{-1}, during 24 hours (Cid et al. 1997). However, cell viability is a good indicator for the selection of microalgal species with bioremediation purposes (González-Barreiro et al., 2006).

Elemental and Biochemical Composition of the Microalgal Biomass

The metabolites can be considered as the end products of cellular regulatory processes and their levels can be interpreted as the ultimate response of biological systems to genetic and environmental changes (Jamers et al., 2009). A very basic index used is the C/N ratio, which has been linked with a growth rate inversely proportional (Laws and Chalup, 1991).

The biochemical composition analysis to characterize the metabolic response of an organism to stimuli or stressors in the environment has been little used so far in microalgae. Different biochemical compounds have been used as study parameters in toxicity tests with microalgae, such as the cellular protein content (Battah et al., 2001), the carbohydrate content of cells (Kobbia et al., 2001), cellular lipid content (Yang et al., 2002) or fatty acids (El-Sheekh et al., 1994). The cellular content of various photosynthetic pigments, mainly chlorophylls and carotenoids (Couderchet and Vernet, 2003), are also related to the photosynthetic activity.

In our investigations we have observed that atrazine exposure induced the process of chlorosis in cyanobacterial cells, given that this herbicide has an effect on photosynthesis, chlorotic subpopulations having low values of chlorophyll *a* autofluorescence (González-

Barreiro et al., 2004). More unpigmented subpopulations (chlorotic) appeared as the atrazine concentration increased and better growth rates resulted.

We have also detected that the herbicide paraquat induces alterations in the elemental and biochemical composition of a non-target microalgal species, *Chlamydomonas moewusii* (Prado et al., 2009b). After 48 h of herbicide exposure, growth rate, dry weight, and chlorophyll *a* and protein content were affected by paraquat concentrations above 0.05 µM. C/N ratio was also affected due to a decrease in nitrogen content in the dry biomass, while the carbon content remained constant for all paraquat concentrations assayed.

The analysis of the photosynthetic pigment content of *C. moewusii*, using a traditional spectrophotometric technique that provides population bulk measurements, indicated us an alteration provoked by paraquat (Prado et al., 2011). By means of flow cytometry, which allowed us characterizing the microalgal response at a single-cell level, we have observed that paraquat concentrations above 50nM induce chlorosis in a percentage of microalgal cells depending on herbicide concentration and exposure time, as reflected by a reduced cell chlorophyll autofluorescence and pigment content of the biomass. The possibility of analyzing chlorotic and non-chlorotic sub-populations separately allowed us the study of morphological properties and physiological status of both cell types, leading to the conclusion that chlorotic cells are non-viable cells. Chlorophyll fluorescence was the most sensitive parameter since even cells exposed to the lowest concentration assayed, 50nM, although not chlorotic,showed a significantly reduced chlorophyll fluorescence with respect to control cells, reflected also by a reduced chlorophyll content of the biomass (Prado et al., 2011).

We have also analyzed the pigment profile of *Phaeodactylum tricornutum* cells exposed to copper (Cid et al., 1995). In this study, the increase of the intracellular pH provoked by the presence of copper induced the alteration of the proportion of the chlorophyll allomers.

Changes in Cell Morphology

Exposure of microalgal cells to pollutants, such as metals or pesticides, can induce changes in cell morphology, both in terms of volume and shape of the cells as produced at the subcellular level changes (changes in the morphology of chloroplasts and mitochondria, appearance of cytoplasmic inclusions, alteration of membrane, and others). These structural and ultrastructural changes can be studied by light and electron microscopy (Torres et al. 2000; Yang et al., 2002) or using flow cytometry (Abalde et al., 1995; Cid et al., 1995; Rioboo et al., 2002).

Cadmium caused ultrastructural changes in *Phaeodactylum tricornutum* cells: deposition of metal on the cell surface, increase of the chloroplast volume, appearance of electrodense granulations, and reduction of lipid inclusions (Torres et al., 2000).

Chlamydomonas moewusii cells exposed to the herbicide paraquat concentrations higher than 0,15 µM, formed palmelloid colonies (clusters of non-flagellated cells closed in a common wall), observed by light microscopy, and the cellular volume and complexity analysed by flow cytometry technique increases, and this fact could be related to palmelloid colony formation, probably due to the incapacity to finish cell division as well as failures of regulation of cell volume because of the attack of oxidative radicals formed to membranes (Franqueira et al.,1999; Rioboo et al., 2008).

Cell Physiology

The toxicity exerted by environmental pollutants may be reflected in different enzyme activities, which can be induced or inhibited by the presence of the toxic agent. Enzyme inhibition measurements in microalgae are becoming increasingly popular indicators of environmental stress because they offer a rapid and sensitive endpoint. Generally tend to study enzymatic activities related to antioxidative mechanisms (catalase, peroxidase, glutathione reductase), as environmental contaminants typically trigger oxidative stress mechanisms may induce antioxidative enzymatic and non-enzymatic, which thus can be used as biomarkers of toxicity (Geoffroy et al., 2002).

Other enzymatic activities were also studied, such as non-specific esterase activity, usually using the fluorogenic substrate fluorescein diacetate, which is transformed into the fluorescent compound fluorescein by esterases, so that measuring the fluorescence resulting from the reaction can be determined the activity of these enzymes (Prado et al. 2009a). On the other hand, there are also studies of particular enzymatic activities, including various enzymes related to nitrogen metabolism. Enzymes involved in nitrogen assimilation were affected by the herbicide paraquat, being nitrate reductase activity more sensitive to paraquat than nitrite reductase, leading a significant alteration of the C/N ratio (Prado et al., 2009b).

In the case of the microalgae *Dunaliella tertiolecta*, whose growth is not a particularly sensitive parameter (Abalde et al., 1995) found that a bioassay of toxicity of various pollutants based on the activity of ß-galactosidase enzyme is more sensitive than traditional tests of growth inhibition, this enzyme bioassay is rapid, sensitive and reproducible, and correlates well with other ecologically relevant parameters such as growth (Peterson and Stauber, 1996).

The photosynthetic activity can be assessed by different methods, two of the more traditional methods of assimilation of ^{14}C-radiolabeled CO_2 and monitoring of the evolution of O_2 through electrodes as the type Clark. Since photosynthesis is one of the most common targets of pollutants, especially herbicides, the use of this physiological parameter in toxicity tests with microalgae is quite common. Inhibition of photosynthesis rapidly reflects the toxic effect of different pollutants on microalgae (Cid et al., 1995; Macinnis-Ng and Ralph, 2003; Strom et al., 2009).

A relatively short time has begun using chlorophyll fluorescence as an effective indicator of the physiological state of the photosynthetic apparatus, providing basic information on the functioning of photosynthesis. When a photosynthetic organism is exposed to light, it produces a fluorescence emission originating mainly from chlorophyll a of photosystem II. Photosynthesis and fluorescence are competing processes, so if microalgal cells adapted to darkness are illuminated, the fluorescence rapidly reaches a maximum and begins to decline with the onset of electron transport (Kautsky effect) (Mallick and Mohn, 2003). Under optimal conditions, most of the light energy absorbed by chlorophyll is dissipated via chemical conversion with a small proportion devoted to the emission of fluorescence and heat, but the photosynthetic capacity of the organism may be reduced under stress conditions, giving lead to an increase in fluorescence emission. This has been observed in the case of microalgae exposed to herbicides inhibiting electron transport at the photosystem II level (Eullaffroy and Vernet, 2003). Compared with the traditional method of assimilation of $^{14}CO_2$, the measurement of chlorophyll fluorescence to evaluate the toxicity of pollutants offers several advantages: it is done quickly, so non-invasive, without incubation, without bottle effects without radioactivity and without destroying the integrity of the cells. Additionally, you can determine various parameters related to fluorescence, which can give an idea of the primary mode of action of pollutant.

The effect of copper on photosynthesis and related parameters of a marine diatom was analyzed in our laboratory; a copper concentration of 0,5 mg l^{-1} reduced in a 50% the photosynthetic rate of *Phaeodactylum tricornutum*, measured as radioactive carbon assimilation, whereas a concentration of 0.10 mg l^{-1} is needed to reduce in a 50% the growth rate, and 0,05 mg l^{-1} of copper provoked a significant decrease in the cellular pool of ATP (Cid et al. 1995). The effect of copper on the pigment profile was mentioned above.

Damage at the DNA Level

It has been observed that many of the toxic agents that contaminate aquatic environments are able to interact with the DNA of living cells, causing genotoxic effects, so that the study of the genotoxic potential of these pollutants has become one of the main objectives of bioassays for pollution control in aquatic systems.

Genotoxic effects of environmental contaminants can be tested using a wide range of tests based on biomarkers (Ali and Kumar, 2008), but recently, and for toxicity tests with microalgae, has gained increasing importance the "comet assay", since it has been revealed as a simple, rapid and sensitive for determining genotoxicity and evaluating level of damage to the structure of DNA (Akcha et al., 2008; Li et al., 2009). This test enables the detection of various DNA lesions, such as the presence of breaks in one strand, induced by physical or chemical agents, allowing the study of each cell separately and enabling the establishment of intercellular differences in the population. In addition, this method is applicable to any eukaryotic cell and is independent of cell proliferation (Erbes et al., 1997).

The comet assay is based on single-stranded DNA breaks induced directly by genotoxic agent or as a result of alkali treatment causes the formation of so-called "comets" after DNA migration in electrophoresis in basic media. After staining the DNA with a fluorophore such as ethidium bromide or SYBR Green, damaged cells appear as comets consist of a head and a tail heavily stained diffuse DNA fragments that have migrated because of its small size. This test can detect DNA damage induced by alkylating agents, intercalating or as a consequence of oxidative stress (Henderson et al., 1998).

Potential DNA damage assayed by the comet assay, provoked by the herbicide paraquat on the freshwater microalga *Chlamydomonas moewusii* was also studied in our laboratory (Prado et al., 2009a). After only 24 h of herbicide exposure significant DNA damage was observed in microalgal cells exposed to all paraquat concentrations assayed, with a 23.67% of comets in cultures exposed to 0.05 µM, revealing the genotoxicity of this herbicide (Prado et al., 2009a).

Other Cytotoxicity Endpoints

Flow cytometry has been applied in the study of the aquatic environment since the 1980s; this technique has achieved extensive use in the study of microalgae and this technique has been introduced as an alternative to the more traditional techniques of analysing cells in culture and from natural populations (Franqueira et al., 2000).

Aerobic organisms produce reactive oxygen species (ROS) in their metabolic process, as oxygen peroxide o superoxide. The level of these ROS increases during several cytotoxic process, leading to the named oxidative stress. The fluorogenic oxidation of hydroethidine (dihydroethidium; HE) to ethidium has been used as a measure of O_2^-. Microalgal suspensions were stained with hydroethidine (HE). HE is a chemically reduced fluorophore able to cross the cell membrane and it is oxidized by superoxide ion in the cytoplasm of cells to a red product (ethidium) (Benov et al., 1998). Ethidium binds to the DNA inside cells and has a red fluorescence when is excited with blue light (Shapiro, 1995). As we have demonstrate, *Chlorella vulgaris* cells exposed to 500 nM of terbutryn, showed an important increase of the oxidative stress level, respect to control cells, after 96 h of exposure (Rioboo et al., 2009b). Membranes could be expected to be highly prone to free radical attack inasmuch as unsaturated fatty acids are major components of most membrane lipid bilayers (Cid et al., 1996). The consequences of free radical attack on membranes are numerous and include the induction of lipid peroxidation (Kellog and Fridovich, 1975), lysis (Goldstein and Weissmann, 1977), and fatty acid deesterification (Niehaus, 1978). Senescence is an active process initiated by some combination of internal and environmental triggers, and membrane deterioration is an early and fundamental feature of this process.

Calcium ion plays an important role as a mediator in the transmembrane signal transduction, being its intracellular concentration increase part of the regulation of several cell processes (Tsien, et al, 1982). The development of fluorescent probes that show a spectral response upon binding Ca^{2+} have enabled researchers to investigate changes in intracellular free Ca^{2+} concentrations (Tepikin, 2001). These fluorescent indicators are derivatives of Ca^{2+} chelators. Fluo-3 is a Ca^{2+} fluorescent indicator excited with visible light and it is essentially non-fluorescent unless bound to Ca^{2+}. After its union to Ca^{2+}, Fluo-3 increases its fluorescence between 100 and 200 times (Burchiel et al., 2000). Using the flow cytometry technique, we have seen that the intracellular calcium level is closely related with the photosynthetic metabolism of control cultures cells, since the $[Ca^{2+}]_i$ increases in darkness and decrease in the light period as be found in *Chlorella vulgaris* (Rioboo et al., 2009b). However, those cells treated with 500 nM of terbutryn showed a drastically different pattern, remaining more or less constant the level of calcium after the first light period (Rioboo et al., 2009b). This terbutryn effect confirms that changes in the intracellular calcium level in this kind of cells are a consequence of the alteration in the photosynthetic process, and an increase in the $[Ca^{2+}]_i$ can be interpreted as an early signal of cell stress.

Microorganisms in general, and microalgae in particular, are the first organisms affected by pollutants discharges in aquatic environments because they are directly in contact with the medium, separated only by the cytoplasmic membrane and the cell wall. Cellular membranes are selective, dynamic barriers that play an essential role in regulating biochemical and physiological events, so any alteration produced in the environment provokes changes in microorganism membranes. Potentiometric optical probes enable researchers to perform membrane potential measurements in cells (or organelles) too small for microelectrodes. The plasma membrane of a cell typically has a transmembrane potential as a consequence of K^+, Na^+ and Cl^- concentration gradients that are maintained by active transport processes. Potentiometric probes offer an indirect method of detecting the translocation of these ions. Increases and decreases in membrane potential -referred to as membrane hyperpolarization and depolarization, respectively- play a central role in many physiological processes, including cell signalling.

DiBAC$_4$(3) (bis-(1,3-dibutylbarbituric acid trimethine oxonol) is a slow-response probe, which exhibits potential-dependent changes in its transmembrane distribution that is accompanied by a fluorescence change. The dye enters depolarized cells where it binds to intracellular lipids. Increased depolarization results in more influx of the anionic dye and thus an increase in fluorescence. Conversely, hyperpolarization is indicated by a decrease in fluorescence. This dye is excluded from mitochondria because its overall negative charge, simplifying the membrane potential measurement in eukaryotic organisms (Jepras et al. 1995). We have observed a decrease of DiBAC$_4$(3) fluorescence respect to control cells can be observed in *Chlorella vulgaris* cells exposed to 500 nM terbutryn after 96 hours (Rioboo et al., 2009b), indicating the hyperpolarization of the cell membrane of affected cells.

CONCLUSION

It has been confirmed that inhibition of growth and photosynthesis, as well as other variables closely related to photosynthesis (ATP formation, radioactive carbon assimilation, oxygen evolution and algal fluorescence induction phenomena) reflect the toxic effects of pollutants on microalgae. Nevertheless, other relevant endpoints are less known because experimental difficulties, especially under *in vivo* conditions. The variety of the results obtained in our laboratory using the flow cytometry technique allowed us to indicate new methods for the detection of several pollutants toxicity, based on the physiological response of microalgae, with the purpose of providing an early warning signal of sublethal levels of pollution.

ACKNOWLEDGEMENT

The authors gratefully acknowledge the funding provided over the years to carry out their research in this area: the Spanish Ministerio de Educación y Ciencia (CGL2004-02037BOS) and Xunta de Galicia (PGIDIT04RFO103946PR and 08MDS-20103PR).

REFERENCES

Abalde, J.; Cid, A.; Reiriz, S.; Torres, E. and Herrero, C. 1995. Response of the marine microalga *Dunaliella tertiolecta* (Chlorophyceae) to copper toxicity in short time experiments. *Bull. Environ. Contam. Toxicol.* 54: 317-324.

Akcha, F.; Arzul, G.; Rousseau, S. and Bardouil, M. 2008. Comet assay in phytoplankton as biomarker of genotoxic effects of environmental pollution. *Mar. Environ. Res.* 66: 59-61.

Ali, D. and Kumar, S. 2008. Long-term genotoxic effect of monocrotophos different tissues of freshwater fish *Channa punctatus* (Bloch) using alkaline single cell gel electrophoresis. *Sci. Total Environ.* 405: 345-350.

Battah, M.; Shabana, E.F.; Kobbia, I.A. and Eladel, H.M. 2001. Differential effects of thiobencarb toxicity on growth and photosynthesis of Anabaena variabilis with changes in phosphate level. *Ecotoxicol. Environ. Saf.* 49: 235-239.

Benov, L.; Sztejnberg, L. and Fridovich, I. 1998. Critical evaluation of the use of hydroethidine as a measure of superoxide anion radical – a simple assay for superoxide dismutase. *Free Radic. Biol. Med.* 25: 826-831.

Burchiel, S.W.; Edwards, B.S.; Kuckuck, F.W.; Lauer, F.T.; Prossnit, E.R.; Ransom, J.T. and Sklar, L.A. 2000. Analysis of free intracellular calcium by flow cytometry: multiparameter and pharmacologic applications. *Methods* 21: 221-230.

Burrows, E.M. 1971. Assessment of pollution effects by the use of algae. *Proc. R. Soc. B.* 177: 295-306

Campanella, L.; Cubadda, F.; Sammartino, M.P. and Saoncella, A. 2001. An algal biosensor for the monitoring of water toxicity in estuarine environments. *Water Res.* 35: 69-76.

Carpenter, K.E. 1924. A study of the fauna of rivers polluted by lead mining in the Aberystwyth district of Cardiganshire. *Ann. Appl. Biol.* 11: 1-23.

Cid, A.; Fidalgo, P.; Herrero, C. and Abalde, J. 1996. Toxic action of copper on the membrane system of a marine diatom measured by flow cytometry. *Cytometry* 25: 32-36.

Cid, A.; Herrero, C.; Torres, E. and Abalde, J. 1995. Copper toxicity on the marine microalga *Phaeodactylum tricornutum*: effects on photosynthesis and related parameters. *Aquat. Toxicol.* 31: 165-174.

Cid, A.; Torres, E.; Herrero, C. and Abalde, J. 1997. Disorders provoked by copper in the marine diatom *Phaeodactylum tricornutum* in short-time exposure assays. *Cah. Biol. Mar.* 38: 201-206.

Couderchet, M. and Vernet, G. 2003. Pigments as biomarkers of exposure to the vineyard herbicide flazasulfuron in freshwater algae. *Ecotoxicol. Environ. Saf.* 55: 271-277.

El-Sheekh, M.M.; Kotkat, H.M. and Hammouda, O.H.E. 1994. Effect of atrazine herbicide on growth, photosynthesis, protein synthesis, and fatty acid composition in the unicellular green alga *Chlorella kessleri*. *Ecotoxicol. Environ. Saf.* 29: 349-358.

Erbes, M.; Webles, A.; Obst, U. and Wild, A. 1997. Detection of primary DNA damage in *Chlamydomonas reinhardtii* by means of modified microgel electrophoresis. *Environ. Mol. Mutagen.* 30: 448-458.

Eullaffroy, P. and Vernet, G. 2003. The F684/F735 chlorophyll fluorescence ratio: a potential tool for rapid detection and determination of herbicide phytotoxicity in algae. *Water Res.* 37: 1983-1990.

Femten, J. and Balls, M. 1993. Replacement of fish in ecotoxicology testing: use of bacteria, other lower organism and fish cells in vitro. In: Richardson, M. (ed.) *Ecotoxicology Monitoring*. VCH, Weinheim. pp: 71-81.

Franqueira, D.; Cid, A.; Torres, E.; Orosa, M. and Herrero, C. 1999. A comparison of the relative sensitivity of structural and functional celular responses in the alga *Chlamydomonas eugametos* exposed to the herbicide paraquat. *Arch. Environ. Contam. Toxicol.* 36: 264-269.

Franqueira, D.; Orosa, M.; Torres, E.; Herrero, C. and Cid, A. 2000. Potential use of flow cytometry in toxicity studies with microalgae. *Sci. Total Environ.* 247: 119-126.

Goldstein, I.M. and Weissmann, G. 1977. Effects of generation of superoxide anion on permeability of liposomes. *Biochim. Biophys. Res. Commun.* 70: 452-458.

González-Barreiro, O.; Rioboo, C.; Cid, A. and Herrero, C. 2004. Atrazine-induced chlorosis in *Synechococcus elongatus* cells. *Arch. Environ. Contam. Toxicol.* 46: 301-307.

Geoffroy, L.; Teisseire, H.; Couderchet, M. and Vernet, G. 2002. Effect of oxyfluorfen and diuron alone and in mixture on antioxidative enzymes of *Scenedesmus obliquus*. *Pest. Biochem. Physiol.* 72: 178-185.

Girling, A.E.; Pascoe, D.; Janssen, C.R.; Peither, A.; Wenzel, A.; Schäfer, H.; Neumeir, B.; Mitchell, G.C.; Taylor, E.J.; Maund, S.J.; Lay, J.-P.; Jüttner, I.; Crossland, N.O.; Stephenson, R.R. and Persoone, G. 2000. Development of methods for evaluating toxicity to freshwater ecosystems. *Ecotoxicol. Environ. Saf.* 45: 148-176.

González-Barreiro, O.; Rioboo, C.; Herrero, C. and Cid, A. 2006. Removal of triazine herbicides from freshwater systems using photosynthetic microorganisms. *Environ. Pollut.* 144: 266-271.

Henderson, L.; Wolfreys, A.; Fedyk, J.; Bourner, G. and Windebank, S. 1998. The ability of the comet assay to discriminate between genotoxins and cytotoxins. *Mutagenesis* 13: 89-94.

Jamers, A.; Blust, R. and De Coen, W. 2009. Omics in algae: Paving the way for systems biological understanding of algal stress phenomena? *Aquat. Toxicol.* 92: 114-121.

Jepras, R.I.; Carter, J.; Pearson, S.C.; Paul, F.E. and Wilkinson, M.J. 1995. Development of a robust flow cytometric assay for determining numbers of viable bacteria. *Appl. Environ. Microbiol.* 61: 2696-2701.

Kellog, E.W. and Fridovich, I. 1975. Superoxide, hydrogen peroxide and singlet oxygen in lipid peroxidation by the xanthine oxidase system. *J. Biol. Chem.* 250: 8812-8817.

Kobbia, I.A.; Battah, M.G.; Shabana, E.F. and Eladel, H.M. 2001. Chlorophyll a fluorescence and photosynthetic activity as tools for the evaluation of simazine toxicity to *Protosiphon botryoides* and *Anabaena variabilis*. *Ecotoxicol. Environ. Saf.* 49: 101-105.

Laws, E.A. and Chalup, M.S. 1991. A microalgal growth model. *Limnol. Oceanogr.* 35: 597-608.

Li, M.; Hu, C.; Gao, X.; Xu, Y.; Qian, X.; Brown, M.T. and Cui, Y. 2009. Genotoxicity of organic pollutants in source of drinking water on microalga *Euglena gracilis*. *Ecotoxicology* 18: 669-676.

Lürling, M. and Roessink, I. 2006. On the way to cyanobacterial blooms: impact of the herbicide metribuzin on the competition between a green alga (*Scenedesmus*) and a cyanobacterium (*Microcystis*). *Chemosphere* 65: 618-626.

Ma, J. 2005. Differential sensitivity of three cyanobacterial and five green algal species to organotins and pyrethroids pesticides. *Sci. Total Environ.* 341: 109-117.

Maccinnis-Ng, C.M.O. and Ralph, P.J. 2003. Short-term response and recovery of *Zostera capricorni* photosynthesis after herbicide exposure. *Aquat. Bot.* 76: 1-15.

Malik, J. and Mohn, F.H. 2003. Use of chlorophyll fluorescence in metal-stress research: a case study with the green microalga *Scenedesmus*. *Ecotoxicol. Environ. Saf.* 55: 64-69.

McCormick, P.V. and Cairns, J.J. 1994. Algae as indicators of environmental change. *J. Appl. Phycol.* 6: 509-526

Nie, X.; Gu, J.; Lu, J.; Pan, W. and Yang, Y. 2009. Effects of norfloxacin and butylated hydroxyanisole on the freshwater microalga *Scenedesmus obliquus*. *Ecotoxicology* 18: 677-684.

Niehaus, W.G. 1978. A proposed role of the superoxide anion as a biological nucleophile in the deesterification of phospolipids. *Bioorg. Chem.* 7: 77-84.

Pascoe, D.; Wenzel, A.; Janssen, C.; Girling, A.E.; Jüttner, I.; Fliedner, A.; Blockwell, S.J.; Maund, S.J.; Taylor, E.J.; Diecrich, M.; Persoone, G.; Verhelst, P.; Stephenson, R.R.;

Crossland, N.O.; Mitchell, G.C.; Pearson, N.; Tattersfield, L.; Lay, J.P.; Peither, A.; Neumeier, B. and Velletty, A.R. 2000. The development of toxicity tests for freshwater pollutants and their validation in stream and pond mesocosms. *Water Res.* 34: 2323-2329.

Petersen, S. and Kusk, K.O. 2000. Photosynthesis tests as an alternative to growth tests for hazard assessment of toxicant. *Bull. Environ. Contam. Toxicol.* 56: 750-757.

Peterson, S.M. and Stauber, J.L. 1996. New algal enzyme bioassay for the rapid assessment of aquatic toxicity. *Bull. Environ. Contam. Toxicol.* 56: 750-757.

Prado, R., García, R., Rioboo, C., Herrero, C., Abalde, J. and Cid, A. 2009a. Comparison of the sensitivity of different toxicity test endpoints in a microalga exposed to the herbicide paraquat. *Environ. Int.* 35: 240-247.

Prado, R., Rioboo, C., Herrero, C. and Cid, A. 2009b. The herbicide paraquat induces alterations in the elemental and biochemical composition of non-target microalgal species. *Chemosphere* 76:1440-1444.

Prado, R.; Rioboo, C.; Herrero, C. and Cid, A. 2011. Characterization of cell response in Chlamydomonas moewusii cultures exposed to the herbicide paraquat: Induction of chlorosis. *Aquat. Toxicol.* 102: 10-17.

Rioboo, C.; González, O.; Herrero, C. and Cid, A. 2002. Physiological response of freshwater microalga (*Chlorella vulgaris*) to triazine and phenylurea herbicides. *Aquat. Toxicol.* 59: 225-235.

Rioboo, C., O'Connor, J.E., Pradro, R., Herrero, C. and Cid, A. 2009a. Cell proliferation alterations in Chlorella cells under stress conditions. *Aquat. Toxicol.* 94: 229-237.

Rioboo, C., Prado, R., Herrero, C. and Cid, A. 2009b. Cytotoxic effects of pesticides on microalgae determined by flow cytometry. In: Kanzantzakis, C.M. (ed.) *Progress in Pesticides Research*. Nova Science Publishers, New York. pp: 491-506.

Shapiro, H.M. 1995. *Practical Flow Cytometry*, 3rd ed. Wiley-Lyss Inc., New York.

Strom, D.; Ralph, P.J. and Stauber, J.L. 2009. Development of a toxicity identification evaluation protocol using chlorophyll-a fluorescence in a marine microalga. *Arch. Environ. Contam. Toxicol.* 56: 30-38.

Tepikin AV (Ed) 2001. *Calcium Signalling. A Practical Approach*, 2nd ed. Oxford University Press, 272 pp.

Torres, E., Cid, A., Herrero, C. and Abalde, J. 2000. Effect of cadmium on growth, ATP content, carbon fixation and ultrastructure in the marine diatom *Phaeodactylum tricornutum* Bohlin. *Wat. Air Soil Pollut.* 117: 1-14.

Tsien, R.Y.; Pozzan, T. and Rink, T.J. 1982. Calcium homeostasis in intact lymphocytes: cytoplasmic free calcium monitored with a new, intracellulary trapped fluorescent indicator. *J. Cell Biol.* 94: 325-334.

Yang, S.; Wu, R.S.S. and Kong, R.Y.C. 2002. Physiological and cytological responses of the marine diatom *Skeletonema costatum* to 2,4-dichlorophenol. *Aquat. Toxicol.* 60: 33-41.

Chapter 12

APPLICATION OF GREEN TECHNOLOGY ON PRODUCTION OF EYES-PROTECTING ALGAL CAROTENOIDS FROM MICROALGAE

Chao-Rui Chen[1], Chieh-Ming J. Chang[1]*, Chun-Ting Shen[1], Shih-Lan Hsu[2], Bing-Chung Liau[1], Po-Yen Chen[1], and Jia-Jiuan Wu[3]

[1]Department of Chemical Engineering, National Chung Hsing University, Taichung 402, Taiwan, ROC
[2]Education and Research Department, Taichung Veterans General Hospital, Taichung 407, Taiwan, ROC
[3]Department of Nutrition, China Medical University, Taichung 404, Taiwan, ROC

ABSTRACT

This study investigated co-solvent modified supercritical carbon dioxide ($SC\text{-}CO_2$) extraction of lipids and carotenoids from the microalgal species of *Nannochloropsis oculata*. The changes in content of zeaxanthin in submicronized precipitates generated from the supercritical anti-solvent (SAS) process were also examined. The effect of operational conditions on amount, recovery of the zeaxanthin and mean size, morphology of the precipitates was obtained from experimentally designed SAS process. The mean size of particles falls within several hundreds of nano meters and is highly dependent on the injection time, the content of zeaxanthin in the particulates ranged from 65 to 71%. Finally, the biological assays including antioxidant and anti-tyrosinase abilities were tested to evidence the bioactivity of zeaxanthin. This study demonstrates that elution chromatography coupled with a SAS process is an environmentally benign method to recover zeaxanthin from *N. oculata* as well as to produce nanosize particles containing zeaxanthin from algal solutions.

---

* E-mail: cmchang@dragon.nchu.edu.tw.

Keywords: *Nannochloropsis oculata*, zeaxanthin, SC-CO$_2$ extraction, supercritical anti-solvent, elution chromatography.

INTRODUCTION

Microalgae are microscopic algae that belong to the eukaryotic microorganisms. More than 50,000 microalgal species are found worldwide, but only about 50 microalgal species have been well studied and a much lower number has been cultivated on a large scale [1]. These microalgae are usually used in food coloring agents, cosmetics and health food supplements due to the fact that they possess an abundant resource for protein, vitamins, minerals, enzymes, fatty acids and carotenoids. Two of the 50 microalgal species, *Dunaliella salina* and *Chlorella pyrenoidosa* are well-cultivated in Taiwan and are a major source of carotenoids, such as lutein and β-carotene [2, 3]. In addition to *D. salina* and *C. pyrenoidosa*, recently *Nannochloropsis oculata* was also considered a high potential resource for carotenoids and thus has been widely cultivated by algal biotechnology companies in Taiwan. The attraction of carotenoids is that they are valuable lipid-soluble component, and play an important role in human health, because the human body does not synthesize carotenoids. Therefore, humans have to absorb the carotenoids from daily food or from health food supplements in order to maintain human health. Until today, more than 400 carotenoids have been discovered and some of them have been suggested to prevent cardiovascular and other chronic diseases [4, 5].

One of these carotenoids, zeaxanthin (β,β-carotene-3,3'-diol), has more specific highly conjugated double bonds than any other carotenoids. Zeaxanthin ($C_{40}H_{56}O_2$) is the principal pigment obtained from yellow corn or from marigold and possesses important physiological functions in the human body. Dietary xanthophylls such as lutein and zeaxanthin play important roles on against of the age-related macular degeneration (AMD) especially notably on the age-related eye disease (AED) [6-8]. Lutein and zeaxanthin have been found to be accumulated in high concentration within human retina [9]. Photoprotective mechanisms are important effects to prevent the harmful reactions in plants (marigold) or in algae. Carotenoids have been suggested as key molecules in the photoprotective mechanisms. For example, β-carotene can act as an effective quencher of singlet oxygen in the photosystem II reaction center [10]. Another example, xanthophylls, especially zeaxanthin, is involved in the process of non-photochemical energy dissipation [11-15]. The algae or marigold containing high zeaxanthin can be cultivated in the environment mentioned above; the content of zeaxanthin was shown to be 2170.3 μg/g in some microalgae [3]. Therefore, in view of the literature, zeaxanthin from microalgae could be a potential substitute for fruits and vegetables as a source for carotenoids.

Nowadays, SC-CO$_2$ has been recognized as a green solvent in recovering valuable compounds from natural materials. Except for supercritical fluid extraction, supercritical anti-solvent (SAS) precipitation is another effective procedure for producing the bioactive compounds, including flavonoids, ginkgolides, lycopene, carotenoids and phenolic acid derivatives [16-23]. This approach has also been extensively applied in pigment dispersion and pharmaceutical recrystallization to produce fine particles with high yield [24, 25]. Additionally, experimental data on the phase equilibrium between SC-CO$_2$ and organic solvent are important to understand the SAS process, as supercritical, superheated, liquid and co-existing phases directly influence the morphology, size and distribution of particles [26]. Three dimensionless parameters, Reynolds number ($\rho v d / \mu$), Weber number ($\rho v^2 d / \sigma$) and super-saturation of solutes (S/S*; S: transient solubility, S*: equilibrium solubility) have been recognized as major parameters that govern the mechanism of particle formation associated with SAS precipitation [27]. Super-saturation of solutes in a high-pressure solution is important to generate micronized particles in an SC-CO_2 anti-solvent precipitation process. However, the transient super-saturation of the solutes in such a high-pressure system that is mixed with SC-CO_2 is difficult to be determined [28]. This article elucidates SC-CO_2 extractions of lipids as well as carotenoids from microalgae *N. oculata*, SAS precipitations of nano- and micro-sized zeaxanthin rich particles from purified algal solutions and the associated bioactivities.

SUPERCRITICAL FLUIDS EXTRACTION OF CAROTENOIDS FROM MICROALGAE

Quantification of Triglycerides and Carotenoids

In this study, the conversion of methyl esterification of oil compounds including the TG, DG, MG and free fatty acids (FFAs) in the extracted oil attains 99% and the amount of the FFAs were almost zero in the extracted oil measured by a HPLC method. Thus, the weight of total triglycerides is equal to that of total FAME. GC quantification of seven triglycerides was performed in a 30 m × 0.5 μm non-polar capillary column (DB-5, JandW, USA) in a gas chromatograph (GC-14B, Shimadzu, Japan). The column temperature was set to 443 K initially and programmed to increase by 58 °C/min to 488 K. Then it was set to increase by 28 °C/min to 496 K. Finally, the column temperature was increased by 18 °C/min until it reached 503 K. The injection volume was 1 μL with 3.4:1 of split ratio. The temperature of flame ionization detector was set at 553 K. The regression coefficients (R^2) of ten calibration curves of methylated triglycerides were all at least 0.99. Figure 1 reveals the GC spectra of extracted oil from microalgae.

High-performance liquid chromatography (HPLC) was performed using a Hitachi 2130 pump and 2400 UV series system (Hitachi, Ltd., Tokyo, Japan). The analysis was carried out with a reverse-phase YMC C-30 (5 μm, 250 mm × 4.6 mm) and a Phenomenex Luna security guard cartridge C-18 (5 μm, 4 mm × 2.0 mm).

The microalgal extracts were eluted using mobile phases of water (A) methanol (B) and methyl t-butyl ether (C). The eluent flow rate was maintained at 1 mL/min, the injection volume was 20 μL, and the detection wavelength and column temperature were set to 450 nm and 303 K, respectively. The elution gradients were as follows: 0 min, 10% A, 90% B, 0% C; 5 min, 4% A, 81% B, 15% C; 25 min, 4% A, 81% B, 15% C; 50 min, 4% A, 31% B, 65% C. The R^2 of carotenoids were all greater than 0.99. Figure 2 shows HPLC chromatograms of algal samples and corresponsive standards which chemical structures were depicted in Figure 3.

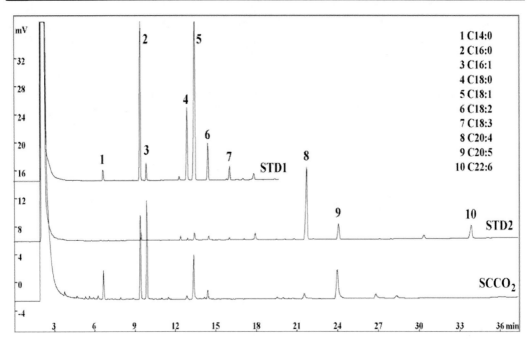

Figure 1. GC spectra of SC-CO$_2$ extract oil and standards by quantification [29].

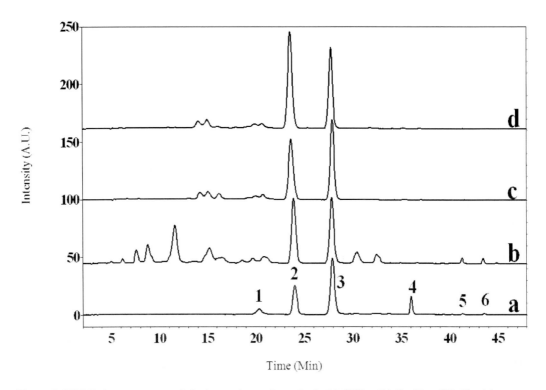

Figure 2. HPLC chromatogram of algal samples and standards (a) STDs; (b) Soxhlet-CH$_2$Cl$_2$; (c) column partition fractions; (d) SAS precipitates. (1. lutein, 2. zeaxanthin, 3. internal standard, 4. β-cryptoxanthin, 5. α-carotene, 6. β-carotene) [32].

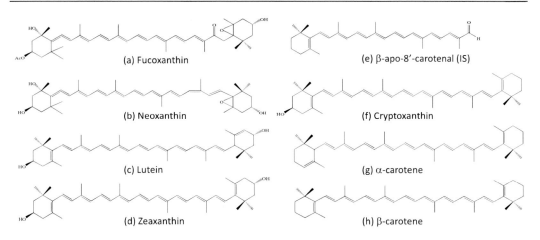

Figure 3. Structures of algal carotenoids and β-apo-8'-carotenal (used as IS) [30].

Soxhlet and Supercritical Fluid Extractions

After 16 h of Soxhlet dichloromethane (DCM) extraction, the content of zeaxanthin in the microalgal extract reached the maximum value of 1.79 mg/g$_{alga}$ that was considered to be 100% recovery of zeaxanthin. In this study, three solvents (ethanol, toluene, n-hexane) were used to evaluate extraction efficiency of zeaxanthin compared to that of the dichloromethane, shown in Table 1.

Table 1 Experimental data concerning Soxhlet extractions of carotenoids from *N. oculata* powder

Exp. #	Solvent	TY (%)	C$_{zea}$ (mg/g)	W$_{zea}$ (mg/g)	R$_{zea}$ (%)	C$_{CAR}$ (mg/g$_{ext}$)	W$_{CAR}$ (mg/g$_{alga}$)	R$_{CAR}$ (%)
1	CH$_2$Cl$_2$	8.91	20.10	1.79	100	25.15	2.24	100
2	n-Haxane	5.33	19.59	1.04	58.1	25.67	1.37	61.2
3	EtOH	23.33	7.55	1.76	98.3	8.49	1.98	88.4
4	Toluene	10.70	9.99	1.07	59.8	12.44	1.33	59.4

TY: total yield of extract = (W$_{ext}$ / W$_{feed}$) × 100%; C$_{zea}$: concentrations of zeaxanthin in extract; W$_{zea}$: the amount of zeaxanthin in microalgae; R$_{zea}$: recovery of zeaxanthin = (W$_{zea}$ / 1.79) × 100%; C$_{CAR}$: concentrations of carotenoids in extract; W$_{CAR}$: the amount of carotenoids in microalgae; R$_{CAR}$: recovery of carotenoids = (W$_{CAR}$ / 2.24) × 100%.

From the table it can be seen that the recovery of zeaxanthin was the highest value when extracted by DCM, but was close to ethanol (R$_{zea}$ = 98%). The recovery of zeaxanthin decreased significantly with decreasing polarity of the solvents, and had no relationship to the boiling point of solvents. This phenomenon was caused by the higher solubility of zeaxanthin dissolved in ethanol and DCM than those of the other solvents. However, dichloromethane is very harmful to living creatures and is a threat to the natural environment. Therefore, the less harmful solvent, ethanol, was adopted as an extraction solvent.

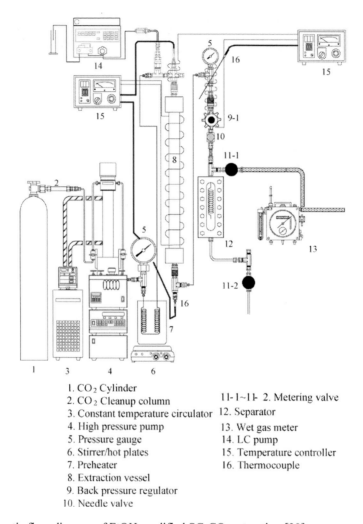

1. CO_2 Cylinder
2. CO_2 Cleanup column
3. Constant temperature circulator
4. High pressure pump
5. Pressure gauge
6. Stirrer/hot plates
7. Preheater
8. Extraction vessel
9. Back pressure regulator
10. Needle valve
11-1~11-2. Metering valve
12. Separator
13. Wet gas meter
14. LC pump
15. Temperature controller
16. Thermocouple

Figure 4. Schematic flow diagram of EtOH modified SC-CO_2 extraction [30].

Figure 4 shows a schematic flow diagram of co-solvent modified SC-CO_2 extraction. In each experiment, 10 g of algae was loaded in a 250 mL stainless steel extraction vessel (8) before extraction. A certain amount of glass wool was packed into the both ends of the extractor to prevent leaking of the sample. Liquid CO_2 flowed from a cylinder (1) which was inserted into a siphon-tube, and passed through a cooling bath (3) at 277 K. The CO_2 was then compressed to the desired pressure using a high pressure pump (100DX, ISCO, USA) (4); after coming out of the pump, the CO_2 flowed through a pre-heater (7) controlled by stirrer/hot plates (6). The flow rate of supercritical CO_2 was maintained at a constant flow rate of 20 mL/min at operative T/P. Meanwhile, the modifier (ethanol, dichloromethane, toluene or soybean oil) was continuously added to the extractor vessel at a constant rate by LC pump (14) during the extraction process. Then, the mixed fluid flowed downward into the extraction vessel where it came to extract the carotenoid from the microalgae. After SC-CO_2 went through the first back-pressure regulator (9), the solute-rich SC-CO_2 was driven into the 130 mL separator (12). Then, the extract was collected in the separator. The volume and rate of CO_2 was measured using a wet gas meter (W-NK-Da-1B, Shinagawa, Japan) (13). At the end of the SFE experiment, extract solution was released from a valve (11-2) followed by concentration of the solution by a vacuum and then it was individually weighed.

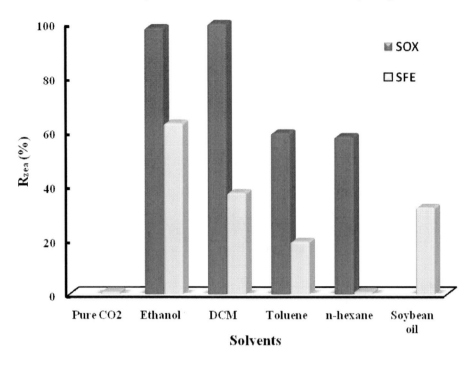

Figure 5. Effects of Soxhlet solvents and SFE modifiers on the recovery of zeaxanthin [30].

In order to improve high extraction efficiency, a few operational parameters were optimized including modifiers, solvent to solid ratio (SSR), CO_2 flow rate, temperature and pressure. 10 mL of each modifiers included ethanol, DCM, toluene and soybean oil were used.

According to the preliminary tests, suitable SFE conditions for recovering carotenoids were 120 in SSR value, 20 mL/min of CO_2 flow rate, 323 K and 350 bar [29]. Figure 5 shows the recovery of zeaxanthin was lowest (R_{zea} = 0.8%) in 99.95% CO_2 extraction [30]. As a result, nonpolar SC-CO_2 was not suitable for extraction of slightly polar carotenoids, such as zeaxanthin.

When co-solvent was added into the SC-CO_2 phase, the recovery of zeaxanthin was highest in ethanol (R_{zea} = 63.2%) due to interaction of polar ethanol and the carotenoids. It is interesting that the SFE method was quite different from that of the Soxhlet extraction. The DCM modified SC-CO_2 extraction seemed to only slightly improve solute solubilities in SC-CO_2.

Soybean oil as a modifier also slightly increased the efficiency to 32.2% of R_{zea}. These results are similar to the study involving extraction of astaxanthin using vegetable oil as a modifier by SFE [26]. Consequently, ethanol can be considered a better modifier in supercritical fluid extraction.

SUPERCRITICAL ANTI-SOLVENT PRECIPITATION OF ALGAL ZEAXANTHIN

Column Chromatography and Fractionation

A medium-pressure normal-phase column partition fractionation was adopted herein to purify zeaxanthin from the algal solution. The freeze-dried microalgae (50.0 g) were exhaustively extracted in dichloromethane with a 300 mL Soxhlet extractor. The algal extracts were concentrated under vacuum to yield a loading sample (3.6 g) for column chromatography. The loading sample was dissolved in the mixed solvents of ethyl acetate (EA) and n-hexane (1:3). The solution was subjected to a 10 cm (ID) × 30 cm (L) glass column which was packed with silica gel as the stationary phase. The feed solution was just introduced on the top of silica gel layer by a positive displacement membrane pump. A similar experimental facility was described in detail in the previous study [31]. Isocratic elution was carried out using EA and n-hexane (1:3) at the flow rate of 7 mL/min. Finally, the fractions from number 24 to 32 were collected and identified as a zeaxanthin-rich portion by HPLC analysis, and the solvent of each fraction was removed under vacuum and then weighed individually. Table 2 presents the concentration of zeaxanthin (C_{zea}) in each fraction. The C_{zea} of 24–32 fractions was enhanced from 121.2 mg/g to 530.9 mg/g. The F24–F32 portions consisting of 56.4% zeaxanthin were mixed to form materials having an average value of the C_{zea} = 436.2 mg/g, these purified samples were stored in a 193 K freezer before HPLC analysis and used for the feed of the SAS precipitation.

Table 2. Experimental data on column partition fractionation of algal extract

Entry	$W_{collected}$ (mg)	C_{zea} (mg/g)	R_{zea} (%)
F1~F22	2270.5	0	0
F23	23.5	121.2	3.9
F24	12.0	348.2	5.8
F25	9.8	443.1	6.0
F26	10.5	468.3	6.8
F27	11.0	462.1	7.0
F28	9.3	522.4	6.7
F29	8.7	530.9	6.4
F30	10.8	484.6	7.2
F31	10.3	462.3	6.6
F32	11.2	252.8	3.9
	93.6	436.2	56.4

W_{feed}: 3600 mg; C_{feed}: 20.1 mg/g_{ext}; Eluent: n-hexane:ethyl acetate = 3:1; $W_{collected}$: weight of each fraction; C_{zea}: concentrations of zeaxanthin in each fraction; R_{zea}: recovery of zeaxanthin = $[(W_{collected} \times C_{zea}) / (3600 \times 20.1)] \times 100\%$.

RSM-Designed SC-CO$_2$ Anti-Solvent Precipitation

Figure 6 shows a conceptual flow diagram of the SAS precipitation developed from the previous works [22, 23]. Liquid CO$_2$ was compressed using a high-pressure pump (Spe-ed SFE, Applied Separations, USA) into a 750 mL surge tank at a constant flow rate after it was preheated using a heat exchanger. Then, CO$_2$ flowed (18, 36, 54 g/min) through a metering valve (SS-31RS4-A, Swagelok, USA) into a visible precipitator (TST, Taiwan), and various concentrations (0.4, 0.6, 0.8 mg/mL) of feed solutions were delivered into the precipitator at a constant flow rate of 2 mL/min via a high-pressure liquid pump (L-6200A, Hitachi, Japan). A coaxial nozzle of 0.007 in. inside diameter was installed in the entrance of the precipitator to spray off the feed solution.

A stainless steel frit (37 µm) and two nylon membrane filters (0.45 µm) were placed at the bottom of precipitator to prevent the penetration of particles. The operating pressure (100, 150, 200 bar) was regulated using a back-pressure regulator (26-1722, Tescom, USA), and the operating temperature (313 K) was controlled using a water bath circulator. The consumption of CO$_2$ was measured using a drum type gas meter (TG10, Ritter, Germany). Additionally, recrystallization time (or called residence time) was also performed in this study. It represents that at constant feed flow rate, the algal solution entered and stayed in the crystallizer before leaving, which is highly related to total injected volume (i.e. injection time).

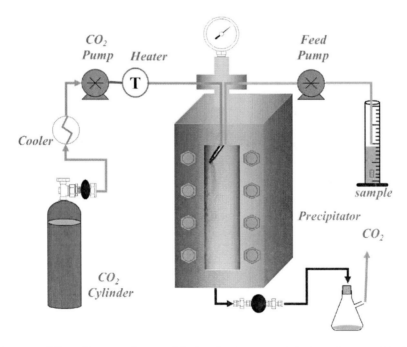

Figure 6. Conceptual flow diagram of supercritical anti-solvent precipitation of zeaxanthin-rich particulates [35].

Prior to the study of a SAS process, solubilities of algal extracts in various solvents were crucial and had to be estimated for selection of suitable feed solutions. Further investigation into these solubilities shows dichloromethane is the best solvent in which to dissolve the zeaxanthin-rich extracts but not suitable for the purification of zeaxanthin in the SAS precipitation process. Ethanol and acetone are better solvents since their solubilities decrease sharply with increasing content of zeaxanthin in the extracts. Acetone was finally selected as a solvent to dissolve the algal extracts to form feed solutions for the SAS process study. Response surface methodology (RSM) based on the center composite scheme for three design variables (pressure, feeding concentration and flow rate of CO_2) with six axial points, eight factor points and one center point was employed for the SC-CO_2 anti-solvent precipitation. Table 3 presents experimental data concerning this RSM-designed. One major response of the C_{zea} was analyzed using a quadratic regression model, which represents the quality of precipitates changed with pressure, solution concentration and flow rate of CO_2 at the center point of P = 150 bar, C_{feed} = 0.6 mg/g and Q_{CO_2} = 36 g/min [32]. The concentration of zeaxanthin increased as pressure and CO_2 flow rate increased, and it decreased as solution concentration increased from 0.4 to 0.8 mg/mL at high pressures. By viewing published literatures, SC-CO_2 extractions modified with co-solvents have increased solubility of the carotenoids [33, 34].

Table 3. RSM-designed SC-CO_2 anti-solvent precipitation at 313 K

RSM #	P (bar)	C_{feed} (mg/mL)	Q_{CO2} (g/min)	C_{zea} (mg/g)	TY (%)	R_{zea} (%)
1(F)	100	0.4	18	486.3	68.3	76.1
2(A)	100	0.6	36	578.4	64.8	85.9
3(F)	100	0.8	18	485.9	80.1	89.2
4(F)	100	0.8	54	653.2	58.6	87.8
5(F)	100	0.4	54	561.1	62.8	80.8
6(A)	150	0.6	54	673.7	58.1	89.7
7(C)	150	0.6	36	602.9	52.4	72.4
8(A)	150	0.4	36	660.7	55.4	83.9
9(A)	150	0.8	36	614.2	55.9	78.7
10(A)	150	0.6	18	597.3	61.0	83.5
11(A)	200	0.6	36	541.1	58.2	72.2
12(F)	200	0.8	18	544.4	65.3	81.5
13(F)	200	0.4	18	655.2	55.2	82.9
14(F)	200	0.4	54	622.3	51.8	73.9
15(F)	200	0.8	54	575.7	58.1	76.7

P: pressure of precipitator; C_{feed}: concentration of feed; Q_{CO2}: flow rate of CO_2; TY: total yield of precipitates = ($W_{crystallization}$ / W_{feed}) × 100%; C_{zea}: concentrations of zeaxanthin in precipitates; R_{zea}: recovery of zeaxanthin = [(10 × TY × C_{zea} / 1000) / (10 × 436.2 / 1000)] × 100%; Q_{feed}: 2 mL/min; $C_{zea, feed}$ = 436.2 mg/g; W_{feed}: 10 mg weight of feed.

However, solubility of zeaxanthin at supercritical carbon dioxide anti-solvent condition (200 bar and 313 K) is quite low so that it is a good reason to precipitate and recover zeaxanthin from the algal solution using the SAS process without adding co-solvents.

Morphology and Particle Size Distribution

Figures 7(a) and (b) display the morphology of precipitate as they vary with pressure. The higher operating pressure produced a smaller size of flake-type precipitates.

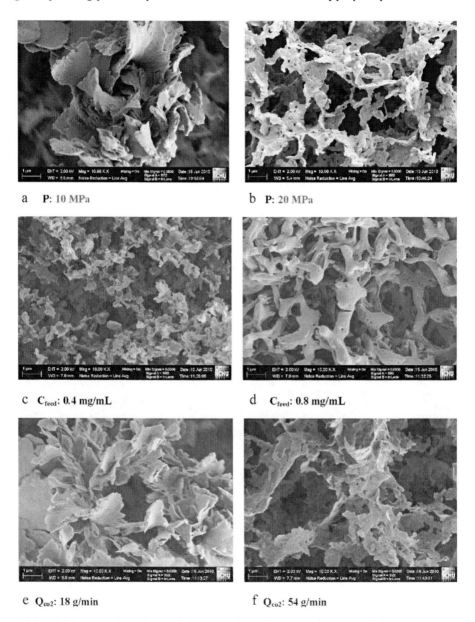

a P: 10 MPa

b P: 20 MPa

c C_{feed}: 0.4 mg/mL

d C_{feed}: 0.8 mg/mL

e Q_{co2}: 18 g/min

f Q_{co2}: 54 g/min

Figure 7. FESEM images of precipitated algal samples at (a) P: 100 bar, C_{feed}: 0.6 mg/mL, Q_{CO2}: 36 g/min; (b) P: 200 bar, C_{feed}: 0.6 mg/mL, Q_{CO2}: 36 g/min; (c) P: 150 bar, C_{feed}: 0.4 mg/mL, Q_{CO2}: 36 g/min; (d) P: 150 bar, C_{feed}: 0.8 mg/mL, Q_{CO2}: 36 g/min; (e) P: 200 bar, C_{feed}: 0.8 mg/mL, Q_{CO2}: 18 g/min; (f) P: 200 bar, C_{feed}: 0.8 mg/mL, Q_{CO2}: 54 g/min [32].

The degree of super-saturation increased as pressures increased above the critical point of the mixture; this was due to a decrease of solute concentration in the solution mixed with CO_2. Figures 7(c) and (d) indicate solution concentrations increasing from 0.4 mg/mL to 0.8

mg/mL result in an increase in the size of the snow flower-type precipitates due to a larger mass input. Figures 7(e) and (f) show a thin size of the snow flower-type precipitates can be found with high CO_2 flow rate. The reason may be that a high CO_2 flow rate is associated with rapid expansion of the feed solution and nucleation of the precipitates occurs in the well mixing of acetone and carbon dioxide. For a given degree of expansion, reducing the solution concentration of zeaxanthin promotes the super-saturation of the solute and the easy formation of narrow size precipitates. In contrast, increasing the solution concentration of zeaxanthin leads the agglomeration of precipitates, such that the precipitates become large.

Figure 8 displays mean size and morphology of the precipitates varied with the recrystallization time of the SAS process [35]. Figure 8(a) shows the smallest particles having tenth of nanometers obtained at 1 min. These small particulates precipitated on the surface of the nylon membrane having 0.45 μm pore size. The mean size of these particles is around 40 nm. However, the numbers of the particles were few since the time was short. The particle size of the precipitates did increase with the time extended to 1.5 min. The precipitates were grown and piled on the seeds that lead to agglomeration and the size increasing. In this experiment, the maximum particle size is about 2 μm. Figure 9 shows that the mean particle size increased with the recrystallization time, especially in the beginning stage of the SAS process, while the purity of zeaxanthin increased slowly with the recrystallization time. The amount of zeaxanthin in the precipitates ranged from 65% to 71%.

Figure 8. SEM images of recrystallized particles by supercritical anti-solvent technique for a recrystallization time of (a) 1; (b) 1.5; (c) 5; (d) 6 min [35].

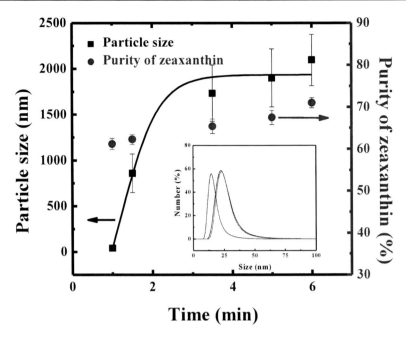

Figure 9. Mean particle size and purity of zeaxanthin of precipitates varied from time. Inset: particle size distribution for recrystallization time of 1 min (triplicate experiments) [35].

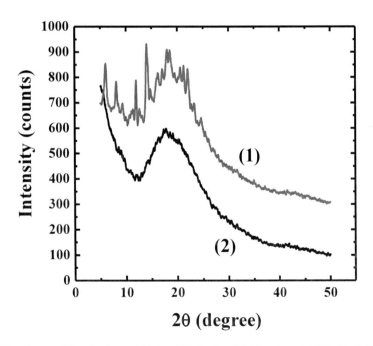

Figure 10. XRD patterns of the algal precipitates. (1): before SAS treatment (40% Zea); (2): after SAS treatment (65% Zea) [35].

The mean particle size was obtained by the following calculation. At least five particles having aspect ratio smaller than 1.5 shown in the SEM image was selected and their average sizes as well as standard deviations were calculated. Figure 10 displays two X-ray diffraction

(XRD) patterns utilized in determining the degree of crystallization of two algal precipitates. One vacuum dried sample contains 40% of zeaxanthin and the other SAS sample contains 65% of zeaxanthin. This XRD study demonstrates that the crystalline sample (40% Zea) might contain some crystalline compounds (*e.g.* polysaccharides), but the SAS process eliminated these compounds because of their high solubilities in the algal solution and resulted in an amorphous sample (65% Zea).

BIOACTIVITY ASSAYS

Antioxidant Capacity

For DPPH radical-scavenging assay, 3.94 mg of DPPH powder was dissolved in the 50 mL methanol to form a 200 μM purple blue solution. 1 mL of the purple blue solution was mixed with 3 mL of the algal extract solution at the concentration ranging from 0.1 mg/mL to 2.5 mg/mL in a quartz tube. After reacting for 30 min and the absorption of this solution was measured at 517 nm by a UV–Vis spectrophotometer (Hitachi, U-3000, Japan). The DPPH scavenging ratio of the extract was calculated to determine a half-effective concentration of the extract (i.e. EC_{50} value), which is a measure of when 50% of the DPPH free radicals in the solution were being scavenged. The scavenging ratio value is defined by $[(ABS_{blank}-ABS_{extract})/ABS_{blank}] \times 100\%$, where ABS_{blank} is the absorption without adding the microalgal extract and $ABS_{extract}$ is the absorption of the microalgal extract sample. Table 4 summarizes the results concerning free radical scavenging activity, as well as the C_{zea} in the extracts.

Table 4, DPPH radical scavenging activities and TEAC anti-oxidant activities of the extracts and purified samples from *N. Oculata*

Sample	EC_{50} (mg sample/mL)	TEAC (mmol TE/g sample)	C_{zea} (mg/g_{ext})
-Tocopherol/Trolox	0.060±0.005	4.000±0.000	-
SFE_{alga}	1.612±0.029	0.313±0.005	13.7
SOX_{alga}	1.103±0.160	0.234±0.029	20.1
29%zea	0.563±0.001	0.480±0.018	290
40%zea	0.416±0.001	0.905±0.014	400
95%zea	0.521±0.003	3.490±0.015	950

All values are represented in terms of mean ± SD (n = 3).

The EC_{50} values of DPPH scavenging of positive control and various samples, and the standard deviations were all below 0.160 mg/mL.

It has been noted that α-tocopherol used as a positive control had the best effect over all the other samples. Experimental data indicated that the DPPH scavenging ability increased with the C_{zea} values ranging from 13.17 to 400 mg/g. However, the worse activity appeared when C_{zea} = 950 mg/g. This phenomenon demonstrated that zeaxanthin does not play a major role in the antioxidant activity of scavenging DPPH free radicals.

For ABTS radical-scavenging assay, the cationic antioxidant activity of algal samples was measured by the ABTS radical decolorization assay according to the method [36] with some modification. The $ABTS^+$ cations were produced by reacting 1.8 mM of ABTS solution with 2.45 mM of potassium persulfate in the ratio of 4:1 (v/v) and allowing the mixtures to stand in the dark at room temperature for 24–28 h before use. The $ABTS^{·+}$ selection was a blue-green solution, and its absorbance was around 0.7 ± 0.05 at 734 nm without further dilution. For preparation of a calibration curve, 0.2 mL of various concentration of Trolox solution (10, 50, 100, 200, 300 and 400 µM) was mixed with 2.0 mL of $ABTS^{·+}$ solution. Water was added to a total volume of 3 mL, and it was allowed to react for 1 min. Absorbance was also measured by a UV–Vis spectrophotometer.

The blank was prepared by the addition of 1 mL of water to 2 mL of the $ABTS^{·+}$ solution. Trolox served as a standard, and the results of the microalgal samples were expressed relatively to Trolox in terms of TEAC. The measurements were made in triplicates with suitable blank solutions each time.

Table 4 also lists the TEAC test expressed as mmol of Trolox equivalent per gram of sample. The TEAC values increased with the C_{zea}, i.e., 95% zea > 40% zea > 20% zea. Nevertheless, this trend is not fit for samples of the SFE_{alga} and SOX_{alga} extracts. This might be the extracts may contain other significant radical scavenging components. The result indicates that zeaxanthin possesses effective ABTS radical scavenging ability close to the standard of Trolox.

Agar-Plate Tyrosinase Inhibition

An agar-plate model was adopted according to the method [37] with some modification to evaluate the correlation between the zeaxanthin contents and tyrosinase inhibition by the algal extracts and the SAS precipitates. The whitening tests were performed using mushroom tyrosinase (200 unit/mL) that was smeared on an agar plate that contained 97.5% distilled water, 0.1% L-Dopa, 2% agar and 0.4% 2-phenoxyethanol. The samples at certain concentrations were loaded on the surface of a piece of no. 2 filter paper which was placed on the agar-plate and then incubated at room temperature for 6 h. The bright area of the sample represents the area of tyrosinase inhibition.

Figure 11 shows that qualitative effect of the 65% zeaxanthin sample on the anti-tyrosinase activity is very close to that of Vitamin C, because that brighten area is almost identical at the dosed concentration of 1 mg/mL. The bright area and brightness representing the ability of anti-tyrosinase are in descending order of the 65% > 40% > 30% zeaxanthin samples. Consequently, the inhibition of tyrosinase activity increased with the content of zeaxanthin in the samples.

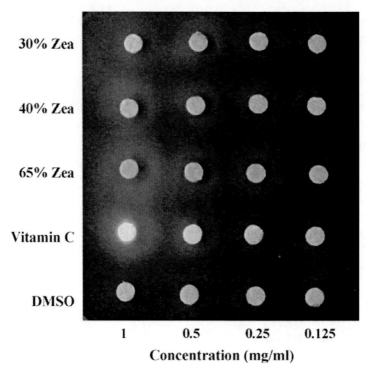

Figure 11. Agar-plate tyrosinase inhibition by column fractions and SC-CO_2 precipitates [32].

CONCLUSION

This study successfully examined the recovery of six carotenoids including bioactive zeaxanthin from *N. oculata* using column chromatography coupled with modified supercritical fluids extraction. This study also demonstrated continuous supercritical carbon dioxide anti-solvent process in producing purified and classified zeaxanthin rich particles with nano- and micro-sized from algal solution is advanced and environmentally benign. Experimental results reveal that the amount of zeaxanthin increased 21 times in the fractions by normal phase liquid column fractionation and further increased 1.5 times in the precipitates by supercritical anti-solvent recrystallization. The concentration of feed solution and the flow rate of carbon dioxide are very important to influence the amount of zeaxanthin, smaller sizes of the round-type particulates can be found in high CO_2 flow rates under a short injection time. Two kinds of chemical anti-oxidant capacities and the anti-tyrosinase activity were applied for the bioactivity assays of several algal samples. This study provides useful information for further exploration of *N. oculata* as health food supplements.

ACKNOWLEDGMENTS

The authors would like to thank the National Science Council of the Republic of China, Taiwan for financially supporting this research (NSC-98-2221-E005-053-MY3 and NSC-99-2622-B-005-005-CC2) and the Taichung Veterans General Hospital and National Chung

Hsing University, Taiwan, for partially supporting under contract no. TCVGH-NCHU 977603. This work is also supported in part by the Ministry of Education, Taiwan under the ATU plan.

REFERENCES

[1] R.L. Mendes, Supercritical fluid extraction of active compounds from algae, in: J.L. Martínez (Eds.), *Supercritical Fluid Extraction of Nutraceuticals and Bioactive Compounds*. CRC Press Taylor and Francis Group, New York, 2008, pp. 189–213.

[2] C.C. Hu, J.T. Lin, F.J. Lu, F.P. Chou, D.J. Yang, Determination of carotenoids in *Dunalilla salina* cultivated in Taiwan and antioxidant capacity of the alga carotenoid extract, *Food Chemistry* 109 (2008) 439–446.

[3] B.S. Inbaraj, J.T. Chien, B.H. Chen, Improved high performance liquid chromatographic method for determination of carotenoids in the microalga *Chlorella pyrenoidosa*, *Journal of Chromatography* A 1102 (2006) 193–199.

[4] H.D. Sesso, J.E. Buring, E.P. Norkus, J.M. Gaziano, Plasma lycopene, other carotenoids, and retinol and the risk of cardiovascular disease in women, *The American Journal of Clinical Nutrition* 79 (2004) 47–53.

[5] A.V. Rao, S. Agarwal, Role of lycopene as antioxidant carotenoid in the prevention of chronic diseases: a review, *Nutrition Research* 19 (1999) 305–323.

[6] N. Krishnadev, A.D. Meleth, E.Y. Chew, Nutritional supplements for age-related macular degeneration, *Current Opinion in Ophthalmology* 21 (2010) 184–189.

[7] S.M. Moeller, P.F. Jacques, J.B. Blumberg, The potential role of dietary xanthophylls in cataract and age-related macular degeneration, *Journal of the American College of Nutrition* 19 (2000) 522–527.

[8] J.P. SanGiovanni, E.Y. Chew, T.E. Clemons, F.L. Ferris, G. Gensler, A.S. Linblad, R.C. Milton, J.M. Seddon, R.D. Sperduto, The relationship of dietary carotenoid and vitamin A, E, and C intake with age-related macular degeneration in a case–control study, *Archives of Ophthalmology* 125 (2007) 1225–1232.

[9] P.S. Bernstein, F. Khachik, L.S. Carvalho, G.J. Muir, D.Y. Zhao, N.B. Katz, Identification and quantitation of carotenoids and their metabolites in the tissues of the human eye, *Experimental Eye Research* 72 (2001) 215–223.

[10] Telfer, S. Dhami, S.M. Bishop, D. Phillips, J. Barber, Beta-carotene quenches singlet oxygen formed by isolated photosystem II reaction centers, *Biochemistry* 33 (1994) 14469–14474.

[11] P. Horton, A. Ruban, Molecular design of the photosystem II light-harvesting antenna: photosynthesis and photoprotection, *Journal of Experimental Botany* 56 (2005) 365–373.

[12] K.K. Niyogi, Photoprotection revisited: genetic and molecular approaches, *Annual Review of Plant Biology* 50 (1999) 333–359.

[13] M. Havaux, K.K. Niyogi, The violaxanthin cycle protects plants from photooxidative damage by more than one mechanism, *Proceedings of the National Academy of Sciences of the United States of America* 96 (1999) 8762–8767.

[14] A.M. Gilmore, N. Mohanty, H.Y. Yamamoto, Epoxidation of zeaxanthin and antheraxanthin reverses non-photochemical quenching of photosystem II chlorophyll a fluorescence in the presence of trans-thylakoid delta pH, *FEBS Letters* 350 (1994) 271–274.

[15] B. Demmigadams, Carotenoids and photoprotection in plants – a role for the xanthophyll zeaxanthin, *Biochimica et Biophysica Acta* 1020 (1990) 1–24.

[16] O.J. Catchpole, J.B. Grey, K.A. Mitchell, J.S. Lan, Supercritical antisolvent fractionation of propolis tincture, *Journal of Supercritical Fluids* 29 (2004) 97–106.

[17] H.L. Hong, Q.L. Suo, L.M. Han, C.P. Li, Study on precipitation of astaxanthin in supercritical fluid, *Powder Technology* 191 (2009) 294–298.

[18] K.X. Chen, X.Y. Zhang, J. Pan, W.C. Zhang, W.H. Yin, Gas antisolvent precipitation of ginkgo ginkgolides with supercritical CO_2, *Powder Technology* 152 (2005) 127–132.

[19] M.J. Cocero, S. Ferrero, Crystallization of beta-carotene by a GAS process in batch – effect of operating conditions, *Journal of Supercritical Fluids* 22 (2002) 237–245.

[20] F. Miguel, A. Martin, T. Gamse, M.J. Cocero, Supercritical anti-solvent precipitation of lycopene – effect of the operating parameters, *Journal of Supercritical Fluids* 36 (2006) 225–235.

[21] F. Mattea, A. Martin, M.J. Cocero, Carotenoid processing with supercritical fluids, *Journal of Food Engineering* 93 (2009) 255–265.

[22] C.R. Chen, C.T. Shen, J.J. Wu, S.L. Hsu, C.J. Chang, Precipitation of 3,5-diprenyl-4-hydroxycinnamic acid in Brazilian propolis from supercritical carbon dioxide solutions, *Journal of Supercritical Fluids* 50 (2009) 176–182.

[23] J.J. Wu, C.T. Shen, T.T. Jong, C.C. Young, H.L. Yang, S.L. Hsu, C.J. Chang, C.J. Shieh, Supercritical carbon dioxide anti-solvent process for purification of micronized propolis particulates and associated anti-cancer activity, *Separation and Purification Technology* 70 (2009) 190–198.

[24] J. Fages, H. Lochard, J.J. Letourneau, M. Sauceau, E. Rodier, Particle generation for pharmaceutical applications using supercritical fluid technology, *Powder Technology* 141 (2004) 219–226.

[25] H.T. Wu, M.J. Lee, H.M. Lin, Nano-particles formation for pigment red 177 via a continuous supercritical anti-solvent process, *Journal of Supercritical Fluids* 33 (2005) 173–182.

[26] C.J. Chang, A.D. Randolph, Precipitation of microsize organic particles from supercritical fluids, *AIChE Journal* 35 (1989) 1876–1882.

[27] M. Rantakyla, M. Jantti, O. Aaltonen, M. Hurme, The effect of initial drop size on particle size in the supercritical antisolvent precipitation (SAS) technique, *Journal of Supercritical Fluids* 24 (2002) 251–263.

[28] C.J. Chang, A.D. Randolph, N.E. Craft, Separation of beta-carotene mixtures precipitated from liquid solvents with high-pressure CO_2, *Biotechnology Progress* 7 (1991) 275–278.

[29] B.C. Liau, F.P. Liang, C.T. Shen, S.E. Hong, S.L. Hsu, T.T. Jong, C.J. Chang, Supercritical fluids extraction and anti-solvent purification of carotenoids from microalgae and associated bioactivity, *Journal of Supercritical Fluids* 55 (2010) 169–175.

[30] B.C. Liau, S.E. Hong, L.P. Chang, C.T. Shen, Y.C. Li, Y.P. Wu, T.T. Jong, C.J. Shieh, S.L. Hsu, C.J. Chang, Separation of sight-protecting zeaxanthin from *Nannochloropsis*

oculata by using supercritical fluids extraction coupled with elution chromatography, *Separation and Purification Technology* 78 (2011) 1-8.

[31] L. Kao, C.R. Chen, C.J. Chang, Supercritical CO_2 extraction of turmerones from turmeric and high-pressure phase equilibrium of CO_2 + turmerones, *Journal of Supercritical Fluids* 43 (2007) 276–282.

[32] C.T. Shen, P.Y. Chen, J.J. Wu, T.M Lee, S.L. Hsu, C.J. Chang, C.C. Young, C.J. Shieh, Purification of algal anti-tyrosinase zeaxanthin from *Nannochloropsis oculata* using supercritical anti-solvent precipitation, *Journal of Supercritical Fluids* 55 (2011) 955–962.

[33] M.G. Sajilata, M.V. Bule, P. Chavan, R.S. Singhal, M.Y. Kamat, Development of efficient supercritical carbon dioxide extraction methodology for zeaxanthin from dried biomass of *Paracoccus zeaxanthinifaciens*, *Separation and Purification Technology* 71 (2010) 173–177.

[34] M. Takahashi, H. Watanabe, J. Kikkawa, M. Ota, M. Watanabe, Y. Sato, H. Inomata, N. Sato, Carotenoids extraction from Japanese persimmon (Hachiyakaki) peels by supercritical CO_2 with ethanol, *Analytical Sciences* 22 (2006) 1441–1447.

[35] P.Y. Chen, C.T. Shen, B.C. Liau, T.M Lee, T.M. Wu, C.J. Shieh, C.J. Chang, Demonstration of continuous supercritical carbon dioxide anti-solvent purification and classification of nano/micro-sized precipitates of algal zeaxanthin from *Nannochloropsis oculata*, *Journal of the Taiwan Institute of Chemical Engineers* (2011), doi:10.1016/j.jtice.2010.11.011.

[36] R. Re, N. Pellegrini, A. Proreggente, A. Pannala, M. Yang, C. Rice-Evans, Antioxidant activity applying an improved ABTS radical cation decolorization assay, *Free Radical Biology and Medicine* 26 (1999) 1231–1237.

[37] J.J. Wu, J.C. Lin, C.H. Wang, T.T. Jong, H.L. Yang, S.L. Hsu, C.J. Chang, Extraction of antioxidative compounds from wine lees using supercritical fluids and associated anti-tyrosinase activity, *Journal of Supercritical Fluids* 50 (2009) 33–41.

Chapter 13

MICROALGAE AS BIODETERIOGENS OF STONE CULTURAL HERITAGE: QUALITATIVE AND QUANTITATIVE RESEARCH BY NON-CONTACT TECHNIQUES

Ana Zélia Miller[1]*, Miguel Ángel Rogerio-Candelera[2], Amélia Dionísio[1], Maria Filomena Macedo,[3] and Cesareo Saiz-Jimenez[2]

[1]Centro de Petrologia e Geoquímica, Instituto Superior Técnico,
Av. Rovisco Pais, 1049-001 Lisboa, Portugal
[2]Instituto de Recursos Naturales y Agrobiología de Sevilla,
IRNAS-CSIC. Av. Reina Mercedes 10, 41012 Sevilla, Spain
[3]Vicarte, Faculdade de Ciências e Tecnologia,
Universidade Nova de Lisboa, Monte de Caparica,
2829-516 Caparica, Portugal

ABSTRACT

Biological colonisation of stone is one of the main problems related to monuments and buildings conservation. It is amply recognised that microalgae have the greatest ecological importance as pioneer colonisers of stone materials, conducting to aesthetic, physical and chemical damages. Their deterioration potential is related with their photoautotrophic nature, using the mineral components of stone substrates and sunlight as energy source without any presence of organic matter.

Stone biodeterioration by microalgae has been assessed by several authors. Most of the employed methodologies for microbial identification and monitoring are time-consuming and require extensive sampling. In addition, the scaffolding and sampling procedures required may also transform the researcher in a biodeteriorating agent itself. In this chapter, non-contact techniques for colonisation detection and monitoring are proposed in order to fulfil the mission of heritage preservation. *In vivo* chlorophyll *a*

---

* Tel.: +351 218419294, E-mail address: azm@fct.unl.pt (A.Z. Miller).

fluorescence and digital image analysis were applied to estimate microalgal biomass and to quantify coverage of limestone samples artificially colonised by algal communities. The results showed that *Ançã* and especially *Lecce* limestones were extensively colonised on their surfaces revealing significant epilithic growth, whereas *Escúzar* and *San Cristobal* limestones were endolithically colonised by photoautotrophic microorganisms.

The easily handled, portable and non-destructive techniques proposed allow the understanding of stone biodeterioration processes avoiding contact and damaging of the objects, which ensures a wide field of application on cultural heritage studies and the design of appropriate conservation and maintenance strategies.

1. INTRODUCTION

Stone materials have been used since the beginning of mankind. Their selection for construction purposes has been driven by questions of durability, availability, workability, cost and appearance. Despite the bewildering variety of stone types, carbonate rocks have been preferentially used as construction material. Among them, limestones were prised for their attractive appearance, ease of quarrying, workability and exceedingly distribution across the Earth's surface. Thus, some of the most remarkable architectural heritage all over the world was built in limestone. Unfortunately, we are confronted with some problems concerning their preservation. Physical and chemical weathering is observed, which induce stone disaggregation and decomposition resulting from material loss (Smith 2003). Several researchers have studied the deterioration processes occurring on limestone surfaces (Maurício et al. 2005; Dionísio 2007; Figueiredo et al. 2007). In many compact limestones, the rate of deterioration may be gradual and, given climatic conditions, largely predictable. However, there are many commonly limestone types which do not decay gradually, but instead experience episodic and sometimes catastrophic breakdown. The problem of understanding the deterioration of limestones is compounded by the large range of intrinsic properties of limestones, and by their varying responses under different climatic and environmental conditions. The interactions between these numerous and synergistically acting factors lead to a dynamic and complex process of physical, chemical and biological deterioration. This last is the cause of many types of deterioration on limestones, through a process referred as biodeterioration. Biological colonisation of cultural heritage assets, especially those exposed to outdoor environment is one of the main problems that curators have to slow down as it constitutes an important risk factor for their conservation. As photoautotrophic organisms, microalgae play an important ecological role integrating the basement of the food chain. Depending on light, carbon dioxide and a few other elements, they are the pioneer colonisers of stone surfaces, forming phototrophic biofilms which can be described as surface attached microbial communities with a clearly present photosynthetic component (Roeselers et al. 2008). Their presence attracts heterotrophic organisms contributing to the development of complex and stratified biofilms, mainly composed of a multilayer of cells embedded in a hydrated extracellular polymeric matrix which hold the cells together (Morton et al. 1998; Warscheid 2000; Roldan et al. 2003). The vital activities of microalgae, as well as those of the other components of the biofilms, have a great biodeteriorating potential which ranges from purely aesthetic to physical and chemical changes and can lead to the total disaggregation and soiling of the surface, a problem especially important in stone cultural heritage elements such as historic buildings and monuments. When microalgae colonise stone materials, they adopt different survival strategies that usually imply the fast development of variously-coloured surface patinas (Ortega-Calvo et al. 1995). Most of these coloured patinas are produced by microbial organic pigments firmly bound to the stone particles (Urzì and Realini 1998; Alakomi et al. 2004; Gorbushina 2007). The consequence is the formation of greenish to blackish biofilm generated patinas, particularly evidenced on light colour limestones (Krumbein 2004). The colour of the biofilm constitutes generally an important aesthetic damage, but this is not the only adverse consequence of microalgae colonisation on cultural assets. As an example, characteristic patterns like crack formation, micropitting and biogenic mineral deposition were detected by Sarró et al. (2006) due to the development of microalgae on the Lions Fountain at the Alhambra Palace (Granada, Spain).

If the environmental conditions are not the most suitable for life, the strategy of microalgae also imply an endolithic growth on the stone substrate, being this euendolithic when they actively dissolve the stone, cryptoendolithic, when the cells find their niche inside the rock pores and structural cavities, or chasmoendolithic, when the cells find protection on fissures and cracks of the rock (Golubic et al. 1981). The development of endolithic biofilms can produce the detachment of stone surface areas, due to the mechanical action of wetting/drying cycles of the extracellular polymeric substances, or due to chemical action of excreted metabolic products on the substrate (Miller et al. 2010a).

The universally recognised value of cultural assets, whether being stone monuments, archaeological remains, paintings, or others works of art, limits the availability of samples for the scientific study of biodeterioration processes. The need for non-contact techniques is of great importance as the researcher can be transformed itself in a biodeteriorating agent since most methodologies employed for microbial identification and monitoring requires extensive sampling.

In the last years, a number of non-destructive characterisation techniques have been developed in order to be applied on cultural heritage assets. The *in situ* application of analytical techniques previously confined to laboratory, is rather demanded, allowing non-destructive studies without sampling procedures.

In general, algal biomass dwelling on stone monuments can be quantified through chlorophyll *a* quantification techniques. However, most of these methods are based on the extraction of chlorophyll from disintegrated cells in an organic solvent and on its subsequent determination by spectrophotometry (Parsons and Strickland, 1963), fluorometry (Yentsch and Menzel, 1963) and high performance liquid chromatography (Goeyens et al., 1982). In fact, these methods are widely used in limnological researches to analyse phytoplankton, periphyton, marine or freshwater algae, as well as in monitoring programs with the purpose of ecosystem management (Macedo et al., 2000, 2001). Nevertheless, they are time-consuming, require large volume of samples to follow the temporal dynamics of photosynthetic communities and do not allow the repeated measurement in time of the same sampling unit, because of their destructive nature.

In recent years, a rapid, reliable and non-destructive chlorophyll determination method based on *in vivo* chlorophyll fluorescence was introduced in the analysis of monuments and historic buildings (Cecchi et al. 2000; Tomaselli et al. 2002; Miller et al. 2006). This method is based on the quantification of chlorophyll *a* on solid substrates through the detection of its natural fluorescence, without sampling procedures. Thus, *in vivo* chlorophyll fluorescence has been used to detect phototrophic microorganisms on monuments and to monitor preventive treatments (Cecchi et al. 2000; Tomaselli et al. 2002; Miller et al. 2006).

The quantification and monitoring of algal biofilms on surfaces can also be performed by digital image analysis techniques, which comprise the set of mathematical operations applied to detect, monitor and quantify different elements included in digital images. As digital image, we understand every pictorial representation of the data obtained by a sensor, i.e., a device capable for detecting electromagnetic radiation, for converting it into a signal and for presenting it in a picture (Chuvieco Salinero 2002). The data obtained by a sensor is directly related to the materials reflectance, which is the percent of reflected radiation in its sensitivity wavelength range. These data are translated to numerical values and ordered in a matrix in which two Cartesian coordinates define the spatial position and a third coordinate defines the reflectance value. Digital images are usually multiband images, as for the Cartesian coordinates, is commonly available more than one reflectance coordinate. This set of two Cartesian coordinates and a third reflectance coordinate receives the name of band. A typical digital photographic image is composed of three bands, each one with encoded reflectance values of the intervals 400-500, 500-600, and 600-700 nm (Blue, Green and Red bands). Thus, digital image analysis is the set of mathematical operations that can be performed with this type of images. These techniques have been used to improve the visualisation of rock art motifs (Rogerio-Candelera 2008), and to record separately different elements (in terms of nature and composition) present in mural paintings (Rogerio-Candelera et al. 2011).

In this chapter, digital image analysis was applied in combination with *in vivo* chlorophyll *a* fluorescence as rapid, reliable and non-contact techniques for the detection and monitoring of biodeterioration processes on different limestone types. With both techniques, qualitative but also quantitative information of microalgal biomass was obtained with the advantages of being low time-consuming and without the need of contact or sampling.

2. MATERIALS AND METHODS

2.1. Colonisation Experiment

The usefulness of quantifying algal biomass by non-contact techniques was illustrated by a laboratory-based stone colonisation experiment, in which five limestone types were inoculated with a culture of microalgae and cyanobacteria and incubated in a climatic chamber (Miller et al. 2010b). The limestone types tested were:

Ançã limestone (CA) – a Portuguese fine-grained, compact to oolitic tendency limestone from the Jurassic;

Lioz limestone (CL) – a Portuguese microcrystalline, very fine-grained limestone from the middle Cretaceous with stylolite joints and recrystallised bioclasts;

San Cristobal stone (SC) – a Spanish coarse-grained calcarenite of the Upper Miocene;

Escúzar stone (PF) – a Spanish heterogeneous and coarse-grained biocalcarenite from the Tortonian;

Lecce stone (PL) – an Italian fine-grained Miocene limestone, almost exclusively composed by sparite bioclasts and scarce cementation.

Before inoculation, replicates of each limestone type (3 cm height and 4.4 cm diameter) were sterilised at 120°C and 1 atm for 20 min. After cooling, the upper surface of the stone samples were inoculated with a multiple-species phototrophic culture composed of microalgae and cyanobacteria, described and tested in a previous study (Miller et al. 2009). The inoculated stone samples were immediately placed in a climatic chamber at 20±2°C and 12h dark/light cycles during 90 days of incubation. Detailed information regarding the laboratory-based colonisation experiment, as well as the petrographic and petrophysical characteristics of each lithotype, is presented in Miller et al. (2010b).

The laboratory-induced colonisation on initially uninhabited limestones presented in this chapter was achieved by inoculating stone samples with a multiple-species community culture since in nature microorganisms involved in stone biodeterioration develop in more or less complex communities because of the diversity of rock ecosystems. Consequently, the choice for the stone inoculation comprised a community of phototrophic microorganisms that are potential deteriorating agents of the selected stone materials. Furthermore, the use of a complex microbial community presents the advantage to simulate the existence of competition and/or synergy between colonising microorganisms, which act singly or in association with other microorganisms, or with physicochemical factors, to deteriorate stones (Koestler et al. 1996). In addition, the stone samples were not re-inoculated and no extra nutrients were added during the experiment. These procedures allowed the comparison of the temporal development of microalgae colonisation on five different limestones throughout two non-destructive photosynthetic biomass quantification techniques.

2.2. Quantification of Photosynthetic Biomass during the Colonisation Experiment

In vivo chlorophyll *a* fluorescence and digital image analysis were applied and compared in order to quantify and monitor the development of photosynthetic growth on the stone samples during the incubation time. Chlorophyll *a* is a photosynthetic pigment present in all photoautotrophic microorganisms, including microalgae and cyanobacteria, used to estimate the amount of photosynthetic biomass present in liquid media, soil and also on rock substrates. Moreover, this pigment is in the origin of green-coloured patinas, which is a good indicator of the presence of algal biofilms. Hence, the temporal dynamic of microalgae colonisations dwelling on stone substrates can be quantified by means of surface areas covered by these green-coloured biofilms using digital image analysis.

In Vivo Chlorophyll a Fluorescence Technique

The growth of phototrophic microorganisms on the stone samples was assessed by *in vivo* chlorophyll *a* fluorescence method. This is a non-destructive, very fast, safe and easy method for the estimation of phototrophic biofilms dwelling on solid substrates, without the extraction of chlorophyll *a* from disintegrated cells. Fluorescence properties of some compounds, such as the natural fluorescence of chlorophyll *a*, are detected with a spectrofluorometer, providing their intensity of fluorescence in counts per second (cps) which give information of their concentration in a sample. A certain excitation wavelength is selected, and a scan is performed to record the intensity versus wavelength, called an emission spectra. Chlorophyll *a* absorbs light in all regions of the visible spectrum, showing maximum absorption in the blue-violet (about 430 nm) and red regions (around 660 nm) and emitting in a wavelength of about 680 nm when light excited at 430 nm.

Stone samples of each lithotype were taken out of the chamber in triplicate at the inoculation time and after each 30 days of incubation (0, 30, 60 and 90 days). Emission spectra were determined using a spectrofluorometer SPEX Fluorolog-3 FL3-22 fitted with a fibre-optic platform (Horiba Jobin Yvon F-3000). For each stone sample five spectrofluorometric measurements were randomly carried out on the surface of the stone samples covered by the biofilm. The fibre-optic end-piece was held steady facing the sample surface at a distance of 2 mm. Measurements were performed with an excitation wavelength of 430 nm (optimum for chlorophyll *a* molecules, APHA/AWWA/WEF, 1992), slits of 4.5 nm, an integration time of 0.3 s and an increment of 1.0 nm.

Digital Image Analysis

Digital image analysis techniques constitute a low-cost and very useful set of tools allowing non-destructive recording and quantification of different elements included in digital images even when they are not recognised by eye. In this chapter, digital image analysis was focused on the detection and quantification of stone surface areas covered by algal biofilms along the incubation time. Three replicates of each lithotype were taken out of the climatic chamber after 0, 45 and 90 days of incubation and placed on millimetric paper under controlled light to ensure fixed conditions for all photographic records. The photographic recording was performed with a digital camera Kodak EasyShare P850. The generated RGB digital images recorded at different incubation times were transferred and processed on a personal computer in order to digitally rectify the geometry of the images, since they should constitute comparable series both geometrically and radiometrically making them consistent for comparative purposes. Radiometric corrections have been performed adjusting the maximum and minimum pixel values. Geometric corrections are necessary due to the different distortions introduced by several factors as the kind and structure of the employed lenses, the focal distance or the relative position of the photographic camera. In most cases, it is necessary the employ of digital photogrammetry techniques to ensure the geometrical consistence of the series. In this chapter, the comparability of the images has been ensured by means of the same lighting conditions, focal length and normal position of the camera respect to the samples. Adobe Photoshop® software was used for the digital rectification of single photographs as shown by Mark and Billo (1999). The result of these geometric corrections is a multi-layer file, in which each layer corresponds to one of the incubation stage recorded.

After radiometric and geometric rectifications, digital decorrelation of RGB images by means of Principal Components Analysis (PCA) technique and simple image classification by the application of a thresholding algorithm were performed.

Image decorrelation by PCA allows the contrast enhancement of digital images (Gillespie et al. 1986), avoiding the loss of information implicit in the methods which redefine the histograms, as the ones known as *linear*, *histogram*, or *special* stretches (Lillesand and Kiefer 2000), and also eliminating the alteration of pixel values obtained by the application of digital filtering. This make these images as suitable for image classification as the original ones, because the transformation experienced by the pixel values is purely geometric, being these linear combinations of the original values (Chuvieco Salinero 2002). One of the main explanations of the spectral differences detected in the different Principal Components bands even in optically homogenous RGB images is that they reflect different compositions. This assumption leads to the application of PCA to issues as mineral survey by means of satellite imagery (Loughlin 1991), the differentiation of phases in rock art paintings (Rogerio-Candelera et al. 2009), or the improvement in the visualisation of rock art panels, even if some figures are not visible at all (Portillo et al. 2008). The Principal Components of a digital image are calculated by means of the expression:

$$PC_j = \sum_{i=1,p} a_{ij} DN_i + R_j \tag{1}$$

where PC_j represent the pixel value corresponding to Principal Component j, a_{ij} is the coefficient applied to the pixel value of the band i in order to generate the component j and R_j a constant introduced in each component in order to avoid negative values.

Therefore, this approach allows the detection of minority elements (of different nature and composition) apparently absent in the initial RGB digital image but masked by the redundant data registered in the Red, Green and Blue bands of the image. With this decorrelation it is possible to choose the most appropriate band corresponding to each PC (PC1, PC2 or PC3) which enhance the visualisation of the photosynthetic biomass present on the stone surfaces. In our case, the colour and texture of some limestones might mask the presence of the algal biofilms on their surfaces, which could not be detected without the digital decorrelation of the images. This approach was performed using the HyperCube v. 9.5 software (*Army Geospatial Centre*, Alexandria, Virginia, USA). The application of an iterative thresholding algorithm was then considered necessary to segment the images into binary, allowing the selection of the colonised areas to be quantified. The binarisation of the images is based in the recognition of the extreme pixel values of the coverage visually identified as biomass. Typically, iterative thresholding algorithms work using the average of the foreground and background class means, establishing a new threshold (T_n) by iteration (Sezgin and Sankur 2004). The algorithm employed displays the threshold according to:

$$T = (g_{max} - g_{min}) \sum_{g=g_{min}}^{g_{mid}^*} p(g) \tag{2}$$

where g_{max} is the highest nonzero grey level, g_{min} is the lowest one, g_{mid} is the midpoint between the two assumed points of the histogram [$g_{mid}=(g_{max}+ g_{min})/2$], and $p(g)$ the probability mass function. For area estimation, the images were scaled, and the selected pixels counted. This allowed obtaining a series of numerical values, permitting the estimation of growth rates along the experimental period. All these image operations were performed using the ImageJ v. 1.38x software (*National Institutes of Health*, Bethesda, MD, USA).

3. RESULTS AND DISCUSSION

3.1. Photosynthetic Biomass Quantified by *In Vivo* Chlorophyll *a* Fluorescence

For the evaluation of the algal colonisation process during the incubation time span, *in vivo* chlorophyll *a* fluorescence was measured on the surface stone samples immediately after inoculation and after 30, 60 and 90 days of incubation. The initial fluorescence intensities obtained for CA and PL abruptly increased during the first 30 days of incubation (Figure 1a). The *in vivo* chlorophyll *a* fluorescence values after 30 days-incubation for SC and PF samples were also about three times higher than those immediately after inoculation. In general, the chlorophyll *a* fluorescence intensities increased during the first 30 days of incubation for all lithotypes, and decreased after 60 days of experimentation, with the exception of SC lithotype (Figure 1a).

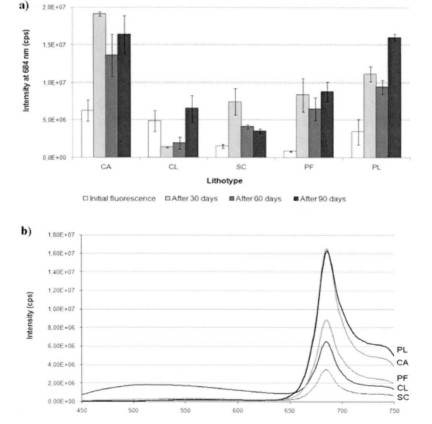

Figure 1. Intensities of chlorophyll *a* fluorescence obtained for each lithotype (excitation wavelength: 430 nm): A) Fluorescence intensity values of chlorophyll *a* at 684 nm measured immediately after inoculation (initial fluorescence), and after 30, 60 and 90 days of incubation. Each column corresponds to the mean value of an average of 15 measurements ± SD. B) Chlorophyll *a* fluorescence spectra measured after 90 days of incubation. Each lithotype spectrum is an average of 15 spectra.

This high development of green biofilms after 30 days of incubation, was conceivably due to residual culture medium (BG11) elements present in the inoculum, providing nutrients for microbial growth. In contrast, on CL surfaces this development was not observed probably due to the very compact nature of this limestone, hindering the inoculum absorption into the samples. In fact, it was verified that after 60 days-incubation a decrease occurred for all lithotypes surfaces except for CL (Figure 1a). This could be attributed to the lack of nutrients provided by total consumption of the elements present in the inoculum and to a negative adaptation to the new type of nutrients supplied by the lithic substrates. However, if growth were only determined by the culture medium elements, similar results would be obtained in all lithotypes and no re-increase of microbial biomass would occur as observed for CA, PF and PL limestones. On the other hand, for CL a great decrease was observed until 60 days of incubation, after which an increase was observed until the end of the incubation experiment. Indeed, CL was the limestone depicting the lowest quantity of chlorophyll a during the firsts 60 days-incubation, which tended to increase during the last 30 days of incubation. This suggests that the phototrophic colonisation would progressively increase if the incubation period were extended. According to Roeselers et al. (2006), the end of exponential growth does not necessarily mean that a stable climax community has established or cessed. The biofilm may be still in an adaptation state, developing slowly towards a final convergence.

The intensity of chlorophyll a fluorescence recorded for the medium-grained SC lithotype decreased after 60 days of incubation until the end of the experiment (Figure 1a), being the least colonised lithotype after 90 days-incubation. This significant decrease noticed to SC samples indicated apparent cessation of epilithic colonisation.

After the inoculum development observed during the first 30 days, PF was the only lithotype where the mean values of chlorophyll a fluorescence remained approximately the same during the 90 days of experimentation. This result suggests an adaptation state of the microalgae colonisation to this stone substrate.

The emission spectra obtained after 90 days for all limestone types showed the typical chlorophyll a fluorescence peak at 684 nm (Figure 1b). After 90 days, high fluorescence intensities were obtained for CA and PL lithotypes which presented visible biofilms formed on their surfaces. High fluorescence intensities represent a high quantity of photosynthetic biomass on the surface of the stone samples. These results obtained for CA and PL lithotypes were probably due to their fine grained textures and petrophysical characteristics (Miller et al. 2010b). As verified in Figure 1b, SC samples showed the lowest quantity of chlorophyll a and thus the lowest algal development on their upper surfaces.

3.2. Stone Coverage Areas Quantified by Digital Image Analysis

For the evaluation of the colonisation process during the incubation time span, the measurement of areas covered by the algal biofilms by means of digital image analysis was also performed. This approach allowed the monitoring of biofilm development on stone samples of the five limestone types throughout the quantification of stone surface areas covered by the green biofilms. By means of the thresholding algorithms applied to bands obtained by PCA (Figure 2), it was possible to isolate the areas covered by the biofilm and quantify the phototrophic cover through the time as represented in Figure 3.

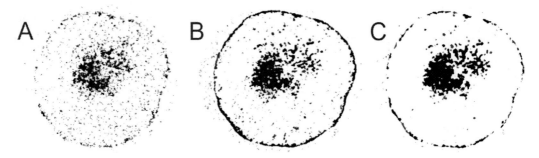

Figure 2. Thresholded areas obtained by ImageJ software for CA: A) After inoculation; B) After 45 days incubation; C) After 90 days incubation. The detected particles are then measured using ImageJ software.

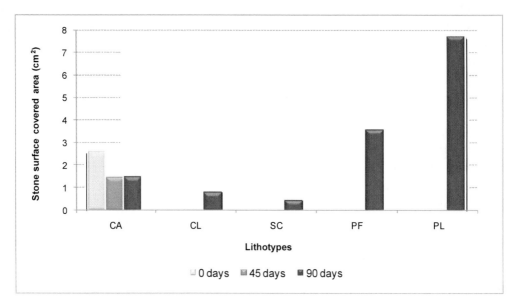

Figure 3. Stone surface areas covered by algal biofilms after 0, 45 and 90 days of incubation, quantified by digital image analysis.

According to this complementary visual monitoring technique, it was possible to assess which samples showed the most extensively colonised surfaces, i.e., a significant epilithic growth. The phototrophic culture, randomly distributed on the lithic surfaces, has grown during the incubation course leading to an increase of surface covered areas. CL showed a very slight increase of surface area covered by growth, also noticed by the *in vivo* chlorophyll *a* fluorescence technique, suggesting the progressive increase of algal colonisation if the incubation were extended. In spite of the difficulties for measuring phototrophic biofilms on PF and SC sample surfaces, strongly masked by the high macroporosity of these lithotypes, the digital image analysis approach was successful since it allowed the quantification of stone surface coverage areas. For SC samples, total surface area covered by the phototrophic biofilm did not show an increase over the course of batch incubation, being the least colonised surface samples among the studied lithotypes, as also observed by *in vivo* chlorophyll *a* fluorescence technique. Distinctively, it was noticeable the epilithic growth registered for PF, which showed a significant increase of phototrophic colonisation on its surfaces; the stone surface covered area was greater than after inoculation, showing a progressive increase during the experiment. As in the case of *in vivo* chlorophyll *a* fluorescence, CA showed a great increase of algal biofilm after the inoculation time, decreasing after 60 days of incubation. A re-increase of biomass was not observed by digital image analysis after 90 days-incubation, as noticed by *in vivo* chlorophyll *a* fluorescence.

According to both approaches, PL samples showed extensive colonised surfaces, revealing significant epilithic growth, followed by PF lithotype. In contrast, SC showed the lowest microalgal biomass. Nevertheless, according to the data presented by Miller et al. (2010) in which *in vitro* chlorophyll *a* quantification technique was combined with *in vivo* chlorophyll *a* fluorescence to analyse the five limestone types, SC and PF were the most colonised stone substrates. The authors concluded that endolithic growth occurred for these lithotypes as revealed by optical and electron microscopy of transversally cut stone samples (Miller et al. 2010b). Gathering all these data together it can be corroborated that the non-destructive techniques used in this chapter can only detect and quantify phototrophic biofilms displayed on the stone surfaces and not growing inside them. Hence, the combination of *in vivo* chlorophyll *a* fluorescence and digital image analysis techniques gave a rather good presentation of algal biomass variation and provided qualitative and quantitative evaluations of epilithic phototrophic growth on the limestones studied. Therefore, it can be concluded that on CA, PL and CL lithotypes algal colonisation occurred epilithically, whereas on SC and PF samples, the microbial growth occurred mainly inside the stone samples.

Both techniques, even though they are non-destructive and produce rapid measurements and quantified observation in less time than conventional methods, are nevertheless insufficient to detect and evaluate endolithic growth without destroying the sample.

CONCLUSION

Experimental simulations investigating stone colonisation are commonly used in ecological studies since they provide a valuable alternative for natural ecological niches by allowing experimental manipulation of the microbial ecosystem. The laboratory-based studies are of great interest for the particular case of cultural heritage materials, as is the case of the study presented in this chapter.

Our results illustrate the suitability of non-destructive methods as digital image analysis and *in vivo* chlorophyll *a* fluorescence to monitor the development of microalgae colonisations on limestone materials, even in an incipient stage when it is difficult to visually appreciate the green colour characteristic of chlorophyll. The most important advantage of the use of these methods is their non-invasivity, which allows obtaining qualitative and quantitative data of repeated samples along time without sampling procedures, and thus to contribute to the elaboration of adequate conservation strategies for cultural heritage assets. Due to the detection of phototrophic microorganisms at an early stage of development on stone surfaces, the use of *in vivo* chlorophyll *a* fluorescence and digital image analysis is also considered as an important tool to control possible relapse. Probably it would not be hazardous to state that the generalisation of the use of these techniques would be of great interest for researchers, conservators and, in last instance, for cultural assets itself. Nevertheless, endolithic growth is not detected by these techniques, which represents a major obstacle when an integral study of the stone phototrophic colonisation is needed.

ACKNOWLEDGMENTS

This study has been partially financed by Centro de Petrologia e Geoquímica do Instituto Superior Técnico (CEPGIST). The authors are grateful to REQUIMTE/CQFB, Departamento de Química, Universidade Nova de Lisboa (Portugal). This is a TCP CSD2007-00058 paper.

REFERENCES

Alakomi, H.L., Arrien, N., Gorbushina, A.A., Krumbein, W.E., Maxwell, I., McCullagh, C., Robertson, P., Ross, N., Saarela, M., Valero, J., Vendrell, M., Young, M.E. 2004. Inhibitors of biofilm damage on mineral materials (Biodam). In: Kwiatkowski, D., Löfvendahl, R. (Eds.) Proceedings of the 10th International Congress on Deterioration and Conservation of Stone, ICOMOS, Stockholm, pp. 399-406.

Alcolea, J.J., Balbín, R. 2007. C^{14} et style. La chronologie de l'art pariétal à l'heure actuelle. *L'Anthropologie* 111: 435-466.

American Public Health Association (APHA), American Water Works Association (AWWA), Water Environmental Federation (WEF), 1992. Standard methods for the examination of water and wastewater, 18th edition, APHA/AWWA/WEF, Washington, D.C.

Cecchi, G., Pantani, L., Raimondi, V., Tomaselli, L., Lamenti, G., Tiano, P., Chiari, R., 2000. Fluorescence lidar technique for the remote sensing of stone monuments. *Journal of Cultural Heritage* 1, 29-36.

Chuvieco Salinero, E., 2002. Teledetección ambiental. La observación de la Tierra desde el Espacio. Ariel, Barcelona.

Dionísio, A., 2007. Stone decay induced by fire on historical buildings: the case of the Cloister of Lisbon Cathedral (Portugal). In: Prikryl, R., Smith, B.J. (Eds), Building Stone decay: from diagnosis to conservation. *Geological Society*, London, Special Publication 271, 87-98.

Figueiredo, C.A.M., Aires-Barros, L., Basto, M.J., Graça, R.C., Maurício, A., 2007. The weathering and weatherability of Basílica da Estrela stones, Lisbon, Portugal. In: Prikryl, R., Smith, B.J. (Eds), Building stone decay: from diagnosis to conservation. *Geological Society*, London, Special Publication 271, 99-107.

Gillespie, A.R., Kahle, A.B., Walker, R.E., 1986. Color enhancement of highly correlated images. I. Decorrelation and HSI contrast stretches. *Remote Sensing of Environment* 20, 209-235.

Goeyens, L., Post, E., Dehairs, F., Vandenhoudt, A., Baeyens, W., 1982. The Use of High Pressure Liquid Chromatography with Fluorimetric Detection for Chlorophyll A

Determination in Natural Extracts of Chloropigments and their Degradation Products. *International Journal of Environmental Analytical Chemistry* 12, 51-63.

Golubic, S., Friedmann, I., Schneider, J., 1981. The lithobiontic ecological niche, with special reference to microorganisms. *Journal of Sedimentary Petrology* 51, 475-478.

Gorbushina, A.A., 2007. Life on the rocks. *Environmental Microbiology* 9, 1613-1631.

Krumbein, W.E., 2004. Life on and in stone – an endless story? In: Kwiatkowski, D., Löfvendahl, R. (Eds), Proceedings of the 10th International congress on deterioration and conservation of stone. ICOMOS, Stockholm, pp. 259-266.

Lillesand, T.M., Kiefer, R.W., 2000. Remote Sensing and Image Interpretation. Fourth Edition. John Wiley and sons, New York.

Loughlin, W.P., 1991. Principal component analysis for alteration mapping. *Photogrammetric Engineering and Remote Sensing* 57, 1163-1169.

Macedo, M.F., Duarte, P., Alves, M., Ferreira, J.G., Costa, V., 2000. Analysis of the deep chlorophyll maximum across the Azores Front. *Hydrobiologia* 441, 155-172.

Macedo, M.F., Duarte, P., Mendes, P., Ferreira, J.G., 2001. Annual variation of environmental variables, phytoplankton species composition and photosynthetic parameters in a Coastal Lagoon. *Journal of Plankton Research* 23, 719-732.

Maurício, A., Pacheco, A., Brito, P., Castro, B., Figueiredo, C., Aires-Barros, L., 2005. An ionic conductivity-based methodology for monitoring salt systems in monument stones. *Journal of Cultural Heritage* 6, 287-293.

Miller, A., Dionísio, A., Macedo, M.F., 2006. Primary bioreceptivity: A comparative study of different Portuguese lithotypes. *International Biodeterioration and Biodegradation* 57, 136-142.

Miller, A.Z., Laiz, L., Dionísio, A., Macedo, M.F., Saiz-Jimenez, C. 2009. Growth of phototrophic biofilms from limestone monuments under laboratory conditions. International *Biodeterioration and Biodegradation* 63, 860-867.

Miller, A.Z., Rogerio-Candelera, M.A., Laiz, L., Wierzchos, J., Ascaso, C., Sequeira Braga, M.A., Hernández-Mariné, M., Maurício, A., Dionísio, A., Macedo, M.F., Saiz-Jimenez, C. 2010a. Laboratory-induced endolithic growth in calcarenites: biodeteriorating potential assessment. *Microbial Ecology* 60, 55-68.

Miller, A.Z., Leal, N., Laiz, L., Rogerio-Candelera, M.A., Silva, R.J.C., Dionísio, A., Macedo, M.F., Saiz-Jimenez, C. 2010b. Primary bioreceptivity of limestones applied on Mediterranean Basin monuments. In: B.J. Smith, M. Gomez-Heras, H.A. Viles, J. Cassar (Eds.) Limestone in the Built Environment: Present Day Challenges for the Preservation of the Past. Geological Society, London, Special Publications 331, 79-92.

Morton, L.H.G., Greenway, D.L.A., Gaylarde, C.C., Surman, S.B., 1998. Consideration of some implications of the resistance of biofilms to biocides. *International Biodeterioration and Biodegradation* 41, 247-259.

Ortega-Calvo, J.J., Ariño, X., Hernandez-Marine, M., Saiz-Jimenez, C., 1995. Factors affecting the weathering and colonization of monuments by phototrophic microorganisms. *Science of the Total Environment* 167, 329-341.

Parsons, T.R., Strickland, J.D.H., 1963. Discussion of spectrophotometric determination of marine pigments with revised equations for ascertaining chlorophylls and carotenoids. *Journal of Marine Research* 21, 155-63.

Portillo, M.C., Rogerio Candelera, M.A., Gonzalez, J.M., Saiz-Jimenez, C. 2008. Estudios preliminares de la diversidad microbiana y análisis de imagen de las manifestaciones parietales en los abrigos de Fuente del Trucho y de Muriecho L (Colungo, Huesca). In: S. Rovira, M. Garcia-Heras, M. Gener, I. Montero (Eds.) Actas del VII Congreso Ibérico de Arqueometría. CSIC, Madrid, pp. 97-107.

Roeselers, G., van Loosdrecht, M.C.M., Muyzer, G. 2008. Phototrophic biofilms and their potential applications. *Journal of Applied Phycology* 20, 227-235.

Rogerio-Candelera, M.A., 2008. Una propuesta no invasiva para la documentación integral del arte rupestre. M Sc Thesis, University of Seville.

Rogerio-Candelera, M.A., Vanhaecke, F., Resano, M., Marzo, P., Porca, E., Alloza Izquierdo, R., Sáiz-Jiménez, C., 2009. Combinación de análisis de imagen y técnicas analíticas para la distinción de diferentes fases en un panel rupestre (La Coquinera II, Obón, Teruel). In: J.A. López Mira, R. Martínez Valle, C. Matamoros de Villa (Eds.) El Arte Rupestre del Arco Mediterráneo de la Península Ibérica. 10 años en la lista del Patrimonio Mundial de la Unesco. Generalitat Valenciana, Valencia, pp. 327-334.

Rogerio-Candelera, M.A., Jurado, V., Laiz, L., Saiz-Jimenez, C., 2011. Laboratory and in situ assays of digital image analysis based protocols for biodeteriorated rock and mural paintings recording. *Journal of Archaeological Science* DOI:10.1016/j.jas.2011.04.020.

Roldan, M., Clavero, E., Hernández-Marine, M., 2003. Aerophytic biofilms in dim habitats. In: Saiz-Jiménez, C. (Ed), Molecular biology and cultural heritage. A.A. Balkema, Lisse, pp. 163-169.

Saiz-Jimenez, C., Ariño, X. 1995. "Biological colonization and deterioration of mortars by phototrophic organisms". *Materiales de Construcción* 45(240): 5-16.

Sarró, M.I., Garcia, A.M., Rivalta, V.M., Moreno, D.A., Arroyo, I., 2006. Biodeterioration of the Lions Fountain at the Alhambra Palace, Granada (Spain). *Building and Environment* 41, 1811-1820.

Sezgin, M., Sankur, B., 2004. Survey over image thresholding techniques and quantitative performance evaluation. *Journal of Electronic Imaging* 13, 146-165.

Smith, B.J., 2003. Background controls on urban stone decay: lessons from natural rock weathering. In: Bimblecombe, P. (Ed), The effects of air pollution on the built environment, *Air pollution reviews, Vol. 2*, Imperial College Press, pp.31-61.

Tomaselli, L., Lamenti, G., Tiano, P., 2002. Chlorophyll fluorescence for evaluating biocide treatments against phototrophic biodeteriogens. *Annals of Microbiology* 52, 197-206.

Urzì, C., Realini, M., 1998. Colour changes of Noto's calcareous sandstone as related to its colonisation by microorganisms. *International Biodeterioration and Biodegradation* 42, 45-54.

Warscheid, T., 2000. Integrated concepts for the protection of cultural artifacts against biodeterioration. In: Ciferri, O., Tiano, P., Mastromei, G. (Eds) Of microbes and Art – The role of microbial communities in the degradation and protection of cultural heritage, Kluwer Academic, New York, pp. 185-201.

Yentsch, C.S., Menzel, D.W., 1963. A method for determination of chlorophyll and phaeophytin by fluorescence. *Deep-Sea Research* 10, 221-31.

In: Microalgae: Biotechnology, Microbiology and Energy
Editor: Melanie N. Johnsen

ISBN 978-1-61324-625-2
© 2012 Nova Science Publishers, Inc.

Chapter 14

ASTAXANTHIN PRODUCTION IN CYSTS AND VEGETATIVE CELLS OF THE MICROALGA *HAEMATOCOCCUS PLUVIALIS* FLOTOW

C. Herrero, M. Orosa, J. Abalde, C. Rioboo, and A. Cid
Laboratory of Microbiology, University of A Coruna, Spain

INTRODUCTION

Carotenoids are isoprenoid polyene pigments widely distributed in nature. They are the main source of the red, orange or yellow colour of many edible fruits (lemons, peaches, apricots, oranges, strawberries, cherries, and others), vegetables (carrots and tomatoes), mushrooms (milk-caps), and flowers. They are also found in animal products: eggs, crustaceans (lobsters, crabs and shrimps) and fish (salmonids) (De Saint Blanquat, 1988).

Carotenoids are synthesized *de novo* by all photosynthetic organisms, including microalgae. Some fungi and non-photosynthetic bacteria can also produce certain carotenoids. Animals lack the ability to synthesize carotenoids *de novo*, but some vertebrates and invertebrates are capable of metabolizing some carotenoids ingested in their diets producing chemical modifications of their structure.

Carotenoids are also found in different organs of higher plant, such as leaves (mainly ß-carotene, lutein, violaxanthin, neoxanthin), fruits (lycopene in the tomato), rhizomes and roots (ß-carotene in carrots), seeds (ß-carotene in wheat), etc. (Goodwin & Britton, 1988).

All algae contain carotenoids; each algal species usually has between five and ten main compounds. Algal carotenoids exhibit high structural diversity, more than those found in higher plants. Although some carotenoids are widely distributed in algae (ß-carotene, violaxanthin, neoxanthin), others are restricted to a few species.

Carotenoids concentration in microalgae ranges from 0.1% to 2% of dry weight; however, some species accumulate much higher quantities under certain conditions. An example of this is *Dunaliella salina*, which accumulates as much as 14% of ß-carotene under conditions of nutritional stress, high salinity, and high luminosity (Borowitzka *et al.*, 1984). Another example is the accumulation of astaxanthin by the green freshwater microalga

Haematococcus pluvialis. The reported astaxanthin conten of this microalgae varied from 1% up to 5% of the dry weight (Zhang *et al.*, 2009).

Primary carotenoids are associated with chlorophylls in the thylakoid; they act as light harvesting molecules with subsequent energy transfer to chlorophylls in photosystems. The light harvesting complex (LHC) consists of carotenoid-chlorophyll-protein complexes. Carotenoids also play a photoprotective role, stabilizing the molecules of chlorophyll against oxygen radicals and extreme radiation levels. Due to this function, as lipid anti-oxidants, carotenoids have been proposed as anti-cancer agents (Peto *et al.*, 1981). Secondary carotenoids may occur in the stigma of algae, where they act as photoreceptors for phototaxis. In chloroplast and thylakoid membranes some carotenoids undergo epoxidation and de-epoxidation induced by light

Other secondary carotenoids are located in the cytoplasm, as is the case of the freshwater microalga *Haematococcus pluvialis,* in which astaxanthin is accumulated in the perinuclear region of the cytoplasm of the aplanospores (Santos & Mesquita, 1984). Astaxanthin accumulataion is carried out under stress conditions, such as nutrient limitation, like chloroplastidic ß-carotene of *Dunaliella.* However, carotenoids synthesis pathways in *Haematococcus* and *Dunaliella* should be quite different, since an excess of ß-carotene in *Dunaliella salina* and *D. parva* leads to the formation of vesicles in the chloroplast (Ben-Amotz *et al.*, 1982 ; Borowitzka *et al.*, 1984), while the astaxanthin of *Haematococcus* forms vesicles surrounding the nucleus (Lang, 1968). It is possible that Golgi bodies are involved in the formation of astaxanthin (Borowitzka & Borowitzka, 1988).

It is remarkable the production of astaxanthin in the yeast *Phaffia rhodozyma,* which this carotenoid appears to protect against oxidative stress; its synthesis is stimulated by oxygen radicals. In *Phaffia*, carotenoids appear to be associated with lipid globules and concentrated near the nuclear envelope (Johnson, 1992).

In photosynthetic organisms carotenoids have two well-defined functions, one in the photosynthesis itself, and another in the protection of the photosynthetic structures from photo-oxidation. In non-photosynthetic tissues of higher plants, in fungi, and in non-photosynthetic bacteria, carotenoids also carry out a photoprotective function, but the mechanism appears to be different from that in photosynthetic tissues (Goodwin, 1980).

In animals, in addition to their anti-oxidant function, carotenoids also affect growth and reproduction (Nagasawa *et al.*, 1989). Thus, juvenile salmonids accumulate pigment in their muscle, and once sexual maturity is reached the pigment is mobilized to the reproductive organs, in which pigment appears to play a role in stimulating fertility and reproduction (Schiedt *et al.*, 1985). On the other hand, ß-carotene is the precursor of retinal, the chromophore of all known visual pigments.

INDUSTRIAL APPLICATIONS OF CAROTENOIDS

Carotenoids have a wide range of industrial applications. They are mainly used as coloring agents on an industrial scale, being used in the feed industry, aquaculture and poultry. Beyond their role in pigmentation, carotenoids benefit in various aspects of animal and human health. They basically protect the cells of the body acting as quencher of reactive oxygen species (Cardozo et al., 2007). Therefore, carotenoids are used in the nutraceutical market as potent antioxidants.

As food coloring agents, the most used are ß-carotene and lycopene, but ß,ε-carotene (α carotene), γ-carotene (ß,ψ-carotene), ß-apo 8′carotenal, ß-apo-8′-carotenic ethylic acid ester are also used (De Saint Blanquat, 1988).

Since animals cannot synthesize carotenoids *de novo*, intensive aquaculture requires a diet including astaxanthin to produce coloration similar to wild fish for market acceptance (Choubert & Heinrich, 1993). The ketocarotenoids astaxanthin (3,3′-dihydroxy-ß-ß-carotene-4,4′-dione) and canthaxanthin (ß-ß-carotene-4,4′-dione) are widely used to obtain the desired flesh coloration of wild salmonids. Although the pigment canthaxanthin can also be beneficial, astaxanthin is preferable because it produces identical pigmentation to that of the flesh of wild salmonids; in addition, astaxanthin is more efficiently deposited (Torrissen, 1986) and remains more stable during the process of pigmentation in the fish (Skrede & Storebakken, 1987).

Therefore, astaxanthin is the most important and expensive colorant in the feed industry for the production of salmonids (salmon, rainbow trout), crustaceans (shrimp, lobster) and poultry (Cardozo et al., 2007). Astaxanthin is one of the most expensive components of salmon farming, accounting for about 15% of total production costs. (Cardozo et al., 2007). In nature the main source of this pigment are crustaceans, but in intensive aquaculture the carotenoids must be added in the diets (Choubert & Heinrich, 1993). Furthermore, precursors of astaxanthin and astaxanthin itself contribute to the characteristic flavor of salmon (An et al., 1989). It has also been demonstrated that these substances increase the survival time of eggs and the percentage of fertilized eggs, by protecting them against extreme conditions (Craik, 1985) and stimulating their growth (Torrissen, 1984).

In addition to its role in the coloration of aquatic animals, astaxanthin possesses several important bioactivities, including antioxidation, enhacement of immune response and anticancer activities (Zhang et al., 2009). The ketocarotenoid astaxanthin is believed to play a key role in the amelioration/prevention of several human pathological processes, such as skin UV-mediated photooxidation, inflammation, prostate and mammary carcinogenesis, ulcers due to *Helicobacter pilori* infection and age-related diseases (Cardozo et al., 2007).

Numerous studies have shown that astaxanthin has health-promoting effects in the prevention and treatment of various diseases, such as cancers, chronic inflammatory diseases, metabolic syndrome, diabetes, diabetic nephropathy, cardiovascular diseases, gastrointestinal diseases, liver diseases, neurodegenerative diseases, eye diseases, skin diseases, exercise-induced fatigue, male infertility, and HgCl2-induced acute renal failure (Yuan et al., 2011).

HAEMATOCOCCUS AS A SOURCE OF ASTAXANTHIN

While astaxanthin is ubiquitous in nature, it exists in low abundance. Natural sources of astaxanthin were mainly dependent on extraction from by-products of crustacean and certain yeast species such as *Phaaffia rodozyma*. These sources were often limited by the availability of natural resources and the low astaxanthin content (0.005%-0.4%) in *Phaffia* yeast. The microalga *Haematococcus pluvialis* (Flotow) has been reported to be the richest source of natural astaxanthin. The organism has since drawn great attention from many researchers, technology development to maximize the astaxanthin content in this alga has also become an attractive research area (Zhang et al., 2009).

H. pluvialis is an ubiquitous single cell biflagellate microalga often found in puddles, eaves and birth bathes. It frequently shows an eye-catching red to deep purple color due a massive accumulation of the secondary carotenoid astaxanthin as well as its fatty acid mono- or diesteres. Cells are in akinete forms within this resting stage and are able to survive unfavorable environmental conditions such as high light, nutrient depletion and eve complete desiccation (Grewe & Griehl, 2008).

H. pluvialis has two types of cell morphology depending on its environmental conditions; green motile and non-motile forms. Under optimal growth conditions the cells are green vegetative cells capable of actively swimming with two flagella and of increasing in number. The pigmentation profile of *Haematococcus* in its green vegetative stage is essentially the same as that of higher plants: chlorophylls *a* and *b*, and the carotenoids ß-carotene, lutein, violaxanthin, neoxanthin and zeaxanthin (Ricketts, 1970).

Under unfavorable conditions, or under various forms of environmental stress, the green vegetative cells cease to be motile, increase their volume drastically, lose their flagella, form a hard, thick cyst-like wall, which may contain sporopollenin (Burczyk, 1987) and enter a resting stage. In this stage astaxanthin is synthesized in the cyst cells, which are then marked by a red color due to astaxanthin accumulation (Kang et al., 2006). Astaxanthin is first deposited around the nucleus and then radially extended until all the protoplast acquires a red coloring. The two processes, encystment and astaxanthin accumulation, are generally coupled, but they are in fact distinct processes and can be experimentally separated in time (Bubrick, 1991).

Resistant forms of this microalga can accumulate as much as 1%-2% astaxanthin in the total dry biomass of microalgae (Borowitzka, 1992), mainly in the form of the monoester fatty acids $C_{16:0}$, $C_{18:0}$, $C_{20:0}$ and $C_{18:1}$, in the early stationary phase, when cells lack a thick cell wall (Renstrom et al., 1981). In later stages, when cells present a thick cell wall, di-esters are predominant, indicating a esterification in the final stages of development (Grung et al., 1992). Mature cysts may contain as much as 3% dry weight in esterified astaxanthin, and values as high as 5% have been reported (Czygan, 1968; Renstrom et al., 1981). Astaxanthin esterification with fatty acids would be a mechanism to concentrate this chromophore in cytoplasmic globules in order to maximize its photoprotective efficiency (Renstrom et al., 1981). These resistant red cells are photosynthetically competent, but their photosynthetic activity is very reduced. This reduction in the photosynthetic activity is mainly due to the lack of cytochrome *f*, and, consequently, the absence of an electron flow from photosystem II (PS II) to photosystem I (PS I), and, to a lesser extent, a decrease in some components of PS II and PS I. The loss and decrease of photosynthetically essential proteins in the aplanaspores may be due to a decrease in the processes of new synthesis and repair, because of the stress imposed by unfavorable environmental conditions (Tan et al., 1995).

Some species of the genus *Haematococcus* can accumulate up to 6-8% (w/w) astaxanthin (Tsavalos et al., 1992); however, there are important problems for the large-scale production of astaxanthin from these microalgae due to their complex life cycles, production process design, and scale-up. Relatively low growth rates and intolerance to high temperatures and high light have limited the use of these microalgae to obtain astaxanthin in open-systems (Lee & Zhang, 1999).

Haematococcus is not the only microalga able to accumulate significant amounts of astaxanthin and other related ketocarotenoids. The capability to synthesize secondary carotenoids under environmental stresses, such as nutrient deficiency (nitrogen, phosphate, etc.), intense light, high temperature, acidic pH, etc, is widely spread over the green microalgae, probably as a defense mechanism against environmental injuries (Borowitzka; Goodwin; Orosa *et al.*, 2001a; Orosa *et al.*, 2000). Astaxanthin accumulated in these microalgae is mainly as isomers 3S and 3´S (Renstrom *et al.*, 1981), with differences in the composition and the quantity of ketocarotenoids among the different species (Tsavalos *et al.*, 1992).

PATHWAY FOR THE SYNTHESIS OF THE CAROTENOID ASTAXANTHIN IN THE GENUS *HAEMATOCOCCUS*

Higher plants and green algae share the same carotenoid biosynthetic pathway to β-carotene. β-carotene ketolase and β-carotene hydroxylase catalyze further steps leading to astaxanthin in *H. pluvialis*. β-carotene ketolase is the only enzyme that exclusively participates in the secondary carotenoid pathway leading to astaxanthin (Lu *et al.*, 2010).

Astaxanthin is synthesized from β-carotene by two different ways (Chumpolkulwong *et al.*; Fan *et al.*, 1995; Fraser *et al.*; Grewe & Griehl, 2008). With the use of diphenylamine, an astaxanthin synthesis inhibitor, a synthesis pathway for *Haematococcus* has been proposed (Fan *et al.*, 1995), via echinenone, cantaxanthin and adonirubin:

β-carotene → echinenone → cantaxanthin → adonirubin → astaxanthin

Astaxanthin biosyntesis also takes place via another route: from β-carotene via β-cryptoxanthin, zeaxanthin and adonixanthin:

β-carotene → cryptoxanthin → zeaxanthin → adonixanthin → astaxanthin

In the case of the yeast *Phaffia rhodozyma* the pathway proposed by Andrewes *et al.* (Andrewes *et al.*) for the synthesis of astaxanthin differs significantly:

β-carotene → echinenone → hydroxy-echinenone → adonirubin → astaxanthin

accumulating in this case the isomer (*3R,3´R*) (Andrewes *et al.*, 1976), whereas in the case of *Haematococcus* is accumulated the stereoisomer (*3S,3´S*) (Andrewes *et al.*, 1974).

FACTORS AFFECTING GROWTH AND ACCUMULATION OF ASTAXANTHIN IN THE GENUS *HAEMATOCOCCUS*

There are two proposed strategies for commercial production. One separates in time the production of biomass (optimal growth, green stage) and pigment (permanent stress, red stage), while the other uses an approach based on continuous culture under limiting stress at steady state.

In general astaxanthin production from *H. pluvialis* is achieved though a two-stage culture: vegetative (green) and aplanospore (red) stages. In the vegetative stage, the slow growth rate, low cell concentration, and susceptibility to contamination are the major problems. In this respect, various studies were performed to improve the growth mainly on the optimization of the culture medium, light intensity and organic carbon nutrition. Although the use of sodium acetate as an organic carbon source under mixotrophic condition seems to be a favourable way of boosting cell concentration and growth rate in *H. pluvialis*, it also increases the contamination risk, particularly with bacteria and *Chlorella* sp., a green microalga able to grow very fast on organic substances. In traditional mixotrophic algal cultures an organic carbon source is present in the medium together with the other inorganic nutrients before inoculation. However, it has been developed a new approach where the cultures were initially grown phototrophically by inorganic nutrients, and sodium acetate addition was done subsequently at the end of the log phase under different light intensities. This alternative mixotrophy has several advantages against traditional mixotrophy such as a much higher cell density in a batch culture period and minimized risks of contamination owing to the shorter exposure of cells to organic carbon sources (Goksan *et al.*).

A two-stage growth process could solve the contradiction between vegetative cell growth and astaxanthin accumulation and culturing *H. pluvialis* in a closed system could reduce the risk of contamination. Different closed bioreactors were used to culture *H. pluvialis* through a two-stage growth process. Efforts were made to simplify the two-stage production process and to develop efficient one-step production process in continuous cultures. But in terms on relative efficiency the two-stage systems performs better than the one-stage systems. The productivities, efficiencies and yields for the pigment accumulation in each case have been compared and analyzed in terms of the algal basic physiology. The two-stage system performs better (by a factor of 2.5-5) than the one-stage system, and the former is best fit in an efficient mass production setup (Aflalo *et al.*, 2007). One-stage production uses an approach based on continuous culture under limiting stress at steady state, the biomass production and astaxanhin production occur simultaneously . Two-stage- production separates in time the production of biomass (green stage) and pigments (red stage). So far, both one-stage and two-stage production of astaxanthin requires closed bioreactors (at least in green stage). Culture in photobioreactors are expensive compared with open culture systems. However Zhang et al. (2009) developed a two-stage growth one-step process for cultivation of *H. pluvialis* in open pond, obtaining an average astaxanthin conten in cyst of 2.10 g per 100g-1 wt.

Many induction methods have been developed for the transformation of green cells to red cysts containing high astaxanthin contents. Induction methods can be divided into two classes according to the role of astaxanthin as an antioxidative storage molecule or as a protoprotective substance. The first class uses various environmental stresses, except for strongh light, to cause retardation of cell multiplication; for example, nitrogen starvation, excess acetate addition, salt stress, or addition of specific inhibitors of cell division (Borowitzka *et al.*, 1991; Cordero *et al.*, 1996; Harker *et al.*, 1996a; Kakizono *et al.*, 1992; Tjahjono *et al.*, 1994a)). A nitrogen deficiency would appear to be the most important factor in the triggering of cellular encystmen. If only carbon sources are available upon cessation of cell division, the cells accumulate astaxanthin as a storage material with an antioxidant function. The other class is the light induction method, which makes use of the photoprotective role of astaxanthin under high-intensity light (Boussiba & Vonshak, 1991).

When *Haematococcus* cells are exposed to high-intensity light, astaxanthin accumulation is accelerated to protect the cells against photodamage and oxidative damage. Moreover, a high-intensity light induction method, using acetate addition, has also been applied to enhance astaxanthin accumulation. Excess acetate addition generates a relative shortage of nitrogen, resulting in a high carbon/nitrogen (C/N) ratio, which triggers cyst formation and astaxanthin accumulation N (Kakizono et al., 1992; Orosa et al., 2001a; Orosa et al., 2005). However, a productive photoautrophic Induction system was established for the production of antioxidative astaxanthin by the green microalga. A favorable CO_2 concentration and controlled specific radiation rate leads to the productive encystement of *H. pluvialis* and enhanced astaxanthin synthesis (Kang et al., 2006).

Recently, it has been reported that methyl jasmonate- and gibberellins A(3) constitute molecular signals in the network of astaxanthin accumulation. Methyl jasmonate- and gibberellins A(3) treatment increased the transcription of three β-carotene ketolase genes (bkts) in *H. pluvialis* enhancing astaxanthin synthesis and accumulation. Methyl jasmonate- and gibberellins A(3) are involved in the stress responses of plants. Induction of astaxanthin accumulation by methyl jasmonate- or gibberellins A(3) without any other stimuli presents an attractive application potential. (Lu et al., 2010).

OBTAINING VEGETATIVE CELLS RICH IN ASTAXANTHIN IN *HAEMATOCOCCUS*

There are several problems for the industrial production of astaxanthin from *Haematococcus* related to the microalgal growth, the synthesis and accumulation of the pigment and the extraction of the astaxanthin from the biomass or the use of entire biomass. As it has been cited above astaxanthin production from *H. pluvialis* generally is carried out in a two-stage culture. The first stage is carried out in optimum conditions to achieve high growth rates and high cellular densities; at this stage there are not nutrient depletion, high light intensity or stress conditions. Once the desired cell density is reached, or when the microalgae have reached the stationary growth phase, different induction methods are used for the transformation of green cells to red cysts containing high astaxanthin contents. *Haematococcus* cysts rich in astaxanthin present a thick cell wall, probably composed of sporopollenin (Burczyk, 1987). The walls of the encysted *Haematococcus* cells are extremely resistant and difficult to break (Johnson & An, 1991), so the extraction of the pigment is difficult and expensive.

However, it has been reported that the synthesis of astaxanthin and the formation of resistant cells in *Haematococcus* are two completely different processes, normally coupled but that can be experimentally separated (Bubrick). Therefore, it seems possible to obtain vegetative flagellated *Haematococcus* cells rich in astaxanthin.

It has also been reported the accumulation of secondary carotenoids in flagellated vegetative *Haematococcus* cells under conditions of nitrogen deficiency and high intensity light (Grünewald et al., 1997; Hagen et al., 2000). However, it has been described that under these conditions flagellated green cells quickly lose their flagella and produce a thick cell wall, becoming first 'webbed' cells and then resistant cells or hematocysts (Borowitzka et al., 1991; Boussiba et al., 1992; Orosa et al., 2001a).

Orosa obtained vegetative cells rich in astaxanthin from *H. pluvialis*. The microalga was cultured in a modified BBM medium (Orosa *et al.*, 2005), at a temperature of 18 ± 1°C under intense light (350 µmol. photon. m^{-2}. s^{-1}), and without nitrate deficiency (12 mM); in these conditions carotenogenesis was induced whereas vegetative cells were not transformed into resistant cells or hematocysts. Astaxanthin was the main accumulated carotenoid, mainly as monoesters. However, the concentration of chlorophylls and primary carotenoids remained relatively high (Abalde *et al.*, 2005). Under the culturing conditions used for the synthesis of astaxanthin in flagellated cells of *Haematococcus pluvialis*, the cells were not subjected to any type of nutrient limitation, an organic carbon source was not needed to induce synthesis, and the temperature was not excessively high.

Regarding the discussion about the most important factor in inducing the synthesis of astaxanthin (nitrogen deficiency or high intensity light), in the case of vegetative cells it is quite clear that this pigment is synthesized and accumulated without any nitrogen limitation in the culturing medium.

High light intensity has been considered the most important factor inducing carotenogenesis in *H. plicatilis* (Harker *et al.*); light intensity of 50-60 µmol. m^{-2}. s^{-1} was the optimal intensity for growth, while the optimal intensity for synthesis of astaxanthin was much higher, about 1600 µmol. m^{-2}. s^{-1}, referring always to the production of astaxanthin in two stages: one for obtaining maximum biomass and then provoking conditions to induce the synthesis of astaxanthin. A light intensity of 350 µmol. m^{-2}. s^{-1} yields a good production of astaxanthin, accompanied by good growth and a high level of biomass production, in a one-stage culture system by Orosa *et al.* (2001 b). Culturing times did not extend beyond the sixth day, in order to ensure that more than 90% of the cells were in a vegetative form; longer time periods lead to the appearance of resistant cells, probably due to a nutritional deficiency.

Obtaining vegetative cells of *Haematococcus pluvialis* rich in astaxanthin supposes important advantages for the commercialization of this product; because these cells do not have thick cell walls indigerible for animals, it is possible used them directly as a food supplement in the diets of salmonids and other organisms, to obtain the desirable flesh color. There are another advantages in the production process itself, since, biomass production and astaxanthin production occurs in a one-stage system of short duration, in contrast to two stage systems currently used. However, the maximum astaxanthin concentration reached is lower to that obtained in two-stage production systems. Astaxanthin accumulation can be improved supplementing the culturie medium with an organic carbon source.

ASTAXANTHIN OVERPRODUCING MUTANTS

Algal strain with improved growth rate and enhanced carotenoid accumulation makes the commercial process of astaxanthin production more feasible. Genetic manipulation techniques have contributed to this purpose, but random screening is the procedure widely used for the improvement of strains (Rowlands, 1984). The main disadvantage of non-spontaneous mutations is the appearance of hidden mutations (Boura-Halfon *et al.*, 1997), that can affect cell growth.

Ultraviolet radiation has been used for the production of mutations but chemical mutagens such as ethyl methane sulphonate (EMS) or N-methyl-N-nitro N nitrosoguanidine (NTG) have also been used. Herbicides affecting carotenoid synthesis pathway are used to select resistant mutants.

Induction and selection of mutants has been a widely employed technique for strain improvement as well as for studying mechanisms of metabolic processes (Tjahjono et al., 1994b). It has been reported astaxanthin formation using norflurazon and floridone in *H. pluvialis* cultures. Mutants of *Phaffia rodozyma* have been obtained by UV exposure and by ethyl methane sulphonate (EMS) or 1-methyl-3-nitro 1 nitrosoguanidine (NTG) treatment for hyperproduction of astaxanthin. It has also been reported mutation *of Haematococcus* by UV and EMS followed by the selection of mutants resistant to compactin, nicotine, diphenylamide, fluridone or norflurazon. However, the information on astaxanthin over-producing mutants is limited.

Orosa (2001) used UV radiation to obtain *H. pluvialis* mutants using diflufenican, norfluazon, fluometuron and amitrol to select astaxanthin hyperproducing mutants. Two astaxanthin overproducing mutants were obtained: RNC-1 (resistant to the herbicide Norflurazon) and RDC-1 (resistant to Diflufenican). RDC1 mutants showed an increase in astaxanthin production of 1.7 (w/w) in flagellated vegetative cells, without differences in the production of biomass. The concentration of total astaxanthin in RNC-1, although lower than that found in RDC-1, was also improved respect to the wild strain (1.5 times higher). These mutations were not different from the wild strain in their carotenoid profile; monoesters and diesters of astaxanthin were the main secondary carotenoids accumulated under conditions of inducing synthesis. The mutants RNC-1 and RDC-1 only showed decreased biomass production in autotrophic cultures; however, in mixotrophic growth there were no significant differences.

Growing culture of green alga *H.pluvialis* was exposed to mutagens such as UV, ethyl methane sulphonate (EMS) and 1-methyl-3-nitro 1 nitrosoguanidine (NTG) and further screened over herbicide glufosinate. The survival rate of cells decreased with increasing concentrations of mutagens and herbicides. The mutants exhibited 23-59% increase in total carotenoids and astaxanthin contents. The NTG treated glufosinate resistant mutants showed increased (2.2 to 3.8 % w/w) astaxanthin content. The NTG treated glufosinate resistant mutants showed increased 82.2 to 3.8 % w/w) astaxanthin content (Kamath et al., 2008).

Another stable astaxanthin overproduction mutant (MT 2877) was obtained by chemical mutagenesis with MNNG (methyl nitro nitrosoguanidina) of a wild type (WT) of the green alga *H. pluvialis*. MT2877 was identical to the WT with respect to morphology, pigment composition and growth kinetics during the early vegetative stage of the life cycle. However, it has the ability to synthesize and accumulate about twice the astaxanthin content of the WT under high light, or under high light in the presence of excess amounts of ferrous sulphate and sodium acetate. MT2877, or other astaxanthin overproduction *Haematococcus* mutants, may offer dual benefits as compared with the wild type, by increasing cellular astaxanthin content while reducing cell mortality during stress-induces carotenogenesis (Hu et al., 2008).

All of these data support the use of mutants for the production of astaxanthin, mainly in the form of vegetative cells rich in this carotenoid.

REFERENCES

Abalde, J., Orosa, M. and Herrero, C. 2005. *Procedimiento para la obtención de células vegetativas (flageladas) de Haematococcus ricas en astaxantina.* Patent ES 2 209 570 B1 Spain.

Aflalo, C., Meshulam, Y., Zarka, A. and Boussiba, S. 2007. On the relative efficiency of two- vs. one-stage production of astaxanthin by the green alga *Haematococcus pluvialis*. *Biotechnology and Bioengineering* 98(1):300-305.

An, G. H., Schuman, D. B. and Johnson, E. A. 1989. Isolation of *Phaffia rhodozyma* mutants with increased astaxanthin content. *Appl Environ Microbiol* 55:116-124.

Andrewes, A. G., Borch, G., Liaaen-Jensen, S. and Snatzke, G. 1974. Animal carotenoids. 9. On the absolute configuration of astaxanthin and actinioerythrin. *Acta Chem Scand Ser B* 28:730-736.

Andrewes, A. G., Phaff, H. J. and Starr, M. P. 1976. Carotenoids of *Phaffia rhodozyma*, a red-pigmented fermenting yeast. *Phytochemistry* 15:1003-1007.

Ben-Amotz, A., Katz, A. and Avron, M. 1982. Accumulation of β-carotene in halotolerant algae: purification and characterization of β-carotene-rich globules from *Dunaliella bardawil* (Chlorophyceae). *J Phycol* 18:529-537.

Borowitzka, L. J., Borowitzka, M. A. and Moulton, T. P. 1984. The mass culture of *Dunaliella salina* for fine chemicals: from laboratory to pilot plant. *Hydrobiologia* 116-117:115-34.

Borowitzka, M. A. 1988. Vitamins and fine chemicals from micro-algae. Pages 153-196 *in* M. A. Borowitzka, L. J. Borowitzka, eds. *Microalgal Biotechnology.* Cambridge Univ. Press, Cambridge.

Borowitzka, M. A. 1992. Compairing carotenogenesis in *Dunaliella* and *Haematococcus*: implications for commercial production strategies. Pages 301-310 *in* T. G. Villa, J. Abalde, eds. *Profiles on Biotechnology.* Servicio de Publicaciones, Universidad de Santiago de Compostela, Santiago de Compostela.

Borowitzka, M. A. and Borowitzka, L. J. 1988. Limits to growth and carotenogenesis in laboratory and large-scale outdoor cultures of *Dunaliella salina*. Pages 371-381 *in* J. Stadler, J. Mollion, M. C. Verdus, Y. Karamanos, H. Morvan, D. Christiaen, eds. *Algal Biotechnology.* Elsevier Applied Science Publisher, Amsterdam.

Borowitzka, M. A., Huisman, J. M. and Osborn, A. 1991. Culture of the astaxanthin-producing green alga *Haematococcus pluvialis*. 1. Effects of nutrients on growth and cell type. *J Appl Phycol* 3:295-304.

Boura-Halfon, S., Rise, M., S., A. and Sivan, A. 1997. Characterization of mutants of the red microalga *Porphyridium aerugineum* (Rhodophyceae) resistant to DCMU and atrazine. *Phycologia* 36:479-487.

Boussiba, S., Fan, L. and Vonshak, A. 1992. Enhancement and determination of astaxanthin accumulation in green alga *Haematococcus pluvialis*. *Methods in enzymology* 213:386-391.

Boussiba, S. and Vonshak, A. 1991. Astaxanthin accumulation in the green alga *Haematococcus pluvialis. Plant Cell Physiol* 32(7):1077-1082.

Bubrick, P. 1991. Production of astaxanthin from *Haematococcus*. *Bioresour Technol* 38:237-239.

Burczyk, J. 1987. Cell wall carotenoids in green algae which form sporopollenins. *Phytochem* 26:121-128.

Cardozo, K. H. M., Guaratini, T., Barros, M. P., Falcao, V. R., Tonon, A. P., Lopes, N. P., Campos, S., Torres, M. A., Souza, A. O., Colepicolo, P. and others. 2007. Metabolites from algae with economical impact. *Comparative Biochemistry and Physiology C-Toxicology & Pharmacology* 146(1-2):60-78.

Cordero, B., Otero, A., Patiño, M., Arredondo, B. O. and Fábregas, J. 1996. Astaxanthin production from the green alga *Haematococcus pluvialis* with different stress conditions. *Biotechnology letters* 18(2):213-218.

Craik, J. C. A. 1985. Egg quality and egg pigment content in salmonid fishes. *Aquaculture* 41:213-226.

Czygan, F. C. 1968. Sekundar-carotinoide in Grünalgen. I. Chemie, vorkommen und faktoren, welche die Bildung dieser polyene beeinflussen. *Arch Mikrobiol* 61:81-102.

Choubert, G. and Heinrich, O. 1993. Carotenoid pigments of the green alga *Haematococcus pluvialis*: assay on rainbow trout, *Oncorhyncus mykiss*, pigmetation in comparison with synthetic astaxanthin and canthaxanthin. *Aquaculture* 112:217-226.

Chumpolkulwong, N., Kakizono, T., Ishi, H. and Nishio, N. 1997. Enzymatic conversion of ß-carotene to astaxanthin by cell-extracts of a green alga *Haematococcus pluvialis*. *Biotechnology letters* 19:443-446.

De Saint Blanquat, G. 1988. Colorantes alimentarios. Pages 275-298 in J. L. Multon, ed. *Aditivos y auxiliares de fabricación en las industrias agroalimentarias.* Ed. Acribia., Zaragoza.

Fan, L., Vonshak, A., Gabbay, R., Hirshberg, J., Cohen, Z. and Boussiba, S. 1995. The biosynthetic pathway of astaxanthin in a green alga *Haematococcus pluvialis* as indicated by inhibition with diphenylamine. *Plant Cell Physiol* 36(8):1519-1524.

Fraser, P. D., Miura, Y. and Misawa, N. 1997. *In vitro* characterization of astaxanthin biosynthetic enzymes. *J Biol Chem* 272:6128-6135.

Goksan, T., Ak, I. and Gokpinar, S. An Alternative Approach to the Traditional Mixotrophic Cultures of *Haematococcus* pluvialis Flotow (Chlorophyceae). *Journal of Microbiology and Biotechnology* 20(9):1276-1282.

Goodwin, T. W. 1980. *The biochemistry of the carotenoids*. Chapman and Hall, London. 377 pp.

Goodwin, T. W. and Britton, G. 1988. Distribution and analysis of carotenoids. Pages 61-132 in T. W. Goodwin, ed. *Plant Pigments*. Academic Press, Padstow, Cornwall.

Grewe, C. and Griehl, C. 2008. Time- and media-pedendent secondary carotenoid accumulation in *Haematococcus pluvialis*. *Biotechnology Journal* 3:1232-1244.

Grünewald, K., Hagen, C. and Braune, W. 1997. Secondary carotenoid accumulation in flagellates of the green alga *Haematococcus pluvialis*. *Eur J Phycol* 32:387-392.

Grung, M., D'Souza, F. M. L., Borowitzka, M. and Liaaen-Jensen, S. 1992. Algal carotenoids 51. Secondary carotenoids, 2. *Haematococcus pluvialis* aplanospores as a source of (3S, 3'S)-astaxanthin esters. *J Appl Phycol* 4:165-171.

Hagen, C., Grünewald, K., Schmidt, S. and Müller, J. 2000. Accumulation of secondary carotenoids in flagellates of *Haematococcus pluvialis* (Chlorophyta) is accompanied by an increase in per unit chlorophyll productivity of photosynthesis. *Eur J Phycol* 35:75-82.

Harker, M., Tsavalos, A. J. and Young, A. J. 1996a. Autotrophic Growth and Carotenoid Production of *Haematococcus pluvialis* in a 30 Liter Air-Lift Photobioreactor. *J Ferment Bioeng* 82:113-118.

Harker, M., Tsavalos, A. J. and Young, A. J. 1996b. Factors responsible for astaxanthin formation in the chlorophyte *Haematococcus pluvialis*. *Bioresource Technology*(55):207 - 214.

Hu, Z. Y., Li, Y. T., Sommerfeld, M., Chen, F. and Hu, Q. 2008. Enhanced protection against oxidative stress in an astaxanthin-overproduction *Haematococcus* mutant (Chlorophyceae). *European Journal of Phycology* 43(4):365-376.

Johnson, E. A. 1992. New advances in astaxanthin production by *Phaffia rhodozyma*. Pages 289-301 *in* T. G. Villa, J. Abalde, eds. Profiles on Biotechnology. Servicio de Publicaciones, Universidad de Santiago de Compostela. Santiago de Compostela.

Johnson, E. A. and An, G.-H. 1991. Astaxanthin from microbial sources. *Crit Rev Biotechnol* 11:297-326.

Kakizono, T., Kobayashi, M. and Nagai, S. 1992. Effect of Carbon/Nitrogen ratio on Encystment Accompanied with Astaxanthin Formation in a Green Alga, *Haematococcus pluvialis*. *J Ferment Bioeng* 74(6):403-405.

Kamath, B. S., Vidhyavathi, R., Sarada, R. and Ravishankar, G. A. 2008. Enhancement of carotenoids by mutation and stress induced carotenogenic genes in *Haematococcus pluvialis* mutants. *Bioresource Technology* 99(18):8667-8673.

Kang, C. D., Lee, J. S., Park, T. H. and Sim, S. J. 2006. Productive encystment of *Haematococcus pluvialis* by controlling a specific irradiation rate in a photoautotrophic induction system for astaxanthin production. *Journal of Industrial and Engineering Chemistry* 12(5):745-748.

Lang, N. J. 1968. Electron microscopic studies of extraplastidic astaxanthin in *Haematococcus*. *Journal of Phycology* 4:12-19.

Lee, Y. K. and Zhang, D. H. 1999. Production of astaxanthin by *Haematococcus*. Pages 196-203 *in* Z. Cohen, ed. *Chemicals from microalgae*. Taylor & Francis, Londres.

Lu, Y. D., Jiang, P., Liu, S. F., Gan, Q. H., Cui, H. L. and Qin, S. 2010. Methyl jasmonate- or gibberellins A(3)-induced astaxanthin accumulation is associated with up-regulation of transcription of beta-carotene ketolase genes (bkts) in microalga *Haematococcus pluvialis*. *Bioresource Technology* 101(16):6468-6474.

Nagasawa, H., Konishi, R., Yamamoto, K. and Ben-Amotz, A. 1989. Effects of carotene rich algae *Dunaliella* on reproduction and body growth in mice. *In vivo* 3:79-82.

Orosa, M. 2001. *Estudio de la producción de astaxantina y otros carotenoides secundarios en microalgas dulceacuícolas, con especial atención a Haematococcus pluvialis Flotow*. PhD Thesis. Universidad de A Coruña, A Coruña. 206 pp.

Orosa, M., Franqueira, D., Cid, A. and Abalde, J. 2001a. Carotenoid accumulation in *Haematococcus pluvialis* in mixotrophic growth. *Biotech letters* 23(5):373-378.

Orosa, M., Franqueira, D., Cid, A. and Abalde, J. 2005. Analysis and enhancement of astaxanthin accumulation in *Haematococcus pluvialis*. *Bioresource Technology* 96(3):373-378.

Orosa, M., Torres, E., Fidalgo, P. and Abalde, J. 2000. Production and analysis of secondary carotenoids in green algae. *Journal of Applied Phycology* 12(3-5):553-556.

Orosa, M., Valero, J. F., Herrero, C. and Abalde, J. 2001b. Comparison of the accumulation of astaxanthin in *Haematococcus pluvialis* and other green microalgae under N-starvation and high light conditions. *Biotechnology Letters* 23(13):1079-1085.

Peto, R., Doll, R., Buckley, J. D. and Sporn, M. B. 1981. Can dietary beta-carotene materially reduce human cancer rates? *Nature* 290:201-208.

Renstrom, B., Borch, G., Skulberg, O. M. and Liaaen-Jensen, S. 1981. Optical purity of (3S, 3'S)-astaxanthin from *Haematococcus pluvialis*. *Phytochem* 20:2561-2564.

Ricketts, T. R. 1970. The pigments of the prasinophyceae and related organisms. *Phytochem* 9:1835-1842.

Rowlands, R. T. 1984. Industrial strain improvement: mutagenesis and random screening procedures. *Enzyme Microb Technol* 6:3-10.

Santos, M. F. and Mesquita, J. F. 1984. Ultrastructural study of *Haematococcus lacustris* (Girod.) Rostafinski (Volvocales). 1.Some aspects of carotenogenesis. *Cytologia* 49:215-28.

Schiedt, K., Leuenberger, F. J., Vecchi, M. and Glinz, E. 1985. Absorption, retention and metabolic transformation of carotenoids in rainbow trout, salmon and chicken. *Pure and Appl Chem* 57:685-692.

Skrede, G. and Storebakken, T. 1987. Characteristics of colour in raw, baked and smoked wild and pen-reared Atlantic salmon. *J Food Sci* 51(3):804-808.

Tan, S., Cunningham, J. F. X., Youmans, M., Grabowski, B., Sun, Z. and Gantt, E. 1995. Cytochrome *f* loss in astaxanthin-accumulating red cells of *Haematococcus pluvialis* (Chlorophyceae): Comparison of photosynthetic activity, photosynthetic enzymes, and thylakoid membrane polypeptides in red and green cells. *J Phycol* 31:897-905.

Tjahjono, A. E., Hayama, Y., Kakizono, T., Terada, Y., Nishio, N. and Nagai, S. 1994a. Hyper-accumulation of astaxanthin in a green alga *Haematococcus pluvialis* at elevated temperatures. *Biotech Lett* 16:133-138.

Tjahjono, A. E., Kakizono, T., Hayama, Y., Nishio, N. and Nagai, S. 1994b. Isolation of resistant mutants against carotenoid biosynthesis inhibitors for a green alga *Haematococcus pluvialis*, and their hybrid formation by protoplast fusion for breeding of higher astaxanthin producers. *J Ferment Bioeng* 77:352-357.

Torrissen, O. J. 1984. Pigmentation of salmonids: Effect of carotenoids in eggs and start-feeding diet on survival and growth rate. *Aquaculture* 43:185-193.

Torrissen, O. J. 1986. Pigmentation of salmonids: a comparison of astaxanthin and canthaxanthin as pigment sources for rainbow trout. *Aquaculture* 53:271-278.

Tsavalos, A. T., Harker, M., Daniels, M. and Young, A. J. 1992. Secondary carotenoids synthesis in microalgae. Pages 47-51 *in* N. Murata, ed. *Research in photosynthesis.* Kluwert, Dordrecht.

Yuan, J. P., Peng, J. A., Yin, K. and Wang, J. H. 2011. Potential health-promoting effects of astaxanthin: A high-value carotenoid mostly from microalgae. *Molecular Nutrition & Food Research* 55(1):150-165.

Zhang, B., Geng, Y., Li, Z., Hu, H. and YG, L. 2009. Production of astaxanthin in open pond by two-stage growth one-step process. *Aquaculture*:275-281.

Chapter 15

NITROGEN SOLUBILITY, ANTIGENICITY, AND SAFETY EVALUATION OF AN ENZYMATIC PROTEIN HYDROLYSATE FROM GREEN MICROALGA *CHLORELLA VULGARIS*

Humberto J. Morris[1]*, Olimpia Carrillo[2], María E. Alonso[2], Rosa C. Bermúdez[1], Alfredo Alfonso[3], Onel Fong[3], Juan E. Betancourt[3], Gabriel Llauradó[1] and Ángel Almarales[4]

[1]Center for Studies on Industrial Biotechnology (CEBI).
Faculty of Natural Sciences, University of Oriente. Ave.
Patricio Lumumba s/n, Santiago de Cuba 5. CUBA CP 90500
[2]Faculty of Biology, University of Havana. 25 e/ J e I. Vedado,
Havana 4. CUBA CP 10400
[3]Center of Toxicology and Biomedicine.
Medical University of Santiago de Cuba. Santiago de Cuba 4.
CUBA CP 90400
[4]Center of Technological Applications for Sustainable Development (CATEDES),
Guantánamo, CUBA

ABSTRACT

Green microalgae biomass would represent in tropical countries an innovative proteinaceous bioresource for developing protein hydrolysates. The proteolytic modification could have special importance for the improvement of solubility of algal protein and for decreasing its residual antigenicity. This chapter examined the nitrogen solubility, residual antigenicity and safety of *Chlorella vulgaris* protein hydrolysate (Cv-PH). A high increase of nitrogen solubility in Cv-PH, with respect to *Chlorella* aqueous extract (Cv-EA) was observed over a wide pH range (2-8). Residual antigenicity of Cv-

---

* Corresponding author. e-mail address: hmorris@cebi.uo.edu.cu Phone number: 53-022-632095, Fax: 53-022-632689.

PH was measured using male guinea pigs sensitized with Cv-EA. Neither mortality nor positive anaphylaxis symptoms were observed in Cv-PH challenged animals. The safety of Cv-PH was evaluated in an oral acute toxicity study (OECD Guideline 423) and in a 28-day repeated dose oral toxicity study (OECD Guideline 407) using mice as an experimental animal model. In the acute toxicity study (at a dose of 2 000 mg/kg) neither mortality nor changes in general condition were observed over a 14-days observation period. In the repeated dose oral toxicity study (a limit test at a dose of 2 000 mg/kg) no clinical changes were found in the experimental animals. The increased hemoglobin levels and leukocyte counts, particularly neutrophils, observed in Cv-PH groups may be related to the hemopoiesis stimulatory effect reported previously in *Chlorella* protein hydrolysates. Organ weights at the end of experimentation and histopathological tests revealed no significant influence of Cv-PH. These findings indicate the safety of Cv-PH in preclinical studies. Since there were no observed adverse effects of Cv-PH in these studies, the NOAEL (no observed adverse effect level) for *Chlorella* protein hydrolysate is 2 000 mg/kg/day administered orally for 28 days. An extended knowledge of the functional properties and safety of microalgae hydrolysates can be useful in understanding their potential use in the food and pharmaceutical industries.

INTRODUCTION

The biotechnology of microalgae, closely related to the biotechnological production, use and application of microalgae, has gained considerable importance in recent decades. It has just started to tap the enormous biological resource and physiological potential of microalgal species growing in all ecological niches (Pulz et al., 2001). With the development of sophisticated culture and screening techniques, microalgal biotechnology can already meet the high demands of both the food and pharmaceutical industries (Pulz and Gross, 2004).

Because their growth requires unexpensive substrates, microalgae can be used as economical and effective biocatalysts to obtain high added-value compounds and during productive processes, the algal biomass formed may be used as a food source such as proteins (Olaizola, 2003; Liang et al., 2004; Shimizu and Li, 2006). Unfortunately, the exploitation of the biological diversity of microalgae and application in human nutrition is constrained by regulations (Görs et al., 2010). Therefore, only a few taxa are utilized for human consumption, e.g. *Dunaliella*, *Spirulina* (*Arthrospira*) and *Chlorella*, whereas the last two ones dominate the microalgal market with several thousand tons biomass per year (Pulz et al. 2001). Currently, *Chlorella* is produced by more than 70 companies (Spolaore et al., 2006) and contains several novel properties.

Chlorella vulgaris is one of the best-studied phototrophic eukaryotes. Even under unfavourable and variable conditions, e.g. light or temperature stress, this microalga is highly productive (Kessler, 1976; Wilson and Huner, 2000) and suitable for protein products sold as health foods and food supplements (Oh-Hama and Miyachi, 1988; Merchant et al., 2002; Iwamoto, 2003).

However, intact green algae *Chlorella* have a low protein digestibility due to their strong wall (Shelef and Soeder, 1980). The enzymatic hydrolysis of cell proteins has been described as a promising method to improving algae protein digestibility, which makes the product useable in human nutrition (Tchorbanov and Bozhkova, 1988; Stoilov et al., 1995; Morris et al., 2008a).

Progress in hydrolysis techniques has led to the production of hydrolysates of many food proteins, using proteolytic enzymes such as pancreatic proteases, bacterial proteases and pepsin (Guadix et al., 2000; Kislukhina, 2002). Enzymatic protein hydrolysates have been reported as suitable sources of protein for human nutrition because of their intestinal absorption, which seems to be more effective than both intact protein and free amino acids. Therefore, protein hydrolysates have been widely used in specific formulations, in order to improve nutritional and functional properties (Siemensma et al., 1993; Mahmoud, 1994). These uses include clinical applications, such as hypoallergenic infant formula, biostimulating preparations, special foods, geriatric products, therapeutic diets and sport drinks (Frokjaer, 1994; Franck et al., 2002; Manninen, 2004). In addition, many food-derived peptides possess bioactive properties (antimicrobial, immunostimulatory, antihypertensive, mineral binding, opiod, etc.) and are often multifunctional (Biziulevicius, 2004). Because of their bioactive properties, some peptides have been claimed to be potential nutraceuticals for food and pharmaceutical applications (Kim and Wijesekara, 2010).

The sources most commonly used in nutritional products are casein and whey proteins and soybean proteins (Clemente et al., 1999). In this sense, green microalgae biomass would represent in tropical countries, e.g. Cuba, an innovative proteinaceous bioresource for developing enzymatic protein hydrolysates suitable for pharmacological nutrition. The utilization of *Chlorella vulgaris* cell biomass derived from outdoor cultivation in tropical conditions for the production of a protein hydrolysate (Cv-PH) has been reported by our research group. Cv-PH enhances host defense activity *in vivo* by stimulating mechanisms involved in both innate and specific immune responses of malnourished mice (Morris et al., 2007). This effect may be due to peptide sequences hidden in *Chlorella* protein, released by enzymatic hydrolysis (Morris et al., 2008b).

Although the consumption of *Chlorella* as health food is well documented, little data are available about the nutraceutical use of *Chlorella* protein hydrolysates. These applications greatly depend on the improvement of protein solubility, a low residual antigenicity and the formulation safety. However, there is no enough evidence to support the usefulness of microalgal protein hydrolysates. This chapter reports the results of a study focused on the evaluation of nitrogen solubility, antigenicity and safety of an enzymatic protein hydrolysate obtained from green microalga *Chlorella vulgaris*. An extended knowledge of the functional properties and safety of microalgae hydrolysates can be useful in understanding their potential use in the food and pharmaceutical industries.

MATERIALS AND METHODS

Microorganism, Cultivation Conditions and Biomass Processing

Algae samples were obtained by autotrophic outdoor cultivation of *Chlorella vulgaris* 87/1 in open circulating cascade systems of 500 m^2. This strain was isolated from *Chalons* dam in Santiago de Cuba in 1987 and is deposited at the Culture Collection of the Solar Energy Research Center (CIES). The growth medium contained (g/L): NH_4NO_3 (1.2), $MgSO_4 \cdot 7H_2O$ (1.0) and a food grade NPK (8:12:12) fertilizer formula (0.9). The algal suspension was bubbled with 1% CO_2. The algae were harvested by continuous-flow centrifugation (separator Alfa Laval, Sweden) up to 10% dry matter in the slurry. The dark-green algae slurry was spray dried in a Niro Atomizer drier (input 200–210°C, output 80–90°C). The powder thus obtained (moisture content 7%) was preserved in plastic boxes for further use. Dry algae samples of 500 g were extracted with ethanol (2 L) at 45°C for 3 h via gentle agitation. Previous results indicated that cell proteins are more readily hydrolysed after extraction with ethanol (Morris *et al.*, 2008a).

Enzymatic Hydrolysis of Cell Biomass

Pancreatin (Merck), having a specific proteolytic activity of 0.47AU/mg of protein, was used for cell protein hydrolysis of the extracted algal biomass. One proteolytic unit was expressed as the amount of enzyme necessary to catalyze at an initial rate the release of 1 μmol tyrosine from a 2% denatured casein solution at pH 7.5 and 37°C within 1 min.

A 10% suspension in water of the ethanol-extracted alga was hydrolysed at an enzyme/substrate ratio of 30AU/g, pH 7.5 and 45°C for 4 h in a 1000 mL reaction vessel, equipped with a stirrer, thermometer and pH electrode. The enzyme reaction was stopped by heat treatment at 85°C for 15 min. The slurry thus obtained was centrifuged and the resultant solutions were spray-dried.

The yields of the technology were of 40–45 g of hydrolysate/100 g of cell biomass. The bulk of the product dry matter consists of soluble hydrolysed protein and free amino acids, accounted for 47.7%. The amino nitrogen/total nitrogen ratio of 26.4% was considered appropriate for protein assimilation. The amino acid pattern was comparable with that of FAO reference protein, except for the low content of sulphur amino acids (Morris *et al.* 2008a). Table 1 shows the mean molecular mass distribution of *Chlorella* protein hydrolysate (Cv-PH) measured by gel filtration chromatography. For comparative purposes, the molecular mass profile of native proteins in *Chlorella* aqueous hot-extract (Cv-EA) is also presented.

Table 1. Comparative distribution (%) of molecular masses of the peptide components in protein hydrolysate and in the aqueous extract of *Chlorella vulgaris* 87/1

Molecular Masses (kDa)	Aqueous extract (Cv-EA)	Protein hydrolysate (Cv-PH)
MM≥ 70	25	9
30 ≤ MM <70	13	6
15 ≤ MM <30	41	15
10 ≤ MM <15	21	17
MM < 10	-	53
4-5		5
2,5-4		37
1-2,5		11

Samples (0.02 g) of *Chlorella* aqueous hot-extract (Cv-EA) or Cv-PH were applied to a column of Sephadex G-100 (1 x 60 cm). The eluent was 0.1 mol/L phosphate buffer pH 6.8 at a flow rate of 12 mL/h. Elution was monitored at 214 nm and the approximate molecular masses were determined using molecular weight standards. The low molecular mass fraction of Cv-PH (< 10 kDa) was further fractionated in a Sephadex G-25 column.

Solubility Curve

Solubility profiles of *Chlorella* aqueous extract (Cv-EA) and the protein hydrolysate (Cv-PH) were obtained following basically the method of Saeed and Cheryan (1988). Samples of both products were extracted by continual stirring (magnetic stirrer) for 1 h with five volumes of distilled water at room temperature and different pHs in the range 2-8. The pH was maintained stable during extraction with 0,5 mol/L NaOH or 0,5 mol/L HCl. Clear supernatants were obtained by centrifugation at 4 000 g for 45 min. The supernatants were filtered through Whatman No. 3 filter paper, and nitrogen was determined in the filtrate by the micro-Kjeldhal method (AOAC, 2005). Solubility was expressed as the percentage of the total nitrogen of the original sample that was present in the soluble fraction.

Antigenicity Study

Antigenicity studies were carried out according to the United States Pharmacopeia specifications for protein hydrolysates (USP 27, 2004). Male guinea pigs, weighing between 420-480 g, were purchased from the National Center for the Production of Laboratory Animals (CENPALAB, Havana, Cuba). The sensitizing solution (*Chlorella vulgaris* aqueous extract, Cv-EA) at a volume of 6 mL was administered intraperitoneally (i.p.) to six animals the second, fourth and sixth days of two consecutive weeks. Thirty days after the last sensitizing dose, five animals were injected intravenously (i.v.) with 3 mL of the protein hydrolysate (Cv-PH, 500 mg/kg) at a flow rate of 2 mL min^{-1}. During the injection and in the 15 minutes later the animals were carefully observed for the apparition of the following symptoms: 1. licking the nose or rubbing the nose with forefeet, 2. ruffling of the fur, 3. labored breathing, 4. sneezing or coughing (three or more times) and 5. retching. The assay requirements (non antigenicity) were complied if none of the inoculated animals showed more than two of the mentioned symptoms and none developed convulsions, prostration or died. The remaining animal was used as a control for evaluating the sensitizing ability of Cv-EA; when injected i.v. with 3 mL of this extract, positive signs of anaphylaxis should be observed.

Safety Evaluation of Cv-PH

In both acute oral toxicity and repeated dose oral toxicity evaluations of Cv-PH, eight week-old male and female Balb/c mice with an average weight of 20-22 g, purchased from the National Center for the Production of Laboratory Animals (CENPALAB, Havana, Cuba) were used. The animals were kept in our laboratory for one week before used, and then, they were housed in an animal room with a temperature of 23 ± 2°C, relative humidity of 60-65% and a 12 h light-dark period. The mice were fed a normal commercial pelleted diet (Ratonina®, CENPALAB, Havana, Cuba). These studies were approved by the institutional Ethical Committee (University of Oriente) and have been performed in accordance with Cuban legislation and the National Research Council Guidelines for the Care and Use of Laboratory Animals.

The *acute oral toxicity* was assessed by a classification test (Acute Toxic Class Method) in accordance with the Guideline 423 of the Organization for the Economic Cooperation and Development (OECD, 1996, revised October 2000). *Chlorella* protein hydrolysate (Cv-PH) was administered at a single dose level of 2000 mg/kg body weight by oral gavage to three animals of each sex (limit test). Careful clinical observations were made daily, as a rule, for 14 days. Observations for general condition included possible changes in skin and fur, eyes, respiratory, circulatory, somatomotor activity and behaviour pattern. Attention was also directed to observations of tremors, convulsions, salivation, diarrhea, lethargy, sleep and coma. All test animals were subjected to gross necropsy and histopathological studies.

The *repeated dose oral toxicity* was evaluated as described in the Guideline 407 (OECD, 1981). Cv-PH was administered daily in the mornings during 28 days to an experimental group formed by 10 Balb/c mice (5 female and 5 male). Animals were caged in groups by sex. A limit test was carried out at a dose level of 2000 mg/kg body weight. A non treated group was used as control. A careful clinical observation was made at least once each day. One day after the conclusion of the administration period, a blood sample was collected from the orbital vein of each animal. The blood specimens were analyzed for hemoglobin, white blood cell counts and differential leukocyte counts.

A portion of the blood sample was allowed to clot and was centrifuged 10 min at 3 000 g under refrigeration to separate the serum. The serum thus obtained was subjected to the following measurements: glucose, alanine aminotransferase (ALT), aspartate aminotransferase (AST) and creatinine with commercial kits (Boehringer Mannheim GmbH Diagnostica, Germany). The animals were autopsied and the following organs were isolated and weighed: liver, spleen, thymus, lungs, heart and kidneys. The weight of each organ relative to the total body weight was calculated. Histopathological studies were performed with hematoxylin-eosine (HandE) stained specimens in organ samples previously fixed in a 10% formalin solution.

Statistical Analysis

Data are expressed as means ± S.E. The Mann-Whitney test at the 5% significance level was used to determine mean differences among the groups for all of the parameters studied. All data were analyzed using the "Statistical Package for Social Sciences" (SPSS) version 12.0/ 2003 for Windows (SPSS Inc. 1989-2003).

RESULTS AND DISCUSSION

In humans, several diseases associated to extensive protein losses and protein-energy malnutrition (PEM) states could be developed. However, the efforts for restoring protein levels by the oral administration of crude proteins or protein extracts generally did not lead to the original protein levels. Therefore, many attempts have been made to supply free amino acids and peptides of different molecular masses through distinct routes (Guildfort and Matz, 2003).

The high protein content of certain microalgae was one of the reasons to select these organisms as unconventional protein sources (Spolaore et al., 2006). The 90-98% of total algae amino acids is contained in proteins (Dortch et al., 1984), so that the enzymatic hydrolysis has been considered as one of the most attractive approaches to improving algae protein digestibility for nutritional purposes.

Enzymatic modification of structure and functional properties of proteins has been a current research subject during the past 20 years. However, microalgae protein hydrolysates find only very marginal use in food and pharmaceutical industries, due in part to problems associated with an incomplete knowledge of their nutritional value, functional capacities, antigenicity and toxicological concerns. In particular, this chapter considers how the proteolytic modification could have special importance for the improvement of solubility of *Chlorella vulgaris* protein and for decreasing its residual antigenicity, thus favouring the safety of *Chlorella* protein hydrolysate (Cv-PH).

Solubility Curve

Proteolytic modification has special importance for the improvement of solubility of protein, e.g. from legumes that are poorly soluble in aqueous media (Clemente et al., 1999). Nitrogen solubility profiles *versus* pH of *Chlorella vulgaris* protein hydrolysate (Cv-PH) and that of aqueous extract (Cv-EA) are shown in Figure 1. The enzymatic treatment with pancreatin, which renders a degree of hydrolysis of 20-22%, modified the solubility curve from and U-shape curve to a nearly straight line over a wide pH range (2-8). A general structural disruption of proteins and a marked increase of interfacial area and charged groups exposed to the aqueous environment are to be expected from such the enzymatic activity.

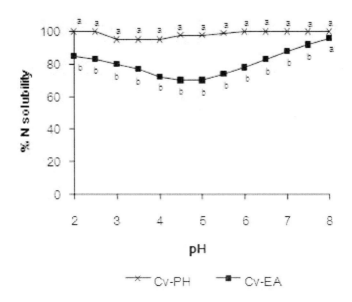

Figure 1. Solubility profile at different pH values of *Chlorella vulgaris* 87/1 aqueous extract (Cv-EA) and the protein hydrolysate (Cv-PH) obtained after enzymatic treatment with pancreatin (30 U/g).

An increase in the extent of enzymatic hydrolysis corresponded to a considerable increase in the nitrogen solubility over the pH range studied, thus indicating a positive relationship. It has been suggested that an increase in the solubility of protein hydrolysates over that of the original protein is due to the reduction of its secondary structure and also to the enzymatic release of smaller peptide units from the protein (Adler-Nissen, 1986; Chobert et al., 1988).

Solubility was expressed as the percentage of the total nitrogen of the original simple that was present in the soluble fraction. Nitrogen was analyzed according to the micro-Kjeldhal method. Means without the same letter at each pH value are significantly different in the Mann-Whitney test ($p < 0,05$).

The improved solubility of Cv-PH compared to Cv-EA was due to its smaller molecular size and a corresponding increase in the number of exposed ionizable amino and carboxyl groups that increase the hydrolysate hydrophilicity. This increasing in solubility is an important factor for the inclusion of protein hydrolysates in several foods (Kong et al., 2007).

Antigenicity Study

The antigenicity of a protein molecule is a function of both its primary structure —the amino acid sequence and the structure of side chains such as attached carbohydrate- and the conformation of the molecule. The primary structure provides the chemical elements for antibody binding, whereas the conformation of the protein molecule determines which chemical elements are involved in the antigen-binding sites, or epitopes. Therefore, enzymatic hydrolysis can reduce antigenicity by changing the primary structure or the conformation of protein (Lee, 1992).

The residual antigenicity of a hydrolysate is estimated by determining its reactivity with antibodies obtained from animals that have been sensitized with the native protein from which the hydrolysate was derived. Although immunoassay tests, e.g. ELISA, are the systems of choice for residual antigenicity analysis because of their inherent sensitivity, animal model tests systems can also be used in non-clinical test programs.

Table 2. Antigenicity evaluation of *Chlorella vulgaris* 87/1 protein hydrolysate (Cv-PH) as determined by sensitization of guinea pigs with *Chlorella* extract

Anaphylaxis symptoms	Animals tested (total No.)	Pathological grade			
		0	1	2	3
Licking the nose or rubbing the nose with forefeet	5		1		
Ruffling of the fur	5	0			
Labored breathing	5	0			
Sneezing or coughing (three or more times)	5	0			
Retching	5	0			

Grade: 0, no abnormality detected; 1, slight; 2, moderate; 3, marked.

The residual antigenicity of *Chlorella vulgaris* protein hydrolysate was assessed in guinea pigs previously sensitized with microalga native proteins contained in aqueous extract (Table 2). The experimental animals did not develop convulsions, prostration, neither died. In only one of the inoculated animals the rubbing of the nose with forefeet was observed. Thus, the requirements for non-antigenicity were accomplished.

Molecular mass profiles of protein hydrolysates with a prevalence of peptides lower than 10 kDa have been associated with a reduced residual antigenicity, estimated by immunoenzymatic methods (Cave and Guilford, 2004). As a result of the hydrolysis of *Chlorella vulgaris* cell proteins, Cv-PH showed a major chromatographic peak of molecular mass lower than 10 kDa, in which three main peptides with masses ranging between 2 and 5 kDa were identified (Morris et al., 2008a). Given this molecular distribution, its uptake from the gut should not cause major problems.

Safety Evaluation of Cv-PH

Since the registration of microalgae products as a drug is expensive and time-consuming, these products are defined as nutraceuticals, which do not easily fall into the legitimated categories of food or drugs but occupy a grey area between both (Gulati and Berry, 2006; Grobbelaar, 2003). Furthermore, the nutraceutical markets are still not well regulated (Bagchi, 2006) and every country has its own food safety regulations. The demand for nutraceuticals made from microalgae has increased during the past decades. These products designed as dietary supplements in human nutrition have to be of supreme quality, especially in terms of the consumer's health (Görs et al., 2010).

While the use of microalgae in functional foods could soon reach the level of mass products, their use in pharmaceutical applications appears to lie more in the future (Pulz and Gross, 2004). Therefore, safety evaluations and regulatory concerns of microalga derived products are relevant.

The *acute oral toxicity* was assessed by a classification test (Acute Toxic Class Method) in accordance with the Guideline 423 of the Organization for the Economic Cooperation and Development (OECD, 1996, revised October 2000). The administration of Cv-PH at a single dose (2000 mg/kg body weight) did not cause alterations in the animal clinical signs, nor deaths were observed. Likewise, no changes in body weight and in food and water consumption were detected (data not shown). The anatomopathological exam did not evidence organ alterations in Cv-PH treated mice. Thus, the product at the assayed dose can be classified as potentially non-toxic.

Vacek et al. (1990) reported that *Chlorella* products are, in general, well tolerated. A lethal mean dose (LD_{50}) of 7.5 g/kg body weight was reached in intraperitoneal administration of mice with Ivastimul –an aqueous extract from *Chlorella kessleri* designed for veterinary use.

The *repeated dose oral toxicity* was evaluated as described in the Guideline 407 (OECD, 1981). Neither modifications in the clinical signs of animal treated during 28 days with Cv-PH at a dose of 2000 mg/kg body weight, nor deaths were observed. The oral administration of Cv-PH led to a significant increase in hemoglobin levels and in leukocyte counts, particularly in the granulocyte pool ($p< 0.05$) (table 3) within the reference interval reported for the specie. These findings could be associated with the hemopoiesis stimulating activity of *Chlorella* products (Vacek et al., 1990; Morris et al., 2007).

The assayed biochemical parameters did not suffer any alteration in Cv-PH administered mice compared to control animals (table 4). As result of Cv-PH treatment neither significant changes were found in organ weights (table 5) nor in anatomopathological studies, both macroscopic and histological.

Table 3. Hematological parameters of mice administered with *Chlorella vulgaris* 87/1 protein hydrolysate (Cv-PH) orally for 28 days (repeated dose oral toxicity study, OECD Guideline 407)

Group	Hemoglobin (g/L)	White blood cell count (x 10^9/L)	Granulocytes (x 10^9/L)	Lymphocytes (x 10^9/L)
Females				
Cv-PH (2 000 mg/kg/day)	133 ± 4 a	8,80 ± 0,28 a	2,75 ± 0,05 a	5,96 ± 0,37 a
Control	117 ± 6 b	6,67 ± 0,18 b	1,15 ± 0,12 b	5,68 ± 0,64 a
Males				
Cv-PH (2 000 mg/kg/day)	128 ± 5 a	7,20 ± 0,13 a	2,30 ± 0,17 a	6,05 ± 0,39 a
Control	112 ± 3 b	5,71 ± 0,08 b	1,10 ± 0,14 b	5,72 ± 0,42 a

Data are expressed as mean ± SE of 10 animals per experimental group.
Different letters indicate significant differences between the Cv-PH administered group and control animals for each sex in the Mann-Whitney test ($p < 0,05$).

Table 4. Serum biochemical parameters of mice administered with *Chlorella vulgaris* 87/1 protein hydrolysate (Cv-PH) orally for 28 days (repeated dose oral toxicity study, OECD Guideline 407)

Group	Glucose (mmol/L)	AST (IU)	ALT (IU)	Creatinine (µmol/L)
Females				
Cv-PH (2 000 mg/kg/day)	6,64 ± 2,58 a	88,01 ± 22,46 a	47,29 ± 16,28 a	47,33 ± 6,47 a
Control	6,68 ± 2,94 a	84,28 ± 26,62 a	45,73 ± 14,64 a	50,17 ± 8,69 a
Males				
Cv-PH (2 000 mg/kg/day)	7,82 ± 3,38 a	86,42 ± 20,19 a	45,87 ± 15,96 a	48,24 ± 10,36 a
Control	7,88 ± 2,96 a	84,90 ± 24,58 a	44,08 ± 18,02 a	50,57 ± 11,55 a

Data are expressed as mean ± SE of 10 animals per experimental group.
Different letters indicate significant differences between the Cv-PH administered group and control animals for each sex in the Mann-Whitney test ($p < 0,05$).

Stoilov et al. (1995) reported the safety of the chronic oral administration to rats of a protein hydrolysate from *Spirulina pacifica* at doses ranging from 2.0-5.0 g/kg body weight during six months, as judged by clinical and morphological parameters.

Our findings indicate the safety of Cv-PH in preclinical studies. Since there were no observed adverse effects of Cv-PH in these studies, the NOAEL (no observed adverse effect level) for *Chlorella* protein hydrolysate is 2 000 mg/kg/day administered orally for 28 days.

Table 5. Relative organ weights (g/100 g) of mice administered with *Chlorella vulgaris* 87/1 protein hydrolysate (Cv-PH) orally for 28 days (repeated dose oral toxicity study, OECD Guideline 407)

Group	Liver	Spleen	Thymus	Lungs	Heart	Kidneys
Females						
Cv-PH (2 000 mg/kg/day)	5,26 ± 0,08 a	0,44 ± 0,06 a	0,35 ± 0,07 a	0,89 ± 0,08 a	0,55 ± 0,03 a	1,41 ± 0,11 a
Control	4,90 ± 0,09 a	0,42 ± 0,09 a	0,33 ± 0,03 a	0,85 ± 0,14 a	0,52 ± 0,07 a	1,52 ± 0,06 a
Males						
Cv-PH (2 000 mg/kg/day)	5,23 ± 0.18 a	0,43 ± 0,04 a	0,23 ± 0,10 a	0,62 ± 0,04 a	0,52 ± 0,02 a	2,15 ± 0,12 a
Control	5,12 ± 0,25 a	0,42 ± 0,04 a	0,23 ± 0,04 a	0,75 ± 0,04 a	0,55 ± 0,03 a	2,12 ± 0,21 a

Data are expressed as mean ± SE of 10 animals per experimental group.
Different letters indicate significant differences between the Cv-PH administered group and control animals for each sex in the Mann-Whitney test ($p < 0,05$).

CONCLUSION

In this chapter, we demonstrated that the proteolytic modification of *Chlorella vulgaris* proteins has special importance for the improvement of solubility of algal protein and for decreasing its residual antigenicity in association with the reduction of its secondary structure and also to the enzymatic release of smaller peptide units from the protein. Given this molecular distribution, its uptake from the gut did not cause major problems, as shown in safety evaluation of Cv-PH. The NOAEL (no observed adverse effect level) for *Chlorella* protein hydrolysate is 2 000 mg/kg/day administered orally for 28 days. An extended knowledge of the functional properties and safety of microalgae hydrolysates can be useful in understanding their potential use in the food and pharmaceutical industries.

REFERENCES

Adler-Nissen, J. (1986). Enzymic hydrolysis of food proteins. New York: Elsevier Applied Science Publisher.
AOAC (2005). Official Methods of Analysis (18th ed.). Washington, DC, USA: Association of Official Analytical Chemists.

Bagchi, D. (2006). Nutraceuticals and functional foods regulations in the United States and around the world. *Toxicology*, 221, 1–3.

Biziulevicius, G.A. (2004). How food-borne peptides may give rise to their immunostimulatory activities: a look through the microbiologist's window into the immunologist's garden (hypothesis). *British Journal of Nutrition*, 92, 1009-1012.

Cave, N.J., and Guilford, W.G. (2004). A method for *in vitro* evaluation of protein hydrolysates for potential inclusion in veterinary diets. *Research in Veterinary Science*, 77, 231-238.

Clemente, A., Vioque, J., Sánchez-Vioque, R., Pedroche, J., Bautista, J., and Millán, F. (1999) Protein quality of chickpea (*Cicer arietinum* L.) protein hydrolysates. *Food Chemistry*, 67, 269-274.

Chobert, J. M., Bertrand-Harb, C., and Nicolas, M. G. (1988). Solubility and emulsifying properties of caseins and whey proteins modified enzymatically by trypsin. *Journal of Agricultural and Food Chemistry*, 36, 883–886.

Dortch, Q., Clayton, J.R., Thorensen, S.S., and Ahmed, S.I. (1984). Species differences in the accumulation of nitrogen pools in phytoplankton. *Marine Biology*, 81, 237-250.

Franck, P., Moneret, D.A., Dousset, B., Canni, G., Nabet, P., and Parisot, L. (2002). The allergenicity of soybean-based products is modified by food technologies. *International Archives of Allergy and Immunology*, 128, 212-219.

Frokjaer, S., (1994). Use of hydrolysates for protein supplementation. *FoodTechnology*, 48, 86–88.

Görs, M., Schumann R., Hepperle, D., and Karsten, U. (2010). Quality analysis of commercial *Chlorella* products used as dietary supplement in human nutrition. *Journal of Applied Phycology*, 22, 265-276.

Grobbelaar, J.U. (2003). Quality control and assurance: crucial for the sustainability of the applied phycology industry. *Journal of Applied Phycology*, 5, 209–215.

Guadix, A., Guadix, E.M., Páez-Dueñas, M.P., González-Tello, P., and Camacho, F. (2000). Procesos tecnológicos y métodos de control en la hidrólisis de proteínas. *Ars Pharmaceutica*, 41, 79-89.

Guilford, W.G., and Matz, M.E. (2003). The nutritional management of gastrointestinal tract disorders. *New Zealand Veterinary Journal*, 51, 284-291.

Gulati, O.P., and Berry, O.P. (2006). Legislation relating to nutraceuticals in the European Union with a particular focus on botanical-sourced products. *Toxicology*, 221, 75–87.

Iwamoto, H. (2003). Industrial production of microalgal cell-mass and secondary products – major industrial species –*Chlorella*. In: Richmond, A. (Ed.), *Handbook of Microalgae Biotechnology*, (pp. 255–263). Oxford: Blackwell Publishing.

Kessler, E. (1976). Comparative physiology, biochemistry, and the taxonomy of *Chlorella* (Chlorophyceae). *Plant Systematics and Evolution*, 125, 129–138.

Kim, S.K., and Wijesekara, I.(2010). Development and biological activities of marine-derived bioactive peptides: a review. *Journal of Functional Foods*, 2, 1-9.

Kislukhina, O.V. (2002). *Enzymes in production of foods and forage*, Moscow: DeLi print.

Kong, X., Zhou, H., and Quian, H. (2007). Enzymatic preparation and functional properties of wheat gluten hydrolysates. *Food Chemistry*, 101, 615–620.

Lee, Y.H. (1992). Food-processing approaches to altering allergenic potential of milk-based formula. *The Journal of Pediatrics*, 121, 47-50.

Liang, S., Liu, X., Chen, F., and Chen, Z .(2004). Current microalgal health food RandD activities in China. *Hydrobiologia*, 512, 45–48.

Mahmoud, M.I. (1994). Physicochemical and functional properties of protein hydrolysates in nutritional products. *Food Technology*, 48, 89–95.

Manninen, A.H. (2004). Protein hydrolysates in sports and exercise: a brief review. *Journal of Sports Science and Medicine*, 3, 60–63.

Merchant, R.E., Andre, C.A., and Sica, D.A. (2002). *Chlorella* supplementation for controlling hypertension: a clinical evaluation. *Alternative and Complementary Therapies*, 8,370–376.

Morris, H.J., Carrillo, O., Almarales, A., Bermúdez, R.C., Lebeque, Y., Fontaine, R. et al. (2007). Immunostimulant activity of an enzymatic protein hydrolysate from green microalgae *Chlorella vulgaris* on undernourished mice. *Enzyme and Microbial Technology*, 40, 456–460.

Morris, H.J., Almarales, A., Carrillo, O., and Bermúdez, R.C. (2008a). Utilisation of *Chlorella vulgaris* cell biomass for the production of enzymatic protein hydrolysates. *Bioresource Technology*, 99, 7723-7729.

Morris, H.J., Carrillo, O., Alonso, M.E., and Bermúdez, R.C. (2008b). Are the peptide sequences encrypted in food *Chlorella* protein a possible explanation for the immunostimulatory effects of microalgal supplements? *Medical Hypotheses*, 70, 896.

Organization for Cooperation and Economical Development (OECD). (1981) *OECD Guideline for the testing of chemicals*. Guideline 407 Repeated dose oral toxicity-rodent: 28-day or 14-day study. Adopted: 12 May 1981.

Organization for Cooperation and Economical Development (OECD). (1996) *OECD Guideline for the testing of chemicals*. Guideline 423 Acute oral toxicity-acute toxic class method. Adopted: 22.03.96, revised document October 2000.

Oh-Hama, T., and Miyachi, S. (1988). *Chlorella*. In: Borowitzka, M.A., and Borowitzka, L.J. (Eds.), *Microalgal Biotechnology*, (pp. 3–26). Cambridge: Cambridge University Press.

Olaizola, M. (2003). Commercial development of microalgal biotechnology: from the test tube to the market place. *Biomolecular Engineering*, 20, 459–466.

Pulz, O., Scheibenbogen, K., and Gross, W. (2001). Biotechnology with cyanobacteria and microalgae. In: Rehm, H.J., and Reed, G. (Eds.), *Biotechnology*, (2nd ed., vol. 10, pp. 105–136). Weinheim: Wiley-VCH.

Pulz, O., and Gross, W. (2004). Valuable products from biotechnology of microalgae. *Applied Microbiology and Biotechnology*, 65, 635–648.

Saeed, M., and Cheryan, M. (1988). Sunflower protein concentrates and isolates low in polyphenols and phytate. *Journal of Food Science*, 53, 1127-1131.

Shelef, G., and Soeder, C.J. (1980). *Algae Biomass production and Use*. Amsterdam: Elsevier.

Shimizu, Y., and Li, B. (2006). Microalgae as a source of bioactive molecules: special problems and methodology. In: Proksch, P., and Mueller, W.E.G. (Eds.), *Frontiers in Marine Biotechnology*, (pp. 145–174). Wymondham: Horizon Biosciences.

Siemensma, A.D.,Weijer, W.J., and Bak, H.J. (1993). The importance of peptide lengths in hypoallergenic infant formulae. *Trends in Food Science and Technology*, 4, 16–21.

Spolaore, P., Joannis-Cassan, C., Duran, E., and Isambert , A. (2006). Commercial applications of microalgae. *Journal of Biosciences and Bioengineering*, 101, 87–96.

Stoilov, I.L., Georgiev, T.D., Taskov, M.V., and Koleva, I.D. (1995). *Oral Preparation for Patients with Chronic Renal Insufficiency and Other Protein Metabolic Diseases.* WO Patent 95/2952, November 2.

Tchorbanov, B., and Bozhkova, M. (1988). Enzymatic hydrolysis of cell proteins in green algae *Chlorella* and *Scenedesmus* after extraction with organic solvents. *Enzyme and Microbial Technology*, 10, 233–238.

U.S.P. (2004). *US Pharmacopeia National Formulary*, USP 27, NF 22 S1. Rockville, MD:United States Pharmacopeial Convention Inc.

Vacek, A., Rotkovská, D., and Bartonícková A. (1990). Radioprotection of hemopoiesis conferred by aqueous extract from Chlorococcal algae (Ivastimul) administered to mice before irradiation. *Experimental Hematology*, 18, 234-237.

Wilson, K.E., and Huner, N.P.A. (2000). The role of growth rate, redox-state of the plastoquinone pool and the trans-thylakoid Delta pH in photoacclimation of *Chlorella vulgaris* to growth irradiance and temperature. *Planta*, 212, 93–102.

In: Microalgae: Biotechnology, Microbiology and Energy
Editor: Melanie N. Johnsen

ISBN 978-1-61324-625-2
© 2012 Nova Science Publishers, Inc.

Chapter 16

HETEROTROPHIC MICROALGAE IN BIOTECHNOLOGY

Niels Thomas Eriksen[*]
Aalborg University, Denmark

ABSTRACT

Heterotrophic microalgal species can be grown in processes and in bioreactors resembling what is used to grow the more common types of industrial microorganisms, bacteria, yeast, and fungi. This opportunity gives heterotrophic microalgal cultures some advantages over phototrophic microalgal cultures in terms of productivity and hygienic standard. Phototrophic microalgal and cyanobacterial cultures are typically 1-2 orders of magnitude less productive than what is often obtained in heterotrophic cultures, partly because only limited amounts of light can be supplied to these cultures, and partly because inhomogeneous light intensities inside the cultures result in low photosynthetic yields near culture surfaces and no photosynthetic activity in central zones too deep to be reached by light. It is also less problematic to maintain cultures axenic in ordinary bioreactors with more compact designs than large-scale photobioreactors, where large surface areas are needed to maximise the collection of light. Heterotrophic cultures are not influenced by climate and weather, in contrast to sunlight dependent, large-scale phototrophic cultures located outdoors.

Cultivation of heterotrophic microalgae is, however, also not unproblematic. Heterotrophic microalgae grow more slowly than many bacteria and yeasts, and heterotrophic microalgae are therefore mainly of interest if they produce something that is not made by other types of microorganisms. Only a limited number of microalgal species will grow heterotrophically, and the number of heterotrophic microalgae synthesising valuable products that cannot be obtained also from other sources is low. Still, heterotrophic species from a phylogenetically highly diverse selection chlorophytes, rhodophytes, cyanidiophytes, diatoms, heterokontophytes, euglenoids, and dinoflagellates have been or are being developed for productions of food, feed, lipids, pigments, and more.

---

[*] Department of Biotechnology, Chemistry and Environmental Engineering, Aalborg University, Sohngaardsholmsvej 49, DK-9000 Aalborg, Denmark, Tel. +45 99408465, e-mail: nte@bio.aau.dk.

A few heterotrophic microalgal processes have also matured to commercialisation. Green algae of the genus *Chlorella* are produced heterotrophically and used as health food, and docosahexaenoic acid, an essential ω-3 poly-unsaturated fatty acid is produced in the dinoflagellate *Crypthecodinium cohnii* and the thraustochytrids *Schizochytrium* sp. and *Ulkenia* sp. and added to infant formula and foods.

HETEROTROPHIC MICROALGAE

Microalgae are a diverse group of non-related organisms that are grouped together only because of their small sizes and photosynthetic capabilities. Several eukaryote phyla as well as cyanobacteria have been classified as microalgae, and the phylogenetic variability among microalgae are therefore enormously large. Many microalgae are able also to grow mixotrophically combining photosynthesis with heterotrophic nutrition, and many have facultative or obligate heterotrophic relatives that do not need light at all in order to grow. These are the heterotrophic microalgae, an artificial classification of organisms based on their heterotrophic capabilities combined with their relatedness to the phototrophic algae.

The majority of microalgal species will not grow heterotrophically (Chen and Chen 2006). Zaslavskaia et al. (2001) demonstrated that one reasons is lack of uptake systems for organic substrates. They transformed the obligate phototrophic diatom, *Phaeodactylum tricornutum* with glucose transporter genes from either humans or from *Chlorella kessleri*, and the transformed strains acquired the ability to take up glucose and grow in darkness. However, when Zhang et al. (1998) introduced a glucose transporter gene from *Synechocystis* into the phototrophic cyanobacterium *Synechococcus*, the cells did acquire the ability to take up and metabolise glucose but a metabolic imbalance was also created and caused cell lethality. Different parts of cell metabolism may therefore be the reason for obligate photoautotrophy. Lack of α–ketoglutarate dehydrogenase activity and incomplete citric acid cycle is one such example (Wood et al. 2004).

Still, heterotrophic species are found in most microalgal phyla. *Chlorella* species have an inducible hexose/H^+ transporter homologous to mammalian glucose transporters (Sauer and Tanner 1989), and several *Chlorella* species also grow well on glucose. Other organic substrates taken up by heterotrophic microalgae include different hexoses and pentoses, glycerol, ethanol and acetic acid (Perez-Garcia et al. 2011). It is actually a microalga, *Galdieria sulphuraria*, that among all microorganisms have shown the ability to use the highest number of organic substances as sole carbon and energy source (Gross and Schnarrenberger 1995).

Predation is also a common feeding strategy among heterotrophic microalgae, but those heterotrophic species that take up dissolved organic substances from the water have received by far the most attention. Although predatory microalgae have been grown and used for feed in aquaculture (Boëchat et al. 2005), their cultivation depends on simultaneous cultivation also of a suitable prey, and since nutrients are passed through two trophic levels, first prey and then predator, the yield of algal biomass compared to the amount of organic substrates fed to the prey culture will be low.

Table 1. Heterotrophic microalgae that have been cultured heterotrophically, their products, major carbon substrates, and types of cultures. The algal species are sorted in phylogenetic groups following the classification scheme available at Algaebase (www.algaebase.org), except Labyrinthulida that is classified according to UniProt Taxonomy (www.uniprot.org/taxonomy). μ_{max}, x_{max}, and r_x represent maximal specific growth rates, maximal biomass concentrations, and biomass production rates in the cultures, respectively. Numbers given in italics are calculated based on information or data read from graphs in the references

Phylum and species	Product	Substrates	Culture type	μ_{max} day^{-1}	x_{max} g L^{-1}	r_x g L^{-1} day^{-1}	Reference
Cyanobacteria							
Arthrospira platensis	biomass	glucose	batch		0.83	*0.17*	Marquez et al. (1993)
Chlorophyta							
Ankistrodesmus braunii	biomass	glucose	batch		3.0	0.7	Burrell et al. (1984)
Chlamydomonas reinhardtii		acetic acid	perfusion		8.9	*1.5*	Chen and Johns (1994)
Chlorella regularis	carotenoids	glucose	batch				Ishikawa et al. (2004)
Chlorella protothecoides	lutein	glucose	batch	0.92	16.4	*2.9*	Shi et al. (1997)
Chlorella protothecoides	lutein	glucose	batch	0.35	19	*2.1*	Zhang et al. (1999)
Chlorella protothecoides	lutein	glucose	fed batch	1.06	48	10.9	Shi et al. (2002)
Chlorella protothecoides	lutein	glucose	batch	1.08	16.9	*3.4*	Shi et al. (2006)
Chlorella protothecoides	biodiesel	corn powder hydrolysate	fed batch		15.5		Xu et al. (2006)
Chlorella protothecoides	biodiesel	glucose, glycine	fed batch		15.5	*2.1*	Li et al. (2007)
Chlorella protothecoides	biodiesel	glucose	fed batch		51.2	*7.7*	Xiong et al. (2008)
Chlorella protothecoides	biodiesel	sugar cane hydrolysate	fed batch		121.3	19.4	Cheng et al. (2009a)
Chlorella protothecoides	biodiesel	Jerusalem artichoke	batch		28	4.1	Cheng et al. (2009b)
Chlorella pyrenoidosa	xanthophyll	glucose	batch		110		Theriault (1965)
Chlorella pyrenoidosa	xanthophyll	glucose	fed batch		302		Theriault (1965)
Chlorella pyrenoidosa		glucose	fed batch	4.80	116.2	1.0	Wu and Shi (2006)
Chlorella pyrenoidosa	lutein	glucose	batch	1.77	24	8.2	Wu et al. (2007)
Chlorella pyrenoidosa	lutein	glucose	fed batch		70	*14*	Wu et al. (2007)
Chlorella vulgaris	biomass	glucose	batch		2.5	0.9	Burrell et al. (1984)
Chlorella vulgaris		glucose, glycerol	batch		1.2	0.15	Liang et al. (2009)
Chlorella zofingiensis	astaxanthin	glucose, lactose, sucrose	batch	0.67	10.3		Sun et al. (2008)
Chlorella zofingiensis	astaxanthin	glucose	fed batch		53		Sun et al. (2008)
Prototheca moriformis	L-ascorbic acid	glucose	batch		13.2		Running et al. (2002)
Scenedesmus obliqus	waste water	glucose	batch	1.28			Abeliovich and Weisman (1978)
Spongiococcum exetricicum	xanthophyll	dextrose	fed batch	3.24	60	*20.9*	Hilaly et al. (1994)

Table 1. (continued)

Phylum and species	Product	Substrates	Culture type	μmax day-1	xmax g L-1	rx g L-1 day-1	Reference
Tetraselmis sp.	aquaculture feed	glucose	batch	0.67	16.3		Day and Tsavalos (1996)
Tetraselmis suecica		glucose, peptone, YE	batch		28.88	4.8	Azma et al. (2011)
Rhodophyta							
Porphyridium cruentum	biodiesel	glucose	batch	0.22	3.2	0.15	Oh et al. (2009)
Cyanidiophyta							
Galdieria sulphuraria	phycocyanin	molasses	fed batch	1.44	116	15.6	Schmidt et al. (2005)
Galdieria sulphuraria	phycocyanin	glucose	fed batch	1.06	109	17.5	Graverholt and Eriksen (2007)
Galdieria sulphuraria	phycocyanin	glucose	continuous	1.04	83.3	50	Graverholt and Eriksen (2007)
Cryptophyta							
Chilomonas parameceum	aquaculture feed	bacteria					Boëchat et al. (2005)
Bacillariophyta							
Cyclotella cryptica	aquaculture feed	glucose		0.91	1.5	0.33	Pahl et al. (2010)
Nitschia alba	sterols, lipids	tryptone					Tornabene et al. (1974)
Nitschia angularis		glutamate, glucose	batch	1.04			Lewin and Hellebust (1976)
Nitschia laevis		glucose	batch	0.69			Lewin and Hellebust (1978)
Nitschia laevis	EPA	glucose	batch	0.34	2.04	0.45	Wen and Chen (2000a)
Nitschia laevis	EPA	glucose	batch	0.65	5.5	0.55	Wen and Chen (2000b)
Nitschia laevis	EPA	glucose	continuous	0.51		2.8	Wen and Chen (2001)
Nitschia laevis	EPA	glucose	perfusion-bleeding		25	6.8	Wen and Chen (2001)
Nitschia laevis	EPA	glucose	perfusion		40	1.6	Wen and Chen (2002)
Nitschia laevis	EPA	glucose	fed batch		22.1	1.5	Wen et al. (2002)
Heterokontophyta/ Crysophyceae							
Ochromonas danica		glucose	batch				Aaronson and Baker (1959)
Ochromonas danica	bioremediation	phenol	batch				Semple (1998)
Heterokontophyta/ Labyrinthulida							
Aurantiochytrium sp.	DHA	glycerol and glutamate	fed batch		100	13.3	Jakobsen et al. (2008)
Aurantiochytrium limacinum	DHA	glucose and more	batch		23.1	4.0	Nagano et al. (2009)
Schizochytrium limacinum	DHA	biodiesel waste glycerol	batch	0.69	22.1	3.1	Chi et al. (2007)
Schizochytrium limacinum	DHA	glucose	batch	0.58	18.5	3.1	Chi et al. (2007)
Schizochytrium limacinum	DHA	biodiesel waste glycerol	batch	0.59	11.5	1.9	Pyle et al. (2008)
Schizochytrium sp.	DHA	glucose	pH auxostat		63	30.2	Ganuza et al. (2008a)

Table 1. (continued)

Phylum and species	Product	Substrates	Culture type	μmax day-1	xmax g L-1	rx g L-1 day-1	Reference
Schizochytrium limacinum	DHA	glycerol	batch		24.6	4.9	Chi et al. (2009)
Schizochytrium limacinum	lipid	sweet sorghum juice	batch	0.91	9.8	1.9	Liang et al. (2010)
Schizochytrium limacinum	DHA	glucose	fed batch		71	2.9	Ren et al. (2010)
Schizochytrium limacinum	DHA	biodiesel waste glycerol	continuous	0.69	18.0	3.9	Ethier et al. (2011)
Thraustochytrium roseum	DHA	starch	fed batch		17.1	2.9	Singh and Ward (1996)
Thraustochytrium aureum	DHA	glucose, glutamic acid	batch		5.7	2.0	Iida et al. (1996)
Ulkenia sp.	DHA, astaxanthin	potato residues	batch		9	3.8	Quilodran et al. (2010)
Euglenozoa							
Euglena gracilis	α-tocoperol	glucose, glutamic acid	fed batch	1.08	48	7.7	Ogbonna et al. (1998)
Euglena gracilis	α-tocoperol	glucose	batch		8.46		Fujita et al. (2008)
Euglena gracilis	α-tocoperol	glucose, ethanol	fed batch		19.69	3.3	Fujita et al. (2008)
Euglena gracilis	β-1,3-glucan	glucose	batch		14	7.3	Santek et al. (2009)
Euglena gracilis	β-1,3-glucan	potato liquor	batch		16	8.8	Santek et al. (2010)
Dinoflagellata							
Crypthecodinium cohnii	DHA	glucose	batch	1.25	27.7	8.9	de Schwaaf et al. (1999)
Crypthecodinium cohnii	DHA	acetic acid	pH auxostat		45.5	5.2	Ratledge et al. (2001)
Crypthecodinium cohnii	DHA	acetic acid	fed batch		109	5.5	de Schwaaf et al. (2003a)
Crypthecodinium cohnii	DHA	ethanol	fed batch	1.13	83	10.0	de Schwaaf et al. (2003b)
Crypthecodinium cohnii	DHA	glucose	batch	1.44	10	6.2	da Silva and Reis (2008)

DIVERSITY OF HETEROTROPHIC MICROALGAL CULTURES

Perez-Garcia et al. (2011) collected at list of 118 microalgal species that can grow heterotrophically and may have commercial potentials. However, the number of species that are actually utilised in heterotrophic processes, or have been characterised in details for heterotrophic cultivation is much lower. Table 1 provides an overview of the diversity of microalgal species that have been examined for heterotrophic cultivation and the products they have synthesised. The species described in Table 1 are grouped according to their phylogeny. The highest number of heterotrophic species in culture is found within Chlorophyta (green algae). Especially different *Chlorella* species have been studied extensively for heterotrophic production of biomass, lipids, and pigments. At least one unicellular red alga (Rhodophyta) has been investigated for heterotrophic synthesis of lipids, and one Cyanidiophyta (often classified as a sub-group within Rhodophyta) has been characterised with respect to heterotrophic production of the pigment, phycocyanin. Members of Cryptophyta (cryptophytes) are mostly phototrophic, but at least one predatory species without chloroplasts has been grown on a bacterial diet and applied as feed in aquaculture. Several diatoms (Bacillariophyta) can grow heterotrophically. Most interest has focussed on their synthesis of eicosapentaenoic acid (EPA). Sometimes diatoms are included within Heterokontophyta, which is a large phylum that includes also crysophytes and macroscopic brown algae. *Ochromonas danica* is a crysophyte which is able to grow on a broad spectrum of organic substrates, including aromatic compounds. The Heterokontophyta also encompasses several orders of non-photosynthetic organisms, among which the Labyrinthulida is particularly important due to their ability to synthesise docosahexaenoic acid (DHA). The labyrinthulids is a class of organisms that some authors have described as microalgae and they have therefore been included in Table 1, although their classification as microalgae does not seem universally accepted. The same may be true for the two protozoan phyla, Euglenozoa (euglenoids) and Dinoflagellata (dinoflagellates) although both phyla contain photosynthetic species. While heterotrophic *Euglena gracilis* is a potential producer of α-tocopherol, the prevalent form of vitamin E, and functional carbohydrates, the heterotrophic dinoflagellate *Cryptechodinium cohnii* is already used commercially for industrial production of DHA.

The products that have been synthesised in heterotrophic microalgal cultures are also listed in Table 1. The most important products are lipids rich in EPA and DHA and other poly-unsaturated fatty acids (PUFAs) that are used for nutritional purposes. Microalgal oils can also be used as starting materials for manufacturing of biodiesel. Photosynthetic or photoprotecting pigments, vitamins and carbohydrates have also been synthesised in heterotrophic microalgae, and microalgal biomass is used as food supplements and in animal feeds. Heterotrophic microalgae have also been examined for potential uses in waste water treatment and for bioremediation purposes.

PRODUCTIVITIES OF MICROALGAL CULTURES

Only few microalgal species are used commercially, and they are economically less important than many other industrial microorganisms. A major reason is the comparatively low productivities found in light-dependent cultures of phototrophic microalgae, compared to the productivities of many heterotrophic microbial cultures. In microbial cultures, cells are the catalyst and sometimes also the product, and high cell densities and volumetric production rates of biomass and products are important variables. Riesenberg and Guthke (1999) defined high cell density cultivations as microbial cultures with biomass dry weight concentrations above 100 g L^{-1}. Only a handful of microalgal species have been grown to such high biomass concentrations, and they have all been heterotrophs (Table 1). Productivity may be even more important than biomass concentration. Biomass productivities above 100 g L^{-1} day^{-1} have been achieved in cultures of e.g. *Escherichia coli* and as high as 229.7 g L^{-1} day^{-1} in fed-batch cultures of the yeast *Candida brassicae* (Riesenberg and Guthke 1999). Such high productivities have never been obtained in microalgal cultures. Still, heterotrophic microalgal cultures can be far more productive than phototrophic ones and therefore heterotrophic microalgae can be produced at lowest cost (Behrens 2005). Since most microalgal products are either whole cells or biomass constituents, biomass productivity is a variable of paramount importance in microalgal cultures.

Phototrophic cultures are limited by the external supply of light while heterotrophic cultures will often be limited by the rate of oxygen supply into the culture. Figure 1 compares 2 laboratory scale bioreactors, one 3 L stirred bioreactor (Applikon, The Netherlands) designed for cultures of heterotrophic microorganisms and one custom made 3 L photobioreactor designed as a bubble column. The average incident light intensity on the surface of the photobioreactor in Figure 1 is 200 μmol photons m^{-2} s^{-1}. It costs a minimum of 8 photons to fix one CO_2 molecule into biomass. If all photons are utilised for CO_2 fixation in the reactor, the maximal rate of biomass production will be 1.5 g L^{-1} day^{-1} (calculations are described in Table 2).

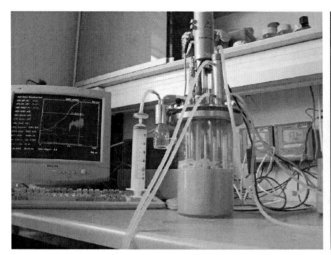 

Figure 1. Left. 3 L Applikon bioreactor for cultivation of heterotrophic microorganisms. The culture in the reactor is the heterotrophic dinoflagellate, *Cryptechodinium cohnii*. Right. Custom made 3 L bubble column photobioreactor illuminated by 6 fluorescent tubes. The culture in the reactor is the green alga, *Chlorella vulgaris* growing photoautotrophically.

In the stirred bioreactor in Figure 1, the maximal oxygen supply rate is determined by the mass transfer coefficient, $k_L a$, and is a function of stirrer speed and aeration rate. If the $k_L a$ value is e.g. 400 h^{-1}, which is a typical $k_L a$ value in this reactor, a maximum of 400 times the amount of oxygen that at any time can be dissolved in the culture medium, can be transferred to the culture in one hour. If this amount of oxygen is utilised to oxidise sugars and generate energy for growth, the biomass production rate can be in the order of 50 g L^{-1} day^{-1} (see Table 2 for details). This productivity is more than 30 times higher than the maximal productivity in the photobioreactor.

The maximal biomass productivities predicted in Table 2 correspond well to the actual biomass productivities that have been measured experimentally. In a 1.7 L bubble-column photobioreactor designed the same way as the photobioreactor in Figure 1, and illuminated by 205 μmol photons m^{-2} s^{-1}, Eriksen et al. (1998) measured the steady-state biomass productivity in a phototrophic continuous flow culture of *Rhodomonas* sp. to be 0.4 g carbon L^{-1} day^{-1} or 0.8 g biomass L^{-1} day^{-1}. This is about half the theoretical maximal productivity at this incident light intensity. In out-door photobioreactors illuminated by the sun, the incident light intensities during day time are considerably higher and the potential biomass productivities are therefore also higher. However, at high light intensities, the photosynthetic efficiency, which is the ratio between energy stored in biomass and absorbed light energy, is low (Ogbonna and Tanaka 2000) and therefore are also the productivities of out-door microalgal cultures normally below 1 g L^{-1} day^{-1} (Eriksen 2008a). Only in specially designed flat panel photobioreactors, where light paths are kept short to minimise dark zones and biomass concentrations high to rapidly attenuate light intensities and create fluctuating light regimes for cells moving around inside the reactors have biomass productivities up to 3-4 g L^{-1} day^{-1} (Zou et al. 2000, Doucha et al. 2005).

Table 2. Estimation of maximal biomass productivities in oxygen limited 3 L bioreactor for heterotrophic microorganisms and light limited 3 L photobioreactor (Figure 1)

Heterotrophic culture	
Biomass composition[1]	$CH_{1.8}O_{0.5}N_{0.2}$
Biomass molecular weight[2]	$MW_x = 26$ g C-mol^{-1}
Biomass yield on oxygen	$Y_{x/O2} = 1$ C-mol mol^{-1}
Oxygen transfer rate	$OTR = k_La \cdot (c_L^* - c_L)$
Oxygen saturation conc.	$c_L^* = 200$ μM
Mass transfer coefficient	$k_La = 400$ h^{-1}
Cell carbon productivity	$r_c = Y_{x/O2} \cdot OTR$
Biomass productivity	$r_x = r_c \cdot MW_x$
Max oxygen transfer rate	$OTR_{max} = k_La \cdot c_L^* = 0.08$ mol L^{-1} h^{-1}
Max cell carbon productivity	$r_{c,max} = Y_{x/O2} \cdot OTR_{max} = 0.08$ C-mol L^{-1} h^{-1}
Max biomass productivity	$r_{x,max} = r_{c,max} \cdot MW_x = 50$ g L^{-1} day^{-1}
Phototrophic culture	
Biomass composition[1]	$CH_{1.8}O_{0.5}N_{0.2}$
Biomass molecular weight[2]	$MW_x = 26$ g C-mol^{-1}
Max photosynthetic efficiency	$PE_{max} = 0.125$ C-mol $photon^{-1}$
Reactor volume	$V = 3$ L
Reactor surface area	$A_s = 0.08$ m^2
Surface photon flux density	$PFFD_s = 200$ μmol photons m^{-2} s^{-1}
Volumetric photon flux	$PFFD_V = (PFFD_s \cdot A_s)/V = 0.0192$ mol photons L^{-1} h^{-1}
Max cell carbon productivity	$r_{c,max} = PE_{max} \cdot PFFD_V = 0.0024$ C-mol L^{-1} h^{-1}
Max biomass productivity	$r_{x,max} = r_{c,max} \cdot MW_x = 1.5$ g L^{-1} day^{-1}

[1]Average biomass composition of microorganisms (Roels 1980)
[2]Molecular weight includes 5% ash in biomass

Biomass productivities in cultures of several heterotrophic microalgae have been above 10 g L^{-1} (Table 1). In continuous flow cultures of the heterotrophic rhodophyte, *Galdieria sulphuraria* grown in the same reactor as shown in Figure 1B, biomass productivities actually reached the theoretical maximal value of 50 g biomass L^{-1} day^{-1} (Graverholt and Eriksen 2007). This is probably the highest biomass productivity seen in a microalgal culture until now.

CULTIVATION OF HETEROTROPHIC MICROALGAE

Heterotrophic microalgae are grown in similar types of cultures as other heterotrophic microorganisms and use the same types of organic nutrients. Many *Chlorella* species and also other microalgae have been grown on glucose. Glycerol and acetic acid have also been frequently used (Table 1). Some species are also supplemented with vitamins, organic nitrogen in the form of amino acids, yeast extract, or other complex substrates. Many of the heterotrophic microalgae have also been grown successfully on cheap, complex substrates such as different plant materials (Cheng et al. 2009a, b, Liang et al. 2010, Quilodrán et al. 2010), waste streams from food industries (Santek et al. 2010), glycerol waste from biodiesel industries (Chi et al. 2007, Pyle et al. 2008, Ethier et al. 2011), and molasses (Schmidt et al. 2005).

Most heterotrophic microalgae have been grown in batch cultures where all nutrients are added the same time as the cultures are inoculated. The final biomass concentrations in batch cultures depend on the initial concentrations of all nutrients. Some microalgae tolerate high nutrient concentrations. *Galdieria sulphuraria* was e.g. only inhibited by glucose concentrations above 200 g L^{-1} (Schmidt et al. 2005). Other microalgae are inhibited by high nutrient concentrations. One example is heterotrophic *Chlamydomonas reinhardtii* growing on acetic acid. This alga obtained highest specific growth rates at acetic acid concentrations below 0.5 g L^{-1} (Chen and Johns 1994). Fed-batch cultures where substrate concentrations are maintained low during growth phase have therefore been used to maximise biomass concentrations in high cell density cultures, and have resulted in biomass concentrations of 50 – 116 g L^{-1} in several heterotrophic microalgae (Table 1). Also continuous flow cultures have been used to grow microalgae in cultures at high biomass concentrations and at low nutrient concentrations (Graverholt and Eriksen 2007, Ethier et al. 2011).

Some microalgal species may also secrete metabolites that inhibit their own growth and is probably why Wen and Chen (2001) were able to triple the biomass productivity of *Nitzschia laevis* in perfusion culture compared to continuous flow culture. Spend medium containing the inhibitory metabolites was separated from cells and removed in a settling devise. Since the cells were retained in the culture, the dilution rate could be higher than in the continuous flow culture. If inhibitory products result from overflow metabolism their production should be avoided if cells are carbon substrate limited. In the green alga, *Spongiococcum exetricicum*, secretion of inhibitory substances was minimised by nutrient limited growth in fed-batch and continuous flow cultures (Hilaly et al. 1994).

NUTRITIONAL OILS FROM HETEROTROPHIC MICROALGAE

A number of oleaginous microalgae accumulate lipids as energy storage materials. In some microalgae, lipids can make up more than half of the biomass and have high contents of long-chain ω-3 and ω-6 PUFAs. Most interest in the production of nutritional oils in heterotrophic microalgae has focussed on eicosapentaenoic acid (EPA, C20:5) and docosahexaenoic acid (DHA, C22:6) synthesis. Both of these fatty acids are mainly obtained from fish oils, where they have been bioaccumulated via marine food chains. They are sometimes referred to as marine lipids, have positive human health effects, also when extracted from algae (Doughman et al. 2007), and are essential ingredients in aquaculture feeds. However, fish oil is a limited natural resource that is already heavily exploited and PUFAs are considered a major constraint to the expansion the mariculture of carnivorous fish (Olsen 2011). Novel applications and increasing demands for PUFAs have therefore resulted in considerably interest in alternative PUFA sources such as microalgal cultures.

Eicosapentaenoic Acid

EPA is synthesised by a variety of microalgae (Vazhappilly and Chen 1998). Diatoms, in particularly species from the genus *Nitzschia*, have received considerable attention as heterotrophic producers of EPA (Barclay et al. 1994, Tan and Johns 1996). EPA production processes have been developed mainly in *N. laevis* (Table 1). In this diatom, EPA may constitute 10-25% of all fatty acids (Tan and Johns 1996, Wen and Chen 2000a, b). Triglycerides constitute roughly 70% of the lipids in *N. laevis*, where also most EPA is bound, but EPA is also found in phospolipids and in other lipid types (Chen et al. 2007). EPA may have specific biological functions in microalgae, but the high EPA content in triglycerides (Chen et al. 2007) and observations of increasing EPA contents relative to other fatty acids at late growth phases in *N. alba* batch cultures (Tornabene et al. 1974) indicate that EPA is also an important energy storage compound.

EPA production processes using *N. laevis* have gone through a series of well-documented optimisation steps with respect to productivity. In batch cultures, heterotrophic *N. laevis* grew to a maximal cell density of 2.04 g L^{-1}, in a medium containing 5 g L^{-1} glucose, while EPA productivities were 6.37 mg L^{-1} day^{-1} (Wen and Chen 2000a). Biomass concentrations could be increased to 5 g L^{-1} at 20-30 g L^{-1} glucose, but decreased again at higher glucose concentrations, and the specific growth rate was maximal at 5 g L^{-1} glucose (Wen and Chen 2000b). In fed-batch cultures were glucose inhibition was avoided by continuous glucose feeding, biomass grew to a concentration of 22.1 g L^{-1} while EPA productivities were 49.7 mg L^{-1} day^{-1} (Wen et al. 2002). Highest EPA productivities of 320.9 mg L^{-1} day^{-1} were, however, obtained in continuous perfusion cultures, where glucose was fed to avoid substrate inhibition while cell separation on the out-flow and partial cell recycling allowed fast dilution rates and removal of inhibitory metabolic products (Wen and Chen 2001). Despite the developments in EPA production processes, fish oil is still the current source to EPA.

Docosahexaenoic Acid

DHA is largely responsible for the nutritional benefits of fish oils in human diets, and fish oil is also still the dominant source to DHA. However, at least 3 heterotrophic microalgae are utilised for commercial DHA productions, the obligate heterotrophic dinoflagellate *Cryptechodinium cohnii* and the labyrinthulids *Schizochytrium* sp. and *Ulkenia* sp. (Ratledge 2004, Spolaore et al. 2006). Microalgal DHA is used as a supplemented to infant formula and various foods for adults (Ward and Singh 2005).

C. cohnii has been grown in batch and fed-batch cultures on glucose, acetic acid, or ethanol (de Swaaf et al. 1999, 2003a, b, Ratledge et al. 2001). The cultures are also supplied an organic nitrogen source, often in the form of yeast extract but several amino acids are also accepted (Pleissner et al. 2011). *C. cohnii* accumulates up to 40% of its dry weight as lipids, and up to 50% of the fatty acids can be DHA. The oil contain only trace amounts of other PUFAs, although strain specific differences do exist (Ratledge et al. 2001). Oleaginous microbes usually accumulate lipids when they are limited in nitrogen or another nutritional component while carbon substrates are in excess (Ratledge 2004). This may also be the case in *C. cohnii* (Mendes et al. 2009), but in addition *C. cohnii* also accumulates lipids as well as floridean starch during nutrient sufficient, exponential growth phases. This may be an adaptation that allows these cells to rapidly store carbon and energy when available nutrients are encountered in the otherwise nutrient deprived oceanic environments (Dauvillé et al. 2009). Lipid accumulation is therefore not restricted to nutrient limited growth phases, although specific lipid contents do increase when cultures get older and some nutrients, most likely components from yeast extract, are depleted (de Swaaf et al. 1999, 2003a, b, Ratledge et al. 2001). *C. cohnii* attain higher lipid contents when it grows on acetic acid or ethanol than when it grows on glucose, and highest biomass concentrations and DHA productivities have been obtained in pH auxostats where acetic acid is used as carbon source as well as acidic titrant to maintain constant pH (Ratledge et al. 2001), and in more traditional fed-batch cultures fed with acetic acid or ethanol (de Swaaf et al. 2003a, b). In these cultures, biomass concentrations have reached 30-83 g L^{-1} with DHA productivities of 0.86-1.3 g L^{-1} day^{-1} (Table 1).

DHA productions have also been described and developed in a number of species belonging to the labyrinthulids, most intensively in *Schizochytrium limacinum*, but also in *Ulkenia* sp., *Thraustochytrium* sp., and *Aurantiochytrium* sp. (Table 1). Batch cultures have in a number of studies been used to characterise growth and lipid production, while fed-batch and continuous flow cultures, where inhibitory effects of nutrients and metabolic by-products can be kept low, have been used to produce high cell densities and high lipid productivities. Labyrinthulids contain wider spectra of differents PUFAs than *C. cohnii* and a variety of fatty acid compositions can be produced depending on species and growth conditions. DHA is abundant in most species, constituting 20-60% of the fatty acids, but also EPA and the ω-6 PUFAs arachidonic acid (C20:4), and docosapentaenoic acid (C22:5) and others may be present (Huang et al. 2003, Ward and Singh 2005). While oils from *C. cohnii* is primarily added to infant formula where other PUFAs are unwanted, oils from *S. limacinum* with their larger numbers of different PUFAs are added as dietary supplements to a number of food products for adults (Ward and Singh 2005). Labyrinthulids can utilise a variety of different carbohydrates and glycerol and in recent years there has been a growing interest in lipid production based on left-over materials from other industries, like agricultural waste (Liang et al. 2010, Quilodrán et al. 2010) and glycerol waste from biodiesel manufacturing (Chi et al. 2007, Pyle et al. 2008, Ethier et al. 2011).

BIODIESEL FROM HETEROTROPHIC MICROALGAE

While the long-chain PUFAs are the interesting components in microalgal oils used for nutritional purposes, oils with lower contents of PUFAs are well suited as starting materials for biodiesel. Such oils are found in green algae and in some microalgae also from other phyla. The lipids are extracted from the biomass and transesterified in methanol and sulphuric acid into fatty acid methyl esters (Miao and Wu 2006, Johnson and Wen 2009). There is widespread interest in the use of phototrophic microalgae as producers of fatty acids for biodiesel manufacturing (Chisti 2007). There is also an emerging research effort in the exploitation of heterotrophic microalgae as producers of fatty acids for biodiesel because the highest lipid productivities in heterotrophic microalgal cultures have been 20 times that of phototrophic cultures (Chen et al. 2011).

The oleaginous green alga, *Chlorella protothecoides* is able to accumulate 45-55% of its biomass dry weight as lipids when grown heterotrophically and has been the main target for studies of microalgal biodiesel production (Miao and Wu 2006, Li et al. 2007, Cheng et al. 2009a, b, Shen et al. 2009). The main carbon source for *C. protothecoides* is glucose, which has been supplemented as pure glucose (Miao and Wu 2006, Shen et al. 2009) or in form of hydrolysed plant materials (Cheng et al. 2009a, b). Microalgal cultures, where lipids are produced for subsequent biodiesel production must obviously be designed for maximal conversion of the energy originally present in the used substrates into energy stored in the produced lipids.

Table 3. Biodiesel production in batch and fed batch cultures of heterotrophic *Chlorella protothecoides* grown on glucose or plant hydrolysates. Maximal biomass concentration x_{max}, total consumption of glucose Δs, yield of biomass on glucose $Y_{x/s}$, specific lipid content in biomass $Y_{p/x}$, yield of lipid on glucose $Y_{p/s} = Y_{x/s} \cdot Y_{p/x}$, energy conversion from glucose to lipid $Y_E = Y_{p/s} \cdot [\Delta H_{c,lipid}/\Delta H_{c,glucose}] \cdot 100\%$ where $\Delta H_{c,lipid}$ and $\Delta H_{c,glucose}$ are heat of combustion of rapeseed oil and glucose, respectively. Numbers given in italics are calculated based on information or data read from graphs in the references

Culture	Substrate	Scale L	x_{max} g L^{-1}	Δs g L^{-1}	$Y_{x/s}$ g g^{-1}	$Y_{p/x}$ g g^{-1}	$Y_{p/s}$ g g^{-1}	Y_E %	Reference
Fed batch	Glucose	5	15.5	*42.4*	*0.37*	0.46	*0.17*	*38*	Li et al. (2007)
Fed batch	Glucose	750	12.8	*32*	*0.40*	0.49	*0.19*	*44*	
Fed batch	Glucose	11,000	14.2	*29.6*	*0.48*	0.44	*0.21*	*48*	
Batch	Glucose	0.2	*7.15*	*23.2*	*0.31*	0.47	*0.14*	*32*	Cheng el at. (2009a)
Batch	Sucrose hydrolysate	0.2	*6.1*	*21.4*	*0.29*	0.44	*0.13*	*28*	
Batch	Sugar cane hydrolysate	0.2	*6.15*	*18.3*	*0.34*	0.43	*0.14*	*33*	
Fed batch	Glucose	2.5	*43.1*	*121.6*	*0.35*	0.48	*0.17*	*38*	
Fed batch	Sugar cane hydrolysate	2.5	*48.5*	*118.6*	*0.41*	0.45	*0.18*	*42*	
Batch	Glucose	0.5	16.5	*24*	0.63	0.45	*0.28*	*64*	Cheng et al. (2009b)
Batch	Jerusalem artichoke hydrolysate	0.5	16	*24*	0.57	0.44	*0.25*	*57*	
Batch	Jerusalem artichoke hydrolysate	0.5	15.5	*28*	0.58	0.43	*0.25*	*56*	
Batch	Jerusalem artichoke hydrolysate	0.5	17	*28*	0.50	0.46	*0.2*	*52*	

From 3 studies on microalgal biodiesel production in *C. protothecoides*, sufficient data on yields of biomass and lipids are available to make calculations on yields of lipid on glucose and energy conversion efficiencies of these processes (Table 3).

In two studies (Li et al. 2007, Cheng et al. 2009a), 28-42% of the combustion enthalpy originally present the consumed glucose was recovered in the produced lipids. In the third study, lipid yields corresponded to 64% recovery of combustion enthalpy (Cheng et al. 2009b). The differences in energy yield is the result of considerably higher yields of biomass in this third study (Table 3).

In all studies in Table 3, the energy recoveries in lipids have been lower than energy recoveries from industrial fuel ethanol fermentations in yeast. Fuel ethanol is mostly produced from sugar cane sugars or hydrolysed corn starch with an average yield of 0.46 kg ethanol per kg glucose (Wheals et al. 1999). With such an ethanol yield, 88% of the enthalpy of combustion originally present in the consumed glucose is recovered in the ethanol. It is questionable whether yields of microalgal lipid on sugars can be sufficiently improved to make microalgal biodiesel energetically competitive to fuel ethanol, as this would imply extremely high lipid yields in the algae. It could therefore become advantageous to base microalgal biodiesel productions on non-fermentable substrates. Johnson and Wen (2009) demonstrated recently that the properties of biodiesel produced by transesterification of lipids from *Schizochytrium limacinum* grown on crude glycerol waste are comparable to other biodiesels. Although biodiesel production in this species is far from being a well characterised process, specific lipid contents can be up to 60% of total dry weight in this and other labyrinthulids (Table 1). A substantial amount of knowledge exists on their productions of oils for nutrition and this knowledge can be directly utilised also for production of oils for fuel.

PIGMENTS FROM HETEROTROPHIC MICROALGAE

Algae synthesise 3 groups of photosynthetic pigments, chlorophylls, carotenoids, and phycobiliproteins. While chlorophyll *a* and *b* are found in plants as well as in algae, chlorophyll *c* and *d* are only found in certain types of algae. Also a number of carotenoids are found only in different algae, and phycobiliproteins are found exclusively in cyanobacteria and algae. Carotenoids and phycobiliproteins are pigments of considerable value. In algae they function as accessory light harvesting pigments. Carotenoids serve also as photoprotective pigments that protect the photosynthetic apparatus from damage by reactive oxygen species. Secondary carotenoids do not work in photosynthesis and can be found also outside chloroplasts and accumulate mainly in response to stress and function as antioxidants (Lemoine and Schoefs 2010). Production of photosynthetic pigments in heterotrophic cultures is particularly challenging as the organisms do not need the pigments when light is absent, and therefore they often degrade their photosynthetic apparatus and contain only low pigment levels when grown heterotrophically (Ahmad and Hellebust 1990, Sloth et al. 2006). Heterotrophic pigment production in microalgae may therefore depend on mutant strains which have partly lost the ability to down-regulate pigment synthesis under heterotrophic conditions (Graverholt and Eriksen 2007).

Heterotrophic production of carotenoids in microalgae has focussed mainly on the xanthophylls lutein and astaxanthin and has been described in greatest details in *Chlorella* species. Carotenoids were among the first microalgal products to be systematically investigated and optimised for production in high cell density heterotrophic cultures. Theriault (1965) describes how *C. pyrenoidosa* batch and fed batch cultures are grown to biomass concentrations of 100 and 302 g L^{-1} in illuminated 30 L bioreactors, while total xanthophyll concentrations reached 512 and 650 mg L^{-1}, respectively. These biomass concentrations are probably still the highest obtained in microalgal batch and fed batch cultures. Also another green alga, *Spongiococcum exetricicum* was early investigated for its production of xanthophylls (Ciegler 1965), and high cell density cultures with biomass concentrations above 50 g L^{-1} combining batch, fed batch and continuous flow stages were developed by Hilaly et al. (1994).

Lutein

Lutein is used as food colourant and feed additive and is one of the important colours of e.g. egg yolk. Lutein is also an essential nutrient to humans. It is found in serum and in the retina and can as such be regarded a vitamin (Semba and Dagbnelie 2003, Fernández-Sevilla et al. 2010). Lutein is presently produced in marigold where it is found in the petals. Lutein production has also been investigated in cultures of several phototrophic microalgae (Fernández-Sevilla et al. 2010) and in heterotrophic *Chlorella* species.

Lutein is a photosynthetic pigment serving a structural role in the light harvesting complex of photosystem II (Kühlbrandt et al. 1994). It can therefore be expected that changes in specific lutein concentrations depend on changes in the photosynthetic apparatus. Ishikawa et al. (2004) used random mutagenesis to develop highly pigmented strains of *C. regularis* with specific lutein contents of up to 5.26 mg g^{-1}, more than twice the value of the parental strain. This increase in specific lutein content was indeed followed by similar increases in specific contents also of other carotenoids and chlorophyll, and also by an increase in chloroplast volume in the mutant strain compared to the parental strain. Highly pigmented strains showed slightly lower specific growth rates than parental strains under heterotrophic conditions (Ishikawa et al. 2004), probably because of the increased metabolic burden related the synthesis of chloroplast components.

Heterotrophic synthesis of lutein has been described in carbon limited batch and fed batch cultures of *C. protothecoides*. Degradation of pigments is a common response to nitrogen limitation in many microalgae (see e.g. Eriksen et al. 2007), and highest specific lutein contents are therefore often found in stationary phases of carbon restricted batch cultures when the nitrogen source is still available (Ishikawa et al. 2004, Wu et al. 2007), and in carbon limited fed batch cultures (Shi et al. 2002). In *Chlorella* species used for heterotrophic production of lutein, specific lutein concentrations have been in the order of 1-5 mg g^{-1} dry weight (Shi et al. 1997, 2002, Wu et al. 2007), while phototrophic microalgal cultures used for lutein production have had slightly higher specific lutein contents of 4-8 mg g^{-1} (Fernández-Sevilla et al. 2010). However, the average lutein productivities of 39-49 mg L^{-1} day^{-1} obtained in 3 heterotrophic fed-batch cultures of *C. protothecoides* (Shi et al. 2002) are still considerably higher than the productivities (0.7-7.2 mg L^{-1} day^{-1}) obtained in outdoor as well as indoor cultures of photoautotrophic microalgae (Fernández-Sevilla et al. 2010).

Astaxanthin

Astaxanthin is mainly used as additive to fish feeds. Salmon and trout obtain their pink flesh from bioaccumulation of astaxanthin taken up via their diet. Astaxanthin has also positive health benefits in humans (Guerin et al. 2003). Astaxanthin can be made synthetically, while natural astaxanthin is extracted from shell fish waste or produced in phototrophic cultures of the green alga *Haematococcus pluvialis* (Olaizola 2003). Astaxanthin does not work in photosynthesis but accumulates in the cell cytoplasm where it may work as an antioxidant and a photoprotective pigment. Under the right conditions, i.e. high light intensity and nitrogen starvation, specific astaxanthin concentration in *H. pluvialis* can be at least up to 58 mg g^{-1} (Suh et al. 2006). Astaxanthin synthesis is not restricted to phototrophic organism, and especially the red yeast, *Phaffia rhodozyma* is also considered a promising candidate for heterotrophic synthesis of natural astaxanthin (Schmidt et al 2011).

Astaxanthin is a common secondary carotenoid also in other green algae than *H. pluvialis*, although in lower concentrations. *H. pluvialis* does not grow well heterotrophically but substantially amounts of astaxanthin (0.44-1.01 mg g^{-1}) have been found in heterotrophic *Chlorella zofingiensis* (Ip and Chen 2005, Sun and Chen 2008, Wang and Peng 2008). This is not so much lower compared to what is found in astaxanthin rich mutant strains of *P. rhodozyma* (0.16-6.6 mg g^{-1}, Schmidt et al 2011). *C. zofingiensis* has so far been grown to biomass concentrations of 53 g L^{-1} while the highest astaxanthin productivity can be estimated to 2.8 mg L^{-1} day^{-1} over an 8 day period based on data presented by Sun and Chen (2008). Unlike photosynthetic pigments like lutein where specific concentrations follow the other photosynthetic pigments and attain maximal values only when nitrogen is in excess (Wu et al. 2007), the specific astaxanthin content in *C. zofingiensis* is highest when grown in media with low contents of nitrogen and when the specific chlorophyll content is lowest (Ip and Chen 2005).

A couple of papers suggest that astaxanthin and other carotenoids can also be efficiently synthesised in the labyrinthulids, *Schizochytrium* sp. (Aki et al. 2003, Yamasaki et al. 2009) and *Ulkenia* sp. (Quilodrán et al. 2010), maybe as a co-product along with the production of PUFAs. Surprisingly high specific astaxanthin concentrations of up to 39 mg g^{-1}, which is similar to astaxanthin contents in *H. pluvialis*, have been found in batch cultures of *Ulkenia* sp. with productivities during 5 day periods of up to 63 mg L^{-1} day^{-1} (Quilodrán et al. 2010).

Phycocyanin

Phycocyanin is the only phycobiliprotein, which production has been described and optimised in heterotrophic microalgal cultures although commercial phycocyanin is still obtained from phototrophic cultures of the cyanobacterium *Arthrospira (Spirulina) platensis* (Eriksen 2008b). Phycocyanin is used as dye in food and cosmetics and as fluorescent marker in diagnostics. Phycocyanin is likely also to find therapeutic applications due to a high degree of structural and functional similarity between the phycocyanobilin chromophores in phycocyanin and bilirubin, a natural antioxidant and NADPH oxidase activity regulator in human plasma (McCarty 2007).

A. platensis can also grow and produce phycocyanin heterotrophically (Marquez et al. 1993) but heterotrophic phycocyanin production has been much more efficient in *Galdieria sulphuraria* (Eriksen 2008b), a unicellular eukaryote that belongs in Cyanidiophyta (Table 1).

G. sulphuraria grows at pH 1-3, has temperature optimum above 40°C, and can utilise a large number of different carbohydrates and polyols as carbon substrates (Gross and Schnarrenberger 1995), as well as sugar beet molasses (Schmidt et al. 2005). Glucose represses phycocyanin synthesis (Rhie and Beale 1994, Stadnichuk et al. 1998, 2000) but the isolation of mutant strains that partly retain their photosynthetic apparatus when grown heterotrophically in darkness have been described several times (Gross and Schnarrenberger 1995, Marquardt 1998). Nitrogen limitation results in almost complete depletion of phycocyanin and other photosynthetic pigments (Sloth et al. 2006). *G. sulphuraria* shows no signs of overflow metabolism or secretion of inhibitory metabolic by-products (Graverholt and Eriksen 2007), and tolerates more than 200 g L^{-1} glucose (Schmidt et al. 2005).

In glucose limited/nitrogen sufficient, heterotrophic, high-cell-density fed-batch and continuous flow cultures with biomass concentrations at 50-110 g L^{-1} biomass productivities have been up to 50 g L^{-1} day^{-1}. The high biomass productivity resulted in phycocyanin productivities as high as 0.86 g L^{-1} day^{-1} (Graverholt and Eriksen 2007). This is at least 10 times higher than what can be reached in *A. platensis* cultures despite the the specific C-PC content in heterotrophic *G. sulphuraria* are considerably lower than in *A. platensis* (Eriksen 2008b).

VITAMINS AND FUNCTIONAL CARBOHYDRATES FROM HETEROTROPHIC MICROALGAE

In addition to lipids and pigments, a few other compounds have also been synthesised in cultures of heterotrophic microalgae. One of these is ascorbic acid (vitamin C). Production of this vitamin was developed and optimised in heterotrophic cultures of *Chlorella pyrenoidosa* by Running et al. (1994). During a 2.5 year period they developed a plate assay that allowed screening of up to 25,000 mutants per week for intracellular ascorbic acid accumulation, developed production strains by mutagenesis followed by selection of high producing mutants, optimised growth medium composition, and improved ascorbic acid productivities 70-fold compared to the parental strain. The cultures were grown to biomass concentrations of 20-40 g L^{-1} resulting in 1-2 g L^{-1} ascorbic acid and the specific ascorbic acid concentrations have therefore been in the order of 5 mg g^{-1} while the parental strain contained just 0.64 mg g^{-1} (Running et al. 1994). A few years later, also processes for extracellular production of ascorbic acid were developed in heterotrophic *C. protothecoides* and several *Prototheca* species (Running et al. 2002). *Prototheca* and *Chlorella* are closely related genera but *Prototheca* species are colourless, obligate heterotrophs that grow a low pH where ascorbic acid is resistant to oxidation. *P. zopfii* was grow to biomass concentrations above 50 g L^{-1}, secreting in the order of 70 mg L^{-1} of ascorbic acid into the growth medium.

A second vitamin and antioxidant that has been produced in heterotrophic microalgae is α-tocopherol, the most active component of vitamin E. Other tocopherols and related compounds are also classified as Vitamin E and synthesised by plants and other photosynthetic organisms. In *Euglena gracilis*, which grows phototrophically as well as heterotrophically α-tocopherol can make up 97% of all tocopherols why α-tocopherol production has been extensively investigated this species (Ogbonna 2009). Heterotrophic fed-batch cultures *E. gracilis* grown on ethanol have been grown to biomass concentrations of up to 39.5 g L^{-1} and specific α-tocopherol concentrations have been 1.2 mg g^{-1} (Ogbonna et al. 1998). *E. gracilis* cultures grown on ethanol produce higher specific α-tocopherol concentrations than cultures grown on glucose. Fujimata et al. (2009) showed that cultures growing on ethanol also have higher intracellular activities of reactive oxygen species and suggested that *E. gracilis* increased its α-tocopherol concentrations to protect against these oxidative species.

Paramylon is a different, but also interesting product from heterotrophic *E. gracilis*. Paramylon is a β-1,3-glucan used as a carbon and energy storage compound in *E. gracilis*. Paramylon may also have biological affects in humans (Santek et al. 2009). In heterotrophic cultures grown on glucose, paramylon can constitute up to 90% of the biomass dry weight (Barsanti et al. 2001). Batch processes for paramylon production have been described in defined media (Santek et al. 2009) as well as in complex media based on glucose enriched potato liquor (Santek et al. 2010). Paramylon is a growth associated product that is produced by *E. gracilis* during growth phases with all nutrients in surplus (Santek et al. 2009). In this respect paramylon synthesis is different from synthesis of most other carbon and energy storage compounds that are produced mainly by nutrient limited cells (Ratledge 2004) but similar to starch synthesis in *C. cohnii* (Dauvillé et al. 2009). Rodriguez-Zavala et al. (2009) investigated co-productions of paramylon, α-tocopherol and also tyrosine in *E. gracilis*, something that could be a possibility, in particular for growth associated products.

HETEROTROPHIC MICROALGAE AS FOOD AND FEED

Pigments and PUFAs extracted from different microalgae are used in foods and feed as described above. A few microalgal species are also eaten directly by humans. *Chlorella* sp. has for long been produced in quantities of hundreds of tonnes per year and marketed as a health food (Lee 1997). A large proportion of the *Chlorella* biomass is produced heterotrophically, and thereby is *Chlorella* the microalga that is produced heterotrophically in largest quantity. Microalgal biomass have also been used as functional components of foods and feed. When rabbits were fed a diet containing 4 g kg^{-1} *Schizochytrium* sp. they obtained higher PUFA contents in their meat compared to rabbits fed other diets and the rabbit meat became possibly a more healthy food source to humans (Mordenti et al. 2010). Also DHA concentrations in milk increased when *Schizochytrium* sp. were include in diets for dairy cows (Franklin et al. 1999).

Most importantly is the use of microalgae in aquaculture feeds. PUFAs, pigments and other cell components from the microalgae are essential nutritional factors needed for correct development of many aquatic animals. Limited availability of fish oil is a severe problem for the continued expansion of mariculture of carnivorous fish (Olsen 2011) and therefore oils from heterotrophic microalgae may become important nutritional supplements. *Cryptechodinium cohnii* and *Schizochytrium* sp. have been tested as substitution for fish oil in fish feed. The PUFA content in catfish meat increases when the fish are fed a diet containing 0.5-2% *Schizochytrium* sp. (Li et al. 2009). Seabreams fed 2-4% *C. cohnii* also incorporated PUFAs from the alga into their own lipids, and showed lower mortality than fish fed fish oil containing diet (Atalah et al. 2007). In a second study, Ganuza et al. (2008b) also found that *C. cohnii* or *Schizochytrium* sp. biomass were suitable ingredients in seabream feed.

However, complete substitution of squid oil by algal biomass caused increased mortality, most likely because of too low EPA contents in the algae. These algae have also been added to feeds for other fish species and can clearly substitute at least part of the fish oils in their feeds. *C. cohnii* or *Schizochytrium* sp. biomass is 10 times more expensive than fish oil but also have higher contents of PUFAs and therefore is the price difference on PUFAs from the two sources much smaller (Hare et al. 2002).

Filter feeding oysters and mussels, as well as rotifers and copepods used as feed for fish larvae depend on small food particles of 5-15 µm in size and many microalgal cells fall into this size interval. Diets based on yeast or microencapsulated nutrients are often preferred by industry for production of rotifers because of the high costs of microalgae (Spolaore et al. 2006), but microalgal diets results in more robust productions, and the rotifers obtain higher nutritional value when fed algal cells. When heterotrophic *Chlorella* cells are enriched in vitamin B_{12} they become a suitable food sources for rotifers *Brachionus* spp. and are used routinely for producing these rotifers (Lubzens et al. 2001). Muller-Feuga (2004) estimated that 50% of all marine fish produced in China, the world's largest producer of marine fish, come from hatcheries that depend on microalgae for their feed production. However, *Chlorella* does not contain long chain PUFAs and the rotifers are therefore supplemented with additional nutrients before fed to the fish. Oleaginous heterotrophic microalgae also show good potentials as feeds and nutritional enrichments for rotifers. PUFAs obtained from their microalgal diets are deposited in the rotifers (Yamasaki et al 2007, Estudillo-del Castillo 2009). The same is true for different copepods (Veloza et al. 2006, Yamasaki et al. 2007), a second group of zooplankton organisms used as aquaculture feed.

Oyster hatcheries typically have their own production of phototrophic microalgae which are needed in the production of larvae and oyster spat. Heterotrophic microalgae may provide an cost-efficient alternative food source if produced heterotrophically in large quantities (Day et al. 1991, Muller-Feuga 2004). Several *Tetraselmis* species have been grown heterotrophically and evaluated as feed for filter feeding bivalves (Glaude and Maxey 1994, Day and Tsavalos 1996), also at industrial scale (Day et al. 1991). *Tetraselmis* sp. grows well heterotrophically and is a commonly used, suitable feed when grown phototrophically. However, when *Tetraselmis* sp. and also a number of other microalgal species used as feed are grown heterotrophically they accumulate more starch and less lipids compared to when they are grown phototrophically, and are therefore of lesser nutritional value to the animals (Day et al. 1991, Glaude and Maxey 1994, Day and Tsavalos 1996). Also the contents of pigments (Day and Tsavalos 1996) and sterols (Jo et al. 2004) are reduced in heterotrophic compared to phototrophic *Tetraselmis* sp. These compounds cannot be synthesised by bivalves and must be supplemented in their diet. Heterotrophic microalgae have therefore not been used successfully to feed filter feeding bivalves as sole food force but used in combination with phototrophic microalgae (Knauer and Southgate 1996, Harel and Place 2004).

CONCLUSION

Heterotrophic microalgal cultures can potentially be much more productive and operated at much higher biomass concentrations than cultures of phototrophic microalgae since the heterotrophic cultures are not limited by external supplies of light. For a number of heterotrophic microalgal species, cultures with productivities of 5-50 g L^{-1} day^{-1} and biomass concentrations above 100 g L^{-1} have been developed. The number of species that have been grown heterotrophically and described in details is limited but they represent great phylogenetic variability with representatives from most of the microalgal phyla. Oils from *Crypthecodinium cohnii* and *Schizochytrium* sp. rich in DHA have been the most successful products made in cultures of heterotrophic microalgae along with *Chlorella* biomass used as health food. EPA and other PUFAs, carotenoids, phycocyanin and vitamins are compounds either found uniquely in microalgae or in higher concentrations than in other organisms. Production processes for all of these compounds have been developed in heterotrophic species, and with much higher productivities compared to what is possible in cultures of their photosynthetic relatives. Microalgae have enormous importance as food for larger organisms in aquatic environments and development of heterotrophic species for use in mariculture may be governed by the steady increasing demand for particular PUFAs and pigments to supplement feeds for fish and shellfish.

ACKNOWLEDGMENT

This paper was prepared as part of the MarBioShell project supported by the Danish Agency for Science, Research and Innovation.

REFERENCES

Aaronson S, Baker H (1959) A comparative biochemical study of two species of *Ochromonas*. *J protozool* 6: 282-284

Abeliowich A, Weisman D (1978) Role of heterotrophic nutrition in growth of the alga *Scenedesmus obliquus* in high-rate oxidation ponds. *Appl Environ Micro*biol 35: 32-37

Ahmad I, Hellebust JA (1990) Regulation of chloroplast development by nitrogen source and growth conditions in a *Chlorella protothecoides* Strain1. *Plant Physiol* 94: 944-949

Aki T, Hachida K, Yoshinaga M, Katai Y, Yamasaki T, Kawamoto T, Maoka T, Shigeta S, Suzuki O, Ono K (2003) Thraustochytrids as a potential source of carotenoids. *J Am Oil Chem Soc* 80: 789-794

Azma M, Mohamed MS, Mohamad Rm Rahim RA, Ariff AB (2011) Improvement of medium composition for heterotrophic cultivation of green microalgae, *Tetraselmis suecica*, using response surface methodology. *Biochem Eng J* 53: 187-195

Barclay WR, Meager KM, Abril JR (1994) heterotrophic production of long chain omega-3 fatty acids utilizing algae and algae-like microorganisms. *J Appl Phycol* 6: 123-129

Barsanti L, Vismara R, Passarelli V, Gualtieri P (2001) Paramylon (β-1,3-glucan) content in wild type and WZSL mutant of *Euglena gracilis*. Effects of growth conditions. *J Appl Phycol* 13: 59-65

Behrens P (2005) Photobioreactors and fermentors: The light and dark sides of growing algae. In: Andersen RA (ed) *Algal cultivation techniques*. Elsevier Academic Press, p. 189-204

Boëchat IG, Schuran S, Adrian R (2005) Supplementation of the protist *Chilomonas paramecium* with a highly unsaturated fatty acid enhances its nutritional quality for the rotifer *Keratella quadrata*. *J Plankton Res* 27: 663-670

Burell ER, Mayfield CI, Inniss WE (1984) Biomass production from the green algae *Chlorella vulgaris* and *Ankistrodesmus braunii*. *Biotechnol lett* 6: 507-510

Chen C-Y, Yeh K-L, Aisyah R, Lee D-J, Chang J-S (2011) Cultivation, photobioreactor design and harvesting of microalgae for biodiesel production: A critical review. *Biores Technol* 102: 71-81

Chen F, Johns MR (1994) Substrate inhibition of *Chlamydomonas reinhardtii* by acetate in heterotrophic culture. *Process Biochem* 29: 145-252

Chen G-Q, Chen F (2006) Growing phototrophic cells without light. *Biotechnol Lett* 28: 607-616

Chen G-Q, Jiang Y, Chen F (2007) Fatty acid and lipid class composition of the eicosapentaenoic acid-producing microalga, *Nitzschia laevis*. *Food Chem* 104: 1580-1585

Cheng Y, Lu Y, Gao C, Wu Q (2009a) Alga-based biodiesel production and optimization using sugar cane as the feedstock. *Energy fuels* 23: 4166-4173

Cheng Y, Zhou W, Gao C, Lan K, Gao Y, Wu Q (2009b) Biodiesel production from Jerusalem artichoke (*Helianthus tuberosus* L.) tuber by heterotrophic microalgae *Chlorella protothecoides*. *J Chem Technol Biotechnol* 84: 777-781

Chi Z, Liu Y, Frear C, Chen S (2009) Study of a two-stage growth DHA-producing marine algae *Schizochytrium limacinum* SR21 with shifting dissolved oxygen level. *Appl Microbiol Biotechnol* 81: 1141-1148

Chi Z, Pyle D, Wen Z, Frear C, Chen S (2007) A laboratory study of producing docosahexaenoic acid from biodiesel-waste glycerol by microalgal fermentation. *Process Biochem* 42: 1537-1545

Chisti Y (2007) Biodiesel from microalgae. *Biotechnol Adv* 25: 294-306

Ciegler A (1965) Microbial carotenogenesis. *Adv Appl Microbiol* 7: 1-34

Dauvillé D, Deschamps P, Ral J-P, Plancke C, Putaux J-L, Devassine J, Durand-Terrasson A, Devin A, Ball SG (2009) Genetic dissection of floridean starch synthesis in the cytosol of the model dinoflagellate *Cryptechodinium cohnii*. *Proc Natl Acad Sci* 106: 21126-21130

Day JD, Edwards AP, Rodgers GA (1991) Development of an industrial-scale process for the heterotrophic production of a micro-algal mollusc feed. *Biores Technol* 38: 245-249

Day JG, Tsavalos AJ (1996) An investigation of the heterotrophic culture of the green alga *Tetraselmis*. *J Appl Phycol* 8: 73-77

da Silva TL, Reis A (2008) The use of multi-parameter flow cytometry to study the impact of *n*-dodecane additions to marine dinoflagellate microalga *Cryptechodinium cohnii* batch fermentations and DHA production. *J Ind Microbiol Biotechnol* 35: 875-887

de Swaaf ME, de Rijk C, Eggink G, Sijtsma L (1999) Optimisation of docosahexaenoic acid production in batch cultivations by *Crypthecodinium cohnii*. *J Biotechnol* 70: 185-192 de Swaaf ME, Pronk JT, Sijtsma L (2003a) High-cell-density fed-batch cultivation of the docosahexaenoic acid producing marine alga *Crypthecodinium cohnii*. *Biotechnol Bioeng* 81: 666-672

de Swaaf ME, Pronk JT, Sijtsma L (2003b) Fed-batch cultivation of the docosahexaenoic-acid-producing marina alga *Crypthecodinium cohnii* on ethanol. *Appl Microbiol Biotechnol* 61: 40-43

Doucha J, Straka F, Lívanský K (2005) Utilization of flue gas for cultivation of microalgae (*Chlorella* sp.) in an outdoor open thin-layer photobioreactor. *J Appl Phycol* 17: 403-412

Doughman SD, Krupanidhi S, Sanjeevi CB (2007) Omega-3 fatty acids for nutrition and medicine: Considering microalgae oil as a vegetarian source of EPA and DHA. *Current Diabetes Rev* 3: 198-203

Eriksen NT (2008a) The technology of microalgal culturing. *Biotechnol. Lett.* 30: 1525-1536

Eriksen NT (2008b) Production of phycocyanin - a pigment with applications in biology, biotechnology, foods, and medicine. *Appl. Microbiol. Biotechnol.* 80: 1-14

Eriksen NT, Poulsen BR, Iversen JJL (1998) Dual sparging photobioreactor for continuous production of microalgae. *J Appl Phycol* 10: 377-382

Eriksen NT, Riisgård FK, Gunther W, Iversen JJL (2007) On-line estimation of O_2 production, CO_2 uptake, and growth kinetics of microalgal cultures in a gas tight photobioreactor. *J Appl Phycol* 19: 161-174

Estudillo-del Castillo, C, Gapasin RS, Leaño EM (2009) Enrichment potential of HUFA-rich thraustochytrid *Schizochytrium mangrovei* for the rotifer *Brachionus plicatilis*. *Aquaculture* 293: 57-61

Ethier S, Woisard K, Vaughan D, Wen X (2011) Continuous culture of the microalgae *Schizochytrium limacinum* on biodiesel-derived crude glycerol for producing docosahexaenoic acid. *Biores Technol* 102: 88-93

Franklin, ST, Martin KR, Baer RJ, Schingoethe DJ, Hippen AR (1999) Dietary marine algae (*Schizochytrium* sp.) increases concentrations of conjugated linoleic acid, docosahexaenoic acid, and transvaccenic acid of milk in dairy cow. *J Nutr* 129: 2048–2052

Fujimata T, Ogbonna JC, Tanaka H, Aoyagi H (2009) Effects of reactive oxygen species on α-tocopherol production in mitochondria and chloroplasts of *Euglena gracilis*. *J Appl Phycol* 21: 185-191

Fujita T, Aoyagi H, Ogbonna JC, Tanaka H (2008) Effect of mixed organic substrate on α-tocopherol production by *Euglena gracilis* in photoheterotrophic culture. *Appl Microbiol Biotechnol* 79: 371-378

Ganuza E, Anderson AJ, Ratledge C (2008a) High-cell-density cultivation of *Schizochytrium* sp. in an ammonium/pH-auxostat fed-batch system. *Biotechnol Lett* 30: 1559-1564

Ganuza E, Benítez-Santana t, Atalah E, Vega-Orellana O, Ganga R, Izquierdo MS (2008b) *Crypthecodinium cohnii* and *Schizochytrium* sp. as potential substitutes to fisheries-derived oils from seabream (*Sparus aurata*) microdiets. *Aquaculture* 277: 109-116

Graverholt OS, Eriksen NT (2007) Heterotrophic high cell-density fed-batch and continuous flow cultures of *Galdieria sulphuraria* and production of phycocyanin. *Appl Microbiol Biotechnol* 77: 69-75

Gross W, Schnarrenberger C (1995) Heterotrophic growth of two strains of the acido-thermophilic red alga *Galdieria sulphuraria*. *Plant Cell Physiol* 36: 633-638

Guerin M, Huntley ME, Olaizola M (2003) *Haematococcus* astaxanthin: applications for human health and nutrition. *Trend Biotechnol* 21: 210-216

Harel M, Koven W, Lein I, Bar Y, Behrens P, Stubblefield J, Zohar Y, Place AR (2002) Advanced DHA, EPA and ArA enrichment materials for marine aquaculture using single cell heterotrophs. *Aquaculture* 213: 347-362

Harel M, Place AR (2004) Heterotrophic production of marine algae for aquaculture. In: Richmond A (ed.) *Handbook of microalgal culture. Biotechnology and applied phycology.* Blackwell Science Ltd, p. 513-524

Hilaly AK, Karim MN, Guyre D (1994) Optimization of an industrial microalgae fermentation. *Biotechnol Bioeng* 43: 314-320

Huang J, Aki T, Yokochi T, Nakahara T, Honda D, Kawamoto S, Shigeta S, Ono K, Suzuki O (2003) Grouping newly isolated docosahexaenoic acid-producing thraustochytrids based on their polyunsaturated fatty acid profiles and comparative analysis of 18S rRNA genes. *Mar Biotechnol* 5: 450-457

Iida I, Nakahara T, Yokochi T, Kamisaka Y, yagi H, Yamaoka M, Suzuki O (1996) Improvement of docosahexaenoic acid production in a culture of *Thraustochytrium aureum* by medium optimization. *J Ferm Bioeng* 81: 76-78

Ip P-F, Chen F (2005) production of astaxanthin by the green microalga *Chlorella zofingiensis*. *Process Biochem* 40: 733-738

Ishikawa F, Sansawa H, Abe H (2004) Isolation and characterization of a *Chlorella* mutant producing high amounts of chlorophyll and carotenoids. *J Appl Phycol* 16: 385-393

Jakobsen AN, Aasen IM, Josefsen KD, Strøm AR (2008) Accumulation of docosahexaenoic acid-rich lipid in thraustochytrid *Aurantiochytrium* sp. strain T66: effect of N and P starvation and O_2 limitation. *Appl Microbiol Biotechnol* 80: 297-306

Johnson MB, Wen Z (2009) Production of biodiesel from the microalga *Schizochytrium limacinum* by direct transesterification of algal biomass. *Energy Fuels* 23: 5179-5183

Knauer J, Southgate PC (1996) Nutritional value of a spray-dried freshwater alga, *Spongiococcum excentricum*, for Pacific oyster (*Crassostrea gigas*) spat. *Aquaculture* 146: 135-146

Lee Y-K (1997) Commercial production of microalgae in the Asia-Pacific rim. *J Appl Phycol* 9: 403-411

Lemoine Y, Schoefs B (2010) Secondary ketocarotenoid astaxanthin biosynthesis in algae: a multifunctional response to stress. *Photosynth Res* 106: 155-177

Lewin J, Hellebust JA (1976) Heterotrophic nutrition of the marine pennate diatom *Nitzschia angularis* var. *affinis*. *Mar Biol* 36: 313-320

Lewin J, Hellebust JA (1978) Utilization of glutamate and glucose for heterotrophic growth by the marine pennate diatom *Nitzschia laevis*. *Mar Biol* 47: 1-7

Li X, Xu H, Wu Q (2007) Large-scale biodiesel production from microalga *Chlorella protothecoides* through heterotrophic cultivation in bioreactors. *Biotechnol Bioeng* 98: 764-771

Li MH, Robinson EH, Tucker CS, Manning BB, Khoo L (2009) Effects of dried algae *Schizochytrium* sp., a rich source of docosahexaenoic acid, on growth, fatty acid composition, and sensory quality of channel catfish *Ictalurus punctatus*. *Aquaculture* 292: 232-236

Liang Y, Sarkany N, Cui Y (2009) Biomass and lipid productivities of *Chlorella vulgaris* under autotrophic, heterotrophic and mixotrophic growth conditions. *Biotechnol Lett* 31: 1043-1049

Liang Y, Sarkany N, Cui Y, Yesuf J, Trushenski J, Blackburn JW (2010) Use of sweet sorghum juice for lipid production by *Schizochytrium limacinum* SR21. *Biores Technol* 101: 3623-3627

Lubzens E, Zmora O, Barr Y (2001) Biotechnology and aquaculture of rotifers. *Hydrobiologia* 446/447: 337–353

Marquardt J (1998) Effects of carotenoid-depletion on the photosynthetic apparatus of a *Galdieria sulphuraria* (Rhodophyta) strain that retains its photosynthetic apparatus in the dark. *J Plant Physiol* 152: 372-80

Marquez FJ, Sasaki K, Kakizono T, Nishio N, Nagai S (1993) Growth characterization of *Spirulina platensis* in mixotrophic and heterotrophic conditions. *J Ferm Bioeng* 76: 408-410

McCarty MF (2007) Clinical potential of *Spirulina* as a source of phycocyanobilin. *J Med Food* 10: 566-570

Mendes A, Reis A, Vasconcelos R, Guerra P, da Silva TL (2009) *Cryptechodinium cohnii* with emphasis on DHA production: a review. *J Appl Phycol* 21: 199-214

Miao X, Wu Q (2006) Biodiesel production from heterotrophic microalgal oil. *Biores Technol* 97: 841-846

Mordenti AL, Sardi L, Bonaldo A, Pizzamiglio V, Brogna N, Cipollini M, Zaghini G (2010) Influence of marine algae (*Schizochytrium* spp.) dietary supplementation on doe performance and progeny meat quality. *Livestock Sci* 128: 179-184

Muller-Feugo A (2004) Microalgae for aquaculture. The current global situation and future trends. In: Richmond A (ed.) *Handbook of microalgal culture. Biotechnology and applied phycology*. Blackwell Science Ltd, p. 352-364

Nagano N, Taoka Y, Honda D, Hayashi M (2009) Optimization of culture conditions for growth and docosahexaenoic acid production by a marine Thraustochytrid, *Aurantiochytrium limacinum* mh0186. *J Oleo Sci* 58: d623-628

Ogbonna JC (2009) Microbiological production of tocopherols: current state and prospects. *Appl Microbiol Biotechnol* 84: 217-225

Ogbonna JC, Tanaka H (2000) Light requirement and photosynthetic cell cultivation – Developments of processes for efficient light utilization in photobioreactors. *J Appl Phycol* 12: 207-218

Ogbonna JC, Tomiyama S, Tanaka H (1998) Heterotrophic cultivation of *Euglena gracilis* Z for efficient production of α-tocopherol. *J Appl Phycol* 10: 67-74

Oh SH, Han JG, kim Y, Ha JH, Kim SS, Jeong MH, Jeong HS, Kim NY, Cho JS, Yoon WB, Lee SY, Kang DH, Lee HY (2009) Lipid production in *Porphyridium cruentum* grown under different culture conditions. *J Biosci Bioeng* 108: 429-434

Olaizola M (2003) Commercial development of microalgal biotechnology: from the test tube to the marketplace. *Biomol Eng* 20: 459-466

Olsen Y (2011) Resources for fish feed in future mariculture. *Aquacult Environ Interact* 1: 187-200

Pahl SL, Lewis DM, Chen F, King KD (2010) Heterotrophic growth and nutritional aspects of the diatom *Cyclotella cryptica* (Bacillariophyceae): Effect of some environmental factors. *J Biosci Bioeng* 109: 235-239

Perez-Garcia O, Escalante FME, de-Bashan LE, Bashan Y (2011) Heterotrophic cultures of microalgae: metabolism and potential products. *Wat Res* 45: 11-36

Pleissner D, Wimmer R, Eriksen NT (2011) Quantification of amino acids in fermentation media by isocratic HPLC analysis of their α-hydroxy acid derivatives. *Analyt Chem* 83: 175-181

Pyle DJ, Garcia RA, Wen Z (2008) Producing docosahexaenoic acid (DHA)-rich algae from biodiesel-derived crude glycerol: Effects of impurities on DHA production and algal biomass composition. *J Agric Food Chem* 56: 3933-3939

Quilodrán B, Hinzpeter I, Hormazabal E, Quiroz A, Shene C (2010) Docosahexaenoic acid (C22:6n-3, DHA) and astaxanthin production by Thraustochytriidae sp. AS4-A1 a native strain with high similitude to *Ulkenia* sp.: Evaluation of liquid residues from food industry as nutrient sources. *Enzyme Microb Technol* 47: 24-30

Ratledge C (2004) Fatty acid biosynthesis in microorganisms being used for single cell oil production. *Biochimie* 86: 807-815

Ratledge C, Kanagachandran K, Anderson AJ, Grantham DJ, Stephenson JC (2001) Production of docosahexaenoic acid by *Cryptechodinium cohnii* grown in a pH-auxostat culture with acetic acid as principal carbon source. *Lipids* 36: 1241-1246

Ren L-J, Ji X-J, Huang H, Qu L, Feng Y, Tong Q-Q, Ouyang P-K (2010) Development of a stepwise aeration control strategy for efficient docosahexaenoic acid production by *Schizochytrium* sp. *Appl Microbiol Biotechnol* 87: 1649-1656

Rhie G, Beale SI (1994) Regulation of heme oxygenase activity in *Cyanidium caldarium* by light, glucose, and phycobilin precursors. *J Biol Chem* 269: 9620-9626

Riesenberg D, Guthke R (1999) High-cell-density cultivation of microorganisms. *Appl Microbial Biotechnol* 51: 422-430

Rodriguez-Zavala JS, Ortiz-Cruz MA, Mendoza-Hernandez G, Moreno-Sanchez R (2009) Increased synthesis of a-tocopherol, paramylon and tyrosine by *Euglena gracilis* under conditions of high biomass production. *J Appl Microbiol* 109: 2160-2172

Roels JA (1980) Applications of macroscopic principles to microbial metabolism. *Biotechnol Bioeng* 22: 2457-2514

Running JA, Huss RJ, Olson PT (1994) Heterotrophic production of ascorbic acid by microalgae. *J Appl Phycol* 6: 99-104

Running JA, Severson DK, Schneider KJ (2002) Extracellular production of L-ascorbic acid by *Chlorella protothecoides*, *Prototheca* species, and mutants of *P. moriformis* during aerobic cultivation at low pH. *J Ind Microbiol Biotechnol* 29: 93-98

Santek B, Felski M, Friehs K, Lotz M, Flashel E (2009) Production of paramylon, a β-1,3-glucan, by heterotrophic cultivation of *Euglena gracilis* on a synthetic medium. *Eng Life Sci* 9: 23-28

Santek B, Felski M, Friehs K, Lotz M, Flashel E (2010) Production of paramylon, a β-1,3-glucan, by heterotrophic cultivation of *Euglena gracilis* on potato flour. *Eng Life Sci* 9: 23-28

Sauer N, Tanner W (1989) The hexose carrier from *Chlorella*. cDNA cloning of a eukaryotic H^+-cotransporter. *FEBS Let* 259: 43-46

Schmidt I, Schewe H, Gassel S, Jin C, Buckingham J, Hümbelin M, Sandmann G, Schrader J (2011) Biotechnological production of astaxanthin with *Phaffia rhodozyma/Xanthophyllomyces dendrorhous*. *Appl Microbiol Biotechnol* 89: 555-571

Schmidt RA, Wiebe MG, Eriksen NT (2005) Heterotrophic high cell-density fed-batch cultures of the phycocyanin producing red alga *Galdieria sulphuraria*. *Biotechnol Bioeng* 90: 77-84

Semba RD, Dagnelie G (2003) Are lutein and zeaxanthin conditionally essential nutrients for eye health? *Med Hypothesis* 61: 464-472

Semple KT (1998) Heterotrophic growth on phenolic mixtures by *Ochromonas danica*. *Res Microbiol* 149: 65-72

Shen Y, Yuan W, Pei Z, Mao E (2009) Heterotrophic culture of *Chlorella protothecoides* in various nitrogen sources for lipid production. *Appl Biochem Biotechnol* 160: 1674-1648

Shi X-M, Chen F, Yuan J-P, Chen H (1997) Heterotrophic production of lutein by selected *Chlorella* strains. *J Appl Phycol* 9: 445-450

Shi X-M, Jiang Y, Chen F (2002) high-yield production of lutein by the green microalga *Chlorella protothecoides* in heterotrophic fed-batch culture. *Biotechnol Prog* 18: 723-727

Shi X, Wu Z, Chen F (2006) Kinetic modelling of lutein production by heterotrophic *Chlorella* at various pH and temperatures. *Mol Nutr Food Res* 50: 763-768

Sing A, Ward OP (1996) production of high yields of docosahexaenoic acid by *Thraustochytrium roseum* ATCC. *J Ind Microbiol* 16: 370-373

Sloth JK, Wiebe MG, Eriksen NT (2006) Accumulation of phycocyanin in heterotrophic and mixotrophic cultures of the acidophilic red alga *Galdieria sulphuraria*. *Enzyme Microb Technol* 38: 168-175

Spolaore P, Joannis-Cassan C, Duran E, Isambert A (2006) Commercial applications of microalgae. *J Biosci Bioeng* 101: 87-96

Stadnichuk IN, Rakhimberdieva MG, Bolychevtseva YV, Yurina NP, Karapetyan NV, Selyakh IO (1998) Inhibition by glucose of chlorophyll a and phycocyanobilin biosynthesis in the unicellular red alga *Galdieria partita* at the stage of coproporphyrinogen III formation. *Plant Sci* 136: 11-23

Stadnichuk IN, Rakhimberdieva MG, Boichenko VA, Karapetyan NV, Selyakh IO, Bolychevtseva YV (2000) Glucose-induced inhibition of the photosynthetic pigment apparatus in heterotrophically-grown *Galdieria partita*. *Russ J Plant Physiol* 47: 585-592

Suh IS, Joo H-N, Lee C-G (2006) A novel double-layered photobioreactor for simultaneous *Haematococcus pluvialis* cell growth and astaxanthin accumulation. *J Biotechnol* 125: 540-546

Sun N, Wang Y, Li Y-T, Huang J-C, Chen F (2008) Sugar-based growth, astaxanthin accumulation and carotenogenic transcription of heterotrophic *Chlorella zofingiensis*. *Process Biochem* 43: 1288-1292

Tan C, Johns MR (1996) Screening of diatoms for heterotrophic eicosapentaenoic acid production. *Appl Phycol* 8: 59-64

Theriault RJ (1965) Heterotrophic growth and production of xanthophylls by *Chlorella pyrenoidosa*. *Appl Microbiol* 13: 402-416

Tornabene TG, Kates M, Volcani BE (1974) Sterols, aliphatic hydrocarbons, and fatty acids of a nonphotosynthetic diatom, *Nitzschia alba*. *Lipids* 4: 279-284

Vazhappilly R, Chen G (1998) Eicosapentaenoic acid and docosahexaenoic acid production potential of microalgae and their heterotrophic growth. *J Am Oil Chem Soc* 75: 393-397

Veloza AJ, Chu F-LE, Tang KW (2006) Trophic modification of essential fatty acids by heterotrophic protists and its effect on the fatty acid composition of the copepod *Acartia tonsa*. *Mar Biol* 148: 779-788

Wang Y, Peng J (2008) Growth-associated biosynthesis of astaxanthin in heterotrophic *Chlorella zofingeniensis* (Chlorophyta). *World J Microbiol Biotechnol* 24: 1915-1922

Ward OP, Singh A (2005) Omega-3/6 fatty acids: Alternative sources of production. *Process Biochem* 40: 3627-3652

Wen Z-Y, Chen F (2000a) Production potential of eicosapentaenoic acid by the diatom *Nitzschia laevis*. *Biotechnol Lett* 22: 727-733

Wen Z-Y, Chen F (2000b) Heterotrophic production of eicosapentaenoic acid by the diatom *Nitzschia laevis*: Effect of silicate and glucose. *J Ind Microbiol Biotechnol* 25: 218-224

Wen Z-Y, Chen F (2001) A perfusion-cell bleeding culture strategy for enhancing the productivity of eicosapentaenoic acid by *Nitzschia laevis*. *Appl Microbiol Biotechnol* 57: 316-322

Wen Z-Y, Chen F (2002) Perfusion culture of the diatom *Nitzschia laevis* for ultra-high yield of eicosapentaenoic acid. *Process Biochem* 38: 523-529

Wen Z-Y, Jiang Y, Chen F (2002) High cell density culture of the diatom *Nitzschia laevis* for eicosapentaenoic acid production: fed-batch development. *Process Biochem* 37: 1447-1453

Wheals AE, Basso LC, Alves DMG, Amorim HV (1999) Fuel ethanol after 25 years. *Trends Biotechnol* 17: 482-487

Wu Z-Y, Shi C-L, Shi X-M (2007) Modelling of lutein production by heterotrophic *Chlorella* in batch and fed-batch cultures. *World J Microbiol Biotechnol* 23: 1233-1238

Wu Z, Shi X (2006) Optimization for high-density cultivation of heterotrophic *Chlorella* based on a hybrid neural network model. *Lett Appl Microbiol* 44: 13-18

Xiong W, Li X, Xiang J, Wu Q (2008) High-density fermentation of microalga *Chlorella protothecoides* in bioreactor for microbio-diesel production. *Appl Microbiol Biotechnol* 78: 29-36

Xu H, Miao X, Wu Q (2006) High quality biodiesel production from a microalga *Chlorella protothecoides* by heterotrophic growth in fermenters. *J Biotechnol* 126: 499-507

Yamasaki T, Aki T, Mori Y, Yamamoto T, Shinozaki M, Kawamoto S, Ono K (2007) Nutritional enrichment of larval fish feed with thraustochytrid producing polyunsaturated fatty acids and xanthophylls. *J Biosci Bioeng* 104: 200–206

Zaslavskaia LA, Lippmeier JC, Shih C, Erhardt D, Grosman AR, Apt KE (2001) Trophic conversion of an obligate photoautotrophic organism through metabolic engineering. *Science* 292: 2073-2075

Zhang C-C, Jeanjean R, Joset F (1998) Obligate phototrophy in cyanobacteria: more than a lack of sugar transport. *FEMS Microbiol Lett* 161: 285-292

Zou N, Zhang C, Cohen Z, Richmond A (2000) Production of cell mass and eicosapentaenoic acid (EPA) in ultrahigh cell density cultures of *Nannochloropsis* sp. (Eustigmatophyceae). *Eur J Phycol* 35: 127-133

Chapter 17

MICROALGAE GROWTH AND FATTY ACID COMPOSITION DEPENDING ON CARBON DIOXIDE CONCENTRATION

C. Griehl, H. Polhardt, D. Müller and S. Bieler
Anhalt University of Applied Sciences,
Department of Applied Biosciences and Process Technology,
Institute of Algae Biotechnology, Köthen, Germany

ABSTRACT

The increase of atmospheric carbon dioxide is considered to be one of the main causes of global warming. Between 1990 and 2008, atmospheric CO_2 rose from 280 ppm (20.541 mill. t) to 400 ppm (29.381 mill. t). At the same time, fossil oil resources are said to be depleted within a few decades if fuel consumption remains at current levels. It is, therefore, crucial to explore alternatives to oil producing sources and also to reduce atmospheric CO_2 concentration.

Microalgae have been suggested as excellent candidates meeting the requirements: they are able to fix large amounts of carbon dioxide and transform it into biomass with high content of lipids.

The lipid content of microalgae is characteristic of their genus and species and also depends on the different growth phases and culture conditions like nutrient supply (especially nitrogen and phosphate amounts in culture medium), light intensity, temperature, pH and carbon dioxide concentration.

Different microalgae species of the division of Chlorophyta (*Scenedesmus* sp. and *Chlorella* sp.) were investigated according to their biomass productivity, lipid content, fatty acid profile and tolerance of high levels of carbon dioxide. The study showed that a higher CO_2 level leads to a decrease in biomass concentration and an increase in lipid content of the analysed species. For the most part, lipids contain saturated and unsaturated fatty acids with a chain length between C14 and C22. With increasing CO_2 concentration, the content of unsaturated fatty acids with 1 and 2 double bonds increases, whereas the content of linolenic acid, an acid with 3 double bonds, decreases.

INTRODUCTION

The greenhouse gas carbon dioxide (CO_2) is considered to be one of the main causes of global warming. It is mainly generated through the combustion of the fossil fuels coal, crude oil and natural gas. Simultaneously fossil oil resources will be depleted in few decades, if the consumption of fuels remains constant. Therefore it is important to investigate new possibilities for oil producing sources and secondly to reduce the CO_2 concentration in atmosphere.

To achieve this, considerable efforts and new approaches are needed. One of these new approaches deals with the introduction of flue-gas CO_2 generated in power plants into large photobioreactors so as to serve as a nutrient for fast-growing microalgae. This also leads to a multiplication of the algae (biomass formation) and, depending on the algae strain used, to a formation of valuable products, such as lipids (for energy production), carotenoids, pharmaceutical drugs, etc. After the valuable products have been extracted, the residual algae biomass is converted into biogas and thus used for energy production.

Compared to terrestrial plants, microalgae can store considerably more CO_2 because of their rapid growth. They can also deliver significantly more biomass, which can then be used for energy or material. Algae can, furthermore, be grown in a variety of climates and on non-arable land, including marginal areas unsuitable for agricultural purposes and do not take up valuable space for food crop cultures. In addition, their production is not seasonal. In summary, microalgae are veritable miniature biochemical factories because of the photosynthetic production of biofuels as well as being providers of biomass feedstock for the production of high value products and electricity.

CO2-Concentrating Mechanism (CCM) and Biosynthesis of Lipids and Fatty Acids in Algae

The efficiency of photosynthesis (PCE - Photo-conversion efficiency) is limited as only about 43% of the entire solar spectrum is used for photosynthesis (PAR = photosynthetic active range, λ = 360-720 nm). Each photon is used with the same stoichiometry, i.e. photons of shorter wavelengths are used with a lower efficiency than those of higher wavelengths (red region of the spectrum) (Rosello Sastre & Posten, 2010). An additional part of the light energy is lost through reflection at the cell. Further energy losses are caused by the relaxation of excited states in the photosynthetic system and energy consumption in the enzymatic steps responsible for the synthesis of organic molecules. Energy is also lost through photorespiration (cellular respiration), an energy-destruction cycle. This process takes place at high light intensities, as it is a means with which the algal cell protects itself against excessive light (Wagner et al., 2005). A PCE value of about 9% is theoretically possible, but in practice not reached by existing technologies.

Photosynthesis can be divided into two partial reactions: the light-dependent and the light-independent reaction.

In the first reaction, the light-dependent reaction, light is absorbed through photosynthetic pigments. These pigments are completely bound to proteins which are either associated to the thylakoid membrane or integrated into it. The light absorption and conversion of light to chemical energy in form of ATP and NADPH take place in the thylakoid membrane (photosynthetic membrane). The light is absorbed by the so-called chlorophyll-protein complexes that include proteins with molecular weights between 20 and 200kDa to which chlorophylls and carotinoids are associated (Wilhelm et al., 1987).

The second reaction of photosynthesis is light-independent (dark reaction) and is also known as Calvin cycle. During the dark reaction, the fixation of CO_2 by the enzyme ribulose bisphosphate carboxylase / oxygenase (Rubisco) is performed. RuBisCo is a very slow enzyme with a low affinity for CO_2. At low CO_2 concentrations (0.04% - atmospheric levels), RuBisCo works at only 25% of its catalytic capacity due to the low concentration of dissolved carbon dioxide and a relatively high concentration of oxygen (~20% - atmospheric level), which compete with CO_2 (Moroney & Ynalvez, 2007). To provide a higher level of carbon dioxide at the active site of RuBisCo, microalgae developed a process for concentrating CO_2, the carbon dioxide concentrating mechanism (CCM). It has by now been established that the CO_2 concentration mechanism can vary among different algae groups (Giordano et al., 2005). Carbonic anhydrase is the key enzyme of this process. It is a zinc-containing metalloenzym, which catalyses the hydration of carbon dioxide and water to bicarbonate and protons [CO_2 + $H_2O \leftrightarrow HCO_3^- + H^+$] (Badger & Price, 1994). Giordano et al. (2005) described that many chlorophyta can use CO_2, HCO_3^- and CO_3^{2-} (DIC-dissolved inorganic carbon) as substrate for RuBisCo. For example the freshwater green algae *Chlamydomonas reinhardtii*, *Scenedesmus obliquus*, different species of *Chlorella* and the diatom *Navicula pelliculosa* can use CO_2 and HCO_3^- simultaneously (Sültemeyer et al., 1991; Rotatore & Colman, 1991 a, b; 1992). Contrary to that the chlorophyte *Eremosphaera viridis* and the eustigmatophyte *Monodus subterraneus* can only use CO_2 as carbon source, because they do not exhibit the specific system for HCO_3^- transport (Huertas et al., 2005).

However, it is assumed that the transport of CO_2 through membranes is increased by carbonic anhydrase. In solution (culture medium of algae) and at a pH of ca. 7, carbon dioxide is principally present as a bicarbonate, but CO_2 gas diffuses about 100 times faster than ions (Bowes, 1968). In the presence of the enzyme carbonic anhydrase, HCO_3^- is converted into CO_2 gas and hence facilitates the passage of CO_2 through the membrane. Once the membrane has been passed, the enzyme converts CO_2 back into bicarbonate ions, thus increasing the diffusion rate in the solution. The CO_2 concentration increases in proximity of RuBisCo. The photosynthesis-affinity of CO_2 is much higher in microalgae cells grown under atmospheric CO_2 levels (0.04% CO_2) than in those grown under higher carbon dioxide concentrations (Tsuzuki & Miyachi, 1989). This might be explained with the fact that low CO_2 algae cells (0.04% CO_2) show a higher activity of carbonic anhydrase than cells grown under high CO_2 levels, because the synthesis of carbonic anhydrase is induced in low CO_2 conditions and stops when the carbon dioxide concentration is increased during algae growth (Yang et al., 1985).

However, carbon dioxide is captured by D-ribulose-1,5 bisphosphate. 2-carboxy-keto-D-arabinit-1,5-bisphosphate is formed as an intermediate, which breaks down into two molecules 3-phosphoglycerate. This reaction is catalysed by the enzyme RuBisCo.

The intermediate 3-phosphoglycerate is a necessary substance for the fatty acid biosynthesis in plants and algae. The biosynthesis of fatty acids takes place in the stroma of the chloroplast.

The new synthesis (de novo biosynthesis) of fatty acids starts with the conversion of 3-phosphoglycerate to acetyl-coenzyme A (acetyl-CoA), onto which more acyl moieties are attached. The acyl groups can only be added in the form of malonyl-CoA, as this is the only material accepted for chain extension. Malonyl-CoA is formed from acetyl-CoA; here, carbonic acid (HCO_3^-) is covalently bound under ATP consumption. This reaction is catalyzed by the enzyme Acetyl-CoA carboxylase (ACCase) (Harwood, 1988). Seven different enzymes, known as fatty acid synthase complex, are involved in the cyclical construction of fatty acid chains. This enzyme complex also includes a soluble polypeptide, the acyl carrier protein (ACP), which binds the substrates acetyl-CoA and malonyl-CoA and the resulting acyl intermediates during the chain extension. At the start of the de novo biosynthesis, acetyl-CoA and malonyl-CoA are separated from the carrier acetyl-CoA and transferred to ACP by specific acyltransferases. This leads to the formation of acetyl-ACP and malonyl-ACP. Together with the enzyme 3-Oxoacyl-ACP synthase, these two molecules react to form acetoacetyl-ACP. Via several intermediate steps, acetoacyl-ACP reacts further to form butyryl-ACP. Butyryl-ACP reacts again with malonyl-ACP, in the same way as during the previous reaction with acetyl-ACP, meaning that the chain extension runs as a cycle reaction. After further synthesis rounds - each extends the chain by another C2-unit - acyl moieties with C16 (palmitoyl-ACP) or C18 (stearoyl-ACP) are synthesized. These two molecules are no longer accepted as the substrate by 3-Oxoacyl-ACP synthase and are consequently no longer included in the chain extension cycle. Further research is needed to establish the exact nature of the subsequent reaction step. This could either be a general reaction step separating ACP from palmitoyl-ACP and stearoyl-ACP through thioesterases, or one that is restricted to the acyl moities released by the chloroplast to the cytoplasm. These contribute to the syntheses of triacylglyceride and membran-lipids. After passage through the chloroplast envelope, the free fatty acids are probably transferred to CoA and passed on to cytosol without modification. This binding reaction is catalyzed by acyl-CoA synthases.

The single steps of the de novo biosynthesis of fatty acids are illustrated in figure 1.

The biosynthesis of triacylglycerols (TAG) and their regulation still offers considerable research potential. TAGs are accumulated in algae cells in larger proportions during stress conditions (for example nutrient limitation). The TAG fraction of microalgae acts as a reserve of fatty acids that either provide energy by oxidation (ATP) or are intended for incorporation into phospholipids. TAGs are synthesized in the cytoplasm or in membranes of the rough endoplasmatic reticulum (rER) through incorporation of the fatty acids into a glycerol backbone. In a first step the CoA-bound fatty acids, which were transferred to the rER, react with sn-glycerol-3-phosphate. The latter is obtained by the hydrogenation of 1,3-dihydroxyacetone and under the influence of the enyzme dihydroxyacetone-reductase. The CoA-linked palmitic acid is then specifically transferred to position 1 of sn-glycerol-3-phosphate by 3-Phosphoglycerol-acyltransferase; lysophosphatidic acid is formed. In the following, another CoA-bounded fatty acid is attached to monoacylglycerol-3-phosphate and diacylglycerol-3-phosphat (phosphatidic acid) is formed. This step is catalyzed by monoacylglycerol-3-phosphate acyltransferase, which prefers unsaturated acyl moieties as substrate. A subsequent dephosphorylation generates diacylglycerol, which crosses over into the rER membrane. Another fatty acid moiety (acyl-CoA) is attached to the resulting hydroxyl group by esterfication and triacylglycerol (TAG) is synthesized. This last implementation of TAG biosynthesis is catalyzed by the enzyme diacylglycerol acyltransferase.

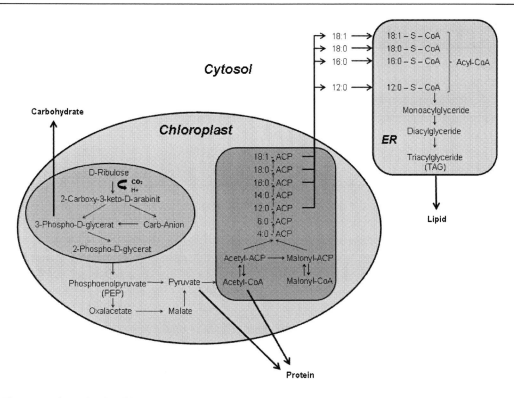

Figure 1. Biosynthesis of fatty acid and TAG-synthesis.

LIPID CLASSES OF ALGAE

Algae are able to survive and to proliferate in a wide range of environmental conditions. This is reflected in their enormous diversity, the unusual structure of their cellular lipids and the ability to modify the lipid metabolism as consequences of environmental changes. The lipids of algae include neutral lipids, polar lipids, wax esters, sterols, hydrocarbons and prenylated derivates such as tocopherol, carotenoids, terpenes, chinones and chlorophylls.

Depending on their chemical structure, algal lipids have various functions (Thompson Jr., 1996). Neutral lipids, especially Triacylglycerols (TAG) are storage substances and energy sources, while polar lipids are structural components of cell membranes, modulators of photosystem efficiency and regulate the energy flow (Harwood & Russell, 1984, Mock & Kron, 2002).

Lipid content and composition of specific lipid classes vary, depending on the specific growth conditions. The inorganic carbon is usually stored as glycolipids during the exponential growth phase, thus resulting in growth and cell division. Under stress conditions, for example nitrogen limitation, carbon is stored by algal cells mainly as TAG and in cytosolic lipid bodies. This process is also accompanied by a break-up of the cell division.

The lipid composition of green algae is dominated by polar glycerolipids (membrane lipid) during growth phase. Glycerolipids of Chlorophyta can be divided into phospholipids, galactolipids and sulfolipids (Goss & Wilhelm, 2009). Phosphatidylethanolamine (PE), phosphatidylcholine (PC), phosphatidylserine and phosphatidylglycerol are the typical phospholipids of green algae. Galactolipids can be classified into monogalactosyl-diacylglycerol (MGDG) and digalactosyldiacylglycerol (DGDG). Sulfoquinovo-syldiacylglycerol (SQDG) is the most common sulfolipid (Goss & Wilhelm, 2009). Phospholipids are constituents of various functional membranes in the cell, whereas MGDG and SQDG can only be found in the thylakoid membrane of chloroplasts. DGDG is also a component of extraplastidial membranes, where it can replace bilayer-forming phospholipids (PC, PE and PG) (Jouhet et al., 2004). The exchange of phospholipids through DGDG can mostly be observed in a phosphate-limitation (Härtel & Benning, 2000).

Neutral lipids (NL) can be classed in monoacylglycerol (MAG), diacylglycerol (DAG) and triacylglycerol (TAG), while MAG and DAG can be regarded as precursors for the TAG biosynthesis. Green algae do not accumulate significant amounts of neutral lipids (storage lipids), unless under certain stress conditions, such as nutrient-limitations (Thompson Jr., 1996). Neutral lipids are formed in stationary phase of algal growth. This is due to the fact that algal cells predominantly synthesize polar lipids during the exponential growth phase. These are required to facilitate growth and incorporated into the algal cell membranes. In the stationary growth phase, algae growth is virtually non-existent; the formation of neutral lipids is the main activity. In addition, hardly any nutrients are left in the stationary phase (especially nitrogen); the accumulation of neutral lipids is thus encouraged. It is to be noted that the content of polar lipids decreases with the age of algae cultures, while the neutral lipid content, especially that of TAG, increases (Brown et al., 1996; Hodgens, 1991), as described above.

Figure 2 shows the distribution of Glycolipids, Phospholipids and neutral lipids depending of different algae classes (Chlorophyceae, Bacillariophyceae, Xanthophyceae, Rhodophyceae and Cyanophyceae).

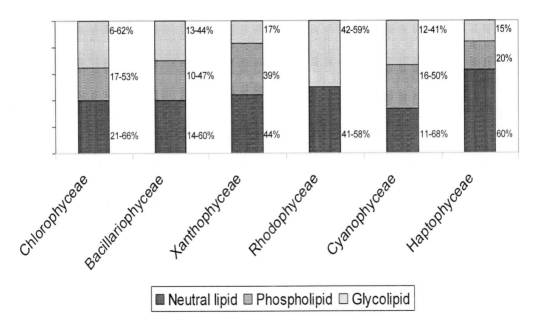

Figure 2. Averages of lipid classes present in different algae class as % of total lipid content (Borowitzka, 1988; Zhu et al., 1997).

The lipid classes of microalgae vary according to genus and species and also depend on the different growth phases and culture conditions like nutrient supply (especially nitrogen and phosphate amounts in the culture medium), light intensity, temperature, pH and carbon dioxide concentration.

INFLUENCE OF CULTURE PARAMETERS ON LIPID CONTENT AND FATTY ACID PROFILE

Algae are able to modify lipid metabolism in response to changes in environmental conditions (Guschina & Harwood, 2006). Under optimal growth conditions algae synthesise fatty acids primarily for integration into glycerol-based membrane lipids (5-20% of their dry cell weight). However, under stress conditions for growth like nitrogen limitation many algae vary their lipid biosynthetic pathway and synthesize mainly neutral lipids (20-50% of their dry cell weight) (Hu et al., 2008).

Table 1. Summary of the influence of different cultivation conditions on lipid classes

Cultivation condition	Influence on lipid class	Statement	Reference
Light intensitiy (high)	triacylglycerids	TAG synthesis requires large amounts of photosynthe-tically produced ATP and NAD(P)H and therefore helps the drainage of excess light energy and prevent photo-chemical damage to the algal cell growth → increasing TAG	Roessler (1990); Khotimchenko & Yakovleva (2005)
Light intensitiy (low)	polar lipids	Structural lipids are synthesized in larger amounts (DGDG, SQDG, PG, PC)	Khotimchenko & Yakovleva (2005); Hu et al. (2008)
Temperature (low)	polar lipids	Increase the levels of PC (Phosphatidylcholin) → regulation of membrane fluidity	Murphy (2004)
	neutral lipids	Content of TAG (triacylglycerol) is lower	Chen et al. (2008)
Temperature (high)	polar lipids	Increase the levels of GL (glycolipids) and MGDG (Monogalactosylglycerol) / decrease in the content of DGDG (Digalactosylglycerol)	Liao et al. (2004); Lynch & Thompson (1982)
Nitrogen-limitation	neutral lipids	TAG rise→ to nitrogen deficiency carbon skeletons not incorporated into proteins, but recovered over the lipid biosynthesis	Gordillo et al. (1998); Rodolfi et al. (2009)
		Degradation of chlorophyll (contain 4 molecules of nitrogen) and the whole chloroplast → chloroplast lipids and fatty acids are suggested to be also catabolised	Piorreck et al. (1983)
		Increasing of the intracellular content of fatty acid acyl-CoA and activation of diacylglycerol acyltransferase (converts fatty acid acyl-CoA to TAG)	Sukenik & Livne (1991); Takagi et al. (2000); Hsieh & Wu (2009)
Salinity (high)	neutral and polar lipids	Lipid content and thus the fatty acid content (especially Polyunsaturated fatty acids) are rising → formation of a fluid cell membrane that prevents the loss of water from the cell	Lee et al. (1989)
Phosphate-limitation	polar lipids	Increase the content of DGDG, which can replace phospholipids, when phosphate is limited	Khozin-Goldberg & Cohen (2006)
Iron (high)	neutral and polar lipids	Peroxidation (oxidative degradation of lipids) is increased → iron complexes elicit different reactions, which represent the beginning of a lipid oxidation	Estevez et al. (2001)
CO_2 (high)	neutral and polar lipids	Peroxidation is reduced by increasing the CO_2 concentration → enhancing lipid content	Yu et al. (2004)

Table. 2. biomass productivity, lipid content and lipid productivity of different microalgae

Microalgae species	Cultivation conditions volume [L]	CO_2 [%]	medium / pH	light [$\mu mol*m^{-2}*s^{-1}$]	T [°C]	lipid extraction-method	biomass-productivity [g/l/d]	Lipid content [% of dry weight]	Lipid-productivity [mg/l/d]	Reference
Amphiprora hyalina	-	-	-	-	-	-	-	22-37	-	Griffiths & Harrison (2008)
Amphora	-	-	-	-	-	-	-	51	160	Griffiths & Harrison (2008)
Amphora coffeaformis	1.5 (Conical flask)	-	f/2	80	25±1	Bligh&Dyer	0.094 (g/l)	19.7	-	Renaud et al. (1999)
Anabaena cylindrica	-	-	-	-	-	-	-	5	-	Griffiths & Harrison (2008)
Anacystis ridulans	8	-	-	800 lux	22	Chloroform / Methanol (2:1)	0.022 (g/l) 13th day	9.33 – 16.8	-	Piorreck & Pohl (1984)
Ankistrodesmus falcatus	-	-	-	-	-	-	0.46	24 – 34	109	Griffiths & Harrison (2008)
Ankistrodesmus sp.	-	10	Chu 13	150	25±1	Bligh&Dyer	0.026	25.79	5.51	Mata et al. (2010)
	-	-	-	-	-	-	0.02	25-75	-	Yoo et al. (2010) Mata et al. (2010)
Botryococcus braunii	3 (tubular reactor)	2	BG 11	150	25	Methanol / Chloroform / Water (2:2:1)	0.084	10.41	-	Ge et al. (2011)
	3 (tubular reactor)	5	BG 11	150	25	Methanol / Chloroform / Water (2:2:1)	0.07	11.21	-	Ge et al. (2011)
	3 (tubular reactor)	10	BG 11	150	25	Methanol / Chloroform / Water (2:2:1)	0.064	12.44	-	Ge et al. (2011)
	3 (tubular reactor)	20	BG 11	150	25	Methanol / Chloroform / Water (2:2:1)	0.096	12.71	-	Ge et al. (2011)
Chaetoceros muelleri	0.1 (Erlenmeyer flask)	5	f (Guillard& Ryther)	100	25	Methanol / Chloroform (2:1)	0.07	33.6	21.8	Rodolfi et al. (2009)
Chaetoceros calcitrans	0.1 (Erlenmeyer flask)	5	f (Guillard& Ryther)	100	25	Methanol / Chloroform (2:1)	0.04	39.8	17.6	Rodolfi et al. (2009)
Chaetoceros sp.	1.5 (conical flask)	-	f/2	80	25±1	Bligh&Dyer	0.125 (g/l)	17	-	Renaud et al. (1999)
Chlamydomonas applanata	-	-	-	-	-	-	-	18 – 33	-	Griffiths & Harrison (2008)

Table 2. (continued)

Microalgae species	Cultivation conditions						lipid extraction-method	biomass-productivity [g/l/d]	Lipid content [% of dry weight]	Lipid-productivity [mg/l/d]	Reference
	volume [L]	CO_2 [%]	medium / pH	light [$\mu mol*m^{-2}*s^{-1}$]	T [°C]						
Chlamydomonas reinhardtii	-	-	-	-	-	-	-	-	21	-	Griffiths & Harrison (2008)
	-	-	-	-	-	-	-	0.03	29 – 63	8	Griffiths & Harrison (2008)
	-	-	-	-	-	-	-	0.036 – 0.04	35 – 63	10.3 – 50	Mata et al. (2010)
	2 (LSL Biolafette)	5	Watanbe	25	25	Bligh&Dyer	0.028	29	8.1	Illman et al. (2000)
	2 (LSL Biolafette)	5	203 mg/l $(NH_4)_2HPO_4$; 2.236 g/l KCl; 2.465 g/l $MgSO_4$; 1.361 g/l KH_2PO_4; 10 mg/l $FeSO_4$; pH=6	25	25	Bligh&Dyer	0.079	63	49.9	Illman et al. (2000)
Chlorella emersonii	230 (tubular PBR)	-	203 mg/l $(NH_4)_2HPO_4$; 2.236 g/l KCl; 2.465 g/l $MgSO_4$; 1.361 g/l KH_2PO_4; 10 mg/l $FeSO_4$; pH=6	130	25	Bligh&Dyer	0.041	25	12.2	Scragg et al. (2002)
	1.8 (conical flask)	0.038	Bristol	3200 lux	30		0.090	-	-	De Morais & Costa (2007)
	1.8 (conical flask)	6	Bristol	3200 lux	30		0.087	-	-	De Morais & Costa (2007)
	1.8 (conical flask)	12	Bristol	3200 lux	30		0.086	-	-	De Morais & Costa (2007)
	1.8 (conical flask)	18	Bristol	3200 lux	30		0.061	-	-	De Morais & Costa (2007)
	-	-	-	-	-	-	0.03	31 – 57	10	Griffiths & Harrison (2008)
Chlorella minutissima	2 (LSL Biolafette)	5	Guillard Medium	25	25	Bligh&Dyer	0.032	31	9.0	Illman et al. (2000)
	2 (LSL Biolafette)	5	Guillard Medium, reduced N (37.5 mg/l $NaNO_3$ instead of 75 mg/l)	25	25	Bligh&Dyer	0.016	57	10.2	Illman et al. (2000)

Table 2. (continued)

Microalgae species	Cultivation conditions						lipid extraction-method	biomass-productivity [g/l/d]	Lipid content [% of dry weight]	Lipid-productivity [mg/l/d]	Reference
	volume [L]	CO_2 [%]	medium / pH	light [$\mu mol*m^{-2}*s^{-1}$]	T [°C]						
Chlorella protothecoides	2 (LSL Biolafette)	5	Watanbe	25	25		Bligh&Dyer	0.003	11	2.5	Illman et al. (2000)
	2 (LSL Biolafette)	5	203 mg/l $(NH_4)_2HPO_4$; 2.236 g/l KCl; 2.465 g/l $MgSO_4$; 1.361 g/l KH_2PO_4; 10 mg/l $FeSO_4$; pH=6	25	25		Bligh&Dyer	0.023	23	5.4	Illman et al. (2000)
	-	-	-	-	-			-	16 – 64	-	Griffiths & Harrison (2008)
	0.8 (Erlenmeyer flask)	0.03	BG 11	180	25±1		Chloroform / Methanol (2:1)	0.065	20.90	-	Tang et al. (2011)
	0.8 (Erlenmeyer flask)	5	BG 11	180	25±1		Chloroform / Methanol (2:1)	0.133	20.65	-	Tang et al. (2011)
	0.8 (Erlenmeyer flask)	10	BG 11	180	25±1		Chloroform / Methanol (2:1)	0.144	24.25	-	Tang et al. (2011)
Chlorella pyrenoidosa	0.8 (Erlenmeyer flask)	20	BG 11	180	25±1		Chloroform / Methanol (2:1)	0.121	25.48	-	Tang et al. (2011)
	0.8 (Erlenmeyer flask)	30	BG 11	180	25±1		Chloroform / Methanol (2:1)	0.075	26.02	-	Tang et al. (2011)
	0.8 (Erlenmeyer flask)	50	BG 11	180	25±1		Chloroform / Methanol (2:1)	0.054	26.75	-	Tang et al. (2011)
	-	-	-	-	-			0.55	18	45 – 97	Griffiths & Harrison (2008)
	-	-	-	-	-			0.23 – 1.47	19 – 22	44.7	Mata et al. (2011)
Chlorella sorokiniana	0.1 (Erlenmeyer flask)	5	BG 11	100	25		Methanol / Chloroform (2:1)	0.23	19.3	44.7	Rodolfi et al. (2009)
	2 (LSL Biolafette)	5	Watanbe	25	25		Bligh&Dyer	0.003	20±1.6	2.7	Illman et al. (2000)

Table 2. (continued)

Microalgae species	Cultivation conditions						lipid extraction-method	biomass-productivity [g/l/d]	Lipid content [% of dry weight]	Lipid-productivity [mg/l/d]	Reference
	volume [L]	CO$_2$ [%]	medium / pH	light [µmol*m^{-2}*s^{-1}]	T [°C]						
	2 (LSL Biolafette)	5	203 mg/l (NH$_4$)$_2$HPO$_4$; 2.236 g/l KCl; 2.465 g/l MgSO$_4$; 1.361 g/l KH$_2$PO$_4$; 10 mg/l FeSO$_4$; pH=6	25	25	Bligh&Dyer	0.005	22	4.8	Illman et al. (2000)
	-	-	-	-	-	-	0.02 – 2.5	10 – 48	42.1	Mata et al. (2010)
	3 (air-lift PBR)	0.3	Bold´s Basal Medium	350	28±2	Bligh&Dyer	0.695 (g/l)	-	0.145 (g/l)	Fulke et al. (2010)
	3 (air-lift PBR)	3	Bold´s Basal Medium	350	28±2	Bligh&Dyer	1.484 (g/l)	-	0.161 (g/l)	Fulke et al. (2010)
Chlorella sp.	3 (air-lift PBR)	10	Bold´s Basal Medium	350	28±2	Bligh&Dyer	0.315 (g/l)	-	0.121 (g/l)	Fulke et al. (2010)
	3 (air-lift PBR)	15	Bold´s Basal Medium	350	28±2	Bligh&Dyer	0.212 (g/l)	-	0.089 (g/l)	Fulke et al. (2010)
	0.8 (cylindrical PBR)	2	f/2	300	26±1	Methanol / Chloroform (2:1)	0.422		143	Chiu et al. (2008)
	0.8 (cylindrical PBR)	5	f/2	300	26±1	Methanol / Chloroform (2:1)	0.393		130	Chiu et al. (2008)
	0.8 (cylindrical PBR)	10	f/2	300	26±1	Methanol / Chloroform (2:1)	0.366		124	Chiu et al. (2008)
	0.8 (cylindrical PBR)	15	f/2	300	26±1	Methanol / Chloroform (2:1)	0.295		97	Chiu et al. (2008)
	1 (rectangular PBR)	2	Artificial seawater + 0.1% (v/v) Walne´s medium+0,1% TMS Urea: 0.200 g/l	~600	30	-	0.464 g/l (after 6 d)	66.1	51	Hsieh&Wu (2009)
Chlorella sp.	1 (rectangular PBR)	2	Artificial seawater + 0.1% (v/v) Walne´s medium+0,1% TMS Urea: 0.200 g/l	~600	30	-	0.849 (g/l) (after 6 d)	60.2	85	Hsieh&Wu (2009)
	1 (rectangular PBR)	2	Artificial seawater + 0.1% (v/v) Walne´s medium+0,1% TMS Urea: 0.200 g/l	~600	30	-	1.422 (g/l) (after 6 d)	52.2	124	Hsieh&Wu (2009)

Table 2. (continued)

Microalgae species	Cultivation conditions						lipid extraction-method	biomass-productivity [g/l/d]	Lipid content [% of dry weight]	Lipid-productivity [mg/l/d]	Reference
	volume [L]	CO$_2$ [%]	medium / pH	light [µmol*m^{-2}*s^{-1}]	T [°C]						
	1 (rectan-gular PBR)	2	Artificial seawater + 0,1% (v/v) Walne's medium+0,1% TMS Urea: 0.200 g/l	~600	30			1.785 (g/l) (after 6 d)	36.5	109	Hsieh&Wu (2009)
	1 (rectan-gular PBR)	2	Artificial seawater + 0,1% (v/v) Walne's medium+0,1% TMS Urea: 0.200 g/l	~600	30			2.027 (g/l) (after 6 d)	32.6	110	Hsieh&Wu (2009)
	-	-			-			0.11	25 – 42	26 – 28	Griffiths & Harrison (2008)
	-	-			-			0.02 – 0.20	5 – 58	11.2 – 40	Mata et al. (2010)
	0.1 (Erlen-meyer flask)	5	BG 11	100	25	Methanol / Chloroform (2:1)		0.20	18.4	36.9	Rodolfi et al. (2009)
	-	10	BG11	150	25±1	Bligh&Dyer		0.105	6.6	6.91	Yoo et al. (2010)
	0.1	0.03 – 15	BG11	80	25±1	Bligh & Dyer		0.339	21-40	825 mg/d	Bhola et al. (2011)
Chlorella vulgaris	1 (polycarbonate flask)	-	0.25 g NaNO3, 0.025 g CaCl2_2H2O, 0.075 g MgSO4_7H2O, 0.075 g K2HPO4, 0.175 g KH2PO4, 0.025 g NaCl, 1 g proteose peptone	-	-	Bligh&Dyer		0.013	4	33	Liang et al. (2009)
	1L (polycarbo-nate flask)	-	0.025 g CaCl2_2H2O, 0.075 g MgSO4_7H2O, 0.075 g K2HPO4, 0.175 g KH2PO4, 0.025 g NaCl, 1 g proteose peptone	-	-	Bligh&Dyer		0.010	4	38	Liang et al. (2009)
	2 (LSL Biolafette)	5	203 mg/l (NH$_4$)$_2$HPO$_4$; 2.236 g/l KCl; 2.465 g/l MgSO$_4$; 1.361 g/l KH$_2$PO$_4$; 10 mg/l FeSO$_4$; pH=6	25	25	Bligh&Dyer		0.037	40	14.9	Illman et al. (2000)

Table 2. (continued)

Microalgae species	Cultivation conditions					lipid extraction-method	biomass-productivity [g/l/d]	Lipid content [% of dry weight]	Lipid-productivity [mg/l/d]	Reference
	volume [L]	CO_2 [%]	medium / pH	light [$\mu mol*m^{-2}*s^{-1}$]	T [°C]					
Chlorella vulgaris	230 (tubular PBR)	-	203 mg/l $(NH_4)_2HPO_4$; 2.236 g/l KCl; 2.465 g/l $MgSO_4$; 1.361 g/l KH_2PO_4; 10 mg/l $FeSO_4$; pH=6	130	25	Bligh&Dyer	0.024	58	13.9	Scragg et al. (2002)
Chlorococcum sp.	0.1 (Erlen-meyer flask)	5	BG 11	100	25	Methanol / Chloroform (2:1)	0.28	19.3	53.7	Rodolfi et al. (2009)
Cryptheco-dinium cohnii	-	-	-	-	-	-	10	20-51.1	-	Mata et al. (2010)
	-	-	-	-	-	-	-	25	-	Griffiths & Harrison (2008)
Cryptomonas sp.	1.5 (Conical flask)	-	f/2	80	25±1	Bligh&Dyer	0.066	22	-	Reanud et al. (1999)
Cyclotella cryptica	-	-	-	-	-	-	-	18 – 38	-	Griffiths & Harrison (2008)
Cylindrotheca	-	-	-	-	-	-	-	27	-	Griffiths & Harrison (2008)
Dunaliella primolecta	-	-	-	-	-	-	-	14 – 23	-	Griffiths & Harrison (2008)
	-	-	-	-	-	-	0.09	23.1	-	Mata et al. (2010)
Dunaliella salina	-	-	-	-	-	-	-	10 – 19	-	Griffiths & Harrison (2008)
	-	-	-	-	-	-	0.22-0.34	6.0-25.0	116.0	Mata et al. (2010)
Dunaliella tertiolecta	-	-	-	-	-	-	-	15 – 18	-	Griffiths & Harrison (2008)
	-	-	-	-	-	-	0.12	16.7 – 71	-	Mata et al. (2010)
	0.5 (Roux-bottle)	3	NORO	150	30	Methanol / Chloroform (2:1)	0.1	63.5	-	Takagi et al. (2006)
Dunaliella sp.	-	-	-	-	-	-	0.17	17.5 – 67	33.5	Mata et al. (2010)
Ellipsoidion sp.	-	-	-	-	-	-	0.17	27.4	47.3	Mata et al. (2010)
Etlia oleoabundans	-	-	-	-	-	-	0.46	36 – 42	136 – 164	Griffiths & Harrison (2008)
Euglena gracilis	-	-	-	-	-	-	-	20 – 35	-	Griffiths & Harrison (2008)
	-	-	-	-	-	-	7.70	14.0-20.2	-	Mata et al. (2010)

Table 2. (continued)

Microalgae species	Cultivation conditions						lipid extraction-method	biomass-productivity [g/l/d]	Lipid content [% of dry weight]	Lipid-productivity [mg/l/d]	Reference
	volume [L]	CO$_2$ [%]	medium / pH	light [μmol*m^{-2}*s^{-1}]	T [°C]						
Fragilaris pinnata	1.5 (Conical flask)	-	f/2	80	25±1	Bligh&Dyer	0.097 (g/l)	14.9	-	Renaud et al. (1999)
	-	-	-	-	-	-	0.05 – 0.06	25	-	Mata et al. (2010)
Haematococcus pluvialis	1	0.06	Bold's Basal	90	24	by Zhekiskeva et al. (2002)		9.2	-	Damiani et al. (2010)
	1	0.06	Bold's Basal	300	24	by Zhekiskeva et al. (2002)		19.80	-	Damiani et al. (2010)
	1	0.06	Bold's Basal N-frei	300	24	by Zhekiskeva et al. (2002)		16.60	-	Damiani et al. (2010)
Hymenomonas carterae	-	-	-	-	-			14 – 20	-	Griffiths & Harrison (2008)
							0.16	25 – 29	38	Griffiths & Harrison (2008)
Isochrysis galbana							0.32 – 1.60	7 – 40	-	Mata et al. (2010)
Isochrysis sp.	1.5 (Conical flask)	-	f/2	80	25±1	Bligh&Dyer	0.054 (g/l)	23.4	-	Renaud et al. (1999)
	-	-	-	-	-		0.08 – 0.17	7.1 – 33	37.8	Mata et al. (2010)
	0.1 (Erlenmeyer flask)	5	f (Guillard & Ryther)	100	25	Methanol / Chloroform (2:1)	0.14–0.17	27.4-29.2	37.7-37.8	Rodolfi et al. (2009)
	1.5 (Conical flask)	-	f/2 salinity: 10 ppt	80	25±1	Bligh&Dyer	0.046 (g/l)	25.5	-	Renaud & Parry (1994)
	1.5 (Conical flask)	-	f/2 salinity: 15 ppt	80	25±1	Bligh&Dyer	0.078 (g/l)	25.8	-	Renaud & Parry (1994)
	1.5 (Conical flask)	-	f/2 salinity: 20 ppt	80	25±1	Bligh&Dyer	0.072 (g/l)	25.6	-	Renaud & Parry (1994)
Isochrysis sp.	1.5 (Conical flask)	-	f/2 salinity: 25 ppt	80	25±1	Bligh&Dyer	0.056 (g/l)	26.5	-	Renaud & Parry (1994)
	1.5 (Conical flask)	-	f/2 salinity: 30 ppt	80	25±1	Bligh&Dyer	0.075 (g/l)	29.30	-	Renaud & Parry (1994)

Table 2. (continued)

Microalgae species	Cultivation conditions volume [L]	CO_2 [%]	medium / pH	light [$\mu mol*m^{-2}*s^{-1}$]	T [°C]	lipid extraction-method	biomass-productivity [g/l/d]	Lipid content [% of dry weight]	Lipid-productivity [mg/l/d]	Reference
	1.5 (Conical flask)	-	salinity: 35 ppt	80	25±1	Bligh&Dyer	0.077 (g/l)	30	-	Renaud & Parry (1994)
Microcystis aeruginosa	8	-	-	800 lux	22	Chloroform / Methanol (2:1)	0.034 (g/l) 46th day	16.5-23.4	-	Piorreck & Pohl (1984)
Monallanthus salina	-	-	-	-	-	-	0.08	20.0-22.0	-	Mata et al. (2010)
Monodopsis subterranea	-	-	-	-	-	-	0.19	13 – 25	30 – 48	Griffiths & Harrison (2008)
Monodus subterraneus	0.1 (Erlenmeyer flask)	5	BG 11	100	25	Methanol / Chloroform (2:1)	0.19	16.1	30.4	Rodolfi et al. (2009)
Monoraphidumminutum	-	-	-	-	-	-	-	22 – 52	-	Griffiths & Harrison (2008)
Nannochloris sp.	-	-	-	-	-	-	0.23	28 – 30	63 – 77	Griffiths & Harrison (2008)
	-	-	-	-	-	-	0.17 – 0.51	20 – 56	60.9 – 76.5	Mata et al. (2010)
Nannochloropsis oculata	0.8 (cylindrical PBR)	2	f/2	300	26±1	Methanol / Chloroform (2:1)	0.480	29.7	142	Chiu et al. (2009)
	0.8 (cylindrical PBR)	5	f/2	300	26±1	Methanol / Chloroform (2:1)	0.441	26.2	113	Chiu et al. (2009)
	0.8 (cylindrical PBR)	10	f/2	300	26±1	Methanol / Chloroform (2:1)	0.398	24.6	97	Chiu et al. (2009)
	0.8 (cylindrical PBR)	15	f/2	300	26±1	Methanol / Chloroform (2:1)	0.372	22.7	84	Chiu et al. (2009)
	1.5 (Conical flask)	-	f/2 salinity: 10 ppt	80	25±1	Bligh&Dyer	0.062 (g/l)	28	-	Renaud & Parry (1994)
Nannochloropsis oculata	1.5 (Conical flask)	-	f/2 salinity: 15 ppt	80	25±1	Bligh&Dyer	0.115 (g/l)	29.5	-	Renaud & Parry (1994)
	1.5 (Conical flask)	-	f/2 salinity: 20 ppt	80	25±1	Bligh&Dyer	0.104 (g/l)	30.5	-	Renaud & Parry (1994)
	1.5 (Conical flask)	-	f/2 salinity: 25 ppt	80	25±1	Bligh&Dyer	0.132 (g/l)	31.5	-	Renaud & Parry (1994)

Table 2. (continued)

Microalgae species	Cultivation conditions volume [L]	CO$_2$ [%]	medium / pH	light [$\mu mol \cdot m^{-2} \cdot s^{-1}$]	T [°C]	lipid extraction-method	biomass-productivity [g/l/d]	Lipid content [% of dry weight]	Lipid-productivity [mg/l/d]	Reference
	1.5 (Conical flask)	-	f/2 salinity: 30 ppt	80	25±1	Bligh&Dyer	0.144 (g/l)	33.5	-	Renaud & Parry (1994)
	1.5 (Conical flask)	-	f/2 salinity: 35 ppt	80	25±1	Bligh&Dyer	0.157 (g/l)	33	-	Renaud & Parry (1994)
Nannochloropsis salina	-	-	-	-	-	-	-	27 – 46	-	Griffiths & Harrison (2008)
Nannochloropsis sp.	-	-	-	-	-	-	0.27	31 – 41	52 – 82	Griffiths & Harrison (2008)
	-	-	-	-	-	-	0.17 – 1.43	12 – 53	37.6 – 90	Mata et al. (2010)
	0.1 (Erlenmeyer flask)	5	f (Guillard& Ryther)	100	25	Methanol / Chloroform (2:1)	0.17-0.20	21.6-35.7	37.6-60.9	Rodolfi et al. (2009)
Navicula acceptata	-	-	-	-	-	-	-	33 – 46	-	Griffiths & Harrison (2008)
Navicula pelliculosa	-	-	-	-	-	-	-	27 – 45	-	Griffiths & Harrison (2008)
Navicula saprophila	-	-	-	-	-	-	-	24 – 51	-	Griffiths & Harrison (2008)
	-	-	-	-	-	-	-	29 – 65	90 – 134	Mata et al. (2010)
	0.8 (columnar glass device)	5	SE-Medium + 3mM NaNO$_3$	360	30±2	Soxhlet (Ethylether)	0.31	-	125	Li et al. (2008)
	0.8 (columnar glass device)	5	SE-Medium + 5mM NaNO$_3$	360	30±2	Soxhlet (Ethylether)	0.40	-	133	Li et al. (2008)
Neochloris oleoabundans	0.8 (columnar glass device)	5	SE-Medium + 10mM NaNO$_3$	360	30±2	Soxhlet (Ethylether)	0.63	-	98	Li et al. (2008)
	0.8 (columnar glass device)	5	SE-Medium + 15mM NaNO$_3$	360	30±2	Soxhlet (Ethylether)	0.58	-	44	Li et al. (2008)
	0.8 (columnar glass device)	5	SE-Medium + 20mM NaNO$_3$	360	30±2	Soxhlet (Ethylether)	0.54	-	38	Li et al. (2008)
Nephroselmis sp.	1.5 (Conical flask)	-	f/2	80	25±1	Bligh&Dyer	0.032 (g/l)	10.5	-	Renaud et al. (1999)
Nitzschia dissipata	-	-	-	-	-	-	-	28 – 47	-	Griffiths & Harrison (2008)

Table 2. (continued)

Microalgae species	Cultivation conditions					lipid extraction-method	biomass-productivity [g/l/d]	Lipid content [% of dry weight]	Lipid-productivity [mg/l/d]	Reference
	volume [L]	CO_2 [%]	medium / pH	light [$\mu mol*m^{-2}*s^{-1}$]	T [°C]					
	-	-	-	-	-	-	-	26	-	Griffiths & Harrison (2008)
Nitzschia frustulum	1.5 (Conical flask)	-	f/2	80	25±1	Bligh&Dyer	0.071 (g/l)	13.9	-	Renaud et al. (1999)
	1.5 (Conical flask)	-	f/2 salinity: 10 ppt	80	25±1	Bligh&Dyer	0.077 (g/l)	33.5	-	Renaud & Parry (1994)
	1.5 (Conical flask)	-	f/2 salinity: 15 ppt	80	25±1	Bligh&Dyer	0.064 (g/l)	32	-	Renaud & Parry (1994)
	1.5 (Conical flask)	-	f/2 salinity: 20 ppt	80	25±1	Bligh&Dyer	0.077 (g/l)	30.5	-	Renaud & Parry (1994)
	1.5 (Conical flask)	-	f/2 salinity: 25 ppt	80	25±1	Bligh&Dyer	0.072 (g/l)	27	-	Renaud & Parry (1994)
Nitzschia frustulum	1.5 (Conical flask)	-	f/2 salinity: 30 ppt	80	25±1	Bligh&Dyer	0.066 (g/l)	28.5	-	Renaud & Parry (1994)
	1.5 (Conical flask)	-	f/2 salinity: 35 ppt	80	25±1	Bligh&Dyer	0.071 (g/l)	26.5	-	Renaud & Parry (1994)
Nitzschia palea	-	-	-	-	-	-	-	40 – 47	-	Griffiths & Harrison (2008)
	-	-	-	-	-	-	-	16 – 47	-	Mata et al. (2010)
Nitzschia sp.	1.5 (Conical flask)	-	f/2	80	25±1	Bligh&Dyer	0.096 (g/l)	16	-	Renaud et al. (1999)
Oscillatoria	-	-	-	-	-	-	-	7 – 13	-	Griffiths & Harrison (2008)
Oocystis pusilla	-	-	-	-	-	-	-	10.5	-	Mata et al. (2010)
Ourococcus	-	-	-	-	-	-	-	27 – 50	-	Griffiths & Harrison (2008)
Pavlova lutheri	-	-	-	-	-	-	0.21	36	50 – 75	Griffiths & Harrison (2008)

Table 2. (continued)

Microalgae species	Cultivation conditions volume [L]	CO$_2$ [%]	medium / pH	light [μmol*m^{-2}*s^{-1}]	T [°C]	lipid extraction-method	biomass-productivity [g/l/d]	Lipid content [% of dry weight]	Lipid-productivity [mg/l/d]	Reference
Pavlova salina	0.1 (Erlenmeyer flask)	5	f (Guillard& Ryther)	100	25	Methanol / Chloroform (2:1)	0.14	35.5	40.2	Rodolfi et al. (2009)
	-	-	-	-	-	-	0.16	31	49	Griffiths & Harrison (2008)
Phaeodactylum tricornutum	0.1 (Erlenmeyer flask)	5	f (Guillard& Ryther)	100	25	Methanol / Chloroform (2:1)	0.16	30.9	49.4	Rodolfi et al. (2009)
	-	-	-	-	-	-	0.34	21 – 26	45 – 72	Griffiths & Harrison (2008)
	-	-	-	-	-	-	0.003 – 1.9	18 – 57	44.8	Mata et al. (2010)
Porphyridium cruentum	0.1 (Erlenmeyer flask)	5	f (Guillard& Ryther)	100	25	Methanol / Chloroform (2:1)	0.24	18.7	44.8	Rodolfi et al. (2009)
	-	-	-	-	-	-	0.36 – 1.50	9 – 18.8/60.7	34.8	Mata et al. (2010)
	0.1 (Erlenmeyer flask)	5	f (Guillard& Ryther)	100	25	Methanol / Chloroform (2:1)	0.37	9.5	34.8	Rodolfi et al. (2009)
Porphyridium purpureum	-	-	-	-	-	-	0.23	11	24 – 35	Griffiths & Harrison (2008)
Prymnesium parvum	-	-	-	-	-	-	-	30	-	Griffiths & Harrison (2008)
Rhodomonas sp.	1.5 (Conical flask)	-	f/2	80	25±1	Bligh&Dyer	0.102 (g/l)	20.4	-	Renaud et al. (1999)
Rhodosorus sp.	1.5 (Conical flask)	-	f/2	80	25±1	Bligh&Dyer	-	4.5	-	Renaud et al. (1999)
Scenedesmus dimorphus	-	-	-	-	-	-	-	26	-	Griffiths & Harrison (2008)
	-	-	-	-	-	-	0.12	21 – 42	25	Griffiths & Harrison (2008)
	-	-	-	-	-	-	0.004 – 0.74 (g/l)	11 – 55	-	Mata et al. (2010)
Scenedesmus obliquus	0.8 (Erlenmeyer flask)	0.03	BG 11	180	25±1	Chloroform / Methanol (2:1)	0.083	15.15	-	Tang et al. (2011)
	0.8 (Erlenmeyer flask)	5	BG 11	180	25±1	Chloroform / Methanol (2:1)	0.158	16.45	-	Tang et al. (2011)

Table 2. (continued)

Microalgae species	Cultivation conditions						lipid extraction-method	biomass-productivity [g/l/d]	Lipid content [% of dry weight]	Lipid-productivity [mg/l/d]	Reference
	volume [L]	CO$_2$ [%]	medium / pH	light [µmol*m^{-2}*s^{-1}]	T [°C]						
	0.8 (Erlenmeyer flask)	10	BG 11	180	25±1	Chloroform / Methanol (2:1)	0.155	19.25	-	Tang et al. (2011)
	0.8 (Erlenmeyer flask)	20	BG 11	180	25±1	Chloroform / Methanol (2:1)	0.134	19.85	-	Tang et al. (2011)
	0.8 (Erlenmeyer flask)	30	BG 11	180	25±1	Chloroform / Methanol (2:1)	0.081	19.90	-	Tang et al. (2011)
	0.8 (Erlenmeyer flask)	50	BG 11	180	25±1	Chloroform / Methanol (2:1)	0.056	24.40	-	Tang et al. (2011)
Scenedesmus quadricauda	0.1 (Erlenmeyer flask)	5	BG 11	100	25	Methanol / Chloroform (2:1)	0.19	18.4	35.1	Rodolfi et al. (2009)
	0.8 (Roux)	40	f/2 with 48 g/l (NH$_4$)$_2$CO$_3$	300	26±1		0.497±0.034			Lin & Lin (2010)
Scenedesmus rubescens	0.8 (Roux)	40	f/2 with 30 g/l urea	300	26±1		0.511±0.062			Lin & Lin (2010)
	0.8 (Roux)	40	f/2 with 85g NaNO$_3$	300	26±1		0.443±0.055			Lin & Lin (2010)
	0.8 (Roux)	40	f/2 with 15g urea and 42.5g NaNO$_3$	300	26±1		0.539±0.040			Lin & Lin (2010)
Scenedesmus sp.	-	-	-	-	-	-	0.03 – 0.26	19.6 – 21.1	40.8 – 53.9	Mata et al. (2010)
	-	10	BG11	150	25±1	Bligh&Dyer	0.217	9.5	20.65	Yoo et al. (2010)
Selenastrum gracile	-	-	-	-	-	-	-	21 – 28	-	Griffiths & Harrison (2008)
	-	-	-	-	-	-	0.08	16 – 25	13 – 17	Griffiths & Harrison (2008)
Skeletonema costatum	0.1 (Erlenmeyer flask)	5	f (Guillard& Ryther)	100	25	Methanol / Chloroform (2:1)	0.08	21.1	17.4	Rodolfi et al. (2009)
	1.5 (Conical flask)	-	f/2	80	25±1	Bligh&Dyer	0.057 (g/l)	13.5	-	Renauld et al. (1999)
Skeletonema sp.	0.1 (Erlenmeyer flask)	5	f (Guillard& Ryther)	100	25	Methanol / Chloroform (2:1)	0.09	31.8	27.3	Rodolfi et al. (2009)

Table 2. (continued)

Microalgae species	Cultivation conditions					lipid extraction-method	biomass-productivity [g/l/d]	Lipid content [% of dry weight]	Lipid-productivity [mg/l/d]	Reference
	volume [L]	CO$_2$ [%]	medium / pH	light [μmol*m^{-2}*s^{-1}]	T [°C]					
	1.5 (Conical flask)	-	f/2	80	25±1	Bligh&Dyer	0.059 (g/l)	13.3	-	Renauld et al. (1999)
Spirulin maxima	-	-	-	-	-	-	-	7	-	Griffiths & Harrison (2008)
	-	-	-	-	-	-	0.21	4.1	-	Mata et al. (2010)
Spirulina platensis	-	-	-	-	-	-	-	10 – 13	-	Griffiths & Harrison (2008)
	-	-	-	-	-	-	0.06-4.3	4.0-16.6	-	Mata et al. (2010)
Synechococcus	-	-	-	-	-	-	-	11	75	Griffiths & Harrison (2008)
	-	-	-	-	-	-	0.59	17 – 26	32 – 99	Griffiths & Harrison (2008)
Tetraselmis suecica	0.1 (Erlenmeyer flask)	5	f (Guillard& Ryther)	100	25	Methanol / Chloroform (2:1)	0.32	8.5	27.0	Rodolfi et al. (2009)
	-	-	-	-	-	-	0.08	16 – 26	13 – 17	Griffiths & Harrison (2008)
Thalassiosira pseudonana	0.1 (Erlenmeyer flask)	5	f (Guillard& Ryther)	100	25	Methanol / Chloroform (2:1)	0.08	20.6	17.4	Rodolfi et al. (2009)
Thalassiosira weisflogii	-	-	-	-	-	-	-	22 – 24	-	Griffiths & Harrison (2008)
Tetraselmis suecica	-	-	-	-	-	-	0.12 – 0.32	8.5 – 23	27 – 36.4	Mata et al. (2010)
Tetraselmis sp.	1.5 (Conical flask)	-	f/2	80	25±1	Bligh&Dyer	0.075 – 0.098 (g/l)	12.3 – 13.8	-	Renaud et al. (1999)
	-	-	-	-	-	-	0.30	12.6 – 14.7	43.4	Mata et al. (2010)
Tribonema	-	-	-	-	-	-	0.51	12 – 16	59	Griffiths & Harrison (2008)

TMS – Tracemetalsolution.

Apart from the dependence of lipid classes on algal species and growth period (Lombardi & Wangersky, 1995), changes in lipid composition are also caused by environmental factors such as temperature, light intensity, pH (CO_2 concentration) and nutrient supply. These parameters are summarized in Table 1.

Researchers usually specify the lipid content of algae as total lipid content and independent of the lipid classes. Under optimal conditions, algae accumulate 5-20% of their dry weight in glycerol-based membrane lipids (Rosello Sastre & Posten, 2010). Under stress and limiting conditions, the triacylglycerol (TAG) content as a favourable substrate for biodiesel production rises which results in a considerable increase of the total lipid content.

Table 2 is a representation of biomass productivity, lipid contents and productivities of different microalgae at different cultivation conditions.

It should be noted that the lipid content of microalgae taken from data in the literature is difficult to compare, as a wide variety of extraction methods is used (for example. Bligh & Dyer, Folch, Guckert). Using the example of *Scenedesmus* sp., a previous study of our research group investigated the effects of different lipid extraction methods. These led to significant differences in the lipid content. A lipid content of 7.31% of dry weight was obtained with the Folch method, whereas 30.48% of lipids were achieved with the Bligh & Dyer method (Griehl et al., 2010). This illustrates that there are substantial discrepancies when it comes to the determination of lipid contents on the basis of different lipid extraction methods.

The lipid classes displayed in figure 2 contain circa 80% of fatty acids with chain lengths of C10-C20 (Rosello Sastre & Posten, 2010). The predominant saturated fatty acids in algae are C12:0 (lauric acid), C14:0 (myristic acid), C16:0 (palmitic acid) and C18:0 (stearic acid). Unsaturated fatty acids with a chain length of C16-C22 and 1-6 double bonds are also present in algae.

Fatty acids of the division Chlorophyta consist mainly of C16-C18 and show a high number of double bonds (Ahlgreen et al, 1992). Compared to other plants, however, the content of C16 fatty acids is higher than that of C18 fatty acids. Polyunsaturated fatty acids (PUFA) with more than 18C atoms can also be found in marine green algae. These display a greater diversity in their fatty acid profiles (Goss & Wilhelm, 2009).

Various cultivation conditions influence the fatty acid profile of algae, such as nutrient supply, light intensity and temperature. As observed in the microalgae species *Chlorella kessleri* and compared to saturated fatty acids, a phosphate limitation during algae cultivation results in higher unsaturated fatty acid content (El-Sheek & Rady, 1995). Nitrogen deficiency in the culture medium leads to an increase in the total lipid content, as described above. Some researchers have described a lower concentration of polyunsaturated fatty acids than in cultivations with normal nitrogen supply (Klein Breteler et al., 2005). The increase of the cultivation temperature also affects the fatty acid profile. A lower temperature induces a higher content of double bonds in the fatty acids. In their investigations, Sushchik et al. (2003) observed a decrease of polyunsaturated fatty acids at an increased temperature during the cultivation of *Chlorella vulgaris*. The light intensity present during the algae growth affects the fatty acid composition in algae, too. It is assumed that high light intensities support an accumulation of TAG with a higher content of saturated fatty acids, while low light intensities lead to an increase of polar lipids with an increase of unsaturated fatty acids (Goss & Wilhelm, 2009).

Table. 3. fatty acid composition of some microalgae

Microalgae species	Culture conditions	Total lipid [% of dw]	C14:0	C16:0	C16:1	C16:2	C16:3	C18:0	C18:1	C18:2	C18:3	C20:0	C20:1	C20:5	saturated	un-saturated	Reference
Amphidinium carterae	-	-	3	24	1	1	-	-	5	1	2	-	-	14	27	28	Harwood et al. (1988)
Anabeana variabilis	0.04% CO$_2$	-	3.4	35.8	21.4	1.9	-	-	7.4	14.3	16.0	-	-	-	39.2	61.0	Tsuzuki et al. (1990)
	5% CO$_2$	-	4.8	35.4	17.0	2.3	-	-	7.5	15.5	17.6	-	-	-	40.2	59.9	Tsuzuki et al. (1990)
Anacystis ridulans	0.04% CO$_2$	-	4.0	49.2	38.7	-	-	-	4.0	-	-	-	-	-	53.1	38.7	Tsuzuki et al. (1990)
	5% CO$_2$	-	2.9	49.9	38.8	-	-	-	5.4	-	-	-	-	-	52.8	38.8	Tsuzuki et al. (1990)
Aphanocapsa sp.	-	-	29-34	-	36-39	-	-	1-2	1-2	1-2	-	-	-	-	29-34	39-45	Hu et al. (2008)
Biddulphia aurica	-	-	32.0	5.0	27.0	2.0	8.0	-	-	-	-	-	-	26.0	37.0	63.0	Hu et al. (2008)
Botryococcus braunii	10% CO$_2$	25.79	-	29.5	3.4	-	-	1.0	44.9	21.2	-	-	-	-	30.5	69.5	Yoo et al. (2009)
Chlamydomonas reinhardtii	0.04% CO$_2$	-	3.0	26.6	3.2	2.9	3.3	-	16.7	10.2	23.1	-	-	-	29.6	59.4	Tsuzuki et al. (1990)
	5% CO$_2$	-	1.5	38.2	0.2	0.2	0.8	-	18.7	11.3	20.1	-	-	-	39.7	51.3	Tsuzuki et al. (1990)
Chaetoceros sp.	-	-	23.6	9.2	36.5	6.9	2.6	-	3	-	1.4	-	-	8.0	32.8	55.4	Hu et al. (2008)
Chlorella minutissima	-	-	12	13	21	-	-	-	1	2	-	-	-	45	25	69	Ratledge & Wilkinson (1988)
	0.03% CO$_2$	20.90	0.89	36.18	1.03	3.48	10.38	1.60	2.19	17.89	24.04	-	-	-	40.99	59.01	Tang et al. (2011)
	5% CO$_2$	20.65	0.80	26.17	1.22	8.55	14.31	1.06	3.33	20.52	23.56	-	-	-	28.50	71.50	Tang et al. (2011)
Chlorella pyrenoidose	10% CO$_2$	24.25	0.69	25.02	1.76	8.57	15.17	0.71	3.44	19.38	24.57	-	-	-	27.11	72.89	Tang et al. (2011)
	20% CO$_2$	25.48	0.83	26.64	1.50	9.01	15.42	0.94	2.10	20.15	22.89	-	-	-	28.93	71.07	Tang et al. (2011)
	30% CO$_2$	26.02	0.92	30.42	3.12	4.04	17.33	0.76	2.83	11.56	26.12	-	-	-	25.01	64.99	Tang et al. (2011)

Table 2. (continued)

Microalgae species	Culture conditions	Total lipid [% of dw]	C14:0	C16:0	C16:1	C16:2	C16:3	C18:0	C18:1	C18:2	C18:3	C20:0	C20:1	C20:5	saturated	un-saturated	Reference
	50% CO_2	26.75	0.66	27.90	0.72	1.07	21.32	0.81	2.23	5.93	35.79	-	-	-	32.94	67.06	Tang et al. (2011)
	20mmol/l N	16.3	2.1	21.2	6.5	14.1	11.1	-	2.1	23.2	19.8	-	-	-	23.3	76.8	Richardson et al. (1969)
Chlorella sorokiniana	10mmol/l N	15.2	-	22.2	4.3	9.1	12.9	-	3.9	23.5	24.2	-	-	-	22.2	77.9	Richardson et al. (1969)
	5mmol/l N	14.6	1.7	24.0	6.0	6.7	12.9	-	3.6	20.4	24.7	-	-	-	25.7	74.3	Richardson et al. (1969)
	-	-	-	40.0	4.0	11.0	17.0	-	5.0	36.0	23.0	-	-	-	-	-	Hu et al. (2008)
	0.04% CO_2	-	3.20	23.8	0.8	2.4	7.4	1.0	1.0	22.2	38.1	-	-	-	28.0	71.9	Tsuzuki et al. (1990)
Chlorella vulgaris	2% CO_2	-	3.0	24.5	0.8	5.8	1.3	1.8	5.1	39.8	17.8	-	-	-	29.3	70.6	Tsuzuki et al. (1990)
	10% CO_2	6.6	-	24.0	2.1	-	-	1.3	24.8	47.8	-	-	-	-	25.3	74.7	Yoo et al. (2009)
	-	-	-	18.0	5.0	12.0	2.1	-	9.2	43.0	10.0	-	-	-	18.0	81.3	Hu et al. (2008)
Crypthecodium cohnii	-	-	47	19	1	5	-	-	-	5	-	-	-	-	67	10	Ratledge & Wilkinson (1988)
Cryptomonas maculata	-	-	5	15	7	1	-	-	3	-	6	-	-	17	20	34	Harwood et al. (1988)
Dunaliella salina	-	-	-	41	15	-	-	-	11	8	19	-	-	-	41	53	Harwood et al. (1988)
Dunaliella tertiolecta	0.04% CO_2	-	0.8	22.9	6.3	2.5	2.4	-	2.8	10.8	35.2	-	-	-	23.7	60.0	Tsuzuki et al. (1990)
	5% CO_2	-	1.4	24.2	2.3	3.4	2.8	-	5.0	12.4	30.9	-	-	-	25.6	56.8	Tsuzuki et al. (1990)
Emiliana huxleyi	-	-	35.1	5.1	-	-	-	1.0	15.3	2.1	7.0	-	-	1.0	41.2	25.4	Hu et al. (2008)
Euglena gracilis	0.04% CO_2	-	-	23.4	11.6	-	-	-	7.7	9.2	24.6	-	-	15.5	23.4	68.6	Tsuzuki et al. (1990)
	5% CO_2	-	-	25.6	9.2	-	-	-	7.6	6.9	30.8	-	-	12.4	25.6	66.9	Tsuzuki et al. (1990)

Table 3. (continued)

Microalgae species	Culture conditions	Total lipid [% of dw]	C14:0	C16:0	C16:1	C16:2	C16:3	C18:0	C18:1	C18:2	C18:3	C20:0	C20:1	C20:5	saturated	un-saturated	Reference
Glossomastrix chrysoplasta	-	-	22.0	4.4	4.0	-	-	-	6.6	3.9	-	-	-	39.2	26.4	68.94	Hu et al. (2008)
Gonyaulax catanella	-	-	13	30	3	-	-	-	7	3	6	-	-	1	43	19	Harwood et al. (1988)
Gymnodinium sanguineum	-	-	6.5	24.8	2.6	-	-	1.9	12.1	0.7	0.4	-	-	14.1	33.2	29.9	Hu et al. (2008)
Hemiselmis brunescens	-	-	2.0	13.3	13.0	3.0	-	-	2.0	-	11.0	-	18.0	11.0	15.3	58.0	Hu et al. (2008)
Isochrysis galbana	-	-	23.1	14.0	2.0	1.0	-	1.1	14.0	5.0	7.0	-	-	5.0	38.2	36.0	Hu et al. (2008)
Isochrysis sp.	25°C	20.7	25.9	12.6	11.3	1.8	-	0.5	9.2	4.1	8.2	-	-	1.5	39.0	36.1	Renaud et al. (1999)
	27°C	21.7	31.7	12.3	6.1	1.6	-	0.5	9.8	4.6	8.7	-	-	0.5	44.5	32.3	Renaud et al. (1999)
	30°C	21.2	28.2	12.9	6.7	1.7	-	0.6	10.1	5.7	9.1	-	-	0.6	41.7	33.9	Renaud et al. (1999)
	33°C	20.2	27.2	13.1	8.4	1.6	-	1.2	9.6	3.6	9.9	-	-	0.8	41.5	33.9	Renaud et al. (1999)
Monochrysis lutheri	-	-	10	13	22	5	7	-	3	1	-	-	-	18	23	52	Harwood et al. (1988)
Monodus subterraneus	-	-	2.3	20.2	30.5	-	-	0.6	4.5	2.0	1.0	-	-	37.1	23.1	75.1	Hu et al. (2008)
Nannochloropsis oculata	100 µmol*m^{-2}*s^{-1}	36.5	5.3	20.5	25.0	0.9	-	0.3	3.7	2.4	0.9	-	4.8	34.8	26.1	72.5	Renaud et al. (1991)
	243 µmol*m^{-2}*s^{-1}	25.8	4.9	14.0	19.3	6.4	-	1.0	3.1	2.6	0.8	-	5.6	40.0	19.9	77.8	Renaud et al. (1991)
	340 µmol*m^{-2}*s^{-1}	22.8	4.4	18.0	22.5	3.8	-	0.7	4.3	2.9	0.7	0.2	6.6	37.0	23.8	77.8	Renaud et al. (1991)
	1100 µmol*m^{-2}*s^{-1}	30.4	5.4	18.4	19.9	7.6	-	1.1	5.0	2.5	1.6	-	5.1	33.8	24.9	75.5	Renaud et al. (1991)
Nannochloropsis sp.	-	-	6.9	19.9	27.4	-	-	-	1.7	3.5	0.7	-	-	34.9	26.8	68.2	Hu et al. (2008)
Olisthodiscus ssp.	-	-	8	14	10	2	2	-	4	4	6	-	-	19	22	47	Harwood et al. (1988)

Table 3. (continued)

Microalgae species	Culture conditions	Total lipid [% dw]	C14:0	C16:0	C16:1	C16:2	C16:3	C18:0	C18:1	C18:2	C18:3	C20:0	C20:1	C20:5	saturated	un-saturated	Reference
Parietochloris incise	-	-	-	9.1	0.7	0.6	-	2.1	15.1	9.3	1.6	-	-	-	11.2	27.3	Hu et al. (2008)
Phaeodactylum tricornutum	-	-	-	10	21	-	-	-	4	3	-	-	-	33	10	61	Ratledge & Wilkinson (1988)
Phaeomonas parva	-	-	18.7	3.7	3.5	-	-	-	1.3	1.8	1.1	-	-	56.0	22.4	63.7	Hu et al. (2008)
Porphyridium cruentum	0.04% CO_2	-	-	46.9	2.1	-	-	-	-	8.8	-	-	-	20.3	46.9	31.2	Tsuzuki et al. (1990)
	5% CO_2	-	-	44.8	3.1	-	-	0.7	1.3	8.1	0.5	-	-	19.4	45.5	35.1	Tsuzuki et al. (1990)
Rhodomonas lens	-	-	18.0	13.1	5.0	6.1	-	2.2	10.1	2.2	16.0	-	-	13.0	33.3	52.4	Hu et al. (2008)
	0.03% CO_2	15.15	0.43	31.50	0.69	1.54	5.25	1.70	0.67	13.00	43.35	0.13	0.32	0.97	34.22	65.78	Tang et al. (2011)
	5% CO_2	16.45	0.24	26.38	0.15	2.82	5.26	1.66	0.74	20.65	40.00	0.14	0.31	1.44	28.61	71.39	Tang et al. (2011)
Scenedesmus obliquus	10% CO_2	19.25	0.21	24.62	0.49	3.19	4.89	2.31	0.87	19.00	41.54	0.23	0.55	1.89	27.58	72.42	Tang et al. (2011)
	20% CO_2	19.85	0.22	25.17	0.13	2.47	5.93	1.97	1.03	19.14	40.67	0.11	0.45	2.41	27.78	72.22	Tang et al. (2011)
	30% CO_2	19.90	0.28	22.33	0.25	0.92	6.07	0.93	1.15	13.43	48.44	0.16	0.47	5.36	23.92	76.08	Tang et al. (2011)
	50% CO_2	24.4	0.28	22.18	0.27	0.91	6.41	0.92	1.20	13.34	48.23	0.16	0.46	5.43	23.76	76.24	Tang et al. (2011)
Scenedesmus sp.	10% CO_2	9.5	-	36.3	4.0	-	-	2.7	25.9	31.1	-	-	-	-	39.0	61.0	Yoo et al. (2009)
Scrippsiella sp.	-	-	3.2	9.4	0.7	-	-	0.5	1.5	1.5	5.4	-	-	1.8	13.1	10.9	Hu et al. (2008)
Spirulina platensis	-	-	1	26	5	-	-	-	23	10	21	-	-	-	27	59	Ratledge & Wilkinson (1988)
Thalassiosira fluviatilis	-	-	6	11	22	3	9	-	3	1	-	-	-	20	17	58	Harwood et al. (1988)
Trichodesmium erythraeum	-	-	7.21	11-17	4-7	-	-	2-6	4-11	6-19	-	-	-	-	20.21-30.21	14-37	Hu et al. (2008)

The carbon dioxide concentration used during cultivation is another parameter influencing the fatty acids composition. Tang et al. (2011) examined the impact of high CO_2 concentrations in the air and observed a decrease in unsaturated fatty acids. Tsuzuki et al. (1990) also observed an increase in saturated fatty acids compared to unsaturated fatty acids in *Chlorella vulgaris* under higher CO_2 concentration in the air. In contrast, Muradyan et al. (2004) suggested that the levels of unsaturated fatty acids rise with an increase in CO_2. The rise in the saturation of the fatty acids could be due to an increase of carbon dioxide content, in turn leading to a decreasing oxygen concentration and affecting the enzymatic desaturases (formation of unsaturated fatty acids from saturated fatty acids) and the β-oxidation of fatty acids (Vargas et al., 1998, Manoharan et al., 1999).

Table 3 provides an overview of the changes in fatty acid composition of different microalgae with regard to their different cultivation parameters.

POTENTIAL OF MICROALGAE FOR EXTRACTION OF BIODIESEL

Over the past two decades, international concern about the negative environmental impacts of fossil fuels has increased. This has also led to focusing on the potential of renewable and environmentally friendly biofuels in replacing petroleum-based fuels (Huntley & Redalje, 2007).

Biodiesel — a common name for acyl-esters — is one of these renewable, biodegradable and non-toxic alternative fuels produced from domestic resources. It consists of fatty acid methyl esters (FAME) derived from plant oils or animal fats through the transesterification of mono-, di-and triacylglycerols (MAG, DAG, TAG) or the esterification of free fatty acids. Fatty acids can also be separated from polar lipids (phospholipids and glycerolipids) and used for biodiesel production (Frühwirth, 2010) (figure 3).

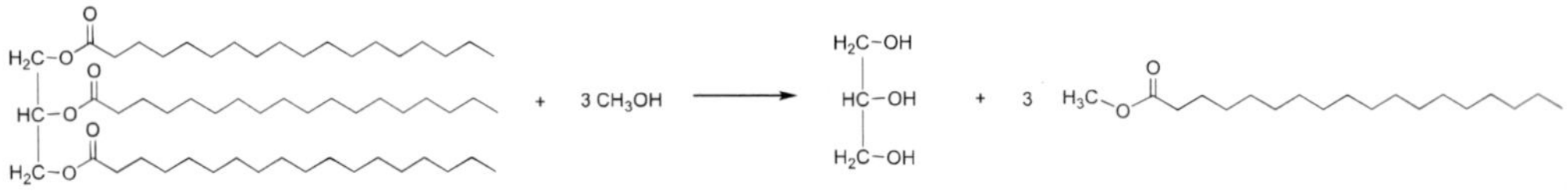

Figure 3. transesterification reaction of a triacylglycerol (tristearin) with methanol.

For the production of biodiesel the crude oil is mixed with 10% alcohol (either methanol or ethanol) and different catalysts (KOH, NaOH, Ethanolates, sulfuric acid, hydrochloric acid). The alkali-catalysed transesterification is much faster than acid catalysed transesterification and is often used for commercial purposes. A temperature of approximately 60°C and ambient pressure the ester bonds of the triacylglycerids occurring in the plant oil is broken and methyl esters (the chemical name for biodiesel) are built from the reaction with the methanol (=transesterification). The transesterification reaction is affected by molar ratio of glycerides to alcohol, used catalysts, reaction temperature, reaction time and free fatty acids and free water content of the oils and fats. The commonly accepted molar ratio alcohol to glycerides is 6:1. Higher reaction temperatures accelerate the reaction.

The resulting glycerol (a valuable byproduct usually sold to be used in soaps and other products) has to be removed from the biodiesel by settling or centrifugation. Then the biodiesel is ready for use.

At present, biodiesel is mainly generated from oil crops which are preferably grown on arable land, such as rapeseed, soybean, palm or sunflower. Recent studies have shown that an increased production of biofuels from oil seeds leads to a rise in deforestation: As a consequence, more CO_2 is emitted into the atmosphere, which runs contrary to the aim of reducing CO_2 emissions through the use of biofuels (Fargione et al., 2008). It is, therefore, crucial to locate an inexpensive source of raw materials for lipids, which does not require arable land and provides biodiesel with a competitive advantage over petroleum diesel.

Potentially, instead of oil crops, microalgae can be used to make biodiesel. They have the potential to displace fossil diesel and plant biodiesel too, as they offer several advantages compared to energy crops: they show higher photosynthetic efficiency, grow extremely rapidly (by tolerating marginal lands, that are not suitable for conventional agriculture) and many are exceedingly rich in neutral lipids.

Among triacylglycerides also fatty acid containing polar lipids (phospholipids and glycerolipids) can be used for biodiesel production (Frühwirth, 2010) (figure 4). In summary microalgal lipids contains up to 80% saponifiable fatty acids demonstrating the potential of algae for alternative feedstock for fuel production.

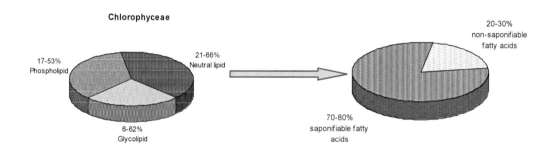

Figure 4.lipid classes of chlorophyceae for saponifiable fatty acids recovery for biodiesel production (Borowitzka, 1988; Frühwirt, 2010).

The first concept for the "biological transformation of solar energy" was described by Oswald & Golueke 1960, using a large scale open pond system to produce algal biomass that was subsequently anaerobically fermented to biogas. The CO_2 present and the digester effluent formed the basis for growing the algae, while the methane generated electricity. Additionally nutrients (N, P, C) were provided by waste waters.

Research in algal biodiesel focuses on the most efficient species in terms of oil production. A lot of research has been done in the field of screening for high lipid containing algae. In the Aquatic Species Program (Sheehan et al. 1998), in particular, 3000 species were screened and over 300 strains with high lipid contents were identified, mostly green algae and diatoms. Between 1978 and 1996, the U.S. National Renewable Energy Laboratory in the Aquatic Species Program (ASP) concentrated on the production of biodiesel from high lipid-content algae grown in outdoor ponds using CO_2 from coal-fired power plants so as to increase the rate of algae growth and reduce carbon emissions.

Considering current technology, yields are still lower than what could be achieved in theory, and the cost-effectiveness needs further improvement, too. It is, therefore, practical to screen those microalgae that provide a straightforward cultivation for a strain which accumulates high lipid contents already in the logarithmic growth phase, so that the yields of lipids can be improved. Hence, it seems worthwhile to examine a variety of strains for enhanced growth performance and oil productivity. In close proximity to an industrial area with CO_2-emissions, algal technology applied for the production of valuable substances such as biodiesel and co-products could facilitate algal growth and reduce the atmospheric partial pressure of carbon dioxide.

Clearly, algae are a superior alternative as a feedstock for large-scale biodiesel production. In practice, however, biodiesel has not yet been produced on a wide scale from algae.

The following steps have to be in place to ensure the large-scale production of biodiesel from algae:

- a sustainable, large-scale production of high-oil-yielding algae strains
- the large-scale extraction of oil from the algae
- the large-scale conversion of algal oil into biodiesel.
- reduction in high production costs (very controversial literature data to the now-term price of biodiesel varying from 2 to 22 $/l (Timilsina & Shrestha, 2011).

It is for economic aspect of the biodiesel production that a large scale application has thus far not been realized. All efforts in biofuel production in the USA and other countries have proceeded beyond rather small laboratory or field testing stage.

Aim of this Study

In this study, different microalgae species of the division of Chlorophyta (*Scenedesmus* sp. and *Chlorella* sp.) were cultivated according to their biomass productivity. The two microalgae species with the highest biomass productivity were chosen for investigation of their biomass growth, lipid content and fatty acid composition under different carbon dioxide levels in order to examine the potential of these microalgae to produce biodiesel.

METHODS

Algae Strain and Culture Conditions

The microalgae strains Scenedesmus sp. (Scenedesmus acuminatus, Scenedesmus obliquus, Scenedesmus pectinatus, Scenedesmus producto-capitatus) and Chlorella sp. (Chlorella kessleri, Chlorella protothecoides, Chlorella sorokiniana, Chlorella vulgaris) were obtained from the 'Culture Collection of Algae' department (Sammlung von Algenkulturen, SAG) at the University of Göttingen, Germany. Scenedesmus sp. and Chlorella sp. were grown in Setlik-medium with the following composition (mg/l): KNO_3 2020; KH_2PO_4 340; $MgSO_4*7H_2O$ 990; Fe-EDTA 18.5; $Ca(NO_3)_2*4H_2O$ 10; H_3BO_3 3.09; $MnSO_4*4H_2O$ 1.2; $CoSO_4*7H_2O$ 1.4; $CuSO_4*5H_2O$ 1.24; $ZnSO_4*7H_2O$ 1.43; $(NH_4)_6Mo_7O_{24}*4H_2O*$ 1.84.

The microalgae were cultivated in 2 L bubble-column glass (working volume 1.5L) reactors at a constant illumination of 100 $\mu E*m^{-2}*s^{-1}$ and at a temperature of 28°C. The cultures for the pre-screening of microalgae were aerated with 3% CO_2. The cultures for the analysis of the influence of different CO_2 concentrations on biomass growth and fatty acid composition were aerated with 200 L/h ambient air enriched with 3%, 10% or 15% CO_2.

The algae cells were harvested by centrifugation after a cultivation time of 14 days.

Determination of Cell Density and Biomass Concentration

The cell density was measured with a Coulter Counter by Beckman Coulter, using the electrical sensing zone method.

For biomass determination 10-mL samples were taken, centrifuged in pre-weight tubes, washed twice with distilled water and dried to weight constancy at 100°C.

Cell Disruption and Lipid Extraction

Cell disruption was performed with a vibration mill with 0.2g freeze-dried biomass and 0.4g sea sand. Lipid extraction was carried out using a solvent mix of n-hexane, isopropanol and distilled water (Guckert et al, 1988). For each species, duplicate samples were analysed with regard to total lipid content and fatty acid composition. The total lipid concentration was determined gravimetrically after extraction. For this purpose, the extracts were evaporated to dryness and then stored in a drying oven at 100°C for one hour. After cooling, the extracts were weighed.

Quantification of FAME by Gas Chromatography

The extracts were evaporated and solved in hexane. The transesterification of the samples was performed with hydrochloric acid / methanol (3M) for 1 hour at 70°C. After cooling, water was added so as to wash out the polar substances and the upper layer was removed. The fatty acid composition was analysed with GC-MS (Shimadzu; GC-MS QP2010) on a SGE BPX 70 GC capillary column.

RESULTS AND DISCUSSION

Growth Analysis

A pre-screening of different Scenedesmus sp. (S. acuminatus, S. obliquus, S. pectinatus, S. producto-capitatus) and Chlorella sp. (C. kessleri, C. protothecoides, C. vulgaris, C. sorokiniana) with an aeration of 3% CO_2 was carried out with the aim of establishing the Scenedesmus and Chlorella species with the highest biomass productivity (figure 5)

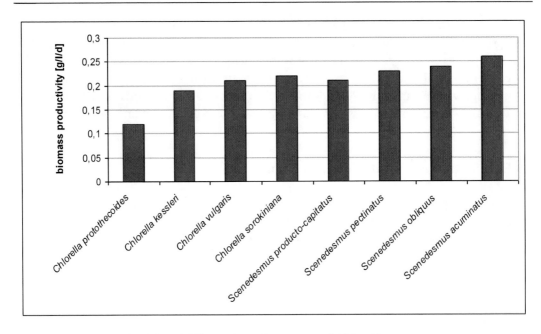

Figure 5. biomass productivity of different *Scenedesmus* sp. and *Chlorella* sp.

The microalgae species *Chlorella sorokiniana* and *Scenedesmus acuminatus* exhibited the highest values of biomass productivity with 0.22 g/l/d and 0.26g/l/d, respectively. These two species were, therefore, selected for test of the influence of different carbon dioxide concentrations (3%, 10% and 15%) on biomass productivity and fatty acid composition (figure 6; tab. 4).

Figure 6 shows that both species achieved the best microalgae growth at an aeration of 3% CO_2. The growth decreased with increasing the carbon dioxide concentration. The maximum biomass concentrations after 14 days of cultivation of *Scenedesmus acuminatus* and *Chlorella sorokiniana* were 2.13 g/l and 2.45 g/l, respectively (at 3% CO_2). The biomass growth decreased with increasing CO_2 in the air stream. The final biomass concentrations and biomass productivities of both microalgae are displayed in table 4.

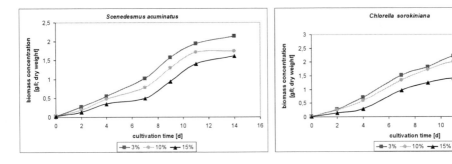

Figure 6. Effects of different carbon dioxide concentrations on growth of *Scenedesmus acuminatus* and *Chlorella sorokiniana*.

Table 4. values of the biomass concentration after 14 cultivation days and biomass productivity of *Scenedesmus acuminatus* and *Chlorella sorokiniana*

	Scenedesmus acuminatus		*Chlorella sorokiniana*	
	biomass concentration (14th d) [g/l]	biomass productivity [g/l/d]	biomass concentration (14th d) [g/l]	biomass productivity [g/l/d]
3% CO_2	2.13	0.26	2.45	0.22
10% CO_2	1.74	0.18	2.15	0.20
15% CO_2	1.61	0.15	1.50	0.17

Schmid-Staiger et al. (2009) investigated the influence of carbon dioxide on the growth of *Phaedactylum tricornutum* and found that the biomass growth decreased to 72% at an aeration of 8% CO_2 compared to a biomass growth of 100% at 2% CO_2. This is consistent with the results recorded for *Scenedesmus acuminatus* and *Chlorella sorokiniana*.

Lipid Extraction

The lipid extraction was performed according to Guckert et al. (1988) using the solvent system hexane / isopropanol / water. The determination of the total lipid content was carried out gravimetrically and the results are shown in figure 7.

The investigation of the lipid content showed that the lipid concentration of the microalgae *Scenedesmus acuminatus* and *Chlorella sorokiniana* was at its peak at an aeration of 15% CO_2 with 22.34% and 27.92% of dry weight, respectively.

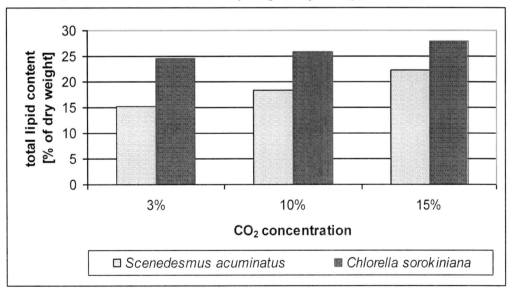

Figure 7. Total lipid content (in % of dry weight) of *Scenedesmus acuminatus* and *Chlorella sorokiniana*.

Our results can be compared to those mentioned in the literature. Widjaja et al. (2009) investigated lipid contents of *Chlorella vulgaris* under different CO_2 concentrations and registered a lipid concentration of approximately 28% at an aeration of 20ml/min CO_2. Illman et al. (2000) detected lipid contents of 20% (5% CO_2) and 22% (low nitrogen medium) in *Chlorella sorokiniana*. These values are lower than those found by us. Rodolfi et al. (2009) analyzed a lipid content of 21% for the microalgae species *Scenedesmus* sp., but without data on CO_2 in the air stream. Tang et al. (2011) also examined the effect of different carbon dioxide concentrations on lipid content of *Scenedesmus* sp. They found that the total lipid content of the two microalgae displayed a rising trend with an increase of the CO_2 concentration.

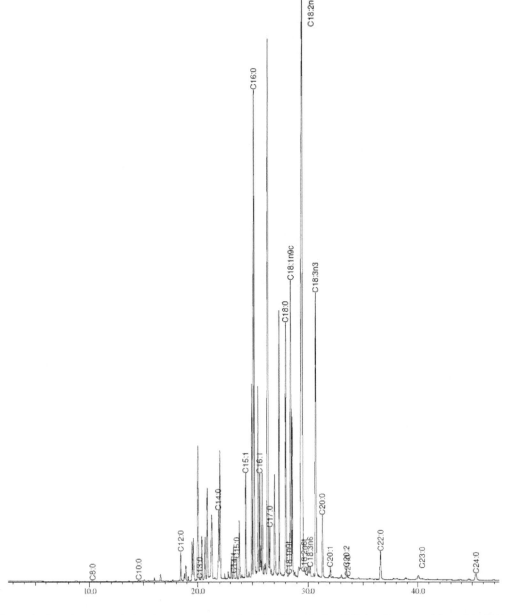

Figure 8. Gas chromatography chromatogram of the FAME analysis of *Chlorella* sp., cultivated at 15% CO_2 and legend of analysed fatty acids.

Analysis of the Fatty Acid Composition

The fatty acid analysis of the microalgae species *Scenedesmus acuminatus* and *Chlorella sorokiniana* was performed using gas chromatography / mass spectrometry (GC-MS). Based on the obtained data, the total fatty acid methyl ester (FAME) concentration and the concentration of single FAMEs was determined. Figure 8 shows the chromatogram of the FAME determination of *Chlorella* sp., cultivated at 15% carbon dioxide.

Table 5. fatty acids detected by GC

abbreviated form	common name	chemical structure
C8:0	Caprylic acid	~~~COOH
C10:0	Capric acid	~~~~COOH
C12:0	Lauric acid	~~~~~COOH
C13:0	Tridecanoic acid	~~~~~~COOH
C14:0	Myristic acid	~~~~~~COOH
C14:1n5 (9c)	Myristoleic acid	~~~=~~COOH
C15:0	Pentadecanoic acid	~~~~~~~COOH
C15:1n5 (10c)	cis-10-Pentadecenoic acid	~~~~=~~COOH
C16:0	Palmitic acid	~~~~~~~~COOH
C16:1n7 (9c)	Palmitoleic acid	~~~~=~~~COOH
C17:0	Heptadecanoic acid	~~~~~~~~~COOH
C17:1n7 (10c)	cis-10-Heptadecenoic acid	~~~~~=~~~COOH
C18:0	Stearic acid	~~~~~~~~~~COOH
C18:1n9 (9t)	Elaidic acid	~~~~~~/~~COOH
18:1n9 (9c)	Oleic acid	~~~~~=~~~COOH
C18:2n6 (9,12t)	Linolelaidic acid	~~~~/~/~~COOH
C18:2n6 (9,12c)	Linoleic acid	~~~~=~=~~COOH
C18:3n3 (9,12,15c)	α-Linolenic acid	~=~=~=~~COOH
C18:3n6 (6,9,12c)	γ-Linolenic acid	~~~=~=~=~COOH
C20:0	Arachidic acid	~~~~~~~~~~~COOH
C20:1n9 (11c)	cis-11-Eicosenoic acid	~~~~~=~~~~COOH
C22:0	Behenic acid	~~~~~~~~~~~~COOH
C23:0	Tricosanoic acid	~~~~~~~~~~~~COOH
C24:0	Lignoceric acid	~~~~~~~~~~~~~COOH

Table 6 illustrates the fatty acid methyl ester (FAME) composition of both cultivated microalgae species. According to the results of the fatty acid methyl ester analysis, palmitic acid (C16:0) is the most abundant fatty acid in *Scenedesmus acuminatus* and *Chlorella sorokiniana* at CO_2 concentrations of 3, 10 and 15%. Palmitoleic acid (C16:1) is the main unsaturated fatty acid in *Chlorella sorokiniana* and Linolenic acid (C18:3) in *Scenedesmus acuminatus*.

Table 6. FAME composition of the most abundant fatty acid methyl esters of the cultivated microalgae *Scenedesmus acuminatus*. and *Chlorella sorokiniana*

microalgae	CO_2 [%]	[% of total fatty acids]						
		C14:0	C16:0	C16:1	C18:0	C18:1	C18:2	C18:3
Scenedesmus sp.	3	1.16	50.10	3.20	9.18	0.18	2.12	25.69
	10	1.67	47.05	4.73	9.27	4.25	8.74	17.10
	15	1.44	44.17	8.30	9.85	6.37	9.41	15.83
Chlorella sp.	3	1.63	43.26	19.60	7.05	6.81	4.48	14.45
	10	1.81	45.45	20.49	7.53	5.12	4.67	12.66
	15	1.31	42.41	21.45	7.99	7.75	4.72	11.55

It appears that increasing CO_2 concentration in airstreams lead to an increase in the degree of unsaturation in fatty acids, but a decrease in the content of saturated fatty acids and linolenic acid (figure 9).

For the microalgae species *Botryococcus braunii*, Yoo et al. (2010) observed an increase of oleic acid (C18:1) from 56% to 60% of total fatty acids through a cultivation with ambient air and real flue gas (5.5% CO_2), but a decrease of linoleic acid (C18:3) from 28% to 12% of total fatty acids. Tsuzuki et al. (1990) investigated the influence of CO_2 on the microalgae species *Chlorella vulgaris* and found that, at the expense of C18:1 and/or C18:2, the content of C18:3 (linolenic acid) was higher in low-CO_2 cells than in high-CO_2 cells. The present study confirms these results, as the content of linolenic acid (C18:3) was reduced from 25.69% to 15.83% in *Scenedesmus* sp. and from 14.45% to 11.55% of total fatty acids in *Chlorella* sp.

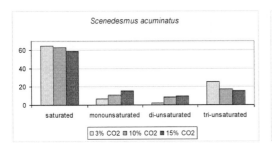
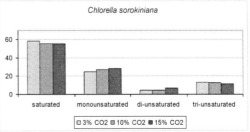

Figure 9. Ratio of saturated and unsaturated fatty acids depending on CO_2 concentration.

Ota et al. (2009) found that the desaturation rate was slower than the elongation rate, which seemed to enhance the production of lower unsaturated fatty acids (Tang et al., 2011). This could be explained with the increase of CO_2 resulting in a decrease of oxygen, influencing the desaturation pathway of fatty acids and thus inducing an increase in the concentration of unsaturated fatty acids (Vargas, 1998). Tang et al. (2011) observed an increase of the unsaturated fatty acids content from 65.78% at 0.03% CO_2 to 76.24% at 50% CO_2 for *Scenedesmus* sp. and from 59.01% to 67.06% of total fatty acids for *Chlorella* sp. It was recorded in the present investigation that an increase of the carbon dioxide concentration from 3% to 15% led to more mono- and di-unsaturated fatty acids. The mono and di-unsaturated fatty acid contents increased from 6.76% and 2.45% to 15.41% and 9.62% for the microalgae species *Scenedesmus acuminatus* and for *Chlorella sorokiniana* from 24.94% and 4.48% to 28.12% and 6.72%, respectively (figure 9). It should, however, be noted that the content of linolenic acid (C18:3) decreased in both species by an increase of CO_2; this is inconsistent with the explanation of an increasing desaturation rate with a rise in CO_2.

In order to assess the suitability of the fatty acids of *Scenedesmus acuminatus* and *Chlorella sorokiniana* for the biodiesel production, the fatty acid composition of these microalgae species have to be consistent with the European standard for *'Fatty acid methyl esters (FAME) for diesel engines'* (EN14214:2008). This regulation states that the main fatty acid methyl esters consist of fatty acids with a chain length of between C14 and C22 and that the content of linolenic acid methyl ester has to be below 12% of the total fatty acid methyl ester content. Figure 10 shows the potential of the analysed microalgae species for the biodiesel production.

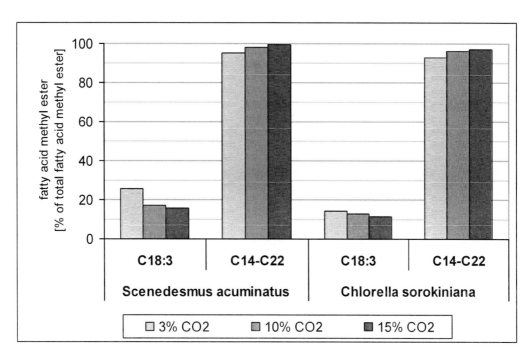

Figure 10. content of linolenic acid (C18:3) and fatty acids with a chain lenght of C14-C22 in order to assess the potential of *Scenedesmus* sp. and *Chlorella* sp. for biodiesel production.

Without further purification, the fatty acid profile of *Scenedesmus acuminatus* is not suitable for the biodiesel production, because, at a cultivation with 15% carbon dioxide in air stream, the linolenic acid methyl ester (C18:3) content lies at at least 15.83%. At 15% CO_2, the linolenic acid methyl ester concentration of *Chlorella sorokiniana* is with at least 11.55% below the limit of 12% C18:3 of the total fatty acid methyl esters. *Chlorella sorokiniana* can, therefore, be considered for biodiesel production. It should, nevertheless, be mentioned that the biodiesel production from microalgae oil is not economically advantageous because of the high costs of microalgae cultivation, harvesting and downstreaming processes. A coproduction of other high valuable products (e.g. proteins, carbohydrates or the removal of polyunsaturated fatty acids prior to further processing) could reduce the costs. The digestion of the waste biomass of microalgae to biogas is promising, too. It can be used as an environmentally friendly energy source.

CONCLUSION

In the present study different Scenedesmus sp. (Scenedesmus acuminatus, Scenedesmus obliquus, Scenedesmus pectinatus, Scenedesmus producto-capitatus) and Chlorella sp. (Chlorella kessleri, Chlorella protothecoides, Chlorella sorokiniana, Chlorella vulgaris) were cultivated at 3% CO_2 so as to determine the species with the highest biomass productivity. This pre-screening revealed the microalgae species Scenedesmus acuminatus and Chlorella sorokiniana to show the highest values for biomass productivity with 0.26g/l/d and 0.22 g/l/d, respectively. These microalgae were, therefore, cultivated at different CO_2 contents in the air stream (3%, 10%, 15% CO_2) with the aim of investigating the influence of a high carbon dioxide concentration on biomass growth, lipid content and fatty acid composition. Scenedesmus acuminatus and Chlorella sorokiniana showed the highest biomass increase at 3% CO_2 after 14 cultivation days with 2.13 g/l and 2.45 g/l, respectively. The biomass growth decreased with increasing CO_2 for both species. On the contrary, the lipid content increased at higher CO_2 levels. The microalgae species reached maximal lipid content at an aeration of 15% CO_2 (Scenedesmus acuminatus: 22.34 % of dry weight; Chlorella sorokiniana: 27.92% of dry weight). The analysis of the fatty acid composition showed that the proportions of the saturated fatty acid palmitic acid (C16:0) decreased with higher CO_2 concentrations during a rise of the stearic acid (C18:0) content. The mono- and diunsaturated fatty acids palmitoleic acid (C16:1) and oleic acid (C18:1) increased under higher CO_2 concentrations. Contrary to that and for both species, the content of linolenic acid (C18:3) decreased with increasing CO_2. In summary, the unsaturated fatty acids (mono- and di-unsaturated fatty acids) contents increased with increasing CO_2 from 3% to 15% CO_2 for both microalgae species from 9.21% to 25.03% (Scenedesmus acuminatus) and from 29.42% to 34.84% (Chlorella sorokiniana). A critical analysis of the microalgae demonstrates the fatty acid methyl ester profil of Chlorella sorokiniana to be suitable for biodiesel production. The content of linolenic acid (C18:3) was below the required limit of 12% C18:3 of total fatty acid content.

The results suggest that the chloropyhta Scenedesmus acuminatus and Chlorella sorokiniana can grow under high CO_2 concentration and thus have great potential for carbon dioxide fixation. Furthermore, the lipid content could be raised through higher CO_2 levels and the resulting fatty acids of Chlorella sorokiniana can be used for biodiesel production as a renewable energy source.

REFERENCES

Ahlgreen, G., Gustafsson, I.-B., Boberg, M. (1992). Fatty acid content and chemical composition of freshwater microalgae. *Journal of Phycology*, Vol. 28, 37-50

Badger M.R., Price G.D. (1994) The role of carbonic anhydrase in photosynthesis, *Annu. Rev. Physiol. Plant Mol. Biol*, Vol 45, pp. 369-392

Bhola, V., Desikan, R., Kumari Santosh, S., Subburamu, K., Sanniyasi, E., Bux, F. (2011). Effects of parameters affecting biomass yield and thermal behaviour of *Chlorella vulgaris. Journal of Bioscience and Bioengineering*, Vol. 111, 377-382

Borowitzka, M.A. (1988). Fats, oils and hydrocarbons. In: M.A. Borowitzka and L.J. Borowitzka, Editors, *Micro-algal biotechnology*, Cambridge University Press, Cambridge, 257–287

Bowes G.W. (1968). Carbonic anhydrase in marine Algae, *Plant Physiol.*, Vol. 44, pp. 726-732

Brown, M.R., Dunstan, G.A., Norwood, S.J., Miller, K.A. (1996). Effects of harvest stage and light on the biochemical composition of the diatom *Thalassiosira pseudonana*. *The Journal of Phycology*, Vol. 32, 64-73

Chen, G.-Q., Jiang, Y., Chen, F. (2007). Fatty acid and lipid class composition of the eicosapentaenoic acid-producing microalgae, *Nitzschia laevis*. *Food Chemistry*, Vol. 104, 1580-1585

Chui, S.-Y., Kao, C.-Y., Chen, C.-H., Kuan, T.-C., Ong, S.-C., Lin, C.-S. (2008).Reduction of CO_2 by a high-density culture of *Chlorella* sp. in a semicontinuous photobioreactor. *Bioresource Technology*, Vol. 99, 3389-3396

Chiu, S.-Y., Kao, C.-Y., Tsai, M.-T., Ong, S.-C., Chen, C.-H., Lin, C.-S. (2009). Lipid accumulation and CO_2 utilization of *Nannochloropsis oculata*. *Bioresource Technology*, Vol. 100, 833-838

Damiani, M.C., Popovich, C.A., Constenla, D., Leonardi, P.I. (2010). Lipid analysis in *Haematococcus pluvialis* to assess ist potential use as a biodiesel feedstock. *Bioresource Technology*, Vol. 101, 3801-3807

De Morais, M.G., Costa, J.A.V. (2007). Isolation and selection of microalgae from coal fired thermoelectric power plant for biofixation of carbon dioxide. *Energy Conversion and Management*, Vol. 48, 2169-2173

El-Sheek, M.M., Rady, A.A. (1995). Effect of phosphorus starvation on growth, photosynthesis and some metabolic processes in the unicellular green alga *Chlorella kessleri*. *Phyton*, Vol. 35, 139-151

Estevez, M.S., Malanga, G., Puntarulo, S. (2001). Iron-dependent oxidative stress in *Chlorella vulgaris*. *Plant Science*, Vol. 161, 9-17

Fargione, J., Hill, J., Tilman, D., Polasky, S., Hawthorne, P. (2008). Land clearing and the biofuel carbon dept. *Science*, Vol. 319, 1235-1238

Frühwirt, H. (2010). Biodiesel production from Algae. *2nd Algae World Europe Brüssel*; 31.05-01.06.2010

Fulke, A.B., Mudliar, S.N., Yadav, R., Shekh, A., Srinivasan, N., Ramanan, R., Krishnamurthi, K., Saravan Devi, S., Chakrabarti, T. (2010). Bio-mitigation of CO_2, calcite formation and simultaneous biodiesel precursors production using *Chlorella* sp.. *Bioresource Technology*, Vol. 101, 8473-8476

Ge, Y., Liu, J., Tian, G. (2011). Growth characteristics of *Botryococcus braunii* 765 under high CO_2 concentration in photobioreactor. *Bioresource Technology*, Vol. 102, 130-134

Giordano, M., Beardall, J., Raven, J.A. (2005). CO_2 concentrating mechanisms in algae: mechanism, environmental modulation, and evolution. *Annual Review of Plant Physiology and Plant Molecular Biology*, Vol. 56, pp. 99-131

Gordillo, F.J.L., Goutx, M., Figueroa, F.L., Niell, F.X. (1998). Effects of light intensity, CO_2 and nitrogen supply on lipid class composition of *Dunaliella viridis*. *Journal of Applied Phycology*, Vol. 10, 135-144

Goss, R., Wilhelm, C. (2009). Lipids in Algae, Lichens and Mosses. In: *Essential and Regulatory Functions*, 117-137

Griehl, C., Polhardt, H., Müller, D., Bieler, S. (2010). *The potential of microalgae to produce lipids for biofuels*. Narossa (In Press)

Griffith, M.J., Harrison, S.T. (2008). Lipid productivity as a key characteristic for choosing algal species for biodiesel production. *Journal of Applied Phycology*, Vol. 21, 493-507

Guckert, J.B., Cooksey, K.E., Jackson, L.L. (1988). Lipid solvent systems are not equivalent for analysis of lipid classes in the microeukaryotic green alga. *Chlorella, Journal of Microbiological Methods* 8, 139-149

Guschina, I.A., Harwood, J.L. (2006). Lipids and lipid metabolism in eukaryotic algae. *Progress in Lipid Research*, Vol. 45, 160-186

Härtel, H., Benning, C. (2000). Can digalactosyldiacylglycerol substitute for phosphatidylcholine upon phosphate deprivation in leaves and roots of *Arabidopsis*?. *Biochemical Society Transaction*, Vol. 28, 729-732

Harwood, J.L. (1988). Fatty-acid metabolism. *Annual Review of Plant Physiology and Plant Molecular Biology*, Vol. 39, 101-138

Harwood, J.L., Russell, N. (1984). *Lipids in Plants and Microbes*. George Allen & Unwin, London, 162

Harwood, J.L., Pettitt, T.P., Jones, A.L. (1988). *Lipid metabolism. In Biochemistry of the Algae and Cyanobacteria*. Oxford Science Publications

Hodgens, P.A., Henderson, R.J., Sargent, J.R., Leftley, J.W. (1991). Patterns in variation in the lipid class and fatty acid composition of *Nannochloropsis oculata* (Eustigmatophyceae) during batch culture. Part I: The growth cycle. *The Journal of Applied Phycology*, Vol. 3, 169-181

Hsieh, C.-H., Wu, W.-T. (2009). Cultivation of microalgae for oil production with a cultivation strategy of urea limitation. *Bioresource Technology*, Vol. 100, 3921-3926

Hu, Q., Sommerfeld, M., Jarvis, E., Ghirardi, M., Posewitz, M., Seibert, M., Darzins, A. (2008). Microalgal triacylglycerols as feedstock for biofuel production: perspectives and advances. *The Plant Journal*, Vol. 54, 621-639

Huertas, I.E., Bhatti, S., Colman, B. (2005). Characterization of the CO_2-concentrating mechanism in the unicellular alga Eustigmatos vischeri. *European Journal of Phycology*, Vol. 40, 409-415

Huntley, M.E., Redalje, D.G. (2007). CO_2 mitigation and renewable oil from photosynthetic microbes: a new appraisal. *Mitigation and Adaption Strategies for Global Change*, Vol. 12, 573-608

Illman, A.M., Scragg, A.H., Shales, S.W. (2000). Increase in Chlorella strains calorific values when grown in low nitrogen medium. *Enzyme and Microbial Technology*, Vol. 27, 631-635

Jouhet, J., Marechal, E., Baldan, B., Bligny, R., Joyard, J., Block, M.A. (2004). Phosphate deprication induces transfer of DGDG galactolipid from chloroplast to mitochondria. *Journal of Cell Biology*, Vol. 167, 863-874

Khotimchenko, S., Yakovleva, I.M. (2005). Lipid composition of the red alga *Tichocarpus crinitus* exposed to different levels of photon irradiance. *Photochemistry*, Vol. 66, 73-79

Khozin-Goldberg, I., Cohen, Z. (2006). The effect of phosphate starvation on the lipid and fatty acid composition of the fresh water eustigmatophyte *Monodus subterraneus*. *Phytochemistry*, Vol. 67, 696-701

Klein Breteler, W.C.M., Schogt, N., Rampen, S. (2005). Effect of diatom nutrient limitation on copepod development: role of essential lipids. *Marine Ecology Progress Series*, Vol. 291, 125-133

Lee, Y. Tan, H., Low, C. (1989). Effect of salinity of medium on cellular fatty acid composition of marine alga *Porphyridium cruentum* (Rhodophyceae). *Journal of Applied Phycology*, Vol. 19-23

Li, Y., Horsman, M., Wang, B., Wu, N. (2008). Effects of nitrogen sources on cell growth and lipid accumulation of green alga *Neochloris oleoabundans*. *Appl. Microbiol. Biotechnol.*, Vol. 81, 629-636

Liang, Y., Sarkany, N., Cui, Y. (2009). Biomass and lipid productivities of *Chlorella vulgaris* under autotrophic, heterotrophic anc mixotrophic growth conditions. *Biotechnology Letters*, Vol. 31, 1043-1049

Liao, F.Y., Li, H.M., He, P. (2004). Effect of high irradiance and high temperature on chloroplast composition and structure of *Dioscorea zingiberensis*. *Photosynthetica*, Vol. 42, 487-492

Lin, Q., Lin, J. (2010). Effects of nitrogen source and concentration on biomass and oil production of a *Scenedesmus rubescens* like microalgae. *Bioresource Technology*, Vol. 101, 1615-1621

Lnych, D.V., Thompson, G.A. (1982). Low temperature-induced alterations in the chloroplast and microsomal membranes of *Dunaliella salina*. *Plant Physiology*, Vol. 69, 1369-1375

Lombardi, A.T., Wangersky, P.J. (1995). Particulate lipid class composition of three marine phytoplankters *Chaetoceros gracilis, Isochrysis galbana* (Tahiti) and *Dunaliella tertiolecta* grown in batch culture. *Hydrobiologia*, Vol. 306, 1-6

Manoharan, K., Lee, T.K., Cha, J.M., Kim, J.H., Lee, W.S., Chang, M., Park, C.W., Cho, J.H. (1999). Acclimation of *Prorocentrum minimum* (Dinophyceae) to prolonged darkness by use of an alternative carbon source from triacylglerides and galactolipids. *Journal of Phycology*, Vol. 35, 287-292

Mata, T.M., Martins, A.A., Caetano, N.S. (2010). Microalgae for biodiesel production and other applications: a review. *Renewable and Sustainable Energy Reviews*, Vol. 14, 217-232

Mock, T., Kroon, B.M.A. (2002). Photosynthetic energy conversion under extrem conditions – II: the significance of lipids under light limited growth in Antarctic sea ice diatoms. *Phytochemistry* 61, 53-60

Manoharan, K., Lee, T.K., Cha, J.M., Kim, J.H., Lee, W.S., Chang, M., Park, C.W., Cho, J.H. (1999). Acclimation of *Prorocentrum minimum* (Dinophyceae) to prolonged darkness by use of an alternative carbon source from triacylglerides and galactolipids. *Journal of Phycology*, Vol. 35, 287-292

Moroney, J.V., Ynalvez. R.A. (2007). Proposed Carbon Dioxide Concentrating Mechanism in *Chlamydomonas reinhardtii. Eukaryotic Cell,* 1251-1259

Muradyan, E.A., Klyachko-Gurvich, G.L., Tsoglin, L.N., Sergeyenko, T.V., Pronina, N.A. (2004). Changes in lipid metabolism during adaption of the *Dunaliella salina* photosynthetic apparatus to high CO_2 concentration. *Russian Journal of Plant Physiology,* Vol. 51, 53-62

Murphy, D. (2004). *Plant lipids: Biology, utilisation and manipulation.* Oxford: Blackwell

Oswald, W.J., Golueke, C.G. (1960): Biological transformation of solar energy. *Advances in Applied Microbiology 2, pp. 223-262*

Ota, M., Kato, Y., Watanabe, H., Sato, Y., Smith, R.L. (2009). Fatty acid production of from a highly CO_2 tolerant alga, *Chlorococcum littorale,* in the presence of inorganic carbon and nitrate. *Bioresource Technology,* 100, pp. 5237-5242

Piorreck, M., Pohl, P. (1984). Formation of biomass, total protein, chlorophylls, lipids and fatty acids in green and blue-green algae during one growth phase. *Phytochemistry,* Vol. 23, 217-223

Piorrek, M., Baasch, K.-H., Pohl, P. (1984). Biomass production, total protein, chlorophylls, lipids and fatty acids of freshwater green and blue-green algae under different nitrogen regimes. *Phytochemistry,* Vol. 23, 207-216

Ratledge C, Wilkinson SG (1988). *Microbial lipids.* Academic Press, London

Reed, M.L., Graham, D. (1968). Control of photosynthetic carbon fixation during an induction phase of Chlorella. *Plant Physiology,* Vol. 43

Renaud, M.S., Thinh, L.V., Parry, D.L. (1999). The gross chemical composition and fatty acid composition of 18 species of tropical Australian microalgae for possible use in mariculture. *Aquaculture,* 170, 147-159

Renaud, M.S., Parry, D.L., Thinh, L.V. (1991). Microalgae for use in tropical aquaculture I:Gross chemical and fatty acid composition of twelve species of microalgae from the Northern Territory, Australia. *Journal of Applied Phycology,* 6, 337-345

Renaud, S.M., Parry, D.L. (1994). Microalgae for use in tropical aquaculture II: Effect of salinity on growth, gross chemical composition and fatty acid composition of three marine microalgae. *Journal of Applied Phycology,* Vol. 6, 347-356

Richardson, B., Orcutt, D.M., Schwertner, H.A., Martinez, C.L., Wickline, H.E. (1969). Effects of Nitrogen Limitation on the Growth and Comosition of Unicellular Algae in Continous Culture. *Pllied Microbiology,* Vol.18, 245-250

Rodolfi, L., Zittelli, G.C., Bassi, N., Padovana, G., Biondi, N., Bovini, G., Tredici, M.R. (2009). Microalgae for Oil: Strain Selection, Induction of Lipid Synthesis and Outdoor Mass Cultivation in a Low-Cost Photobioreactor. *Biotechnology and Bioengineering,* Vol. 102, No. 1

Roessler, P. (1990). Environmental control of glycerolipid metabolism in microalgae: commercial implications and future research directions. Journal of Phycology, Vol. 26, 393-399

Rotatore, C., Colman, B. (1991a). The actice uptake of carbon dioxide by the unicellular green algae *Chlorella saccharophila* and *C. ellipsoidea. Plant, Cell & Environment,* Vol. 14, 371-375

Rotatore, C., Colman, B. (1991b). The acquisition and accumulation of inorganic carbon by the unicellular green alga *Chlorella ellipsoidea. Plant, Cell & Environment,* Vol. 14, 377-382

Rotatore, C., Colman, B. (1992). Active uptake of CO_2 by the diatom *Navicula pelliculosa*. *Journal of experimental botany*, Vol. 43, 571-576

Rosello Sastre, R., Posten, C. (2010). Die vielfältigen Anwendungen von Mikroalgen als nachwachsende Rohstoffe. *Chemie Ingenieur Technik*, Vol. 82, 1925-1939

Schmid-Staiger, U., Preisner, R., Marek, P., Trösch, W. (2009). Kultivierung von Mikroalgen im Photobioreaktor zur stofflichen und energetischen Nutzung. *Chemie Ingenieur Technik*, Vol. 81, 1783-1789

Scragg, A.H., Illman, A.M., Carden, A., Shales, S.W. (2002). Growth of microalgae with increased calorific values in a atubular bioreactor. *Biomass and Bioenergy*, Vol. 23, 67-73

Sheehan, J., Dunahay, T., Benemann, J., Roessler, P. (1998): A look back at the U.S. Department of Energy's aquatic species program—biodiesel from algae. *NREL/TP-580-24190. U.S. Department of Energy's Office of Fuels Development*

Sukenik, A., Livne, A. (1991). Variations in lipid and fatty acid content in relation to acetyl-CoA carboxylase in the marine prymnesiophyte *Isochrysis galbana*. *Plant Cell Physiology*, Vol. 32, 371-378

Sültemeyer, D.F., Fock, H.P., Canvin, D.T. (1991). Active uptake of inorganic cabon by *Chlamydomonas reinhardtii*: evidence for simultaneous transport of HCO_3^- and CO_2 and characterization of active CO_2 transport. *Canadian Journal of Botany*, Vol. 69, 995-1002

Sushchik, N.N., Kalacheva, G.S., Zhila, N.O., Gladyshev, M.I., Volova, T.G. (2003). A temperature dependence of the intra- and extracellular fatty acid composition of green algae and cyanobacteria. *Russian Journal of Plant Physiology*, Vol. 50, 374-380

Takagi, M., Watanabe, K., Yamaberi, K., Yoshida, T. (2000). Limited feeding of potassium nitrate for intracellular lipid and triglyceride accumulation of *Nannochloris* sp. UTEX LB1999. *Applied Microbiology and Biotechnology*, Vol. 54, 112-117

Takagi, M., Karseno, Yoshida, T. (2006). Effect of salt concentration on intracellular accumulation of lipids and Triacylglyceride in Marine Microalgae *Dunaliella* cells. *Journal of Bioscience and Bioengineering*, Vol. 101, 223-226

Tang, D., Han, W., Li, P., Miao, X., Zhong, J. (2011). CO_2 biofixation and fatty acid composition of *Scenedesmus obliquus* and *Chlorella pyrenoidosa* in response to different CO_2 levels. *Bioresource Technology* 102, pp. 3071-3076

Thompson Jr., G.A. (1996).Lipids and membrane function in green algae. *Biochimica et Biophysica Acta*, Vol. 1302, 17-45

Timilsina, G.R., Shrestha, A. (2011). How much hope should we have for biofuels?. *Energy*, Vol. 36, 2055-2069

Tsuzuki, M., Miyachi, S. (1989). The function of carbonic anhydrase in aquatic photosynthesis. *Aquatic photosynthesis*, Vol. 34, 85-104

Tsuzuki, M., Ohnuma, E., Sato, N., Takaku, T., Kawaguchi, A. (1990). Effects of CO_2 concentration during grwoth on fatty acid composition in microalgae. *Plant Physiol.*, Vol. 93, 851-856

Vargas, M.A. (1998). Biochemical composition and fatty acid content of filamentous nitrogen-fixing cyanobacteria. *Journal of Phycology* 34, pp. 812-817

Wagner, H., Jakob, T., Wilhelm, C. (2005). Balancing the enrgy flow from capture light to biomass under fluctuating light conditions. *New Phytologist*, Vol. 169, 95-108

Widjaja, Chien, C.-C., Ju, Y.-H. (2009). Study of increasing lipid production from fresh water microalgae *Chlorella vulgaris*. *Journal of Taiwan Institute of Chemical Engineers*, Vol. 40, pp. 13-20

Wilhelm, C., Krämer, P., Wiedemann, I. (1987). Die Lichtsammelkomplexe der verschiedenen Algenstämme. *Biologie in unserer Zeit*, Vol. 5, 138-143

Yang, S.-Y., Tsuzuki, M., Miyachi, S. (1985). Carbonic anhydrase of *Chlamydomonas*: purification and studies on ist induction using antiserum against *Chlamydomonas* carbonic anhydrase. *Plant Cell Physiology*, Vol. 26, 25-34

Yoo, C., Jun, S.-Y., Ahn, C.-Y., Oh, H.-M. (2010). Selection of microalgae for lipid production under high levels carbon dioxide. *Bioresource Technology*, Vol. 101, S71-S74

Yu, J., Tang, X., Zhang, P., Tian, J., Cai, H. (2004). Effects of CO2 Enrichment on Photosynthesis, Lipid Peroxidation and Activities of Antioxidative Enzymes of Platymonas subcordiformis Subjected to UV-B Radiation Stress. *Acta Botanica Sinica* (Chinese Plant Science), Vol. 46, 682-690

Zhu, C.J., Lee, Y.K., Chao, T.M. (1997). Effects of temperature and growth phase on lipid and biochemical composition of *Isochrysis galbana* TK1. *Journal of Applied Phycology*, Vol. 9, 451-457

INDEX

2

20th century, 173
21st century, 23, 43

A

abatement, 38, 238, 287
absorption spectra, 259, 262, 265
abstraction, 70
access, x, 227, 239, 241
accounting, 87, 361
acetic acid, 155, 388, 389, 391, 395, 396, 410
acetone, 21, 74, 93, 253, 334, 336
acetylcholine, 19
acidic, 62, 66, 75, 175, 179, 301, 363, 397
action potential, 19
activated carbon, 170
active compound, vii, 1, 40, 43, 150, 341
active oxygen, 130, 136
active site, 415
active transport, 319
acute renal failure, 361
adaptation, 31, 128, 353, 397
adaptations, viii, 102, 103, 119
additives, ix, 6, 35, 141, 142, 179, 180, 240
adenocarcinoma, 16
adjustment, 210
adsorption, 119, 144, 147, 297, 304
adults, 8, 396, 397
adverse effects, xiii, 103, 118, 374, 383
aesthetic, xii, 345, 346
Africa, 5, 6, 149, 150, 154
agar, 339
age, 52, 150, 158, 207, 220, 265, 326, 341, 361, 418
age-related diseases, 361
aggregation, 272
agonist, 58
agriculture, ix, xi, 4, 24, 104, 141, 142, 219, 277, 297, 312, 439

AIDS, 49, 150, 161
air pollutants, 27
alanine, 45, 378
alcohols, viii, 38, 61, 93, 94, 97
alfalfa, 243
algorithm, 268, 350, 351
alimentation, 143, 162
alkaloids, 18, 20
alkenes, 184
ALT, 378, 382
alternative energy, 279
alternative medicine, 6
ambient air, 302, 441, 446
amine, 189, 297
amine group, 189
amines, 19, 156, 306
amino, ix, 13, 15, 45, 118, 141, 147, 149, 189, 193, 309, 375, 376, 378, 379, 380, 395, 397, 410
amino acids, 13, 147, 189, 193, 309, 375, 376, 378, 379, 395, 397, 410
ammonia, 33, 47, 84, 157, 306
ammonium, 111, 208, 219, 407
amplitude, 256
amylase, 235
anaerobic bacteria, 157
anaerobic digestion, 24, 29, 97, 156, 157, 159, 168, 229, 238, 288
anaphylaxis, xiii, 374, 377
anatomy, 2
ancestors, 202
anemia, 150
aneuploidy, 133
anhydrase, 415, 449, 453, 454
antagonism, 116
anthocyanin, 273
antibiotic, 14, 17, 21, 36, 48, 53
antibiotic resistance, 36
antibody, 37, 43, 49, 53, 380
anti-cancer, 342, 360

antigen, 38, 47, 380
antigenicity, xiii, 373, 375, 377, 379, 380, 381, 383
antioxidant, vii, xii, 1, 7, 11, 12, 17, 48, 116, 118, 128, 132, 136, 149, 150, 151, 152, 246, 325, 339, 341, 364, 401, 402
antitumor, 14
apoptosis, 16
appetite, 172
aquaculture, vii, ix, 1, 4, 7, 10, 13, 33, 52, 131, 142, 149, 161, 162, 201, 203, 207, 208, 210, 211, 219, 224, 229, 238, 240, 241, 253, 274, 296, 360, 361, 388, 390, 392, 396, 403, 404, 408, 409, 452
aquatic environment, viii, 33, 40, 101, 103, 115, 134, 312, 313, 318, 319, 405
aquatic habitats, viii, 101
aquatic systems, 312, 318
aqueous solutions, 138
aquifers, 103, 128, 306
aromatic compounds, 135, 137, 392
aromatic hydrocarbons, viii, 102, 112, 121, 124, 130, 131, 133, 134, 135, 137, 138, 139, 231
aromatics, 157
arrest, 19
arsenic, 131
arteriosclerosis, 153
ascorbic acid, 118, 134, 137, 389, 402, 410
aseptic, 144
Asia, 5, 7, 52, 149, 408
aspartate, 378
assessment, vii, xi, 9, 53, 61, 64, 78, 82, 83, 86, 87, 91, 92, 128, 132, 135, 138, 139, 167, 243, 244, 245, 272, 273, 296, 311, 313, 323, 357
assets, 346, 347, 355
assimilation, 54, 137, 236, 240, 288, 308, 317, 318, 320, 376
atmosphere, viii, 3, 29, 141, 142, 146, 160, 172, 228, 229, 239, 278, 296, 302, 303, 414, 439
atmospheric deposition, 103
atoms, 158, 433
ATP, 28, 51, 104, 109, 131, 318, 320, 323, 415, 416, 419
Austria, 247
authorities, xi, 311
automobiles, 283
autotrophic organisms, vii, 1
awareness, 312

B

Bacillus subtilis, 21
backscattering, 262
bacteria, viii, xiii, 9, 17, 20, 21, 28, 29, 33, 44, 51, 101, 103, 123, 129, 136, 147, 149, 157, 174, 235, 252, 289, 292, 303, 304, 312, 321, 322, 359, 360, 364, 387, 390

banks, 149
barriers, 240, 292, 319
base, 23, 24, 27, 64, 231, 253, 255, 278, 280, 287, 291, 357, 399, 419
basement membrane, 17
batteries, 313
beams, 257
beer, 190
beet molasses, 402
behaviors, 183
beneficial effect, 22, 150, 152, 299
benefits, 7, 11, 12, 22, 23, 49, 145, 146, 149, 150, 163, 278, 279, 288, 290, 291, 297, 367, 396, 401
benign, xii, 325, 340
benzene, 109, 278
beta-carotene, 57, 151, 152, 342, 370, 371
beverages, 10
bicarbonate, 415
bilirubin, 401
binarisation, 351
bioaccumulation, 102, 103, 115, 124, 131, 135, 137, 147, 401
bioassay, 115, 131, 314, 317, 323
bioavailability, 111, 122
biocatalysts, 374
biochemistry, x, xi, 45, 227, 241, 252, 369, 384
biodegradation, 33, 111, 115, 123, 129, 133, 135
biodiesel production, vii, 25, 26, 34, 41, 44, 48, 53, 54, 57, 60, 61, 62, 63, 64, 65, 66, 68, 73, 76, 78, 79, 80, 84, 96, 97, 157, 166, 231, 233, 241, 242, 245, 246, 248, 249, 271, 281, 292, 293, 305, 306, 398, 399, 406, 408, 412, 433, 438, 439, 440, 447, 448, 450, 451
biodiversity, vii, 2, 33, 239, 279
bioenergy, 1, 30, 41, 44, 57, 230, 293
biofuel, vii, x, xi, 25, 32, 38, 39, 41, 44, 48, 50, 52, 56, 59, 62, 79, 80, 82, 96, 146, 156, 157, 158, 159, 162, 163, 168, 227, 229, 230, 233, 235, 236, 237, 240, 242, 243, 244, 245, 254, 269, 277, 278, 279, 280, 282, 283, 284, 286, 288, 289, 290, 291, 292, 293, 296, 304, 305, 440, 449, 450
biogas, vii, 1, 29, 40, 41, 54, 60, 156, 158, 160, 166, 238, 288, 292, 414, 439, 448
biological activity, 13, 14, 15, 159
biological control, 51
biological processes, viii, 45, 101, 103
biological samples, 258
biological systems, 257, 315
biologically active compounds, vii, 1, 43, 150
biomarkers, 128, 138, 317, 318, 321
biomass cultivation, viii, 61, 82, 83, 95
biomass growth, 145, 440, 441, 442, 443, 448
biomass materials, ix, 93, 171, 172, 173, 196, 197

biomaterials, 306
biomolecules, 1, 34
biopolymer, 42, 151, 155
biopolymers, ix, 29, 141, 156, 280, 281, 282, 289
bioremediation, viii, x, 101, 103, 128, 129, 146, 202, 229, 246, 251, 280, 289, 307, 315, 390, 392
biosphere, viii, 141
biosynthesis, 35, 38, 39, 58, 155, 157, 264, 274, 287, 371, 408, 410, 411, 412, 415, 416, 418, 419
biotechnological applications, vii, 5
biotechnology, iv, vii, viii, 1, 4, 9, 35, 40, 42, 49, 54, 56, 57, 131, 141, 160, 161, 162, 164, 165, 169, 198, 202, 224, 240, 247, 248, 252, 267, 268, 269, 274, 305, 306, 309, 326, 374, 385, 407, 409, 449
biotic factor, 122
birds, 13, 314
bisphenol, 116, 117, 124, 135
bleaching, 272
bleeding, 390, 412
blends, 231, 235, 247, 286
blood, 22, 42, 378, 382
blood vessels, 22
body weight, 9, 378, 381, 382
boilers, 83, 85, 94
bonding, 186
bonds, xiv, 326, 413, 433, 438
bone, 167
brain, 42
branching, 231
Brazil, 26, 141, 143, 145, 146, 158, 162, 166, 169, 283
breakdown, 129, 189, 285, 289, 346
breathing, 19, 377, 380
breeding, 13, 307, 371
Britain, 46, 221
brominated flame retardants, 131
budding, 174
Butcher, 223
butyl ether, 187, 194, 327
by-products, 66, 79, 102, 166, 278, 309, 361, 397, 402

C

Ca^{2+}, 319
cables, 115
cadmium, 34, 116, 130, 136, 139, 147, 167, 323
calcium, 17, 22, 49, 54, 68, 81, 84, 159, 319, 321, 323
calibration, 255, 269, 327, 339
campylobacter, 294
cancer, 14, 53, 57, 58, 342, 360, 371
candidates, xiv, 14, 28, 32, 174, 269, 280, 288, 304, 413
cane sugar, 229, 399

capillary, 327, 441
capsule, 149
carbohydrate, 27, 158, 191, 193, 208, 209, 211, 254, 280, 284, 285, 287, 315, 380
carbohydrates, 13, 22, 27, 29, 39, 147, 156, 158, 159, 174, 192, 193, 194, 212, 217, 218, 238, 279, 283, 287, 288, 301, 304, 392, 397, 402, 448
carbon atoms, 158
carbon emissions, 439
carbon monoxide, 172, 278
carbon neutral, 172, 183, 186, 197, 278
carbon tetrachloride, 59
carbonization, ix, 171, 172, 173, 175, 176, 179, 180, 181, 182, 184, 188, 191, 194, 196, 197
carboxyl, 186, 380
carboxylic acid, 180, 188
carcinogenesis, 17, 361
carcinoma, 16
cardiovascular disease, 341, 361
carotene, 7, 10, 17, 41, 42, 57, 120, 142, 150, 151, 152, 160, 177, 229, 274, 275, 296, 326, 328, 341, 342, 359, 360, 361, 362, 363, 365, 368, 369, 370, 371
carotenoids, vii, x, xii, 1, 6, 7, 11, 17, 45, 142, 147, 149, 150, 151, 152, 163, 166, 221, 251, 253, 265, 273, 288, 315, 325, 326, 327, 329, 331, 334, 340, 341, 342, 357, 359, 360, 361, 362, 363, 365, 366, 367, 368, 369, 370, 371, 389, 399, 400, 401, 405, 408, 414, 417
Carotenoids, vi, xii, 151, 263, 274, 325, 326, 327, 342, 343, 359, 360, 368, 399, 400
case study, 322
casein, 375, 376
catabolism, 39
catalysis, 62, 63
catalyst, 40, 62, 66, 67, 68, 72, 73, 75, 84, 85, 175, 231, 282, 392
cataract, 341
catfish, 403, 408
cation, 19, 343
cattle, 190
C-C, xii, 294, 325, 412
cDNA, 410
cell culture, 50, 268
cell cycle, 314
cell division, 33, 36, 105, 133, 137, 236, 314, 316, 364, 417
cell line, 16, 17
cell membranes, 417, 418
cell metabolism, 388
cell size, 105, 111, 259, 262, 315
cell surface, 119, 147, 316

cellulose, ix, 171, 173, 174, 190, 197, 235, 283, 285, 290
cellulosic biofuel, 283
Chad, 5, 41, 149, 160
Chaetoceros, 12, 52, 149, 203, 209, 222, 223, 224, 225, 232, 420, 434, 451
challenges, xi, 38, 43, 53, 128, 144, 147, 160, 172, 244, 248, 292, 295
changing environment, 128
channel blocker, 18
cheese, 193
chemical, vii, x, xi, xii, 1, 5, 7, 8, 10, 24, 25, 26, 31, 38, 78, 79, 89, 96, 102, 116, 139, 147, 148, 156, 166, 176, 186, 189, 190, 192, 193, 197, 201, 203, 208, 209, 211, 212, 215, 218, 219, 220, 222, 224, 227, 230, 244, 247, 249, 280, 287, 296, 297, 304, 308, 311, 313, 317, 318, 327, 340, 345, 346, 347, 359, 367, 380, 415, 417, 419, 438, 445, 449, 452
chemical properties, 249
chemical reactions, 304
chemical structures, 327
chemicals, xi, 4, 81, 102, 103, 104, 130, 132, 134, 135, 137, 147, 156, 162, 164, 169, 186, 203, 240, 243, 287, 297, 303, 304, 311, 313, 314, 368, 385
chemoprevention, 53
chicken, 371
children, 8, 48, 150, 162, 163
China, 6, 7, 10, 56, 101, 103, 132, 164, 177, 202, 228, 325, 340, 385, 404
chitosan, 151, 164, 165
Chlorella protein hydrolysates, xiii, 374, 375
chlorine, 53
chloroform, 74, 75, 253
chlorophyll, xii, 2, 6, 52, 105, 110, 115, 116, 134, 139, 147, 152, 161, 252, 260, 264, 265, 267, 270, 271, 272, 273, 293, 301, 312, 314, 315, 316, 317, 321, 322, 323, 342, 345, 347, 348, 349, 350, 352, 353, 354, 355, 357, 358, 360, 369, 399, 400, 401, 408, 411, 415, 갬419
chloroplast, 6, 7, 51, 52, 59, 134, 257, 316, 360, 400, 405, 415, 416, 419, 451
cholera, 59
cholesterol, 162
cholinesterase, 19
chromatograms, 327
chromatography, xii, 165, 187, 252, 253, 325, 326, 327, 332, 340, 343, 347, 376, 444, 445
chromium, 34, 54, 146
chronic diseases, 326, 341
chymotrypsin, 20
circulation, 22, 153
civilization, 149

classes, 19, 46, 255, 364, 417, 418, 419, 433, 439, 450
classification, 2, 343, 350, 378, 381, 388, 389, 392
cleaning, 144
climate, xiii, 23, 40, 59, 143, 163, 228, 279, 387
climate change, 40, 59, 163, 228, 279
climates, 4, 172, 280, 286, 414
clinical application, 375
clinical symptoms, 150
clone, 224
cloning, 410
clusters, 259, 316
coal, 27, 51, 52, 80, 85, 94, 96, 135, 145, 146, 164, 165, 172, 181, 182, 183, 185, 186, 194, 197, 198, 228, 244, 288, 290, 302, 307, 309, 414, 439, 449
coding, 39
coefficient of variation, 193
coenzyme, 157, 304, 416
coffee, 192
coke, 186, 197
collagen, 22
colonisation, xii, 345, 346, 348, 349, 352, 353, 354, 355, 356, 358
colonization, 2, 12, 357, 358
color, 362, 366
colorectal adenocarcinoma, 16
coma, 378
combined effect, 128, 129, 132, 139
combustion, vii, xi, 1, 24, 27, 60, 79, 83, 84, 85, 91, 96, 102, 145, 146, 159, 160, 172, 182, 183, 184, 185, 188, 193, 197, 198, 228, 231, 239, 244, 249, 277, 278, 282, 288, 398, 399, 414
combustion processes, 102
commercial, x, 4, 6, 7, 11, 13, 19, 26, 35, 42, 56, 64, 66, 68, 71, 96, 142, 143, 144, 145, 177, 188, 189, 201, 202, 203, 204, 211, 214, 215, 216, 217, 222, 227, 229, 232, 240, 241, 242, 252, 287, 298, 302, 363, 366, 368, 377, 378, 384, 391, 396, 401, 438, 452
commodity, 6, 270, 296, 297, 298, 305
communication, 99
communities, xii, 132, 186, 313, 346, 347, 349, 358
community, xi, 7, 111, 252, 311, 312, 349, 353
comparative analysis, 57, 185, 408
compensation, 259, 262
competition, 132, 235, 239, 281, 283, 322, 349
competitive advantage, 230, 439
competitors, 288
complement, 202
complexity, 175, 189, 316
complications, 253
composites, 173

compounds, viii, ix, x, 2, 4, 8, 10, 12, 13, 14, 15, 17, 18, 19, 20, 22, 33, 38, 39, 40, 50, 58, 102, 104, 111, 114, 115, 117, 118, 120, 122, 123, 124, 129, 132, 135, 137, 142, 144, 146, 148, 149, 156, 157, 168, 172, 189, 192, 202, 210, 213, 240, 242, 245, 251, 253, 255, 257, 278, 287, 298, 313, 315, 326, 327, 338, 341, 343, 349, 359, 374, 392, 402, 403, 404, 405
compression, 164, 230, 241, 246, 249, 297
computation, 192
computer, 350
condensation, 154, 155
conditionally essential, 411
conditioning, 22
conductivity, 181, 357
conference, 245, 246
configuration, 368
congress, 357
Congress, 42, 139, 356
conjugation, 118
connective tissue, 22
consensus, 279
conservation, xii, 131, 189, 345, 346, 355, 356, 357
constant rate, 236, 330
constituent materials, 195
constituents, 21, 57, 132, 181, 210, 221, 268, 393, 418
construction, 25, 76, 174, 346, 416
consumers, 11, 230
consumption, xiv, 8, 9, 28, 42, 77, 80, 111, 147, 202, 228, 230, 238, 242, 247, 277, 287, 296, 297, 333, 353, 374, 375, 381, 398, 413, 414, 416
containers, 180, 203, 215
contaminant, 128, 236
contaminated water, viii, 101, 128
contamination, 4, 143, 144, 152, 235, 241, 286, 301, 312, 364
contour, 179, 191
contradiction, 364
controversial, 190, 279, 440
convergence, 353
conversion rate, 27, 239
cooking, 192, 199
cooling, 64, 77, 78, 236, 301, 330, 349, 441
copper, 128, 130, 137, 138, 147, 315, 316, 318, 320, 321
coproduction, 448
copyright, iv
Copyright, iv
correlation, 263, 266, 267, 339
correlation coefficient, 267
correlations, 254
cosmetic, 6, 22

cosmetics, vii, 1, 22, 35, 46, 151, 152, 229, 240, 326, 401
cost, vii, viii, ix, x, xi, 1, 24, 26, 29, 31, 33, 57, 62, 68, 69, 85, 90, 92, 93, 94, 95, 96, 101, 102, 141, 143, 145, 148, 155, 157, 158, 160, 172, 174, 175, 177, 187, 189, 197, 201, 214, 227, 228, 229, 235, 237, 239, 241, 247, 252, 268, 279, 283, 285, 286, 287, 288, 289, 297, 298, 301, 303, 304, 309, 311, 346, 350, 392, 404, 439
coughing, 377, 380
crabs, xiii, 359
cracks, 2, 347
creatinine, 378
credentials, 288, 289
critical analysis, 448
criticism, 25
crop, x, 4, 23, 29, 227, 240, 278, 279, 280, 414
crop production, 278
crops, x, xi, 4, 23, 24, 25, 38, 45, 63, 156, 157, 160, 162, 163, 227, 229, 233, 234, 235, 239, 240, 242, 244, 277, 278, 279, 280, 283, 286, 291, 439
crude oil, 84, 90, 92, 131, 228, 243, 414, 438
crystalline, 338
crystallization, 338
crystals, 10
Cuba, 373, 375, 377
cultivation conditions, 158, 159, 269, 419, 433
cultural heritage, xii, 346, 347, 355, 358
culture conditions, xiv, 25, 26, 54, 57, 142, 158, 242, 253, 291, 409, 413, 419
culture media, 36, 144, 145, 223
culture medium, xiv, 20, 21, 28, 33, 36, 144, 148, 153, 205, 207, 208, 214, 215, 220, 223, 236, 313, 353, 364, 393, 413, 415, 419, 433
cyanide, 34, 49, 146, 163
cyanobacteria, vii, viii, ix, 1, 2, 3, 6, 9, 10, 13, 14, 16, 17, 18, 19, 23, 25, 27, 28, 37, 40, 41, 44, 45, 49, 51, 52, 53, 56, 58, 103, 104, 123, 141, 142, 153, 154, 157, 161, 162, 164, 165, 167, 168, 171, 175, 202, 256, 257, 260, 261, 271, 273, 283, 314, 348, 349, 385, 388, 399, 412, 453
cycles, 15, 52, 155, 245, 247, 347, 349, 362
Cylindrotheca closterium, 110, 121
cyst, 274, 362, 364, 365
cytochrome, 118, 362
cytomegalovirus, 15
cytometry, 252, 254, 271, 314, 315, 316, 318, 319, 320, 321, 323, 406
cytoplasm, 157, 319, 360, 401, 416
cytotoxicity, 16, 54, 136, 164

D

damages, xii, 33, 345
database, 128

deaths, 381
decay, 346, 356, 358
decomposition, 70, 71, 75, 76, 111, 285, 346
deconvolution, 256
defence, 18, 118
deficiency, 6, 7, 150, 153, 155, 158, 166, 168, 265, 363, 364, 365, 366, 419, 433
deficit, 245
deforestation, 439
degradation, viii, 15, 39, 47, 101, 103, 111, 116, 119, 123, 128, 129, 133, 138, 149, 274, 285, 292, 358, 419
dehydration, ix, 171, 188
dehydrochlorination, 123
Delta, 103, 132, 386
denaturation, 194
dengue, 38
denitrification, 158
Denmark, 387
Department of Energy, 43, 46, 58, 99, 145, 169, 248, 294, 453
dephosphorylation, 416
depolarization, 19, 319, 320
deposition, 102, 103, 143, 261, 265, 274, 296, 316, 347
deposits, xi, 311
deprivation, 450
depth, xi, 129, 138, 203, 216, 217, 295, 301
derivatives, 240, 319, 326, 410
desiccation, 362
destruction, 151, 286, 414
detachment, 347
detection, xii, 253, 254, 269, 312, 318, 320, 321, 327, 345, 347, 348, 350, 351, 355
detergents, 15
detoxification, 56, 118, 123
developing countries, 102, 244
deviation, 192, 260, 267
diabetes, 361
diabetic nephropathy, 361
diacylglycerol, 416, 418, 419
diaphragm, 19
diarrhea, 378
diatoms, xiii, 46, 60, 123, 136, 222, 224, 229, 283, 387, 392, 411, 439, 451
dibenzo-p-dioxins, 102
dielectric strength, 37
diesel engines, 79, 235, 244, 282, 447
diesel fuel, 39, 84, 85, 94, 96, 187, 231, 247
diet, 8, 9, 162, 210, 361, 371, 377, 392, 401, 403, 404
dietary supplementation, 409
diffraction, 337
diffuse reflectance, 262
diffusion, 119, 235, 415
digestibility, 8, 149, 374, 379
digestion, 24, 29, 30, 58, 97, 156, 157, 159, 164, 168, 229, 238, 288, 293, 448
dilation, 22
discharges, xi, 103, 311, 319
discomfort, 8
diseases, 14, 17, 18, 19, 38, 153, 326, 341, 361, 378
disinfection, 293
dispersion, 326
displacement, 189, 278, 332
disposition, 196
dissolved oxygen, 236, 406
distillation, 64, 69, 70, 71, 72, 73, 75, 78, 87, 172, 190, 285
distillation processes, 73
distilled water, 177, 180, 181, 339, 377, 441
distortions, 350
distribution, xi, 10, 22, 38, 103, 136, 137, 187, 188, 210, 224, 236, 240, 257, 295, 320, 327, 337, 346, 376, 381, 383, 418
diversity, 2, 3, 40, 56, 293, 349, 359, 374, 391, 417, 433
DNA, 9, 36, 37, 43, 111, 150, 284, 318, 319, 321
DNA damage, 318, 321
DNA lesions, 318
DNAs, 36
DOC, 123
docosahexaenoic acid, xiv, 12, 13, 60, 153, 388, 392, 395, 406, 407, 408, 409, 410, 411
dogs, 13
DOI, 98, 99, 264, 265, 267, 358
domestic resources, 438
dominance, 241
double bonds, xiv, 326, 413, 433
dough, 11, 12
down-regulation, 282
drainage, 419
dream, 298
dressings, 11
drinking water, 322
drugs, 17, 18, 20, 381, 414
dry matter, 10, 139, 154, 165, 193, 232, 235, 241, 376
drying, viii, 29, 61, 63, 67, 82, 83, 84, 94, 95, 96, 97, 149, 154, 156, 177, 178, 184, 185, 190, 193, 242, 288, 289, 347, 441
durability, 346
dyes, vii, 1, 240, 253
dynamic control, 14

E

early warning, xii, 312, 320

ecology, 308
economic downturn, xi, 295
economic performance, 245
economics, 25, 62, 163, 190, 243, 245
ecosystem, 104, 139, 242, 279, 312, 313, 347, 355
editors, 161, 270, 271, 292
effluent, 33, 40, 136, 144, 147, 153, 439
effluents, 34, 44, 49, 60, 103, 145, 146, 147, 158, 160, 163, 290
egg, 149, 369, 400
eicosapentaenoic acid, 13, 153, 167, 244, 306, 392, 395, 406, 411, 412, 449
elaboration, 355
elastase inhibitors, 20
elastin, 22
election, 241, 243, 296
electric current, 37
electric field, 242
electricity, viii, 29, 61, 76, 81, 82, 83, 84, 85, 86, 88, 89, 90, 91, 92, 93, 94, 95, 96, 97, 228, 240, 241, 414, 439
electrodes, 37, 317
electrolysis, 27
electromagnetic, 348
electron, 104, 118, 130, 138, 316, 317, 355, 362
electron microscopy, 316, 355
electrons, 24, 230
electrophoresis, 318, 320, 321
electroporation, 37, 43, 58, 59, 242
ELISA, 380
elongation, 115, 157, 446
e-mail, 373, 387
emission, 23, 115, 157, 249, 286, 296, 297, 303, 317, 350, 353, 440
emphysema, 19
employment, 181
EMS, 367
emulsions, 11, 12, 48, 56
encapsulation, 12
endocrine, viii, 102, 115, 117, 124, 135
endocrine system, 115
energy consumption, 77, 228, 242, 414
energy input, ix, 81, 82, 83, 87, 171, 185, 313
energy recovery, 64, 74
energy supply, 228
energy transfer, 360
engineering, x, 1, 14, 25, 34, 38, 39, 40, 41, 43, 46, 49, 55, 56, 62, 76, 96, 131, 142, 222, 223, 228, 241, 242, 243, 249, 282, 283, 285, 287, 289, 290, 304, 305, 412
England, 138
entrapment, 144, 164

environment, vii, viii, xi, 4, 13, 18, 22, 33, 78, 79, 101, 102, 104, 116, 128, 131, 133, 134, 139, 157, 181, 186, 210, 241, 248, 279, 295, 297, 312, 313, 315, 318, 319, 326, 330, 346, 358, 379
environmental change, xi, 311, 315, 322, 417
environmental conditions, xi, 36, 203, 207, 218, 229, 230, 252, 280, 311, 346, 347, 362, 417, 419
environmental effects, 102, 312
environmental factors, 143, 169, 201, 203, 210, 409, 433
environmental impact, 4, 160, 244, 287, 438
Environmental Protection Agency, 130, 138, 313
environmental quality, 313
environmental stress, 119, 120, 236, 317, 362, 363, 364
environmental stresses, 120, 363, 364
environmental sustainability, vii, 1
environmental variables, 357
enzymatic activity, 135, 379
enzyme, 14, 20, 39, 40, 62, 138, 154, 155, 235, 282, 317, 323, 363, 376, 415, 416
enzymes, 13, 22, 26, 28, 33, 118, 132, 136, 279, 285, 289, 304, 317, 322, 326, 369, 371, 375, 416
EPA, 13, 115, 153, 296, 303, 390, 392, 395, 396, 397, 404, 405, 407, 408, 412
epidermis, 21
EPS, 164, 303
equilibrium, 63, 78, 91, 103, 138, 162, 231, 326, 343
equipment, 115, 144, 175, 269
erosion, 279
essential fatty acids, 224, 411
ester, x, 62, 70, 76, 187, 227, 231, 242, 314, 361, 438, 445, 446, 447, 448
ester bonds, 438
estriol, 123
estuarine environments, 321
ethanol, 21, 26, 27, 49, 90, 93, 156, 157, 158, 189, 193, 229, 235, 240, 278, 282, 283, 284, 285, 286, 288, 290, 291, 292, 293, 294, 304, 329, 330, 331, 343, 376, 388, 391, 396, 399, 402, 407, 412, 438
ethers, viii, 102, 132, 133, 134, 136
ethyl acetate, 332
Euglena gracilis, 27, 29, 31, 37, 45, 110, 111, 121, 130, 232, 298, 308, 322, 391, 392, 402, 406, 407, 409, 410, 435
eukaryote, 388, 401
eukaryotic, vii, 1, 2, 28, 40, 56, 123, 142, 151, 153, 163, 222, 229, 257, 262, 279, 283, 291, 318, 320, 326, 410, 450
eukaryotic cell, 318
Europe, 7, 162, 270, 306, 449
European Community, 313
European Parliament, 57

European Union, xi, 9, 311, 313, 384
evaporation, ix, 143, 147, 171, 172, 185, 211, 231, 235, 240, 286, 301
evidence, xii, 2, 7, 54, 129, 149, 202, 325, 375, 381, 453
evolution, vii, 1, 28, 53, 56, 57, 133, 247, 275, 291, 317, 320, 450
excitation, 111, 253, 349, 350, 352
excretion, 118
exercise, 312, 361, 385
exopolysaccharides, 303, 309
experimental condition, 63, 190, 263
exploitation, 7, 10, 14, 157, 374, 398
exposure, xi, 19, 106, 110, 111, 112, 115, 116, 117, 118, 128, 130, 133, 136, 203, 311, 315, 316, 318, 319, 321, 322, 364, 367
extinction, 255
extraction, ix, xii, 12, 29, 62, 63, 70, 74, 75, 76, 83, 89, 90, 93, 96, 151, 152, 154, 160, 168, 171, 186, 187, 191, 193, 194, 196, 202, 228, 233, 238, 240, 242, 243, 253, 254, 270, 282, 285, 287, 288, 289, 296, 303, 304, 309, 325, 326, 329, 330, 331, 340, 341, 342, 343, 347, 349, 361, 365, 376, 377, 386, 420, 421, 422, 423, 424, 425, 426, 427, 428, 429, 430, 431, 432, 433, 440, 441, 443
extracts, 9, 10, 13, 17, 18, 21, 22, 46, 57, 58, 163, 168, 187, 253, 256, 327, 332, 333, 338, 339, 369, 378, 441

F

factories, 7, 10, 52, 161, 244, 283, 414
families, 153
fantasy, 60
farmers, 241, 247, 278
farms, 161, 203, 204
fat, 136, 191, 193, 244, 281, 284
fat soluble, 136
fatty acids, vii, ix, x, xiv, 1, 2, 11, 13, 20, 26, 35, 39, 65, 141, 149, 150, 151, 153, 157, 169, 171, 174, 186, 187, 196, 197, 210, 221, 224, 232, 240, 242, 246, 251, 253, 264, 265, 267, 268, 270, 275, 281, 287, 296, 304, 315, 319, 326, 327, 362, 392, 395, 396, 397, 398, 405, 407, 411, 412, 413, 415, 416, 419, 433, 438, 439, 444, 445, 446, 447, 448, 452
fauna, 321
FDA, 149
feed additives, 6, 35
feedstock, 25, 48, 62, 63, 65, 66, 74, 75, 84, 158, 162, 187, 229, 230, 235, 240, 242, 252, 278, 281, 283, 285, 287, 288, 291, 406, 414, 439, 440, 449, 450
fermentation, viii, ix, 14, 24, 26, 27, 29, 33, 48, 49, 50, 55, 59, 61, 74, 90, 93, 94, 96, 97, 158, 166, 171, 173, 189, 196, 197, 279, 283, 284, 285, 288, 289, 290, 291, 294, 304, 406, 408, 410, 412
fertility, 139, 154, 239, 360
fertilizers, ix, 141, 142, 145, 158, 172, 203, 219, 220, 240, 279
fever, 38
fiber, 166, 193, 199, 258, 261, 262, 265, 267, 268
films, 286
filters, 148, 261, 262, 264, 265, 267, 273, 333
filtration, ix, 67, 68, 72, 73, 74, 75, 148, 149, 171, 172, 177, 181, 185, 186, 241, 376
financial support, 269
fingerprints, 255
first generation, 158, 279
fish, xiii, 13, 33, 38, 42, 46, 115, 130, 132, 149, 153, 163, 201, 210, 312, 314, 320, 321, 359, 361, 395, 396, 401, 403, 404, 405, 409, 412
Fish and Wildlife Service, 132
fish oil, 395, 396, 403
fisheries, 407
fixation, vii, viii, 1, 4, 31, 32, 49, 50, 52, 54, 60, 101, 103, 104, 142, 144, 146, 165, 210, 244, 246, 247, 248, 249, 296, 297, 298, 302, 304, 306, 307, 308, 323, 393, 415, 448, 452
flame, 102, 115, 131, 139, 253, 327
flame retardants, 131, 139
flavonoids, 22, 326
flavor, 361
flex, 283, 286
flocculation, 147, 148, 186, 238, 241, 303, 309
flotation, 148, 235
flour, 410
flowers, xii, 359
fluctuations, 286
flue gas, xi, 31, 32, 33, 46, 51, 52, 60, 161, 165, 238, 280, 286, 287, 295, 298, 299, 301, 302, 305, 307, 308, 309, 407, 446
fluid, 282, 298, 309, 326, 330, 331, 341, 342, 419
fluid extract, 282, 309, 326, 331, 341
fluidized bed, 170
fluorescence, xii, 52, 104, 106, 112, 115, 137, 253, 254, 271, 272, 275, 312, 315, 316, 317, 319, 320, 321, 322, 323, 342, 346, 347, 348, 349, 352, 353, 354, 355, 358
foams, 304
food additives, ix, 35, 141, 142, 240
food chain, 13, 40, 103, 162, 312, 314, 346, 395
food industry, ix, 141, 142, 152, 410
food products, 10, 11, 12, 174, 397
food safety, 11, 381
food security, 239, 280
food web, 103, 119, 128, 134
force, 9, 35, 67, 296, 404

foreign exchange, 279
formation, ix, 17, 22, 52, 68, 70, 142, 154, 155, 157, 160, 162, 164, 171, 172, 182, 183, 186, 187, 188, 189, 194, 253, 254, 304, 316, 318, 320, 327, 336, 342, 347, 360, 365, 367, 370, 371, 411, 414, 416, 418, 419, 438, 449
formula, xiv, 375, 384, 388, 396, 397
fossils, 278
fragments, 318
France, 9, 10, 22
free radicals, 263, 338, 339
freezing, 151
freshwater, vii, xii, 3, 7, 21, 25, 45, 51, 52, 54, 103, 104, 105, 123, 131, 133, 139, 169, 202, 232, 254, 270, 280, 299, 304, 312, 313, 314, 318, 320, 321, 322, 323, 347, 359, 360, 408, 415, 449, 452
fruits, xii, 326, 359
FTIR, 45, 252, 254, 270
FTIR spectroscopy, 45, 270
fuel consumption, xiv, 277, 413
functional food, x, 10, 12, 227, 296, 381, 384
funding, 320
fungi, viii, xiii, 12, 17, 21, 51, 101, 103, 149, 289, 359, 360, 387
fungus, 17, 154
fusion, 15, 37, 371

G

gametophyte, 50
gasification, 24, 27, 53
gastrointestinal tract, 384
GC-FID, 253
gel, 21, 22, 42, 144, 164, 320, 332, 376
gene pool, 2
genes, 35, 36, 37, 39, 59, 274, 365, 370, 388, 408
genetics, 14
genome, 35, 36
genotype, 36
genus, xiv, 142, 149, 158, 159, 176, 263, 274, 362, 388, 396, 413, 419
geometry, 258, 350
Germany, 4, 5, 9, 10, 48, 138, 142, 202, 333, 378, 413, 440
germination, 115
gland, 22
global warming, ix, xi, xiv, 23, 40, 160, 172, 175, 227, 228, 229, 242, 278, 295, 413, 414
glucose, 26, 106, 145, 152, 155, 161, 168, 304, 378, 388, 389, 390, 391, 395, 396, 398, 399, 402, 403, 408, 410, 411, 412
glucoside, 118
glutamate, 309, 390, 408
glutamic acid, 391
glutathione, 118, 120, 134, 136, 317

glycerol, 7, 62, 66, 67, 70, 71, 72, 73, 74, 76, 80, 81, 85, 145, 177, 282, 388, 389, 390, 391, 395, 397, 399, 406, 407, 410, 416, 418, 419, 433, 438
glycogen, 285
glycol, 37
Golgi bodies, 360
gout, 9
governments, 295
gracilis, 27, 29, 31, 37, 45, 110, 111, 121, 123, 130, 136, 223, 232, 298, 308, 322, 391, 392, 402, 403, 406, 407, 409, 410, 425, 435, 451
granules, 151
grass, 41, 182, 297
grasses, 196, 278
gravity, 186
grazing, 241
Great Britain, 46, 221
green alga, 2, 6, 7, 9, 15, 17, 21, 22, 28, 29, 35, 38, 41, 45, 46, 49, 51, 52, 53, 54, 58, 59, 131, 132, 133, 134, 135, 138, 139, 161, 162, 224, 229, 245, 254, 255, 258, 265, 270, 271, 272, 274, 275, 283, 284, 285, 294, 314, 321, 322, 363, 367, 368, 369, 370, 371, 374, 376, 386, 391, 393, 395, 397, 398, 400, 401, 406, 415, 417, 433, 439, 449, 450, 451, 452, 453
green revolution, 269
greenhouse, vii, 5, 43, 145, 156, 172, 210, 218, 230, 247, 292, 293, 296, 414
greenhouse gases, 156, 293
growth factor, 22
growth modes, 296
growth rate, 4, 29, 31, 111, 116, 146, 152, 201, 205, 206, 210, 211, 212, 213, 215, 216, 217, 219, 220, 224, 230, 235, 242, 243, 252, 297, 299, 301, 314, 315, 316, 318, 351, 362, 364, 365, 366, 371, 386, 389, 395, 396, 400
growth temperature, 236, 308
Guangdong, 101
Guangzhou, 101
Guinea, 41

H

HAART, 169
habitat, 57, 201, 210
habitats, viii, 2, 101, 358
harbors, 33
harmful effects, 312
harvesting, x, 5, 25, 29, 40, 42, 80, 81, 82, 83, 84, 147, 148, 186, 229, 233, 238, 241, 251, 252, 253, 266, 269, 272, 280, 281, 286, 287, 288, 296, 303, 304, 305, 309, 341, 360, 399, 400, 406, 448
Hawaii, 4, 6, 10
hazards, 132, 278
healing, 22

health, vii, viii, xiv, 1, 4, 6, 9, 10, 11, 12, 22, 40, 43, 48, 101, 102, 104, 115, 150, 169, 176, 252, 278, 312, 326, 340, 360, 361, 371, 374, 375, 381, 385, 388, 396, 401, 403, 405, 408, 411
health effects, 396
heat loss, 66
heavy metals, 4, 9, 33, 34, 103, 116, 119, 120, 128, 129, 130, 137, 138, 144, 146, 147, 230, 298
heavy oil, 27
height, 148, 349
heme, 410
heme oxygenase, 410
hemicellulose, ix, 171, 173, 194, 235, 280
hemoglobin, xiii, 374, 378, 381
hepatitis, 38, 47
hepatotoxicity, 60
herbicide, 116, 119, 314, 315, 316, 317, 318, 321, 322, 323, 367
herpes, 49
herpes simplex, 49
heterotrophic microalgae, xiii, 280, 387, 388, 392, 394, 395, 396, 398, 402, 403, 404, 405, 406
heterotrophic microorganisms, 393, 394, 395
hexachlorobenzene, 102, 136
hexane, 240, 253, 304, 329, 332, 441, 443
histogram, 350, 351
history, 2, 47, 57, 301, 312
HIV, vii, 1, 14, 15, 42, 150, 161, 168
HIV/AIDS, 150
HIV-1, 15, 42, 168
HIV-2, 15
homeostasis, 150, 323
homolytic, 138
Hong Kong, 101
hormone, 38
hormones, 33, 134
host, 38, 375
hot springs, 2, 146, 168, 301
House, 199
hue, 13
human, viii, xi, 8, 9, 10, 15, 16, 19, 21, 22, 37, 40, 42, 49, 101, 102, 104, 142, 150, 164, 202, 238, 253, 287, 295, 296, 326, 341, 360, 361, 371, 374, 375, 381, 384, 396, 401, 408
human body, 326
human development, 287
human health, viii, 9, 22, 101, 102, 104, 326, 360, 396, 408
human immunodeficiency virus, 49
humidity, 143, 377
Hunter, 41, 45
hybrid, 236, 283, 371, 412

hydrocarbons, 2, 40, 131, 138, 159, 160, 230, 243, 293, 303, 304, 309, 411, 417, 449
hydrogen, 24, 27, 28, 29, 35, 40, 42, 43, 45, 47, 49, 50, 51, 52, 53, 56, 60, 118, 134, 144, 156, 159, 165, 166, 172, 184, 186, 229, 240, 245, 293, 294, 322
hydrogen gas, 53
hydrogen peroxide, 118, 134, 322
hydrogenase, 28, 40, 53, 56, 159
hydrogenation, 24, 416
hydrolysis, ix, 26, 171, 187, 285, 374, 375, 376, 379, 380, 381, 383, 386
hydrophilicity, 380
hydroxide, 231
hydroxyl, 118, 186, 416
hypercholesterolemia, 153
hypertension, 385
hypothesis, 384

I

ideal, 28, 36, 79, 80, 144, 287, 304
identification, xii, 64, 147, 254, 323, 345, 347
identity, 154
illumination, x, 176, 203, 212, 215, 236, 251, 253, 268, 275, 303, 441
image, xii, 43, 181, 337, 346, 348, 349, 350, 351, 353, 354, 355, 358
image analysis, xii, 346, 348, 349, 350, 353, 354, 355, 358
imagery, 351
images, 181, 183, 194, 335, 336, 348, 350, 351, 356
imitation, 45, 153, 155, 158, 270, 281, 294, 366, 400, 416, 417, 419, 450, 451
immobilization, 14, 144, 165
immune response, 12, 43, 361, 375
immune system, 18, 37, 150, 151
immunodeficiency, 15, 49
immunologist, 384
immunostimulant, 20
immunostimulatory, 375, 384, 385
improvements, 62, 96, 97, 181, 243, 272
impurities, 29, 285, 410
in vitro, 14, 17, 43, 49, 150, 161, 321, 355, 384
in vivo, xii, 14, 43, 150, 161, 255, 270, 312, 314, 320, 347, 348, 349, 352, 354, 355, 375
incidence, 143, 148, 160
income, 288
incomplete combustion, 102
incubation period, 20, 353
incubation time, 349, 350, 352, 353
India, 7, 10, 23, 58, 59, 142, 228, 277, 305
induction, 14, 57, 119, 174, 175, 247, 263, 274, 297, 319, 320, 364, 365, 370, 452, 454
induction methods, 364, 365

industrial chemicals, 186
industrial wastes, 156
industrialization, 312
industrialized countries, 279
industries, xiii, 7, 102, 151, 185, 189, 283, 374, 375, 379, 383, 395, 397
industry, ix, 10, 13, 26, 27, 34, 35, 38, 86, 141, 142, 145, 152, 174, 177, 182, 184, 186, 187, 188, 189, 193, 197, 199, 203, 230, 241, 244, 249, 279, 288, 289, 295, 296, 360, 361, 384, 404, 410
infancy, 14, 288
infection, 15, 43, 150, 361
infertility, 361
inflammation, 57, 153, 361
inflammatory disease, 361
inflation, 240
infrastructure, 240, 287
ingestion, 9
ingredients, 9, 10, 22, 47, 56, 57, 309, 396, 403
inhibition, 19, 40, 104, 105, 107, 108, 109, 110, 115, 117, 130, 133, 137, 167, 189, 236, 299, 301, 312, 317, 320, 339, 340, 369, 396, 406, 411
inhibitor, 19, 49, 363
initial state, 79, 88, 90
injuries, 363
inoculation, 164, 205, 217, 223, 349, 350, 352, 354, 355, 364
inoculum, 205, 206, 207, 213, 215, 217, 218, 223, 265, 353
insects, 104
insulation, 185
integration, 36, 63, 76, 77, 85, 89, 95, 287, 350, 419
integrity, 315, 318
intensity values, 352
interaction effect, 116
interferon, 164
intravenously, 377
invertebrates, 132, 359
investments, 203
investors, 230
ionization, 253, 327
ions, 111, 162, 167, 210, 211, 319, 415
Iran, 21, 47
iron, 27, 49, 150, 419
irradiation, 263, 370, 386
irrigation, 246, 280, 283
isolation, 40, 128, 165, 189, 296, 402
isomerization, 123
isomers, 363
isoprene, 40, 52
isoprenoid polyene pigments, xii, 359
isotope, 308
Israel, 9, 10, 58, 142, 202, 251, 273, 280

issues, 23, 34, 172, 239, 241, 247, 279, 290, 297, 312, 351
Italy, 244
iteration, 351

J

Japan, 6, 9, 10, 50, 53, 60, 142, 146, 169, 202, 302, 307, 327, 331, 333, 338
joints, 348
Jordan, 55

K

K^+, 319
Keynes, 46
kidney, 9, 42
kidney stones, 9
kidneys, 378
kinetics, 14, 42, 134, 142, 144, 147, 162, 243, 367, 407
KOH, 438
Korea, 294, 302
Kyoto protocol, 296

L

lactose, 193, 194, 389
lakes, 5, 6, 7, 145, 202
land based plants, vii, 1
landscapes, 292
larvae, 13, 38, 204, 209, 210, 213, 222, 224, 404
larval development, 224
laws, 78
leaching, xi, 311
lead, x, 17, 20, 38, 39, 63, 69, 96, 116, 228, 241, 251, 253, 279, 303, 312, 314, 317, 321, 336, 346, 366, 378, 433, 446
leakage, 103, 297
legend, 444
legislation, 377
lens, 437
lesions, 318
lethargy, 378
life cycle, 36, 51, 245, 247, 288, 296, 362, 367
light conditions, 7, 267, 371, 453
light cycle, 349
light scattering, 256, 272
lignin, ix, 157, 171, 173, 194, 235, 279, 280, 283
limestone, xii, 84, 346, 348, 349, 353, 355, 357
linoleic acid, 150, 407, 446
lipases, 242
lipid metabolism, 39, 163, 282, 417, 419, 450, 452
lipid oxidation, 419
lipid peroxidation, 118, 319, 322
liposomes, 321
liquid chromatography, 327, 347

liquid fuels, 24, 174, 228, 230, 278, 293
liquid phase, 72
liquids, 10, 67
live feed, ix, 201
liver, 18, 36, 42, 137, 361, 378
livestock, 238, 240
localization, 150, 263
lower prices, 151
luminosity, 359
Luo, 132, 133, 138
lutein, 17, 152, 165, 168, 296, 326, 328, 359, 362, 389, 400, 401, 411, 412
lycopene, 17, 326, 341, 342, 359, 361
lymphocytes, 323
lysis, 147, 154, 319

M

machinery, 285, 289
macroalgae, 12, 13, 23, 57, 136, 243, 291
macular degeneration, 326, 341
magnesium, 22, 159
magnetic resonance, 252, 271
magnitude, xiii, 105, 119, 179, 203, 262, 279, 387
Maillard reaction, 192, 193
majority, 23, 228, 388
malaria, 38
Malaysia, 291
malnutrition, 150, 378
mammals, 19, 115, 150, 314
man, 169
management, 205, 207, 214, 245, 268, 293, 312, 347, 384
manganese, 22
manipulation, 44, 129, 142, 181, 355, 366, 452
mantle, 175
manufacturing, 174, 186, 189, 239, 392, 397, 398
manure, 29, 51, 97, 160
mapping, 179, 357
marine diatom, 56, 60, 115, 116, 134, 135, 136, 139, 167, 315, 318, 321, 323
marine environment, 13, 18, 139, 314
marine fish, 404
marine species, 252
marker genes, 35, 36
marketplace, 409
Marx, 57, 293
Maryland, 44
mass, 4, 9, 29, 31, 33, 44, 57, 63, 66, 67, 68, 69, 70, 71, 72, 73, 75, 76, 77, 85, 87, 160, 164, 172, 174, 177, 179, 184, 187, 194, 196, 202, 207, 214, 222, 223, 224, 225, 242, 243, 246, 247, 248, 253, 255, 269, 270, 281, 286, 292, 293, 302, 304, 336, 351, 364, 368, 376, 381, 384, 393, 412, 445
mass spectrometry, 253, 445

materials, viii, ix, xii, 27, 33, 51, 61, 68, 72, 81, 85, 87, 92, 93, 103, 156, 157, 171, 172, 173, 174, 175, 177, 179, 183, 184, 186, 191, 192, 193, 194, 195, 196, 197, 230, 235, 245, 284, 287, 326, 332, 345, 346, 348, 349, 355, 356, 392, 395, 397, 398, 408, 439
matrix, 129, 164, 346, 348
matter, iv, xii, 10, 34, 36, 60, 102, 139, 154, 156, 160, 165, 182, 183, 193, 205, 206, 207, 213, 224, 232, 235, 241, 278, 304, 345, 376
measles, 15
measurements, x, 115, 128, 181, 251, 254, 255, 258, 259, 262, 271, 273, 316, 317, 319, 339, 350, 352, 355, 378
meat, 149, 403, 409
media, 14, 36, 144, 145, 146, 148, 189, 223, 225, 230, 233, 238, 261, 262, 265, 309, 318, 349, 369, 379, 401, 403, 410
median, 314
medicine, ix, 6, 14, 34, 35, 141, 407
Mediterranean, 56, 357
medium composition, x, 251, 253, 402, 405
melanoma, 17
melting, 174
membranes, 147, 211, 256, 257, 263, 296, 316, 319, 360, 415, 416, 417, 418, 451
mercury, 148, 298
messages, 19
meta-analysis, 133
Metabolic, v, 43, 52, 134, 141, 289, 386
metabolic pathways, 35, 38, 39, 154, 287, 289
metabolic syndrome, 361
metabolism, 2, 23, 38, 39, 40, 54, 57, 59, 119, 134, 137, 145, 163, 167, 210, 211, 247, 263, 265, 270, 279, 282, 314, 317, 319, 388, 395, 402, 410, 417, 419, 450, 452
metabolites, ix, 2, 14, 18, 19, 32, 35, 55, 57, 58, 60, 104, 111, 118, 123, 129, 135, 142, 151, 163, 169, 232, 245, 315, 341, 395
metabolized, 123, 147
metabolizing, 174, 186, 359
metal salts, 179
metals, viii, 4, 9, 33, 34, 60, 101, 103, 116, 118, 120, 128, 129, 130, 133, 137, 138, 144, 146, 147, 148, 230, 298, 316
metastasis, 17, 54
meter, 272, 331, 333
methanol, viii, 21, 61, 63, 66, 67, 68, 69, 70, 71, 72, 73, 74, 75, 76, 78, 81, 84, 85, 86, 87, 88, 89, 90, 91, 93, 94, 231, 238, 253, 278, 327, 338, 398, 438, 441
methodology, 29, 168, 255, 334, 343, 357, 385, 405

Mexico, 5, 6, 48, 142, 149, 202, 203, 214, 215, 216, 301, 307
Miami, 28
mice, xiii, 17, 49, 370, 374, 375, 377, 378, 381, 382, 383, 385, 386
microalgae species, ix, xiv, 26, 55, 81, 201, 210, 211, 212, 230, 232, 233, 237, 239, 240, 308, 413, 433, 440, 442, 444, 445, 446, 447, 448
microbial communities, 186, 346, 358
microbial community, 349
microcosms, 130, 133
microcrystalline, 348
micro-Kjeldhal method, 377, 380
micronucleus, 167
micronutrients, 147, 238
microorganism, 174, 181, 285, 319
microorganisms, viii, xii, xiii, 2, 4, 43, 54, 56, 57, 101, 104, 131, 134, 141, 143, 154, 160, 174, 177, 186, 196, 229, 247, 252, 255, 256, 258, 292, 293, 297, 304, 312, 313, 322, 326, 346, 348, 349, 355, 357, 358, 387, 388, 392, 393, 394, 395, 405, 410
microscopy, 315, 316, 355
microstructure, 164, 194
migration, 318
military, 37
mineralization, 129
miniature, 414
Ministry of Education, 269, 341
Minneapolis, 187
Miocene, 348
mission, xii, 345
missions, 142, 159, 169, 228, 278, 286
Mississippi River, 291
mitochondria, 316, 320, 407, 451
mixing, 66, 72, 78, 85, 194, 236, 336
modelling, vii, 61, 64, 65, 67, 68, 71, 72, 74, 96, 411
models, 65, 66, 76, 77, 85, 96, 255
modifications, 25, 35, 62, 63, 64, 78, 91, 97, 203, 207, 214, 359, 381
moisture, 22, 63, 83, 96, 177, 179, 190, 376
moisture content, 63, 83, 376
molar ratios, 66, 94
molasses, 145, 160, 283, 390, 395, 402
molecular mass, 376, 378, 381
molecular weight, 128, 129, 133, 173, 184, 376, 394, 415
molecules, vii, 1, 36, 37, 47, 129, 147, 154, 155, 256, 304, 326, 350, 360, 385, 414, 415, 416, 419
mollusks, ix, 13, 201
Mongolia, 202
monomers, 29
morphology, xii, 2, 164, 301, 308, 316, 325, 327, 335, 336, 362, 367

mortality, xiii, 110, 112, 117, 314, 367, 374, 403
Moscow, 251, 384
Moses, 271
mosquitoes, 38
mother cell, 314
multiple sclerosis, 150
multiplication, 364, 414
mussels, xi, 311, 404
mutagenesis, 367, 371, 400, 402
mutant, 36, 52, 59, 265, 266, 267, 268, 270, 367, 370, 399, 400, 401, 402, 406, 408
mutation, 367, 370
mutations, 366, 367
Myanmar, 10
mycelium, 154

N

Na^+, 18, 319
NaCl, 136, 152, 160, 424
NAD, 419
NADH, 155
Nannochloropsis oculata, xii, 22, 122, 232, 244, 274, 275, 280, 325, 326, 343, 427, 436, 449, 450
nanometers, 336
naphthalene, 123, 137
narcotic, 298
National Institutes of Health, 351
National Research Council, 377
national security, 172
natural food, 5, 6, 7, 10, 13
natural gas, 27, 84, 85, 94, 96, 172, 184, 230, 302, 414
natural resources, 361
nausea, 8
Navicula, 12, 37, 46, 111, 117, 127, 135, 415, 428, 453
necrosis, 38
negative effects, 115
nephropathy, 361
nerve, 18
nervous system, 104
Netherlands, 98, 393
neural network, 412
neurodegenerative diseases, 18, 361
neutral, ix, 157, 172, 183, 186, 189, 193, 197, 227, 252, 253, 254, 270, 271, 278, 288, 417, 418, 419, 439
neutral lipids, 270, 417, 418, 419, 439
neutrophils, xiii, 374
New Zealand, 61, 98, 292, 384
NH2, 82
nickel, 34, 128
nicotine, 367
Nile, 38, 252, 253, 270, 271

NIR, 252, 254, 259, 263, 270
NIR spectra, 255
nitric oxide, 186
nitrification, 158
nitrite, 317
nitrogenase, 28, 53, 154
Nitzschia, 26, 109, 111, 112, 121, 125, 126, 153, 210, 233, 395, 396, 406, 408, 411, 412, 428, 429, 449
NMR, 187, 252, 254, 271
non-polar, 253, 256, 327
non-renewable resources, viii, 27, 61, 79, 80, 84, 96
Norway, 167
nuclear magnetic resonance, 252, 271
Nuclear Magnetic Resonance, 254
nucleation, 336
nucleic acid, 2, 9, 134
nucleus, 36, 360, 362
nuisance, 21
nutraceutical, 41, 361, 375, 381
nutrient, viii, xiv, 5, 7, 25, 33, 34, 51, 55, 97, 101, 103, 144, 147, 148, 155, 163, 167, 169, 189, 197, 220, 221, 230, 241, 274, 301, 307, 360, 362, 363, 365, 366, 395, 397, 400, 403, 410, 413, 414, 416, 418, 419, 433, 451
nutrient concentrations, 144, 167, 221, 395
nutrients, viii, 7, 25, 33, 36, 40, 60, 79, 81, 82, 97, 101, 103, 144, 145, 146, 147, 148, 151, 154, 155, 159, 160, 176, 203, 210, 219, 230, 235, 237, 238, 240, 249, 280, 301, 307, 314, 349, 353, 364, 368, 388, 395, 397, 403, 404, 411, 418, 439
nutrition, vii, 1, 10, 33, 38, 40, 42, 53, 149, 163, 167, 193, 202, 210, 240, 253, 280, 364, 374, 375, 381, 384, 388, 399, 405, 407, 408

O

obstacles, 29, 186, 189
oceans, 172, 174
oil production, x, 25, 54, 227, 239, 241, 247, 252, 294, 306, 307, 410, 439, 450, 451
oil spill, 103
oilseed, 23, 25, 245
oleic acid, 65, 66, 155, 242, 274, 304, 446, 448
omega-3, 10, 405
operating costs, 240
operations, 12, 50, 148, 151, 172, 174, 177, 184, 189, 197, 348, 351
opportunities, 23, 44, 56, 161, 244, 293
optical density, 106, 112, 256, 257, 258, 260, 262, 265
optical parameters, 262
optical properties, x, 251, 256, 258, 262, 264, 265, 267, 272, 273

optimization, 35, 37, 49, 53, 168, 285, 296, 364, 406, 408
organ, 378, 381, 382, 383
organelles, 319
organic chemicals, 134
organic compounds, 33, 118, 122, 132, 157, 172, 192, 210, 213
organic matter, xii, 34, 60, 102, 156, 160, 205, 207, 213, 224, 345
organic pollutants, viii, xi, 101, 102, 118, 119, 128, 132, 133, 134, 136, 311, 322
organic polymers, 186
organic solvents, 15, 103, 194, 253, 304, 386
organism, 2, 13, 26, 34, 36, 37, 40, 52, 54, 166, 201, 210, 315, 317, 321, 361, 401, 412
organochlorine pesticides (OCPs), viii, 102, 106
organs, 359, 360, 378
osmosis, 147
osmotic pressure, 176, 213
overproduction, 367, 370
ox, 158
oxidation, 12, 60, 104, 111, 123, 151, 153, 184, 188, 299, 319, 360, 402, 405, 416, 419, 438
oxidative damage, 118, 274, 365
oxidative stress, 7, 116, 118, 135, 137, 138, 317, 318, 319, 360, 370, 449
oxygen, vii, viii, 3, 7, 17, 28, 53, 111, 118, 129, 130, 133, 136, 141, 142, 147, 155, 173, 231, 236, 263, 278, 319, 320, 322, 326, 341, 360, 361, 393, 394, 399, 403, 406, 407, 415, 438, 446
oxygen consumption, 111
oyster, 404, 408
oysters, 149, 404

P

Pacific, 52, 224, 305, 408
palm oil, 157, 230, 278, 281
Parliament, 57
particle bombardment, 37, 59
partition, 120, 256, 328, 332
pasta, 10, 11, 46
patents, 15
pathogens, 14
pathways, 35, 38, 39, 123, 154, 157, 287, 289, 304, 360
PCA, 350, 354
PCBs, viii, 102, 103, 105, 110, 116, 119, 121, 122, 124, 125, 130, 136, 138
PCP, 130
Pearl River Delta, 103, 132
pepsin, 375
peptide, 44, 189, 375, 376, 380, 383, 385
peptides, 13, 18, 20, 375, 378, 381, 384
perfusion, 389, 390, 395, 396, 412

permeability, 321
permeation, 254
permission, iv, 257, 260, 261, 265, 267
peroxidation, 118, 319, 322
peroxide, 118, 134, 319, 322
persistent organic pollutants (POPs), viii, 102, 121, 124
Peru, 48
pesticide, 131, 139, 314
pests, 104
petroleum, ix, 27, 80, 133, 146, 159, 172, 185, 227, 229, 231, 232, 240, 241, 242, 243, 278, 280, 281, 282, 287, 438, 439
Petroleum, 99, 293, 305
pH, xiii, xiv, 4, 20, 25, 34, 155, 157, 175, 176, 187, 189, 207, 210, 211, 213, 231, 241, 298, 299, 301, 302, 303, 316, 342, 363, 373, 376, 377, 379, 380, 386, 390, 391, 397, 402, 407, 410, 411, 413, 415, 419, 420, 421, 422, 423, 424, 425, 426, 427, 428, 429, 430, 431, 432, 433
pharmaceutical, xiii, 13, 14, 18, 20, 21, 22, 23, 37, 55, 152, 326, 342, 374, 375, 379, 381, 383, 414
pharmaceuticals, vii, 1, 4, 14, 43, 132, 142, 229, 240, 287
pharmaceutics, ix, 41, 141
pharmacology, 52
PHB, 154, 155
phenol, 390
phenolic compounds, 135
phenotype, 130
phenotypes, 2
phenylalanine, 189
Philadelphia, 138
phosphate, xiv, 82, 155, 159, 208, 214, 219, 320, 363, 376, 413, 416, 418, 419, 433, 450, 451
phosphates, 240
phosphatidylcholine, 417, 450
phospholipids, 255, 416, 417, 419, 438, 439
phosphorous, 33, 176, 189, 192, 213, 230, 301
phosphorus, 27, 33, 42, 51, 58, 144, 145, 146, 147, 155, 158, 164, 168, 211, 243, 247, 449
photoautotrophy, 388
photographs, 350
photolysis, 24, 27, 111
photons, 256, 264, 393, 394, 414
photooxidation, 139, 361
photosynthesis, vii, viii, xi, 1, 6, 24, 28, 31, 40, 43, 49, 58, 104, 105, 107, 116, 118, 130, 132, 133, 134, 137, 141, 142, 143, 147, 152, 157, 159, 164, 202, 228, 230, 236, 237, 255, 271, 273, 275, 278, 284, 297, 308, 312, 315, 317, 318, 320, 321, 322, 341, 360, 369, 371, 388, 399, 401, 414, 415, 449, 453

photosynthetic systems, 272
phototaxis, 360
phycobilin, 410
phycocyanin, vii, 1, 6, 10, 17, 43, 59, 150, 152, 161, 162, 166, 168, 301, 390, 391, 401, 402, 405, 407, 411
phycoerythrin, 150, 152, 162
phylum, 392
physical environment, 25
physical properties, 167
physicochemical properties, 111
physics, 61
Physiological, 139, 323
physiology, x, xi, 2, 12, 28, 122, 137, 227, 241, 252, 270, 287, 364, 384
phytoplankton, 13, 105, 132, 133, 135, 202, 210, 213, 222, 273, 291, 313, 320, 347, 357, 384
pigmentation, 7, 302, 360, 361, 362
pigs, xiii, 12, 149, 153, 163, 374, 377, 380
plankton, xi, 3, 201, 255, 261, 262, 311
plants, vii, ix, xi, 1, 2, 4, 7, 17, 31, 39, 40, 43, 57, 76, 104, 111, 118, 130, 131, 133, 134, 135, 137, 145, 146, 150, 154, 157, 159, 162, 171, 174, 189, 229, 232, 235, 238, 239, 244, 252, 262, 271, 272, 273, 274, 278, 279, 280, 283, 286, 295, 297, 298, 299, 301, 304, 311, 326, 341, 342, 359, 360, 362, 363, 365, 399, 402, 414, 415, 433, 439
plasma membrane, 37, 319
plastics, 154, 172
plastid, 2
platform, 52, 350
playing, vii, 1
PLS, 252, 255
PM, 292, 308
polar, 118, 253, 256, 315, 327, 331, 417, 418, 419, 433, 438, 439, 441
polarity, 329
policy, 292
pollutants, viii, xi, xii, 23, 27, 33, 101, 102, 103, 116, 118, 121, 124, 128, 129, 130, 132, 133, 134, 135, 136, 137, 230, 311, 312, 313, 314, 316, 317, 318, 319, 320, 322, 323
pollution, xi, xii, 138, 238, 240, 278, 279, 287, 311, 312, 318, 320, 321, 358
poly(3-hydroxybutyrate), 164
polybrominated diphenyl ethers, viii, 102, 132, 134, 136
polycarbonate, 424
polychlorinated biphenyl, viii, 102, 110, 121, 124, 132, 133, 135
polychlorinated biphenyls (PCBs), viii, 102, 110
polycyclic aromatic hydrocarbon, viii, 102, 112, 121, 124, 130, 131, 133, 134, 135, 137, 138, 139, 231

polyesters, 154, 168
polyether, 175
polymer, 254
polymerization, 155
polymers, 154, 186, 197, 286
polypeptides, 189, 371, 416
polyphenols, 22, 385
polysaccharide, 15, 17, 49, 54
polyunsaturated fat, vii, x, 1, 11, 13, 44, 149, 150, 153, 240, 246, 251, 253, 287, 296, 408, 412, 433, 448
polyunsaturated fatty acids, vii, x, 1, 11, 13, 149, 150, 153, 240, 246, 251, 253, 287, 296, 412, 433, 448
polyurethane, 144, 304
polyurethane foam, 144, 304
ponds, 4, 6, 8, 43, 133, 161, 167, 174, 177, 216, 235, 236, 237, 238, 240, 241, 246, 286, 288, 293, 302, 309, 405, 439
pools, 2, 203, 215, 217, 306, 384
population, xi, 8, 38, 104, 109, 110, 112, 117, 133, 160, 167, 169, 172, 176, 271, 277, 279, 311, 313, 314, 316, 318
population density, 109, 167, 169, 176
population growth, 133, 314
portfolio, 287
Portugal, 1, 46, 345, 356
positive correlation, 266
positive relationship, 380
potassium, 159, 231, 339, 453
potassium persulfate, 339
potato, 167, 391, 403, 410
poultry, 7, 12, 360, 361
power generation, 51, 164, 172, 183, 186
power plants, 31, 43, 145, 146, 238, 239, 299, 304, 414, 439
precipitation, 147, 194, 326, 332, 333, 334, 342, 343
predation, 174
predators, 119, 288
prediction models, 255
preparation, 11, 37, 138, 198, 339, 384
preservation, xii, 345, 346
prevention, 15, 240, 269, 341, 361
Principal Components Analysis, 350
principles, 45, 258, 410
probability, 351
probe, 270, 320
probiotic, 12
producers, viii, xi, 3, 14, 18, 25, 28, 49, 101, 103, 219, 232, 311, 313, 314, 371, 396, 398
production costs, 34, 38, 148, 160, 202, 235, 289, 307, 361, 440
production technology, 243

project, 34, 38, 246, 269, 307, 405
prokaryotes, 43
proliferation, 23, 103, 137, 298, 314, 318, 323
proline, 138
promoter, 59
propane, 138
propylene, 90, 93
prostration, 377, 380
protease inhibitors, 19, 20
protection, 9, 15, 22, 59, 263, 347, 358, 360, 370
protective role, 263
protein hydrolysates, xiii, 373, 375, 377, 379, 380, 381, 384, 385
protein structure, 189
protein synthesis, 22, 158, 321
proteins, x, 22, 29, 35, 38, 46, 56, 147, 149, 150, 174, 189, 193, 194, 196, 202, 217, 227, 238, 279, 280, 281, 287, 296, 301, 304, 362, 374, 375, 376, 378, 379, 380, 381, 383, 384, 386, 414, 419, 448
proteolytic enzyme, 375
protons, 24, 230, 415
Pseudomonas aeruginosa, 21
Pseudo-nitzschia, 210, 224
psoriasis, 19
public concern, 102
public health, 312
Puerto Rico, 44
pulp, 196
pumps, 76, 94, 95, 96, 97
purification, 37, 50, 65, 70, 71, 72, 151, 152, 165, 167, 168, 186, 197, 270, 275, 285, 334, 342, 343, 368, 447, 454
purines, 9
purity, 70, 71, 72, 73, 75, 76, 152, 336, 337, 371
PVC, 143
pyrolysis, 24, 54, 243, 306

Q

quality control, 162
quality improvement, 269
quality of life, 23
quanta, 256
quantification, 64, 253, 262, 268, 270, 327, 328, 347, 348, 349, 350, 353, 354, 355
quartz, 338
quinone, 136

R

race, 303
radiation, 22, 111, 130, 133, 138, 150, 255, 263, 273, 274, 348, 360, 365, 367
Radiation, 138, 454
radicals, 111, 114, 263, 316, 338, 339, 360
rainfall, 143

Raman spectroscopy, 254, 271
raw materials, viii, 61, 68, 72, 81, 85, 92, 93, 157, 235, 439
reactant, 87, 190, 194
reactants, 79, 186, 193
reaction center, 326, 341
reaction temperature, ix, 63, 66, 73, 75, 171, 174, 175, 178, 191, 196, 197, 438
reaction time, 66, 73, 178, 179, 182, 187, 191, 192, 196, 197, 438
reactions, 62, 94, 104, 118, 123, 174, 175, 192, 193, 194, 304, 326, 414, 419
reactive oxygen, 17, 118, 319, 361, 399, 403, 407
reactivity, 111, 114, 281, 380
real time, 255, 269
recognition, 9, 103, 351
recombination, 36, 54, 229
recommendations, iv
reconciliation, 2
recovery, xii, 29, 37, 50, 64, 68, 70, 71, 72, 73, 74, 75, 76, 81, 83, 87, 91, 94, 97, 148, 151, 165, 166, 242, 273, 289, 304, 309, 312, 322, 325, 329, 331, 332, 334, 340, 399, 439
recovery process, 76
recrystallization, 326, 333, 336, 337, 340
rectification, 350
recycling, 33, 42, 68, 97, 166, 169, 314, 396
redistribution, 172
refractive index, 256
regeneration, 22, 36, 79, 306
regression, 17, 179, 190, 252, 255, 327, 334
regression analysis, 179
regression equation, 179, 190
regression method, 255
regression model, 334
regulations, 11, 374, 381, 384
rehydration, 177
reinforcement, 186
relatives, 36, 283, 388, 405
relaxation, 414
relevance, 19
REM, 68
remediation, viii, 101, 103, 238, 280
remote sensing, 356
renal failure, 361
renewable energy, vii, 1, 40, 49, 59, 168, 228, 230, 245, 246, 248, 293, 305, 448
renewable fuel, ix, 24, 227, 240, 281, 303
repair, 150, 362
replacement rate, 144
replication, 42, 49, 150, 201, 210, 217, 220
reproduction, 115, 203, 315, 360, 370
reproductive organs, 360

requirements, vii, xiv, 28, 36, 61, 62, 63, 64, 65, 70, 71, 76, 77, 81, 82, 85, 87, 92, 145, 187, 201, 203, 210, 220, 237, 239, 242, 280, 304, 377, 381, 413
researchers, 2, 39, 103, 159, 202, 252, 301, 303, 319, 346, 356, 361, 433
reserves, xi, 228, 230, 264, 277
residuals, 166
residues, ix, 29, 67, 68, 72, 73, 74, 75, 81, 95, 97, 145, 171, 173, 196, 197, 278, 283, 391, 410
resistance, viii, 12, 36, 102, 103, 146, 157, 273, 296, 357
resolution, 2, 271, 272
resources, viii, x, xi, xiv, 10, 25, 27, 30, 32, 56, 61, 79, 80, 84, 89, 92, 96, 141, 166, 172, 228, 251, 252, 277, 279, 283, 292, 361, 413, 414, 438
respiration, 104, 105, 109, 118, 414
respiratory arrest, 19
respiratory problems, 158
response, viii, xii, 8, 12, 43, 45, 101, 111, 116, 118, 128, 137, 138, 152, 168, 170, 203, 205, 208, 210, 219, 244, 270, 273, 274, 290, 312, 315, 316, 319, 320, 322, 323, 334, 361, 399, 400, 405, 408, 419, 453
restoration, viii, 61, 79, 80, 84, 86, 90, 91, 92, 93, 94, 96
retardation, 364
reticulum, 39, 416
retina, 326, 400
retinol, 7, 341
revenue, 289
Rhodophyta, 46, 47, 390, 391, 409
ribosomal RNA, 2, 3
Richland, 305
rings, 174
risk, viii, 4, 9, 101, 115, 129, 152, 301, 313, 341, 346, 364
risk assessment, 129, 313
RNA, 2, 3, 9, 110
room temperature, 183, 339, 377
roots, 2, 235, 239, 359, 450
rotifer, 208, 406, 407
rotifers, 13, 208, 404, 409
routes, 63, 64, 65, 76, 77, 97, 103, 288, 293, 378
routines, 203, 205, 215, 218
rowing, 374
Royal Society, 42, 56
runoff, xi, 103, 311
Russia, 251

S

safety, xiii, 8, 9, 11, 63, 136, 174, 297, 373, 375, 379, 381, 382, 383

salinity, 4, 25, 152, 153, 158, 159, 160, 207, 210, 212, 213, 222, 224, 281, 293, 359, 426, 427, 428, 429, 451, 452
salmon, 7, 149, 361, 371
salt concentration, 59, 453
salts, 28, 145, 150, 179, 180, 213, 238
saltwater, 313
saturated fat, 433, 438, 446, 448
saturated fatty acids, 433, 438, 446
saturation, 39, 236, 327, 335, 394, 438
savings, 77, 148, 279
scatter, 258, 259, 262, 265, 267
scattering, x, 251, 255, 256, 257, 258, 259, 260, 261, 262, 263, 271, 272, 273
science, ix, 133, 141, 142, 199
sclerosis, 150
scope, 102
SCP, 8, 9, 149
seafood, 30
second generation, 158, 168, 279
secrete, 18, 395
secretion, 395, 402
security, xi, 172, 229, 230, 239, 242, 277, 279, 280, 327
sediment, 111, 136, 139
sedimentation, 235, 236, 238, 241, 259
sediments, 115, 116, 131, 136
seed, 38, 115, 232, 248, 252
senescence, 26
sensing, 255, 356, 441
sensitivity, viii, 17, 82, 83, 90, 102, 103, 128, 129, 131, 135, 163, 313, 314, 321, 322, 323, 348, 380
sensitization, 380
sensors, 269
sequencing, 2, 35, 142
serine, 19, 20
serum, 8, 9, 378, 400
services, 292
sewage, 33, 34, 47, 55, 115, 167
sex, 378, 382, 383
shape, 111, 256, 259, 262, 316, 379
shear, 143
shellfish, 18, 49, 405
shortage, xi, 33, 160, 295, 365
shortfall, 14, 77
showing, 32, 67, 73, 74, 350, 355
shrimp, 47, 54, 149, 162, 204, 209, 210, 213, 218, 222, 361
side chain, 380
side effects, 8, 9
signal transduction, 319
signalling, 319
signals, 18, 254, 365

significance level, 378
signs, 2, 22, 301, 377, 381, 402
silica, 21, 332
silicon, 37, 46
simulation, 64, 76, 154, 248
simulations, 355
siphon, 330
skeleton, 155
skin, 12, 19, 22, 53, 58, 59, 153, 361, 378
skin diseases, 153, 361
sludge, 158, 303, 309
society, 296
sodium, 18, 19, 52, 155, 159, 160, 181, 214, 231, 364, 367
sodium hydroxide, 231
software, 64, 65, 67, 68, 69, 71, 96, 350, 351, 354
soil erosion, 279
solar collectors, 236, 237
solid tumors, 16
solubility, xiii, 102, 103, 111, 188, 210, 298, 301, 327, 329, 334, 373, 375, 379, 380, 383
solution, 151, 156, 160, 187, 192, 229, 242, 253, 254, 255, 256, 282, 289, 297, 303, 306, 327, 331, 332, 333, 334, 335, 338, 339, 340, 376, 377, 378, 415
solvents, 15, 63, 64, 65, 72, 74, 78, 81, 89, 91, 94, 103, 194, 253, 269, 304, 329, 331, 332, 333, 334, 342, 386
somatomotor, 378
South Africa, 154, 165
Soxhlet extractor, 332
soybeans, 62, 229, 247
Spain, 4, 311, 345, 347, 358, 359, 368
speciation, 2, 34
specifications, 70, 71, 72, 73, 76, 377
spectrophotometry, 255, 269, 314, 347
spectroscopy, xi, 45, 252, 253, 254, 255, 269, 270, 271, 273
spending, 145
spleen, 378
Spring, 218
sprouting, 239
Sri Lanka, 227
stability, 11, 12, 60, 144, 151, 154, 164, 165, 231, 242, 282
stabilization, 55, 114, 133, 293
stack gas, 297
standard deviation, 179, 192, 211, 260, 337, 338
starch, 11, 23, 24, 26, 27, 39, 46, 50, 55, 93, 166, 230, 235, 283, 284, 285, 290, 391, 397, 399, 403, 404, 406
starch polysaccharides, 11

starvation, 25, 29, 158, 254, 263, 265, 266, 270, 272, 273, 274, 275, 364, 371, 401, 408, 449, 451
state, 9, 40, 79, 80, 81, 83, 87, 88, 90, 91, 133, 243, 247, 248, 253, 254, 263, 269, 271, 302, 314, 317, 353, 356, 363, 364, 386, 393, 409
states, 167, 215, 378, 414, 447
statistics, 199, 281
steel, 32, 174, 186, 330, 333
sterile, 189
sterols, 2, 42, 45, 390, 404, 417
stigma, 360
stoichiometry, 66, 414
Stone biodeterioration, xii, 345
storage, x, 2, 11, 38, 39, 119, 151, 163, 164, 186, 232, 251, 252, 254, 263, 265, 282, 297, 364, 395, 396, 403, 417, 418
strain improvement, 367, 371
stress, 6, 7, 25, 59, 116, 118, 129, 132, 135, 137, 138, 143, 158, 236, 247, 255, 259, 263, 265, 270, 274, 317, 318, 319, 322, 323, 359, 360, 362, 363, 364, 365, 367, 369, 370, 374, 399, 408, 416, 417, 418, 419, 433, 449
stressful events, 269
stressors, 119, 315
stroma, 415
structure, vii, 1, 2, 15, 16, 20, 114, 134, 158, 168, 181, 189, 192, 196, 203, 229, 237, 245, 259, 282, 318, 350, 359, 379, 380, 383, 417, 445, 451
style, 356
substitutes, 253, 407
substitution, 138, 403
substrate, 118, 123, 155, 156, 157, 159, 160, 173, 175, 178, 189, 193, 197, 298, 317, 347, 353, 376, 395, 396, 407, 415, 416, 433
substrates, xii, 34, 54, 93, 156, 158, 166, 173, 174, 175, 179, 184, 186, 191, 197, 345, 347, 349, 353, 355, 374, 388, 389, 392, 395, 397, 398, 399, 402, 416
sucrose, 179, 389
sugar beet, 26, 196, 278, 283, 402
sugar industry, 145, 184
sugarcane, x, 158, 169, 227, 278, 280
sulfate, 148, 181, 244, 299
sulfur, 146, 183, 240, 278, 299
sulphur, 27, 84, 231, 286, 376
Sun, v, 37, 38, 44, 47, 56, 59, 61, 83, 98, 99, 199, 291, 371, 389, 401, 411
supercritical anti-solvent (SAS), xii, 325, 326
supervision, 6
supplementation, 17, 162, 384, 385, 409
suppression, 154, 283
surface area, xiii, 236, 301, 347, 349, 350, 353, 354, 387, 394

surfactant, 116, 139
surfactants, 136
surplus, 281, 403
surrogates, 39
survival, 115, 210, 222, 347, 361, 367, 371
susceptibility, xi, 104, 111, 235, 311, 364
suspensions, x, 251, 253, 256, 257, 258, 259, 260, 261, 263, 265, 266, 267, 269, 271, 272, 273, 319
sustainability, vii, x, 1, 43, 227, 228, 229, 239, 242, 244, 384
sustainable development, 228, 291
sustainable energy, 47
Sweden, 9, 292, 376
switchgrass, 229, 283
Switzerland, 22, 138
symbiosis, 154
symptoms, xiii, 8, 9, 150, 374, 377, 380
syndrome, 361
synergistic effect, 116
synthesis, 18, 22, 24, 25, 40, 57, 60, 82, 110, 153, 154, 155, 157, 158, 183, 186, 210, 213, 230, 247, 274, 286, 289, 321, 360, 362, 363, 365, 366, 367, 371, 391, 395, 399, 400, 401, 402, 403, 406, 410, 414, 415, 416, 417, 419
synthetic fuels, 135

T

T cell, 15
Taiwan, 6, 10, 294, 325, 326, 333, 340, 341, 343, 454
tanks, 43, 143, 151, 161, 162, 203, 205, 214, 216, 217, 288
target, x, 34, 36, 78, 104, 115, 116, 235, 251, 253, 268, 296, 316, 323, 398
taxa, 45, 374
taxonomy, 2, 169, 274, 384, 389
techniques, x, xii, 9, 29, 35, 40, 89, 149, 152, 185, 227, 233, 241, 243, 251, 253, 269, 271, 273, 282, 304, 318, 345, 346, 347, 348, 349, 350, 355, 356, 358, 366, 374, 375, 406
technologies, vii, 1, 43, 63, 95, 157, 161, 233, 241, 279, 287, 290, 296, 305
technology, viii, xi, 26, 29, 30, 34, 35, 37, 47, 61, 64, 82, 95, 96, 101, 128, 203, 235, 243, 244, 268, 277, 278, 279, 280, 283, 286, 287, 288, 289, 292, 294, 342, 362, 376, 407, 439
temperature dependence, 453
tension, 150
terpenes, 417
testing, 105, 303, 321, 385, 440
textiles, 115
texture, 351
Thailand, 10, 99

Thalassiosira, 35, 108, 109, 110, 124, 135, 149, 211, 212, 221, 222, 223, 233, 432, 437, 449
therapeutic benefits, 149
thermal energy, 83, 172, 184
thermal expansion, 174
thermal properties, 65
thermal treatment, 189
thermochemical cycle, 79
thermodynamics, 78
thermogravimetric analysis, 246
thin films, 286
thoughts, 103
thrombin, 19
thymus, 378
time periods, 366
time series, 271
tincture, 342
tissue, 22, 258, 294
tocopherols, 402, 409
toluene, 278, 329, 330, 331
total cholesterol, 162
total product, 6, 7, 361
toxic effect, xii, 105, 111, 115, 118, 130, 147, 312, 314, 317, 320
toxic substances, 8
toxicity, viii, ix, xi, xiii, 102, 103, 104, 105, 111, 114, 115, 116, 119, 128, 130, 131, 132, 133, 134, 135, 136, 137, 138, 139, 141, 167, 231, 311, 312, 313, 314, 315, 317, 318, 320, 321, 322, 323, 374, 377, 378, 381, 382, 383, 385
toxin, 18, 44, 52, 59, 116
traits, 288
transcription, 116, 137, 365, 370, 411
transduction, 319
transesterification, vii, 24, 41, 61, 62, 63, 64, 65, 66, 67, 68, 69, 70, 71, 72, 73, 74, 75, 76, 77, 78, 80, 81, 83, 84, 85, 86, 87, 88, 89, 90, 91, 92, 93, 94, 95, 96, 97, 187, 231, 233, 281, 282, 289, 291, 292, 399, 408, 438, 441
transesterification processes, vii, 61, 64, 65, 66, 76, 77, 78, 80, 81, 83, 84, 85, 91, 92, 93, 94, 95, 96, 97
transformation, viii, 34, 36, 37, 39, 42, 45, 46, 50, 51, 52, 54, 55, 57, 58, 59, 79, 101, 103, 118, 119, 128, 129, 192, 350, 364, 365, 371, 439, 452
transformation processes, 129
transformations, 79
translation, 42
translocation, 319
transmission, 15, 18, 52, 150, 262, 273
transport, xi, 27, 102, 104, 118, 130, 138, 242, 277, 278, 280, 283, 288, 289, 317, 319, 412, 415, 453
transport processes, 319

transportation, 172, 174, 183, 186, 197, 228, 244, 248, 278, 297
treatment, vii, 1, 4, 14, 15, 23, 30, 32, 33, 34, 40, 54, 55, 60, 71, 97, 130, 133, 136, 142, 144, 145, 146, 147, 153, 156, 157, 164, 165, 166, 172, 189, 202, 208, 238, 254, 259, 279, 280, 283, 285, 289, 297, 302, 318, 337, 361, 365, 367, 376, 379, 382, 392
triacylglycerides, 196, 439
trial, 205, 213
tricarboxylic acid, 155
triggers, 319, 365
triglycerides, x, 25, 65, 75, 157, 227, 231, 281, 282, 327, 396
trypsin, 19, 38, 384
tumor, 16, 17
tumor invasion, 17
tumors, 16, 17
Turkey, 21
turnover, viii, 23, 101, 103
tyrosine, 376, 403, 410

U

UK, 44, 98, 133
ultrasound, 63, 64, 65, 66, 71, 73, 78, 86, 87, 89, 91, 92, 93, 94, 96, 97, 151, 282
ultrastructure, 2, 169, 323
UN, 247, 306
unicellular structure, vii, 1
uniform, 143, 192, 235, 239, 257, 303
United, 43, 54, 59, 138, 142, 158, 168, 169, 228, 248, 283, 291, 292, 293, 341, 377, 384, 386
United Nations, 138, 248
United Nations Development Programme, 248
United States, 43, 59, 142, 158, 168, 169, 228, 283, 291, 292, 293, 341, 377, 384, 386
up-scaled process, vii, 61
uranium, 228, 230
urban, 53, 358
urea, 82, 220, 431, 450
uric acid, 8, 9
urine, 8, 33
USA, 4, 7, 9, 10, 32, 49, 51, 57, 64, 199, 202, 221, 224, 248, 295, 298, 301, 327, 330, 333, 351, 383, 440
UV, 22, 111, 120, 133, 136, 143, 274, 327, 338, 339, 361, 367, 454
UV radiation, 22, 111, 133, 274, 367

V

vaccine, 38
vacuole, 118
vacuum, 67, 70, 71, 76, 331, 332, 338
Valencia, 12, 60, 358
validation, 323

valuation, 198, 323, 383, 384
valve, 70, 75, 331, 333
vanadium, 29
vapor, 156
variables, 143, 178, 179, 191, 219, 320, 334, 357, 392
variations, 210, 212
varieties, 174
vascular endothelial growth factor (VEGF), 22
vector, 104
vegetable oil, 233, 244, 247, 281, 331
vegetables, xii, 149, 157, 326, 359
vegetation, 2
vehicles, 27, 286
vein, 378
velocity, 46
vertebrates, 359
vessels, 22, 285
vibration, 441
viral infection, 43
virus replication, 49, 150
viruses, 15, 235
viscosity, 210, 231, 240, 245, 281
vision, 138, 294
visualization, 253
vitamin A, 150, 168, 341
vitamin B1, 6, 150, 404
vitamin B12, 6, 150, 404
vitamin B12 deficiency, 150
vitamin C, 7, 22, 402
Vitamin C, 339
vitamin E, 7, 392, 402
vitamins, vii, ix, 1, 7, 13, 22, 136, 141, 145, 149, 202, 287, 326, 392, 395, 405
volatility, 281
volatilization, 111
vomiting, 8

W

Washington, 44, 46, 50, 169, 221, 308, 356, 383
waste, xi, 4, 32, 33, 44, 51, 71, 97, 133, 142, 145, 153, 156, 157, 166, 229, 232, 238, 245, 277, 288, 293, 294, 298, 302, 311, 389, 390, 391, 392, 395, 397, 399, 401, 406, 439, 448
waste treatment, 4, 97, 133, 302
waste water, 4, 145, 389, 392, 439
wastewater, vii, 1, 23, 30, 32, 33, 34, 40, 41, 42, 44, 47, 48, 50, 51, 52, 53, 55, 56, 60, 71, 103, 128, 129, 134, 136, 142, 144, 145, 146, 147, 157, 164, 165, 168, 169, 221, 229, 238, 243, 249, 280, 287, 289, 290, 297, 307, 356
water quality, 218
water resources, xi, 252, 277, 279
water supplies, 279
water vapor, 83, 156
wavelengths, 414
web, 128, 134
weight reduction, 109
wetting, 347
white blood cell count, 378
wild type, 28, 29, 268, 367, 406
wildlife, 115, 119, 132
Wisconsin, 130
wood, 83, 85, 94, 95, 96, 97, 283
wool, 330
workers, 12, 197, 262
World War I, 104
worldwide, xi, 104, 116, 201, 229, 277, 279, 282, 289, 326

X

xanthophyll, 342, 389, 400
X-ray diffraction, 337
X-ray diffraction (XRD), 338
XRD, 337, 338

Y

yeast, xiii, 27, 196, 285, 290, 360, 361, 363, 368, 387, 392, 395, 397, 399, 401, 404
yield, ix, x, 4, 14, 24, 33, 40, 54, 62, 67, 70, 72, 73, 76, 142, 148, 171, 172, 173, 177, 178, 179, 180, 181, 185, 188, 191, 193, 194, 197, 218, 227, 228, 233, 234, 235, 237, 241, 245, 253, 275, 279, 280, 281, 282, 283, 285, 287, 301, 306, 326, 329, 332, 334, 388, 394, 398, 399, 411, 412, 449
yolk, 400

Z

zinc, 22, 34, 116, 137, 138, 146, 147, 167, 415
zooplankton, 13, 208, 241, 404